Fundamentals of Entomology

Fundamentals of Entomology

Fourth Edition

RICHARD J. ELZINGA
Department of Entomology
Kansas State University

PRENTICE HALL
Upper Saddle River, New Jersey 07458

Library of Congress Cataloging-in-Publication Data
Elzinga, Richard J., 1931–
Fundamentals of entomology / Richard J. Elzinga. — 4th ed.
p. cm.
Includes bibliographical references (p. 454) and index.
ISBN 0-13-508037-1
1. Entomology. I. Title.
QL463.E48 1997 96-8206
595.7—dc20 CIP

Production Editor: *Cindy Trickel, Carlisle Publishers Services*
Production Liaison: *Eileen O'Sullivan*
Managing Editor: *Mary Carnis*
Acquisitions Editor: *Charles Stewart*
Manufacturing Buyer: *Ilene Sanford*
Director of Manufacturing & Production: *Bruce Johnson*
Marketing Manager: *Debbie Yarnell*
Compositor: *Carlisle Communications, Ltd.*
Manuscript Editor: *Karen Bankston*
Cover Designer: *Miguel Ortiz*
Cover Photo: *Richard J. Elzinga*

A Simon & Schuster Company
Upper Saddle River, New Jersey 07458

Cover photo description: An American cockroach *(Periplaneta americana)* preening an antenna.

Printed in the United States of America

10 9 8 7 6 5 4 3 2 1

ISBN 0-13-508037-1

Prentice-Hall International (UK) Limited, *London*
Prentice-Hall of Australia Pty. Limited, *Sydney*
Prentice-Hall Canada, Inc., *Toronto*
Prentice-Hall Hispanoamericana, S.A., *Mexico*
Prentice-Hall of India Private Limited, *New Delhi*
Prentice-Hall of Japan, Inc., *Tokyo*
Simon & Schuster Asia Pte. Ltd., *Singapore*
Editora Prentice-Hall do Brasil, Ltda., *Rio de Janeiro*

To those seeking knowledge of the basic insect plan, I dedicate this fourth edition. Hopefully, you will find the study of insects as exciting as I do.

Contents

Preface

Insects are one of the most remarkable animal groups on the earth, and they merit study for at least two major reasons. First, insects have unsurpassed diversity and niches; because of this extensive variation, these animals can provide an in-depth understanding of nature and the many ways that biological problems have been met. Approximately 70 to 75 percent of the known species of animals are classified as insects, and 31 orders and about 600 families are found in North America north of Mexico (Borror, Triplehorn, and Johnson, 1989). Insects fly, they jump, they hide, they see ultraviolet light, they produce and molt an extraordinary exoskeleton, and they possess magnificent colors and shapes. Few habitats exclude insects. In withstanding harsh environments they are unparalleled. Some insects live in the arid deserts, some in hotsprings up to 80° C, others on mountain peaks as high as 6,096 m, some in tropical rain forests, and there are insects that live in arctic temperatures that reach below −20° C.

A second major reason to study insects is that a knowledge of insects is essential as we manipulate ecosystems for increased food production and better health. Back in the early 1900s, many entomologists were concerned about the competition for food between humankind and insects, and some entomologists believed that insect control was imperative for survival of the human race. Although such a position may seem somewhat extreme, insects do consume or spoil sufficient crops and products, causing millions of people to starve each year. And insect-transmitted diseases, both to humans and their crops, remain a threat to health and civilizations.

In the mid-1970s students and professionals encouraged me to write an introductory text that would condense the diversity of insects and their influence upon the ecosystem into a basic insect plan. The first edition of this text presented the topics of the fundamental structure-function concept of both external and internal structures, the development and impact of the environment upon insects, and the major interactions between insects, plants, and human. Control or management aspects were only discussed briefly because of the many detailed texts on these subjects and because of the flux in the status of insecticide usage and other methodology. The first edition also included a chapter on classification and simplified keys, primarily to adult insects, to permit identification of the more common orders and families.

The second edition expanded all chapters, especially the Arthropod Plan; Behavior; Insects, Plants, and Humans; and Classification. In the latter, 3 orders and more than 60 families were added. Additions included keys for identifying common immature insects to family, tables that summarized the predominant diet and habitat preferences of holometabolous orders (both adult and immature), and a chapter on insect parasitism.

The third edition updated previous material, principally in the areas of physiology, development, behavior, social insects, and the impact of insects and pest management on our ecosystem.

This edition updates previous material and covers new material in the chapters titled Predation by Insects and Household Insects. The subject of insect pest management has been expanded into a separate chapter. Also, major changes in insect classification have been made and keys have been updated.

My gratitude is expressed to the many students, colleagues, and to my understanding and supportive family, who have contributed to and encouraged this publication. Acknowledgment is extended to those who loaned or gave permission to use illustrations and photographs. Scanning electron microscope photographs were taken in the SEM Laboratory, Kansas State University.

RICHARD J. ELZINGA
Manhattan, Kansas

Fundamentals of Entomology

1 The Arthropod Plan

It has been said, "To understand a person, one must also understand his family." Similarly, to understand insects (entomology) requires at least a cursory knowledge of other animals and especially those classified as the phylum Arthropoda (from *arthros* for joint, *poda* for foot). Arthropods differ from many other invertebrates by having the following characteristics:

1. Bilateral symmetry
2. An exoskeleton with chitin
3. Externally segmented bodies and embryologically one pair of appendages per segment
4. Appendages modified for feeding
5. A hemocoel instead of a coelom
6. No cilia
7. A ventral nerve cord and dorsal brain

Arthropods are an ancient group of diverse animals that first appeared from the Precambrian to the Cambrian divisions of the earth's history. Two major lines have survived to this day and are classified as subphyla: Mandibulata, those that have well-developed *mandibles,* and the Chelicerata, those that utilize *chelicerae.* Some classifications subdivide the Mandibulata into the Crustacea (see subsequent discussion) and Uniramea (all other mandibulate arthropods); however, the text will not follow this latter system because it is the subject of some controversy.

Arthropods represent an extremely successful group of animals. They live in a wide variety of habitats, exhibit diverse types of locomotion, have a great range of

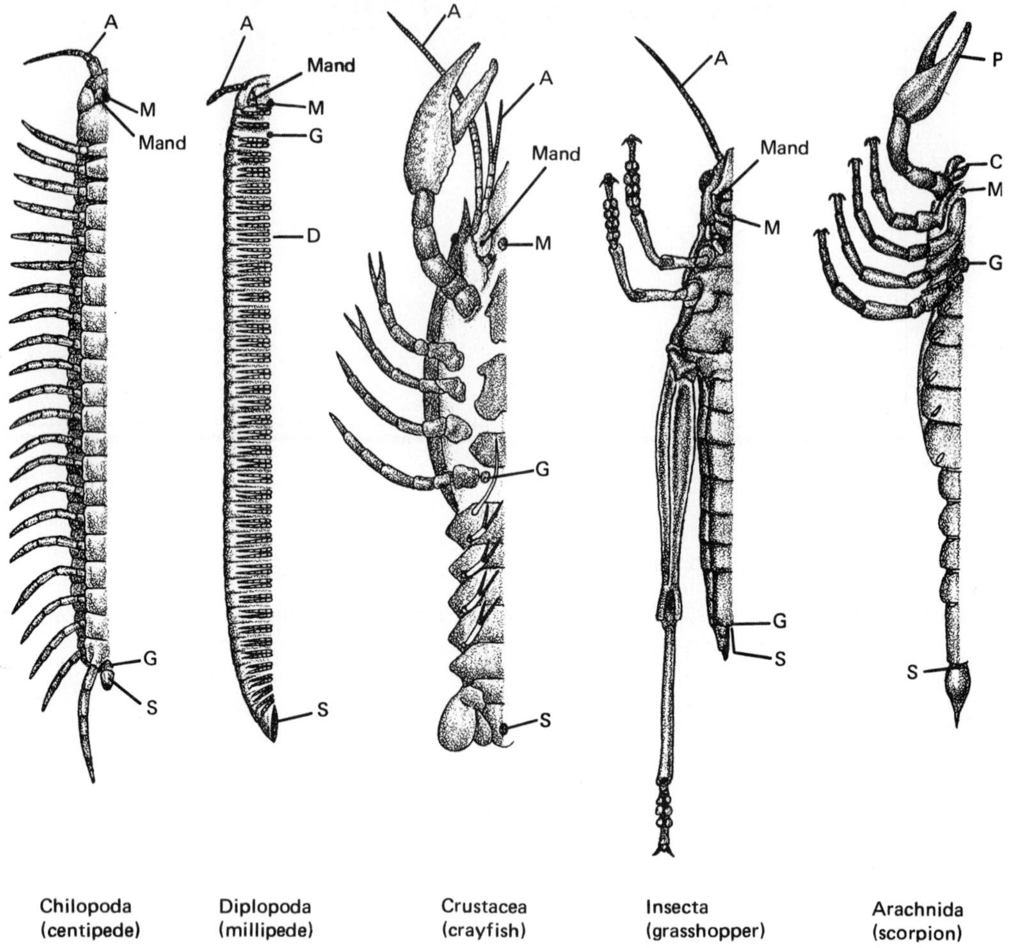

Figure 1. Examples of the classes of arthropods and their body plan. *A*, antenna; *C*, chelicera; *D*, diplosegment; *G*, genital pore; *M*, mouth; *Mand*, mandible; *P*, pedipalp; *S*, anus.

structural variations, eat many varieties of food, and include the greatest number of species. To understand this success, a limited discussion on some general concepts and adaptations is in order.

SEGMENTATION AND TAGMOSIS

The ancestors of arthropods undoubtedly were bilaterally symmetrical and had their major sensory structures located at the anterior end of the body to perceive the forward environment. Their body probably consisted of either 20 or 21 *metameres* or segments. Segmentation did not represent complete divisions of the body, but only

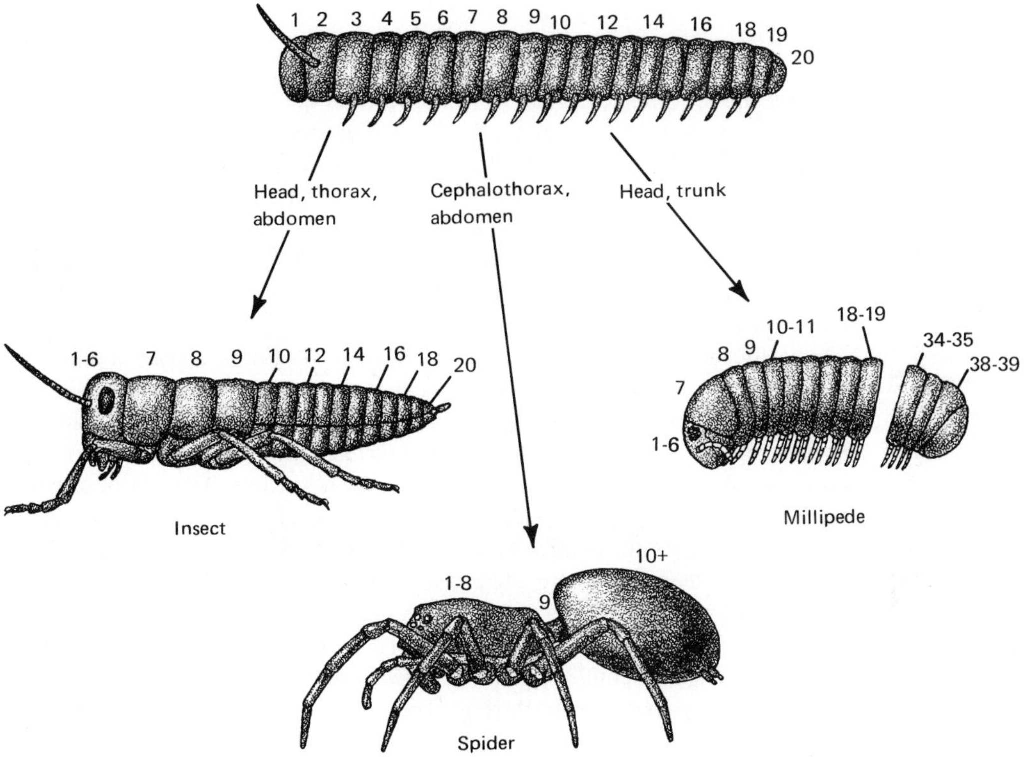

Figure 2. Segmentation and tagmosis of three arthropods from theoretical ancestor.

muscle organization accompanied by divisions in the exoskeleton and nervous system for efficient locomotion. Each segment also possessed a pair of short lobelike appendages. The alimentary canal had two terminal openings, the anus and mouth.

From this archetype evolved the many diverse shapes and forms of present-day arthropods. As seen in Figures 1 and 2, arthropod bodies are specialized into functional regions or *tagma,* a process termed *tagmosis.* In one type of tagmosis, the anterior six segments evolved into the *head* (sensory, feeding, and coordination centers), and the remaining segments, or *trunk,* retained their generalized role including locomotion (centipedes and millipedes). In a second pattern, the anterior 8 (spiders) to 14 segments (crayfish) became modified into a *cephalothorax* or prosoma (sensory, feeding, coordination, and locomotor center), and the remaining segments became the *abdomen* or opisthosoma and commonly lost most of their appendages and role in movement. A third variation, found in insects and many crustaceans, resulted in three body regions, the *head* (sensory, feeding, and coordination center), the *thorax* (locomotion), and the *abdomen.* The localization of the locomotor area into the thorax or cephalothorax reduced undulation tendencies, such as those in centipedes, because the propulsion force of long legs is applied to a small area.

EXOSKELETON

The outer surface of terrestrial animals must protect the body and slow down either water uptake or loss from the body; this is often accomplished by the production of a slime or mucoprotein that covers the body surface. In arthropods, the covering has been modified into a solid structure, the many-layered exoskeleton or *integument.* Not only does this structure cover and protect the entire external surface, but it also is invaginated into the body to line the tracheal system, the anterior and posterior sections of the digestive system, portions of the reproductive system, and to provide internal braces and muscle attachment bases.

Basically, the integument consists of a *basement membrane,* a layer of *epidermal cells,* and an externally secreted layer, the *cuticle,* which contains up to one-half the dry weight of an insect. One of the major compounds within the arthropod cuticle is *chitin,* a polymer of N-actyl-D-glucosamine and closely related to cellulose. This nitrogenous polysaccharide has a tannish color and is flexible. The suit of armor so characteristic of this phylum results from the addition of certain hardening material to form a chitin-protein complex of microfibers in the cuticle. In insects and arachnids, for example, quinones are added that cross-link the chitin-protein microfibers into a plasticlike *sclerotin,* a material that is both strong and light. This tanning or hardening process is termed *sclerotization.* This contrasts with the Crustacea and Diplopoda that add calcium, a process of *calcification.*

Cross sections through the cuticle (Fig. 3) reveal a laminate condition. The outermost, multilayered, yet extremely thin *epicuticle,* consisting of lipids and polyphenols, provides waterproofing and also protection for the epidermal cells when the new cuticle is deposited during molting (see Fig. 105). Inside this thin epicuticle is the hardened *exocuticle,* the layer produced by sclerotization, and the innermost flexible and recyclable *endocuticle.* The finalized suit-of-armor is the result of localized areas becoming specialized in the presence and thickness or absence of these layers. Areas that have a thick exocuticle are termed *sclerites* and may be braced further by creases or *sulci* (Fig. 3). Flexible regions between sclerites, which permit movement in the suit of armor, consist primarily of endocuticle and are referred to as *membrane. Pore canals* extend through the procuticle and aid in the deposition of portions of the epicuticle. Dermal gland ducts also extend from the cellular layer to the external environment through the integument.

The deposition of an exoskeleton results in modifications to the sensory systems, much of which must now develop as part of the integument. Many sensory receptors extend through the integument into hairlike projections, the *setae* (Fig. 3), and others form platelike pressure pads. Most setae are solid and are innervated at their base to detect tactile movement; however, others are hollow and detect chemicals entering them through lateral pores (Fig. 4). Still others include tympanic organs (hearing), temperature-sensitive organs, and photoreceptors (light). Many of these receptors will be discussed in subsequent chapters.

The benefits derived from the integument include protection from most chemicals except strong acids and bases, retardation of water movement both out of and into the body, and high protection from physical damage and abrasion. The integument is a structure that can form concealing colors and shapes for avoiding detection by preda-

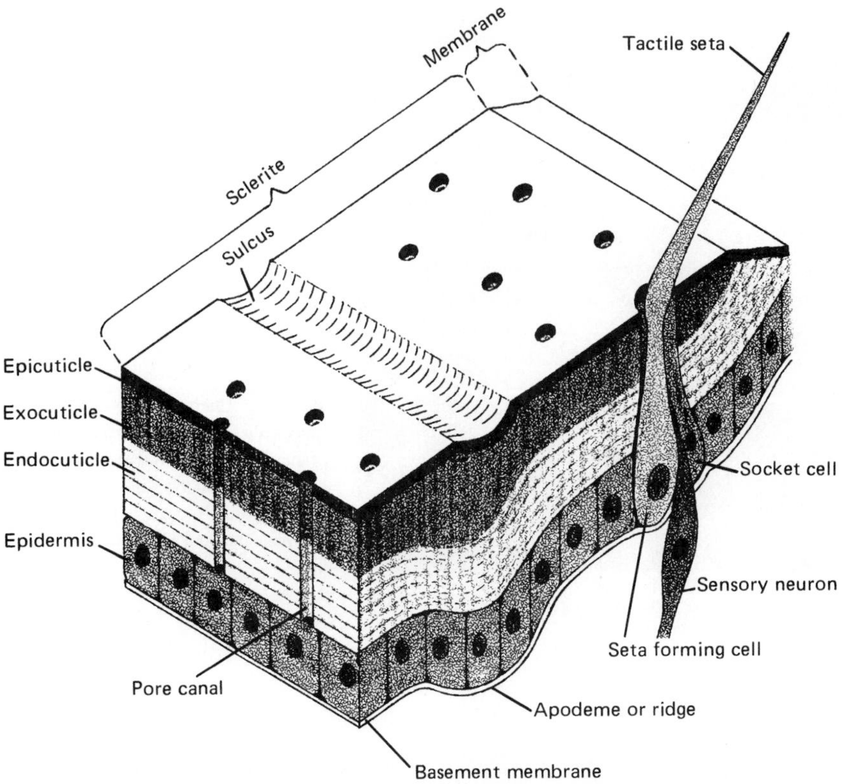

Figure 3. Diagrammatic representation of insect exoskeleton.

tors, a barrier to pathogens and many predators, a reservoir for some waste products, and an excellent structure for attaching a musculature system with good leverage. The cuticle also resists the harmful effects of ultraviolet rays with which terrestrial organisms must cope.

Disadvantages of the exoskeleton include the need for special modifications for gaseous exchange and sensory pickup. The presence of the cuticle also places a major restriction on growth, for only a limited amount of protoplasm can be added until it becomes restrictive; the shedding or *molting* of the cuticle exposes the individual to the physical and chemical forces from which it was previously protected, and water loss becomes critical during this period (Figs. 5, 142). The new exoskeleton also must be larger than the old one or the process would be self-defeating. The actual mechanics of these intriguing paradoxes will be discussed in Chapter 4.

Humans seem to interest themselves in the grandiose things in nature. Dinosaurs, large snakes, and mammals are studied in preference to the average or minute forms. Similarly, no text would be complete without mentioning the large and bizarre examples of arthropods. As with many animals, these unusual species are usually extinct and are found only in the fossil record. The giant of the arthropods appears to have been an

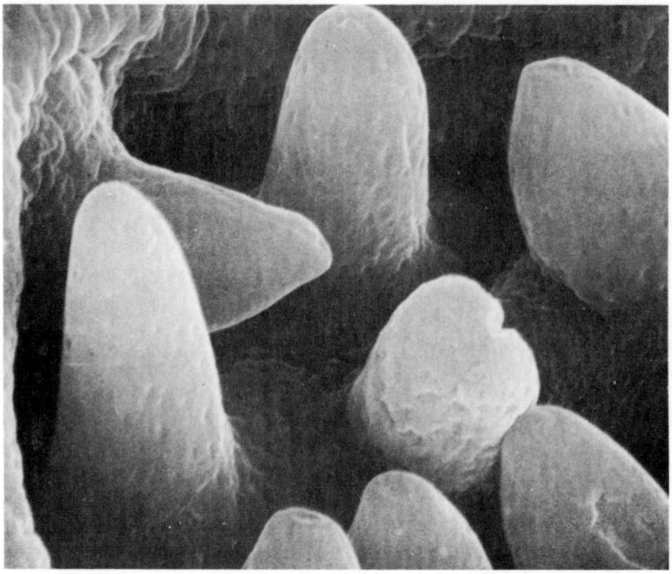

Figure 4. Chemoreceptor setae on maxillary palp of a cutworm caterpillar. The scanning electron micrograph permits magnification to the point of viewing the minute pores through which molecules can pass and contact the internal sensory cells.

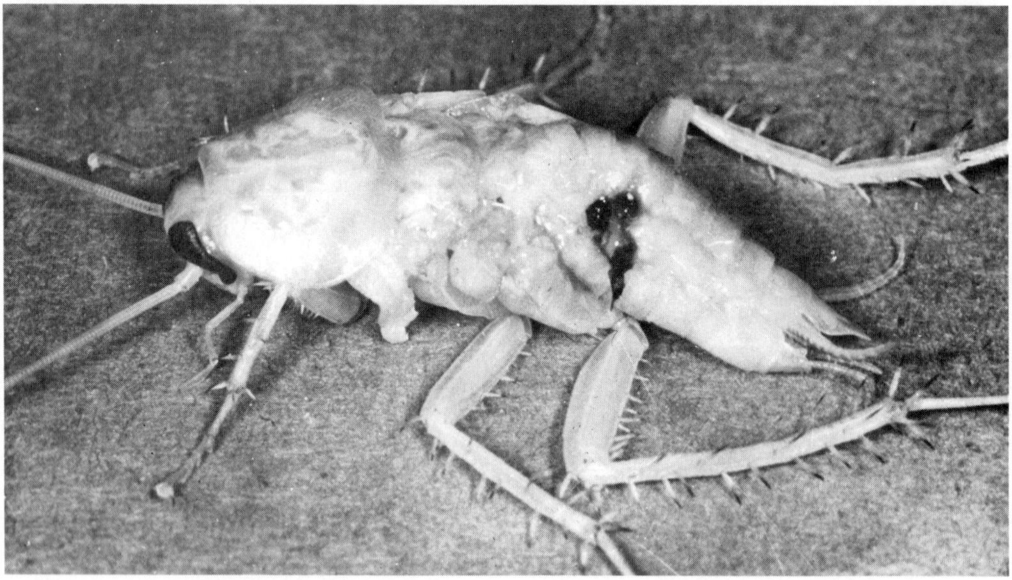

Figure 5. Newly molted arthropods are extremely vulnerable to a dry and hostile environment. This newly molted and vulnerable cockroach, *Periplaneta americana,* has been severely injured by other American cockroaches prior to its hardening of the exoskeleton.

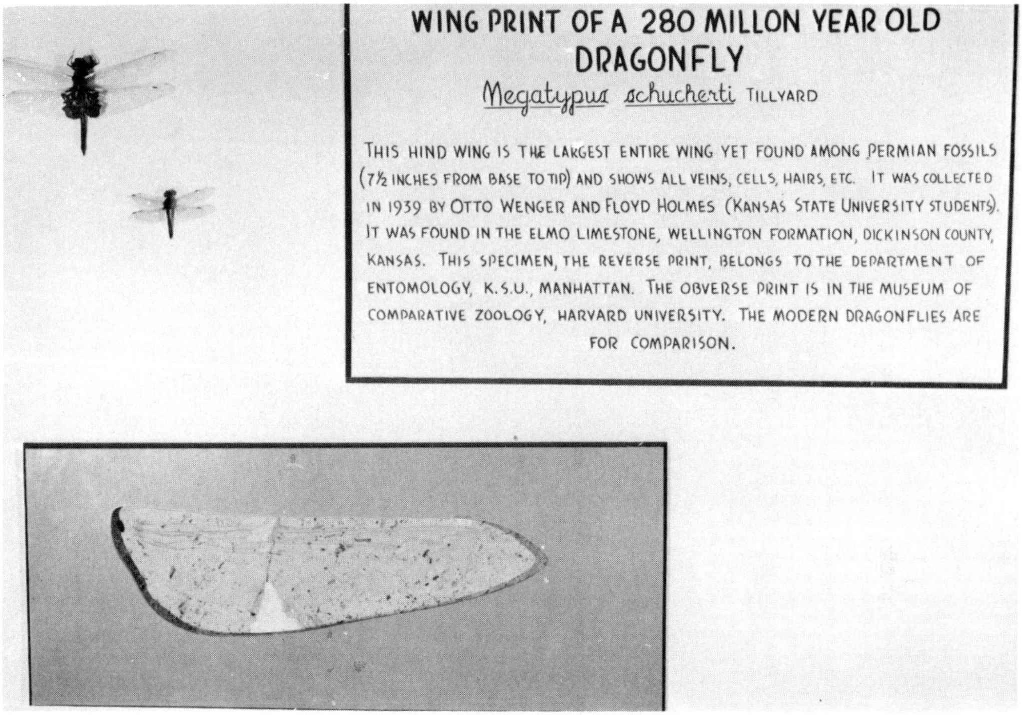

Figure 6. Dragonfly reconstruction based on a fossil wing print from Permian fossils in Elmo, Kansas. The hind wing print is 7 1/2 in. from the tip to base. Compare the reconstruction with the modern-sized dragonflies to the upper left.

aquatic arachnid, resembling a modern scorpion, which attained the length of nearly 2 m. Another formidable fossil is the dragonfly with a wingspan of more than 700 mm, as illustrated in Figure 6. Contrast these examples with the more moderate yet still spectacular modern arthropods such as moths with a 320 mm wingspan, 330 mm walking sticks, the 600 mm Atlantic lobster, the 270 mm centipede, and the 300 mm millipede. The heaviest living insect is probably the Goliath beetle, *Goliathus regius,* from Africa which weighs up to 100 g.

As large as these arthropods are, they are very small compared to such organisms as vertebrates. What has determined the size of these invertebrates? There are many advantages in being small: it is usually the small- to medium-sized species of most animal groups that have survived the long and rigorous geological history of the earth. The major advantages of being small are:

1. Individuals require less energy and time to complete development.
2. Less energy is needed to sustain life both as individuals and as populations.
3. It is easier to find protection from predators and other environmental extremes.
4. A greater number of ecological habitats are available for exploitation than is true for large animals.

Figure 7. An ant trapped within a droplet of water. Size isn't always advantageous.

5. Muscular action is much more efficient.
6. Solar heat can be used to heat the body because of the high surface/volume ratios.
7. Random dispersal by wind action is accomplished with great ease.
8. Individuals can fall great distances without serious injury because gravity has less effect.

As in most instances, advantages often come at a price. Liabilities include such obvious difficulties as predation, physical factors such as being trapped in rain droplets (Fig. 7) and of being captured and borne great distances by wind, and the less obvious, yet equally critical, factors such as the great potential for water loss that comes with small size. Transpiration increases as a function of an increased surface/volume ratio. In other words, the smaller the animal, the less its volume and the greater the ratio. Because small volumes can maintain little water reserves and the evaporative surface is high, serious difficulties can result unless countered by waterproofing to slow down transpiration. With this in mind, let us see how size affects the ratio (without waterproofing) by comparing a 2 cm cube with one that is 1 cm. The surface of one side of the first cube is 2 cm wide by 2 cm high, or 4 cm^2; and the total surface becomes 4 cm^2 times the total number of sides (6), or 24 cm^2. The volume is obtained by multiplying the height by the width by the depth, or 2 cm $\times$ 2 cm $\times$

PHYLUM	
*Arthropoda	945,500
Mollusca	50,000
Chordata	47,000
Protozoa	30,000
Plathyhelminthes	13,000
*Nematoda	15,000
Annelida	12,000
Coelenterata	9,600
Echinodermata	6,000
Porifera	5,000
Ectoprocta	4,000
Misc. Invertebrates	4,000

ARTHROPOD CLASSES	
*Insecta	814,000
*Arachnida	70,000
Crustacea	42,000
Diplopoda	10,000
Chilopoda	3,000
Misc.	6,500

MAJOR INSECT ORDERS	
*Coleoptera	350,000
Lepidoptera	120,000
*Hymenoptera	110,000
*Diptera	100,000
Homoptera	40,000
Hemiptera	25,000
Orthoptera	19,000
Misc.	50,000

Figure 8. Approximate numbers of animal species known based on May (1988) and others. Some estimate the number of insect species as over 20 million, but most lists are more conservative. Asterisk (*) indicates areas where the greatest number of new species probably will be discovered.

2 cm, or 8 cm^3. The surface/volume ratio of this 2 cm cube therefore becomes 24/8 or 3/1 when reduced to the lowest common denominator. In contrast, the 1 cm cube surface is 1 cm $\times$ 1 cm $\times$ 6, or 6 cm^2; and the volume is 1 cm $\times$ 1 cm $\times$ 1 cm, or 1 cm^3 with a ratio of 6/1. The reduction of volume from the 2 cm cube to one of 1 cm was 8 times (8 to 1 cm^3), but the surface area decreased only 4 times (24 to 8 cm^2), resulting in a higher (2$\times$) surface/volume ratio in the smaller cube.

In summary, with the exception of those species that inhabit water and do not have to combat the severity of gravity and high water loss, arthropods are small because of factors operating on their sclerotized exoskeletons.

SPECIES NUMBERS

Arthropods have great adaptability and have radiated into most aquatic and terrestrial habitats. Trilobites, followed by the Crustacea, illustrate the dominant role of this phylum in the marine environment. Insecta and Arachnida have done the same in the terrestrial realm. Although we are far from a knowledge of the total number of species, Figure 8 summarizes current knowledge of animal speciation and shows, to a great degree, the versatility of this group. More than 80 percent of the known animal species are arthropods, and an even higher figure is expected once all species have been classified, depending on the number of nematodes that remain undescribed. Stork (1988) estimated that the number of arthropods is in the 10 to 80 million range; although such a figure may be too high, it illustrates the hyperdiversity of this group. Also, when the 900,000-plus presently described arthropod species are compared to the 300,000 to 400,000 plant and algal species and about 70,000 fungi, the adaptive radiation of arthropods becomes more than sheer estimation.

A cursory view of modern arthropods reveals five major subgroupings or *classes* (see Chapter 13 for classification system). Species in each grouping are more closely related to one another than to those in other groupings because of their tagmosis, origin of major feeding structures, modifications of appendages, location of genital openings, life cycles, and respiratory apparatus. A key is reproduced here to aid in the identification of these classes.

KEY TO THE COMMON ADULTS OF THE ARTHROPOD CLASSES

1. Antennae 2 pairs, 1 pair may be greatly reduced (Fig. 1) . . . . CRUSTACEA
   Antennae either absent or 1 pair 2
2. Antennae absent (Figs. 1, 9) and 4 pairs of walking legs present . . . . ARACHNIDA
   One pair of antennae; leg numbers variable but not 4 pairs 3
3. Many pairs of legs; body divided into head and many-segmented trunk (Fig. 2). 4
   Only 3 pairs of ambulatory legs present; body divided into head, thorax, abdomen; wings may be present . . . . INSECTA
4. Most trunk segments with 2 pairs of legs (Figs. 1, 27) . . . . DIPLOPODA
   Each trunk segment with 1 pair of legs. 5
5. First pair of legs modified into poison claws (Fig. 26); gills lacking; spiracles present. . . . . CHILOPODA
   First pair of legs similar to others; gills may be present; spiracles absent . . . . CRUSTACEA*

*If you arrived here, you failed to see the first pair of antennae and should return to couplet 1 to check your error.

ARTHROPOD CLASSES

Arachnida

Included in the arachnids are the spiders, mites, ticks, scorpions, pseudoscorpions, harvestmen, whipscorpions, and sunspiders (Fig. 9). All are characterized by possessing *chelicerae* and *pedipalpi,* lacking antennae, and having four pairs of legs as adults and a cephalothorax and abdomen. Except for the mites and ticks, specialization only occurs for a carnivorous diet. Most arachnids are terrestrial.

Chelicerae (Fig. 10) are believed to be appendages of the third body segment, usually modified into predatory organs. Chelicerae have two or three segments and most have an opposable "thumb" (*chelate* condition) whereby the prey may be grasped and torn apart (Fig. 11); the exceptions are the many parasitic mites and ticks and spiders in which the thumb is lacking (*unchelate* condition).

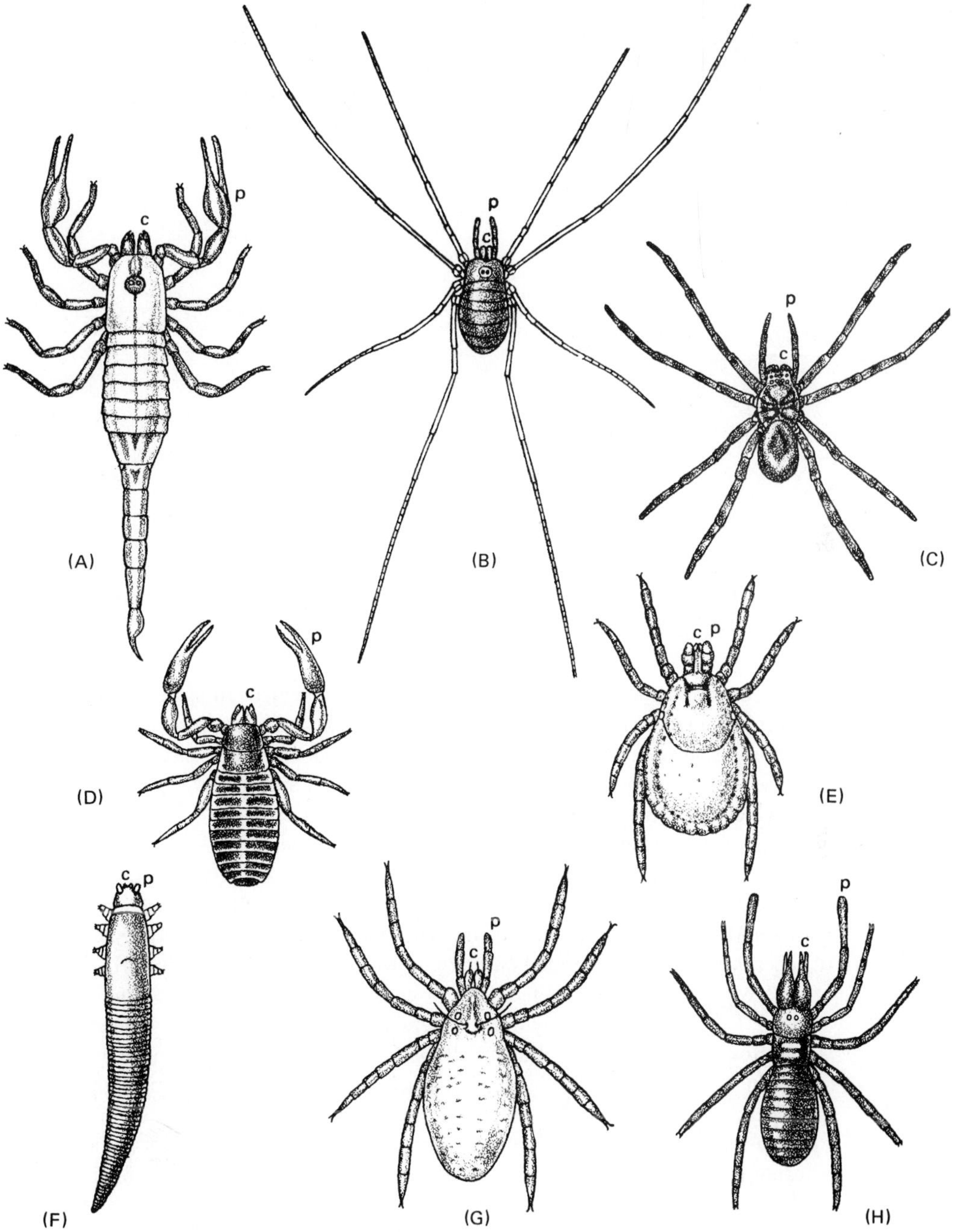

Figure 9. Diagrams of the more common arachnids. (A) scorpion; (B) harvestman; (C) spider; (D) pseudoscorpion; (E) tick; (F) follicle mite; (G) predaceous mite; (H) sunspider; *c*, chelicerae; *p*, pedipalpi.

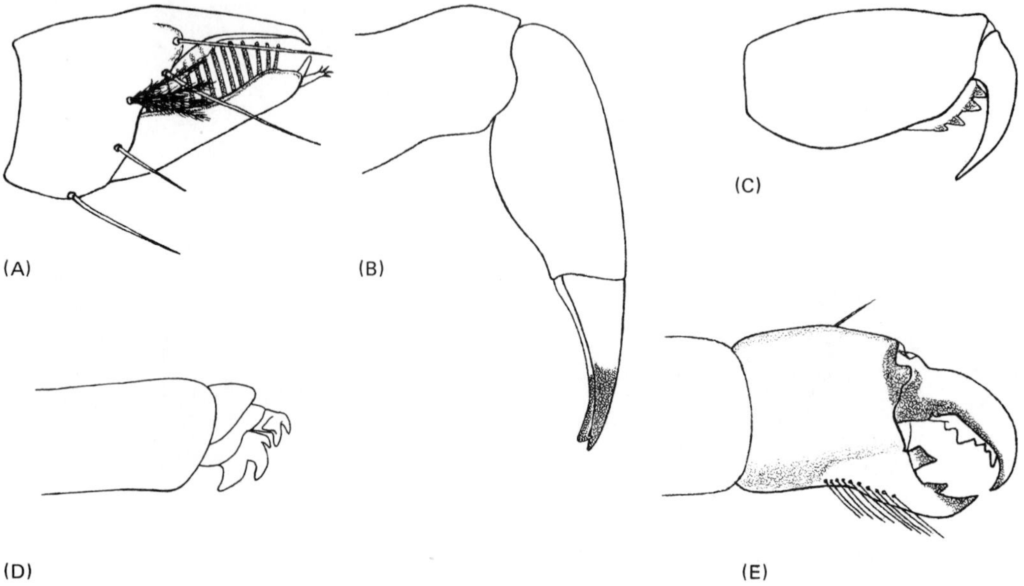

Figure 10. Modifications of some arachnid chelicerae. (A) pseudoscorpion; (B) harvestman; (C) spider; (D) tick; (E) scorpion.

Pedipalpi are variously modified and multisegmented. In some species, such as scorpions and pseudoscorpions, pedipalpi are chelate and are greatly enlarged for capturing prey. In male spiders they become copulatory organs. In most arachnids, however, pedipalpi are sensory structures and also form the base of a preoral cavity in which the chelicerae function.

Silk is produced by many arachnids and has multiple uses, especially for spiders. Silken chambers are spun to cover eggs and form protective retreats for many mites, pseudoscorpions, and spiders (Fig. 12). Young spiders, soon after hatching, draw out silken strands that are captured by the wind and transport them from one habitat to another, a phenomenon termed *ballooning.* Male spiders spin sperm webs, structures onto which sperm is deposited and subsequently withdrawn by the copulatory pedipalpi. Draglines are spun by most spiders and often prevent injury when the spiders fall from a precarious site. Last of all, but certainly of vital importance, is the use of silk in capturing prey. Although most arachnids actively seek or ambush prey, many spiders spin webs whereby sticky silk from abdominal *spinnerets* is formed into a snare for entangling unsuspecting invertebrates (Fig. 13). Regardless of the means of capture, much extraoral digestion of the prey occurs. Digestive fluids are pumped into the wounds (Fig. 14) and mixed with tissues as the chelicerae tear them apart, and then the mixture is sucked into the body by a muscular pump located in the cephalothorax.

Although most arachnids are carnivorous, exceptions may be seen in the ticks and many mites (Fig. 15). Hard-bodied ticks, for example, hatch in great numbers (Fig. 16) and attach themselves to a host by their mouthparts (Fig. 17). After completing a feeding (engorgement), they drop to the ground, molt into a nymphal stage, and reattach

Figure 11. Sunspiders or solpugids have greatly enlarged chelicerae and pedipalpi. The central chelicerae extend from the central eyes outward and are chelate, whereas the pedipalpi are leglike. *C*, chelicera; *P*, pedipalp.

to a new host for another blood meal. Once this has been obtained, the nymph drops to the ground and molts, and the adult subsequently seeks a blood meal. Some species of ticks can transmit disease to humans, wildlife, and domesticated animals. The most common human diseases in the United States are Rocky Mountain spotted fever and Lyme disease.

Many mites also are parasitic during some stage of their life cycle. Although mites have been collected internally from the respiratory systems of both vertebrates and invertebrates, as well as from cloacas of birds and reptiles, most are located on the outer covering of the host's body. The "bite" of the chigger, the larval stage of a specific group of mites, is well known, but less obvious are the skin mites (*Demodex*) that inhabit pores in approximately 50 percent of unsuspecting Americans today [Fig. 9(F)]. These skin mites may also be a scourge among pets and livestock. Most mites, however, are not parasitic but feed on plants and on other arthropods. Phytophagous mites penetrate plant tissues and use their chelicerae to withdraw such liquids as sap, interstitial fluids, or lysed cells. The detrimental effect of these species upon our crops and shrubs by feeding and disease transmission is of economic concern. Less obvious are the mites

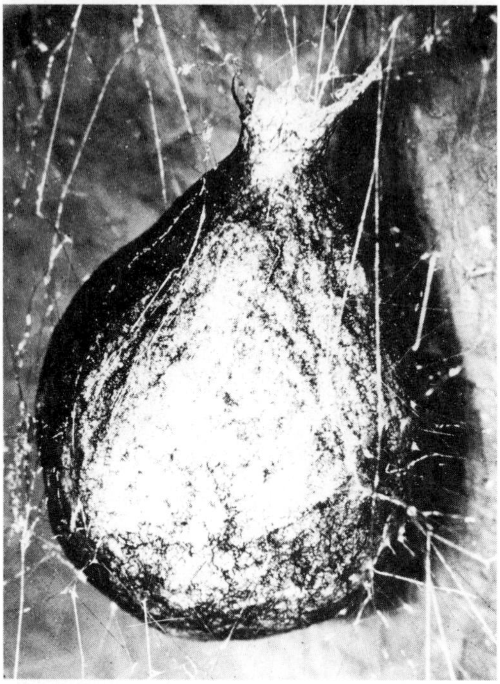

Figure 12. One use of silk by spiders is in covering eggs. This egg sac will moderate the influence of freezing temperatures, provide some protective concealment from predators, and hold the eggs together.

Figure 13. Silk is often used to form a snare by some spiders. The circular orbs are often sticky, and they trap insects flying or jumping into the web. The spider avoids the sticky webbing and moves about on the nonsticky spokes of the web.

Figure 14. Wolf spider (*Lycosa*) feeding on a cockroach. Large spiders such as this do not spin silken snares or webs but must capture their prey by sheer force.

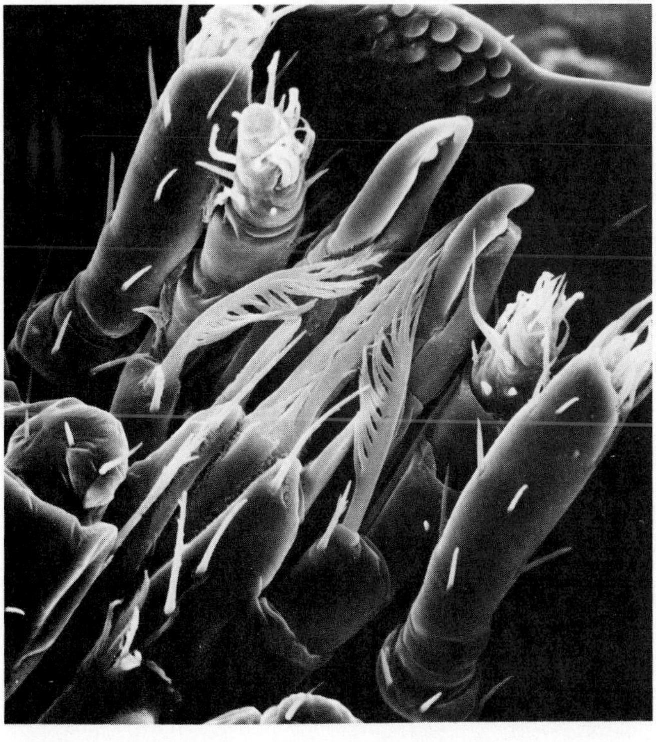

Figure 15. Scanning electron microscope photograph of the mouthparts of a mite (*Planodiscus*) found on army ants.

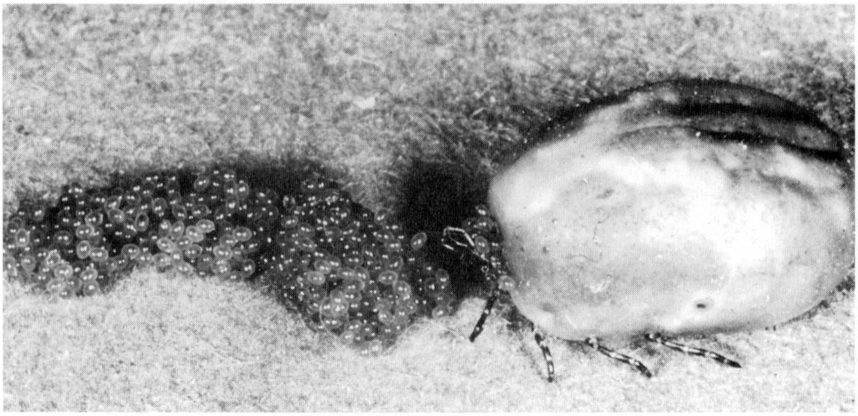

Figure 16. This fully engorged tick (*Dermacentor parumapertus*) is in the process of oviposition. Almost double the number of eggs already seen will be deposited.

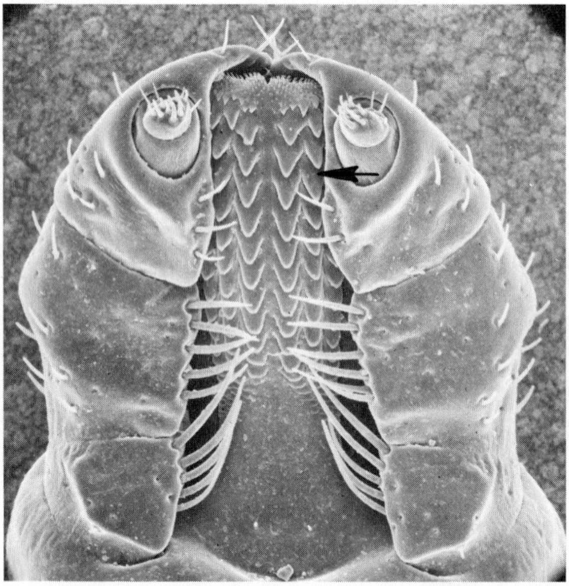

Figure 17. The toothed structure is the hypostome. Once this structure is in place and a "cement" secreted, the tick is difficult to remove.

that feed upon other mites, insects, and arthropod eggs and are beneficial to humans (Fig. 18). In addition to the previously mentioned modes of feeding, many mites are scavengers. One such mite, the house dust mite, feeds on organic material that collects around the home and is a common source of allergies among humans.

Small arachnids obtain oxygen through their exoskeleton and/or minute tracheal tubes. However, large species such as scorpions have *book lungs,* internal sacs with many leaflike membranes, to increase the surface area for gaseous exchange and yet minimize water loss by maintaining this expansion internally. A few species are aquatic

Figure 18. The appearance of giant velvet red mites (*Dinothrombium* sp.) following a desert rainstorm is a dramatic occurrence. A predator, this mite feeds on small arthropods, especially termites.

(Fig. 19) and utilize *gills*. In most arachnids, sperm is transferred inside a *spermatophore,* an often sclerotized "suitcase." After fertilization, development takes two paths. Many, such as spiders and scorpions, hatch from the eggs resembling the adult, and the successive molts produce only minor changes (Fig. 20). In others, such as the mites, more drastic changes take place as the young pass through successive larval (three pairs of legs) and nymphal (four pairs of legs) stages. In both strategies the life cycles vary, but duration is normally a function of size and temperature. Many mites expire after a few months, but a large tarantula may survive for more than 20 years in the tropics.

In addition to the bites and stings of mites and ticks, those by other arachnids may be painful and are sometimes dangerous to humans. Spiders possess poison glands that empty through a duct into the terminal segment of the chelicera. These toxins are normally used to kill prey; but a few species of spiders, including the black widow (Fig. 21) and brown spider species (Fig. 22), also contain enzymes that are harmful to humans and other vertebrates. There are two species of black widow spiders in the United States. One, *Latrodectus variolus,* is primarily a northern form, whereas *L. mactans* is normally found in warmer regions. Both species produce irregular webs in out-of-the-way habitats such as rodent burrows, rock piles, outhouses, piles of

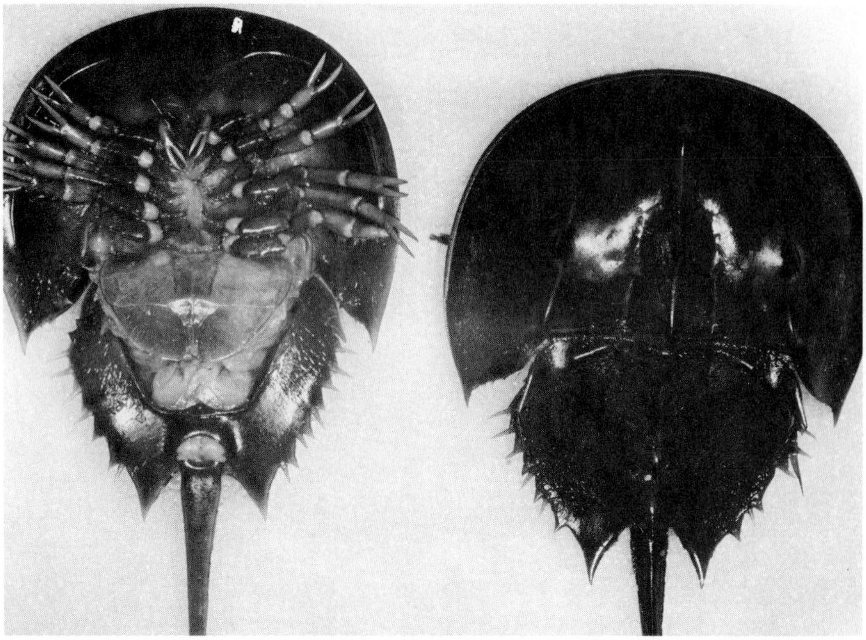

Figure 19. Some arthropods, particularly those that are aquatic and have physical support from water, reach a large size as illustrated by this marine horseshoe crab (*Limulus polyphemus*).

SPIDER	Eggs in silken cocoon	→ Hatch as larva (immobile)	→ Molt and leave cocoon	→ Normal habits as nymph or spiderling	→ 4–20 + molts	→ Adult
SCORPION	Embryos in female	→ Born as larva (ride on female)	→ Molt and leave female	→ Normal habits as nymph	→ 7–8 molts	→ Adult
HARVESTMAN	Eggs in soil	→ Hatch as nymph (nonfeeding)	→ Molt	→ Normal habits as nymph	→ 6–8molts	→ Adult
PSEUDO-SCORPION	Eggs in external female chamber	→ Hatch as larva (attached to female)	→ Molt and leave female	→ Normal habits as nymph	→ 3 molts	→ Adult
MITE	Eggs deposited variously	→ Hatch as larva (nonfeeding, usually)	→ Molt	→ Normal habits as nymph	→ 1–3 molts	→ Adult
TICK	Eggs deposited in soil	→ Hatch as larva (feeding)	→ Molt	→ Normal habits as nymph	→ 1–8 molts	→ Adult

Figure 20. Representative life cycles of the more common arachnids.

firewood, and window wells. The mature female is much larger than the male and hangs upside down in the web exposing a characteristic red or orange hourglass marking on the abdomen. Bites are neurotoxic and can be serious, particularly if the bitten individual is either very young or aged and if serum is not available. The other group of serious biters, the brown spiders, consist of five or six species in the genus

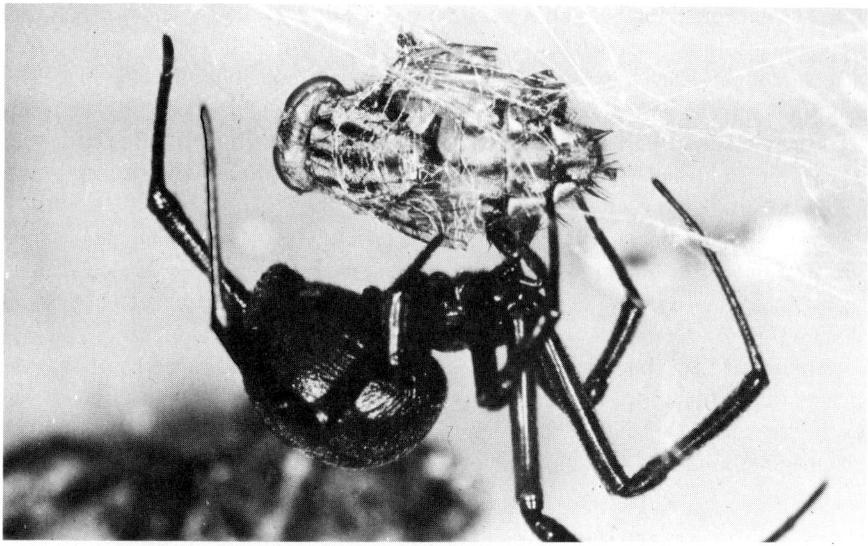

Figure 21. This black widow spider, *Latrodectus mactans,* has combed out sticky silk onto her prey using her hind legs and is about to immobilize the flesh fly, *Sacrophaga* sp., by injecting a toxin.

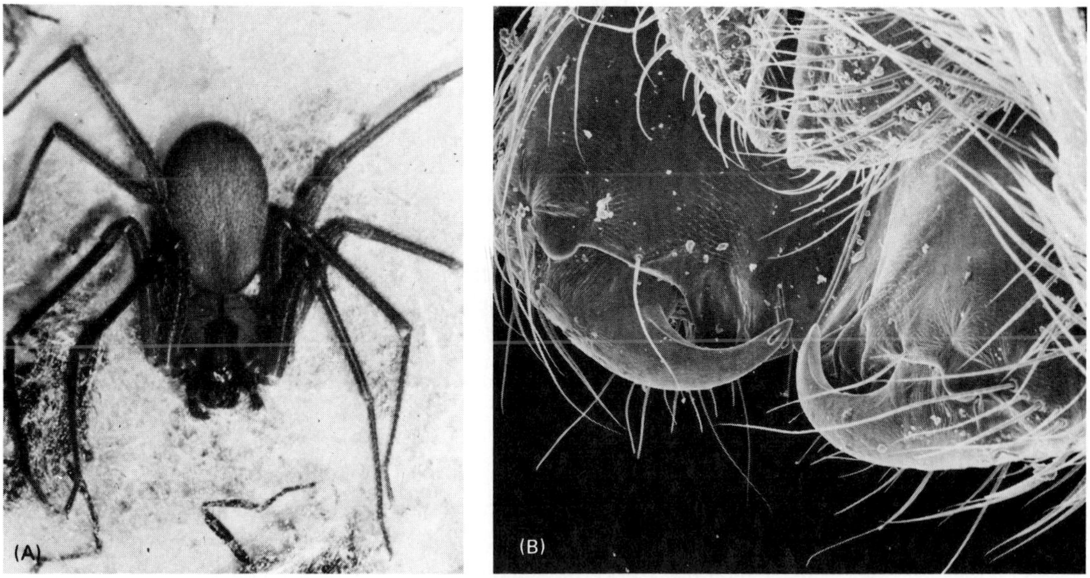

Figure 22. The medically important brown recluse spider (*Loxosceles reclusa*). (A) an overall view indicating the characteristic violin-shaped design on the dorsum of the cephalothorax; (B) posterior view of the chelicerae with the poison duct pore and semi-chelate modification.

Loxosceles. Found mainly in the southern states, all are secretive under rocks, logs, or other undisturbed habitats and may be recognized by their brown color and a characteristic violin-shaped marking on the dorsum of the cephalothorax. One species, *L. reclusa,* may also be found in houses and on occasion will bite humans. The resulting reaction is swelling and localized necrosis with a sloughing of affected tissues. No serum is available for treatment, but early injection of corticosteroids may prevent sloughing if administered within 12 hours.

In addition to spiders, some scorpions also have toxins harmful to humans, but because of their distribution in arid regions of the United States, few humans contact them. Scorpion toxins are injected by the sting located at the terminal end of the abdomen (Fig. 23).

Arachnids are subdivided into orders on the basis of modifications of the chelicerae and pedipalpi and the presence or absence of abdominal segmentation, as well as the presence of specialized adaptations such as stings. The following key will assist the unfamiliar student in identifying the major orders found in the United States.

KEY TO THE COMMON ARACHNID ORDERS

1. Abdomen bearing sting at terminal end of elongate six-segmented tail (Fig. 23). . . . . (scorpions) SCORPIONES
   Tail and sting absent 2

2. Pedipalpi chelate [Fig. 10(E)]. . . . . (pseudoscorpions) PSEUDOSCORPIONES
   Pedipalpi unchelate [Fig. 10(C)]. 3

3. Abdomen distinctly segmented. 4
   Abdomen apparently unsegmented [Fig. 9(G)] 5

4. Abdomen distinct; chelicerae greatly enlarged (Fig. 11) . . . . (sunspiders) SOLIFUGAE
   Abdomen fused indistinguishably with cephalothorax; often legs elongate [Fig. 9(B)] . . . . (harvestmen) OPILIONES

5. Abdomen narrowly united with cephalothorax. . . . . (spiders) ARANEAE
   Abdomen broadly joined to cephalothorax [Fig. 9(E) and (G)] . . . . (ticks, mites) ACARI

Crustacea

Crustacea had an ancient origin, showing up at the beginning of the Cambrian division. This class of arthropods consists of approximately 42,000 species. They are mandibulate, possessing as their major structures appendages called *mandibles.* At least during some stage in their life history, their appendages are branched or *biramous.* Anterior to the mandibles are two pairs of antennae, often used in swimming by some minute species, and posterior to the mandibles are two pairs of *maxillae* that aid in feeding.

When viewing the class Crustacea, one is immediately amazed at the many diverse shapes and sizes (Fig. 24). In addition to the variation in shape, tagmosis is extreme but is usually comprised of either a head, thorax, and abdomen or a cephalothorax

Figure 23. Scorpions coil their abdomen and sting into this typical striking position when attacked. Note also the enlarged pedipalpi that grasp and hold prey.

and abdomen, although some species have little or no divisions of their body into body regions [Fig. 24(C)]. All Crustacea harden their exoskeleton by the addition of calcium salts, hence the origin of the name Crustacea.

Although most Crustacea are marine, some inhabit fresh water (Fig. 25), and a few are terrestrial. Because of their aquatic dependence, oxygen often must be obtained from water where it is in relatively low concentration. Small individuals, with their high surface/volume ratio, are able to take in oxygen through their body surface. Large forms, however, have gills, expansions from their legs, which constantly circulate water across the surface area of the gills to aid in oxygen uptake.

The majority of crustaceans hatch from eggs in a shape very unlike the adult and pass through a series of larval stages. The simplest larva is the *nauplius,* consisting of an unsegmented body, three pairs of appendages, and a single median eye. Species that initiate life with this stage usually progress to the adult through a series of progressive steps separated by molts. At each molt, segments and appendages are added at a "budding zone" near the posterior end of the body until the predetermined adult number is reached. Such a process is termed *anamorphic development.* The three swimming appendages of the original nauplius eventually become the two pairs of antennae and the mandibles of the adult.

Some of the more advanced Crustacea, however, hatch as specialized larvae and bypass the nauplius through comparatively large changes at each molt. As an example, crabs have a *zoea* larva in which many of the abdominal segments are already present at hatching and the mouthparts are functional.

Food often varies according to the size of the individual. Larval stages and small adults are usually dependent on other plankton such as diatoms. The larger crustaceans are usually carnivores or scavengers, although a few are parasites.

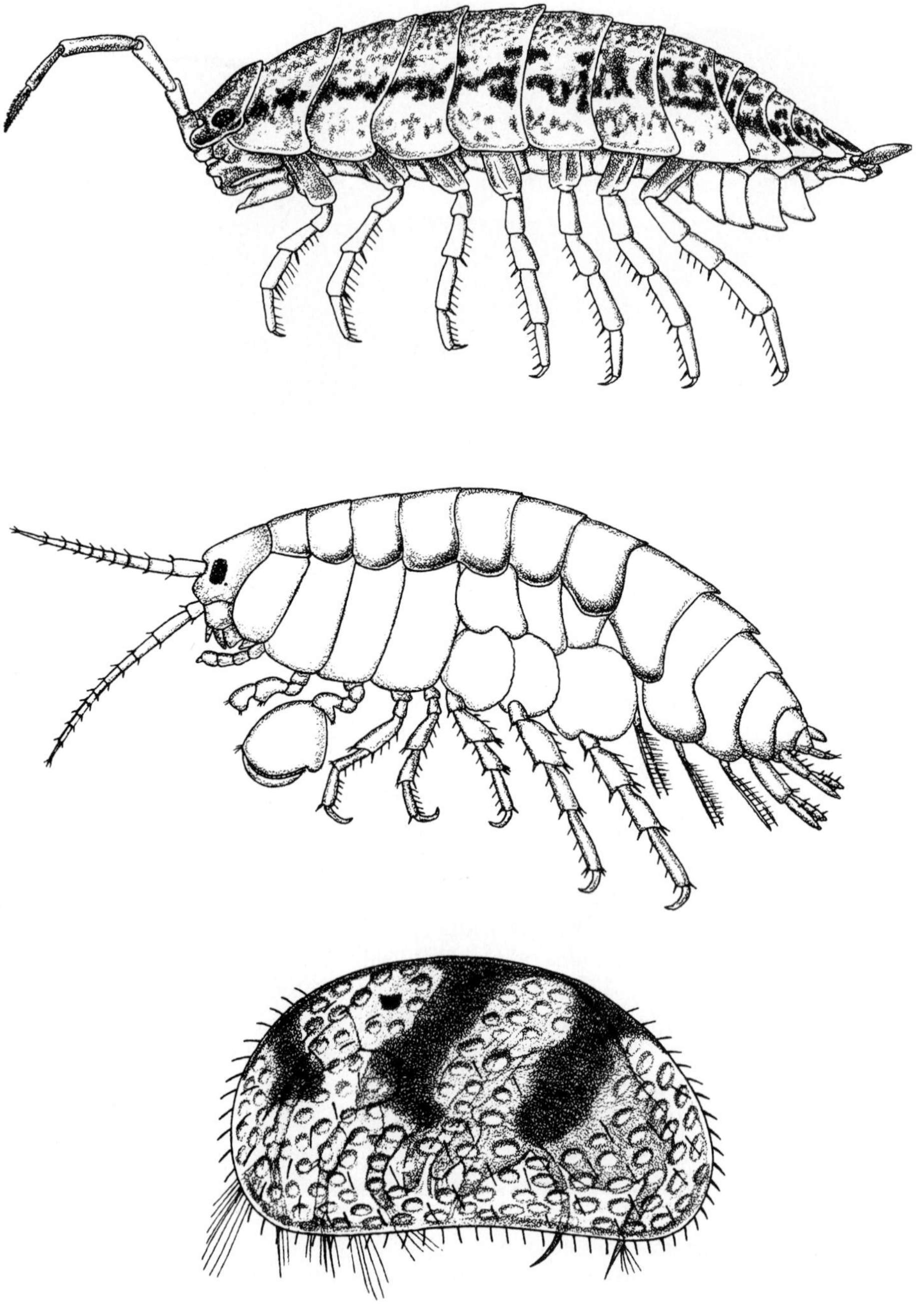

(A)

(B)

(C)

Figure 24. Variations within the Crustacea. (A) terrestrial isopod; (B) aquatic side-swimmer; (C) aquatic ostracod.

Figure 25. This crayfish (*Orconectes* sp.) is contracting its abdomen, a means of rapidly propelling itself posteriorly away from danger.

Chilopoda

Centipedes have a long body divided into a *head* and a multisegmented *trunk.* The head has a pair of moderately long antennae, a pair of mandibles, and two pairs of maxillae. With the exception of two terminal segments, the remaining segments each have a single pair of legs. Some species have short legs that are very efficient in transporting the body through narrow burrows and cracks as they chase prey, but other species, such as the house centipede, have elongated appendages for running in more open situations. The first pair of legs behind the head is modified (Fig. 26) into poison claws or *toxicognaths* used in capturing invertebrates upon which the centipedes feed.

Eggs are deposited either singly or in groups; if they are oviposited in groups, the female usually guards the eggs. Primitive species hatch with few segments and add segments during successive molts, a process mentioned previously as *anamorphic development.* The maximal number of leg-bearing segments is 15 in these primitive species.

The more advanced centipedes are born with the same number of segments as the adult, a condition termed *epimorphic development.* Curiously, these centipedes include the species with the greatest number of segments (up to 190) and appendages. Molting in these organisms results only in a size increase, with the maximal size of approximately 275 mm recorded for one species in South America.

Centipedes usually hide during the daytime but are active at night. Their food consists of other arthropods, earthworms, and slugs. Although efficient in killing their prey, the toxin from their poison claws is rarely dangerous to humans.

Figure 26. A centipede head and poison claw (*P*).

Diplopoda

The millipedes represent another group that has a body differentiated into a head and trunk. The head contains a single pair of short antennae, a pair of *mandibles,* followed by a single pair of fused maxillae, the *gnathochilarium.* Unlike the centipede, most of the trunk segments are actually *diplosegments* (Fig. 27), a condition in which a pair of segments fuse dorsally but not ventrally so that each "segment" appears to have two pairs of appendages. The genital pore is located anteriorly (Fig. 1).

Eggs are deposited in groups of 25 to 250 in soil or moist humus. Larvae possess only three pairs of legs at hatching, but diplosegments are added by anamorphosis in groups of up to three or more at a molt until the predetermined number is reached. Although the name millipede infers 1,000 legs, a maximum number of legs known is 375 pairs, and most have fewer than 50. The longest millipede is 300 mm. From 1 to 7 years are required to complete a life cycle.

Molting often occurs in earthen chambers (Fig. 28) in which water loss is minimized and in which protection from predators is maximized. The exoskeleton, unlike

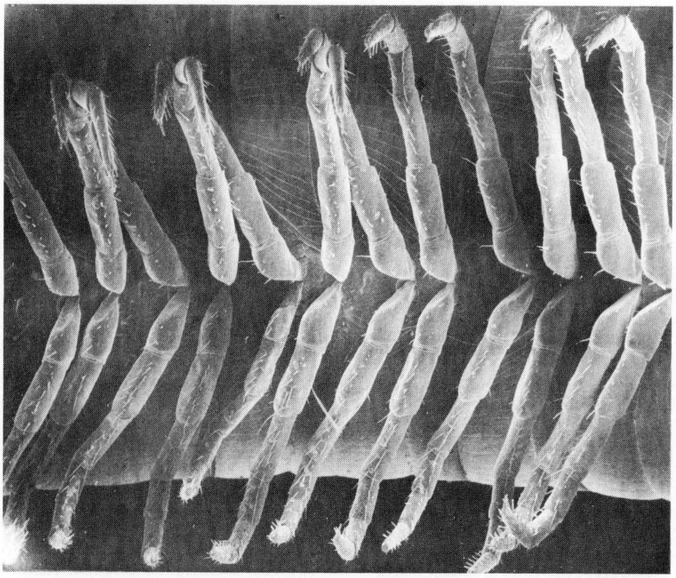

Figure 27. A scanning electron microscope photograph revealing the diplosegments of a millipede.

Figure 28. Earthen cell in which a millipede is estivating during the Costa Rican dry season. Such behavior reduces water loss to a minimum.

that of the centipedes but like that of the crustaceans, is hardened by incorporating calcium carbonate.

Most millipedes are found in rotting logs, leaf litter, humus, and under stones where they feed mainly on decaying vegetative material and fungi. Because their legs are numerous, close together, and short, locomotion is relatively slow and occurs by waves of leg movement down the trunk in synchronous fashion. When disturbed, millipedes often roll into tight coils to gain maximal protection from their hard exoskeleton. Many secrete noxious fluid such as hydrogen cyanide from specialized glands to repel predators.

Insecta

Insecta (insects) are arthropods that, as adults, have three body regions or tagma, three pairs of thoracic legs, and a pair of antennae, and they feed by mandibles. The number of body segments consists of from 19 to 20 in most species. Insects are the only invertebrates with wings. Wings have had a marked influence upon the success of these species. Most insects are terrestrial, although a few occupy aquatic habitats during part of their existence, and only the ocean has eluded extensive exploitation.

Development is epimorphic, except in the order Protura (which may not be a true insect order), and no new segments are added after hatching from the egg. Changes vary from minor (incomplete metamorphosis) to drastic (complete metamorphosis). Most insects complete development in a year or less, but a few require up to 17 years to mature.

Special attention to the great number of insectan species should be made. Why have they become so successful? All the reasons are obviously not known, but some of the major ones are probably:

1. Their small size.
2. Modification and exploitation of appendages into the many types of mouthparts and locomotion-food gathering legs.
3. Extensive development of complete metamorphosis in which the immatures and adults have evolved to feed on different foods and, hence, ephemeral sources can be exploited with little competition among individuals of the same species.
4. Rapid life cycles wherein mutations can be rapidly selected for and incorporated into the population gene pool.
5. The many different species-isolating mechanisms involving genital, hormonal, and behavioral modifications.
6. Seasonal variations wherein one generation can vary from another.
7. Possession of wings during the adult reproductive stage, particularly the type that can be folded (flexed) when not in use.

The success story of this group is one that all biologists should be exposed to, and it will unfold in subsequent chapters.

QUESTIONS

1. What is an *arthropod*? How does it differ from other invertebrates?
2. What are the advantages and disadvantages of an exoskeleton? How are the disadvantages overcome by arthropods?
3. What is *tagmosis*? Of what advantage is this process? What are the evolutionary trends as seen in Crustacea, Insecta, and Arachnida?
4. How does the number of species represent evolutionary success? What classes have the most species?
5. Is small size advantageous or disadvantageous? What are the selective pressures that interplay and affect size in arthropods?
6. Which are the most closely related: spiders and scorpions or centipedes and millipedes? What evidences are available that you can use to determine which are most closely related? How are relationships determined?
7. Compare the classes of arthropods according to morphological, developmental, and habitat preference.
8. What is the difference between anamorphic development and epimorphic development? In which arthropod class is each type of development found?
9. Are any arthropods of potential danger to humans? In what way? How does such knowledge affect your life?
10. How many pairs of appendages are believed to be in each embryonic body segment in the ancestors of arthropods?

2 The Insect Externally

THE HEAD

As indicated in Chapter 1, the most anterior region or tagma of the insect body is the head. What is a head? Are all animal heads the same? Embryologists have noted that heads are formed differently in various animal groups, but, in each instance, an assemblage of structures carries out (1) ingestion of food, (2) major sensory perception, (3) coordination of bodily activities, and (4) protection of the coordinating centers. Sensory structures also are located near the mouthparts and coordinating centers to ingest food more effectively and to enable more rapid responses to incoming stimuli. The head, therefore, is defined as a *functional unit* responsible for carrying out the above four processes.

In insects the individual segments of the head have fused so completely that external evidence of their presence or origin is lost. How then can we determine the head's segmentation? As with most animals, primary evidence of origin comes from studying embryology and the nervous system. Research has found that these two factors are the least likely to be modified by adaptive selection, but we expect and find in as diversified a group as insects that even these factors have undergone sufficient change to prevent decisive conclusions. However, without going into supportive data as found in Matsuda (1965), most entomologists currently believe that the insect head consists of the first six metameres or segments; this assumption will serve as the basis for further discussions.

As in most animals, the insect head is covered by hardened elements to assist both in the protection of the coordinating centers and in the process of feeding. As expected, the least modifications occur in primitive or the least specialized species in which linear invaginations or *sulci* (sometimes referred to as *sutures*) may be seen along the base near the mouthparts and extend dorsally. These folds are braces to prevent collapse of

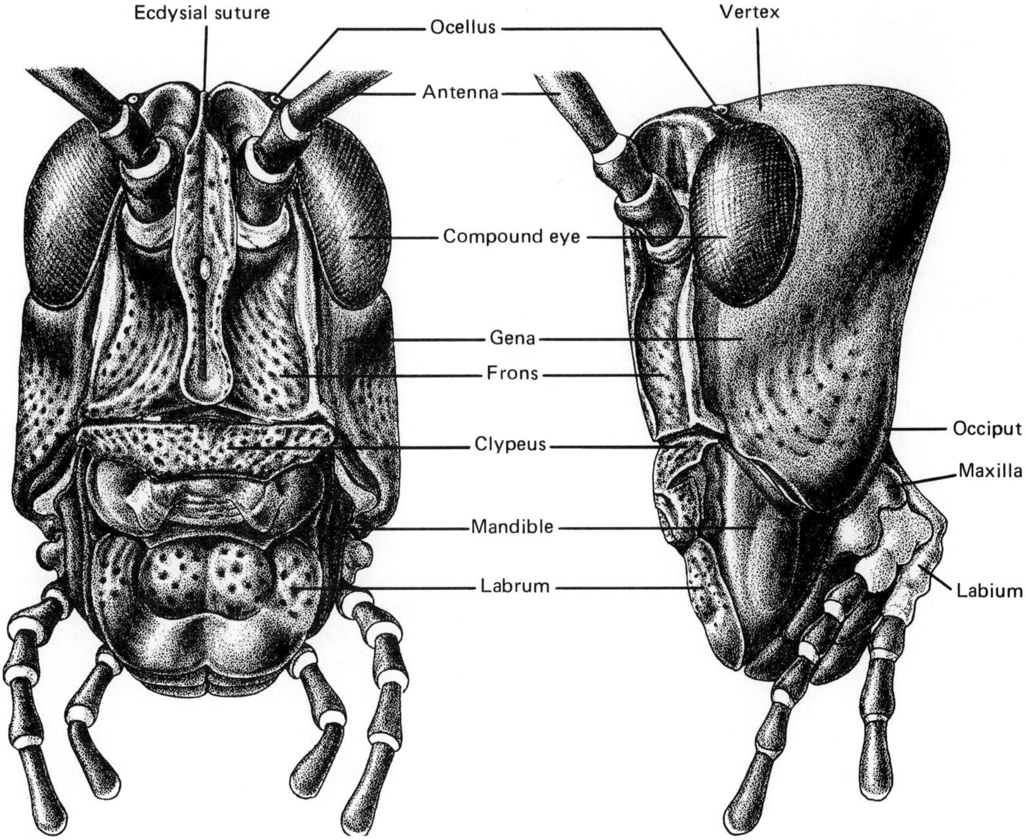

Figure 29. The insect head illustrating the major sclerites, mouthparts, and sensory structures.

the head during feeding or when subjected to external physical stresses. If one creases paper or any flattened yet semirigid material, increased strength is gained. Similarly, sulci can brace heads without substantial thickening. Because evolutionary selection, however, has resulted in harder and thicker head skeletons in the advanced insects, the importance of sulci has diminished accordingly. This will be explained shortly.

The exoskeletal plates between the sulci are termed *sclerites,* which have no segmental value and are useless in elucidating head segmentation. As seen in Figure 29, the sclerite located in the "forehead" region is termed the *frons.* Below this plate is the *clypeus* and the flaplike *labrum.* Beneath the compound eye is the *gena* and above is the *vertex.* A sulcus separates the gena from posterior *occiput.* Only the ventral (for feeding) and posterior (for communication with the remaining portion of the body) parts of the head capsule remain uncovered by sclerites.

Are all external linear lines on the cranium sulci? The answer is yes for the adult stage but no for immatures. Without some weakened area, molting would not be possible, the exoskeleton of the head would remain in place, and no growth could continue. In immature insects there exists between the antennae a λ-shaped *ecdysial suture* that

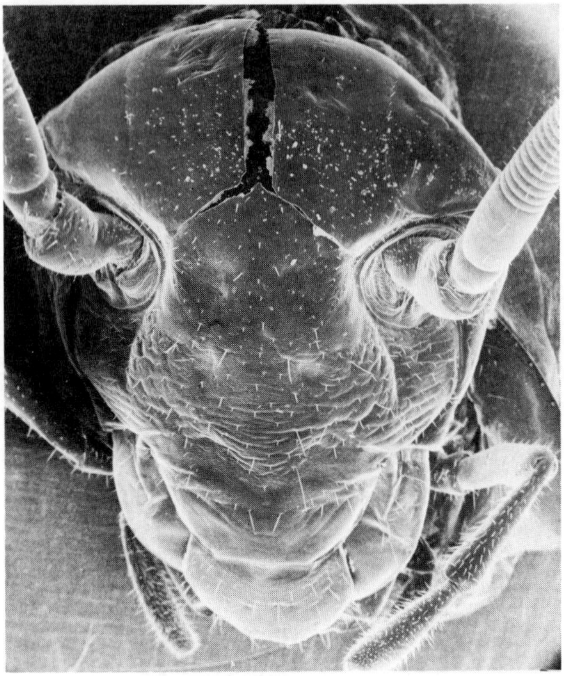

Figure 30. Scanning electron microscope photograph of a cockroach molting. Note the split along the ecdysial suture, which will permit the newly formed cranium to escape.

extends posteriorly to the membranous neck or *cervix.* Because this suture has no exocuticle and hence cannot serve as a brace, it fractures during molting (Fig. 30) and permits the old head capsule to be shed.

Bracing of the head occurs internally by means of the *tentorium* (Fig. 31). This structure is formed by localized invaginations from certain sulci at the base of the head capsule, and because of its double triangular shape, the tentorium has great strength and assists in preventing collapse of the head.

As indicated previously, a progressive loss of sulci occurs with the concurrent thickening of the head capsule and evolution of the various mouthparts. The thickening or increased sclerotization now takes the stresses once imposed on the sulci, and the once obvious sclerites become poorly delineated regions as the sulci become indistinct. Also, the head position may become modified from the early *hypognathous* condition in which the mouthparts hang at right angles to the body axis, to the more advanced *opisthognathous* condition in which they project backward between the legs. Another modification, the *prognathous* condition in which the mouthparts project forward, is commonly seen in carnivorous species and forms that burrow in wood or soil. When the insect is prognathous, the prey can be grasped easily in cramped quarters with little danger to the predator, or the burrow can be easily extended because of the projecting mouthparts.

Feeding in multicellular organisms often requires abrasion or chewing of the food. In mollusks this is accomplished by means of a tonguelike radula, in chordates by jaws and teeth, and in arthropods by modified appendages (refer to Chapter 1). The primary parts in insects are the *mandibles* [Fig. 32(A), (B), and (C)] or paired

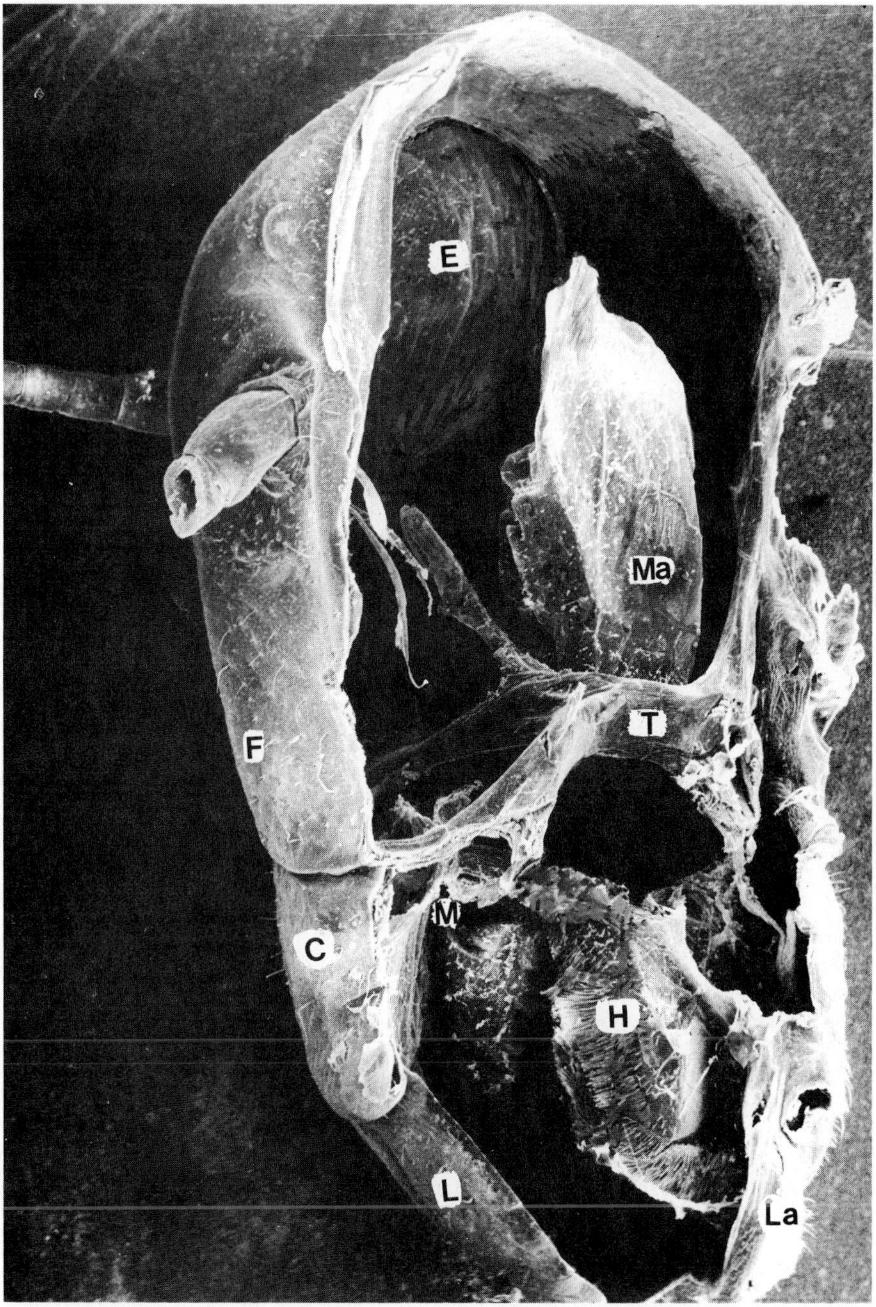

Figure 31. Internal view of a grasshopper cranium after removal of soft tissues, as recorded by the scanning electron microscope. *C*, clypeus; *E*, compound eye; *F*, frons; *H*, hypopharynx; *L*, labrum; *La*, labium; *M*, mouth; *Ma*, mandibular apodeme; *T*, tentorium.

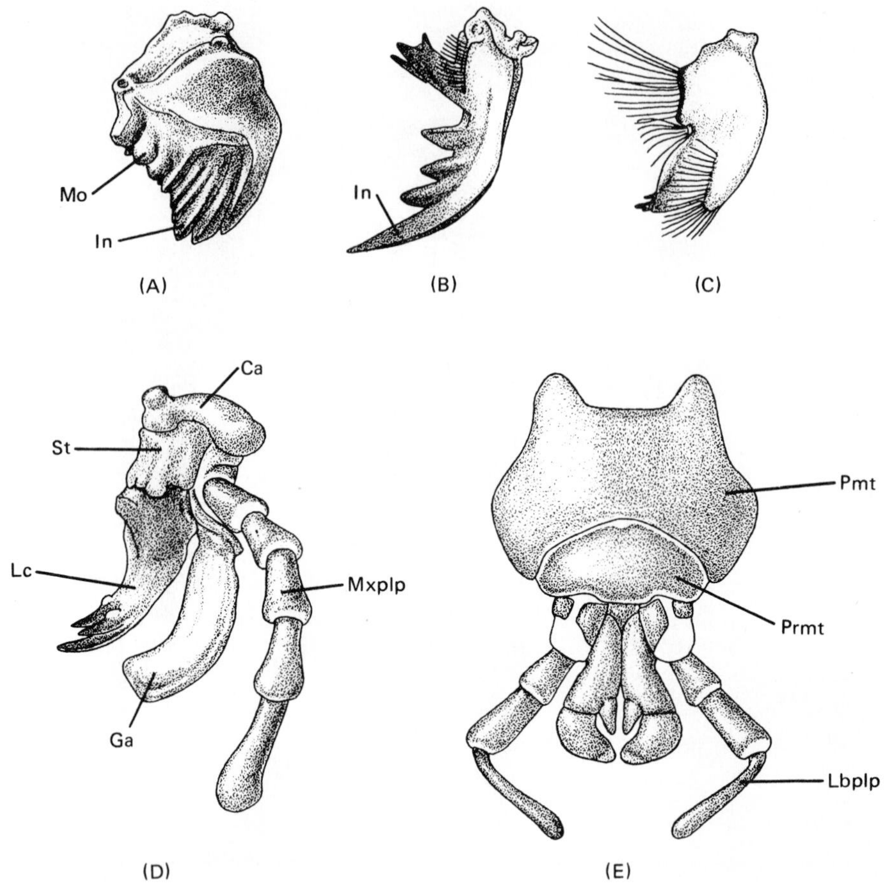

Figure 32. Some modifications of insect mouthparts. (A) mandible for feeding on plant material; (B) mandible of carnivorous insect; (C) mandible of scavenger modified for filtering food from water; (D) maxilla of herbivore; (E) labium of herbivore. *Ca*, cardo; *Ga*, galea; *In*, incisor; *Lc*, lacinia; *Lbplp*, labial palp; *Mo*, molar; *Mxplp*, maxillary palp; *Pmt*, prementum; *Prmt*, postmentum; *St*, Stipes. [Redrawn with slight modifications from Snodgrass, 1935.]

appendages of the fourth segment. Supporting their action are the *maxillae* [Fig. 32(D)] or paired appendages from the fifth segment and the *labium* [Fig. 32(E)] or fused appendages of the sixth segment. Other parts may be involved and include the *labrum* and *hypopharynx* (Matsuda, 1965).

Are all insectan mouthparts similar? This can be answered by asking a second question, "Are all diets of insects similar?" One of the evolutionary marvels has been the numerous modifications that have permitted these invertebrates to feed upon nearly all organic materials. For simplicity, let us view some of these specializations under six major categories: chewing, cutting-sponging, sponging, siphoning, piercing-sucking, and chewing-lapping. Remember that there are many variations of each to exploit the many food sources available in nature. Note that only the chewing type

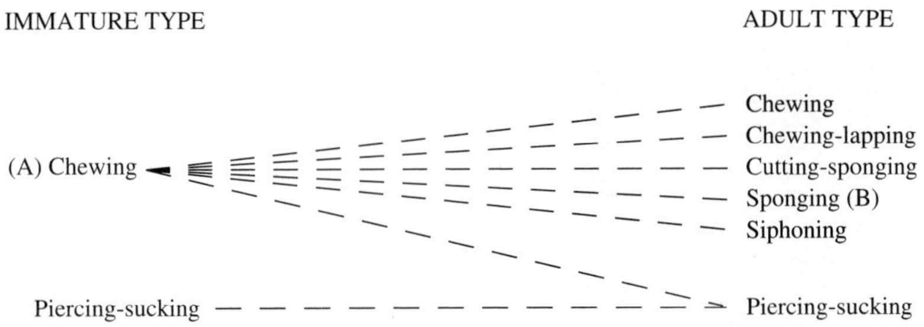

Figure 33. Metamorphosis of types of larval mouthparts into those seen in adults. A, chewing involves some highly specialized mouthparts in Neuroptera and Diptera; B, sponge may be reduced, leaving only scraping or cutting teeth in some species.

is specialized for uptake of solid materials, although many of the others may require penetration of solids to contact the liquid food.

Chewing type. The chewing type is the least specialized and is considered to be the basic insectan type. It is also the most common and is found in most immature stages (Fig. 33). Let's use the June beetle larva (Fig. 34) as an example and view the structure and the feeding process together. The distal portions of the heavily sclerotized mandibles have cutting edges, and their bases have grinding surfaces. Note the arrangement of distal cutting surfaces and that the proximal grinding is similar to the tooth position and function in the jaws of mammals. The mandibles move across one another scissorlike, first cutting with the tips (incisor area) and then grinding with the basal *molar* [Fig. 32(A)]. This is similar to what occurs in the human mouth with the incisor and molar teeth. The muscles operating the mandibles are the largest in the body and normally originate from the vertex. From this dorsal position they extend ventrally to attach to large mandibular apodemes (Figs. 31, 82). Because of their chewing action, the head must either be braced by sulci or have thickened sclerotization.

The anterior labrum, the mandibles and maxillae at the sides, and the posterior labium form a *preoral cavity* between the mouthparts in which chewing occurs (Fig. 35). Food is pulled into this cavity by the maxillae and cutting action of the mandibles. Saliva is secreted between the labium and hypopharynx (Fig. 35) and is mixed with the food during chewing to assist in swallowing. The food bolus is then forced by the maxillae and hypopharynx up into the mouth for swallowing.

Are all mouthparts the same in the chewing type? The answer is a hearty no. Predaceous insects usually have mandibles that are long, pointed, and very sharp [Fig. 32(B)]. These are used similarly to canine teeth in mammals in killing the prey. Because little grinding is required, the molar area is noticeably reduced or absent. Just as mammal herbivores have different tooth modifications resulting from eating different foods of various consistencies, so also do insectan herbivores; those that feed on grasses or seeds have greater grinding surfaces on the mandibles than do those that chew soft succulent leaves.

In all instances, feeding is actually the culmination of a series of behavioral responses to stimuli. Using a phytophagous insect as an example, we see that the sequence typically

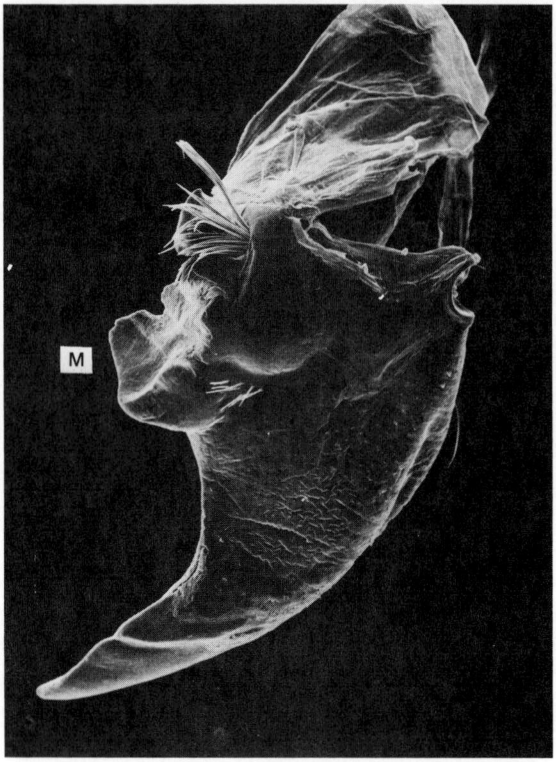

Figure 34. The mandible of a June beetle larva as seen through the scanning electron microscope. Note the enlarged molar area for extensive grinding (*M*).

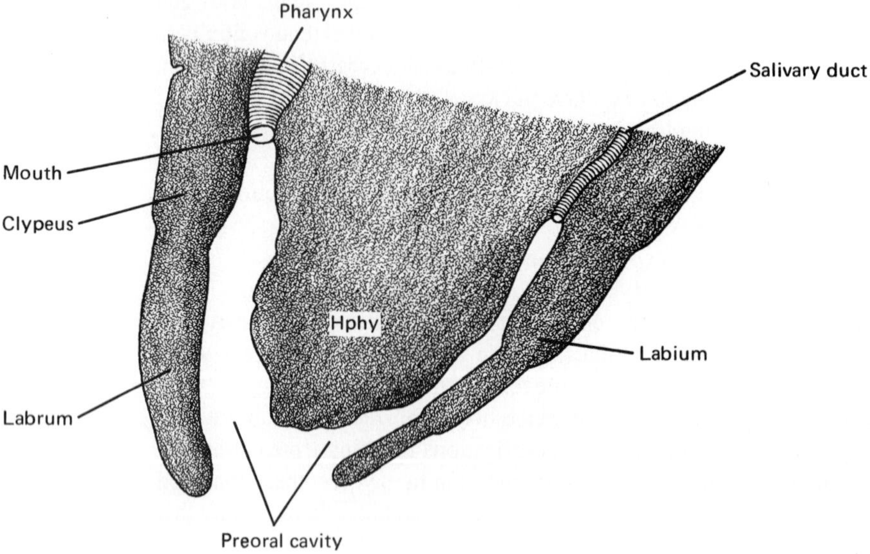

Figure 35. Cross section through the insect head indicating the preoral or chewing cavity and openings into the alimentary canal and salivary duct. (Redrawn with slight modifications from Snodgrass, 1935.)

follows the pattern of first being attracted to the food by odor, sensing the plant through contact receptors on the tarsi or antennae, taking a test bite whereby the taste receptors, especially those on the palpi (Fig. 4), can further discriminate, and finally feeding if the food is acceptable.

Cutting-sponging. The cutting-sponging type [Fig. 36(A)] is restricted to a limited number of adult flies feeding as parasites upon blood from mammal hosts. Black flies and horse flies are good examples. The mandibles and maxillae are elongated and pointed and function as *stylets* to pierce the skin. Once the capillary networks are pierced and disrupted, blood is released, sponged up into the labium with its many canallike pseudotracheae, and sucked into the body through a specialized food canal between the labrum and hypopharynx. Anticoagulants, found in the insect's saliva, are pumped into the wound to prevent blood clotting and may keep the blood flowing from the wound for some time subsequent to the departure of the fly.

Sponging. Only adults of the more specialized flies, such as the house fly, belong to the highly modified sponging type [Fig. 36(B)]. The proboscis consists primarily of the labium with its spongelike *labellum;* the mandibles and maxillae, except for the maxillary palpi, are either lost or incorporated into the basal elements. During feeding, the proboscis is lowered, and salivary secretions are pumped out onto the food. The food may be abraded mechanically by means of small centrally located prestomal teeth or by other teeth forming the *pseudotracheae,* the sponging canals (Fig. 37). The dissolved or suspended food is then moved by capillarity through the numerous pseudotracheae to a median reservoir, the *prestomum.* Certain dipteran parasites, such as the stable flies and tsetse flies (Fig. 38), have lost the pseudotracheae and have elongated the labium into a beak terminated with sharp teeth for piercing tissues and liberating blood. Elzinga and Broce (1986) discuss labellar modifications in great detail. In any of these modifications, the sucking pump now draws up the mixture through the labral-hypopharyngeal tube into the alimentary canal.

Tasting is carried out by sensory hairs on the feet or tarsi and on the labellum. The scanning electron microscope photograph (Fig. 4) has recorded receptors similar to these.

Siphoning. Almost every naturalist has observed a butterfly or moth land upon a flower, uncoil, and extend its proboscis. If nectar is present, this fluid is sucked into the body [Figs. 36(C), 39]. The proboscis then coils up because of its natural elasticity, and the next flower is visited.

The proboscis is highly specialized, consisting of only the greatly elongated galea of the two maxillae held together medially by interlocking spines and hooks. Traversing this tube is the food canal and salivary duct. Uncoiling the proboscis is the result of the contraction of tiny oblique muscles, and coiling results from relaxation of these same muscles plus the natural stiffening of the cuticle (somewhat similar to a watch spring). Nectar is sucked into the alimentary canal by the sucking pump. The other potential mouthparts, the mandibles and labium, are reduced or lost and have no role in food ingestion.

Piercing-sucking. A wide variety of distantly related insects have the piercing-sucking type of mouthpart [Fig. 36(D), (E)]. Herbivores such as the cicada, parasites

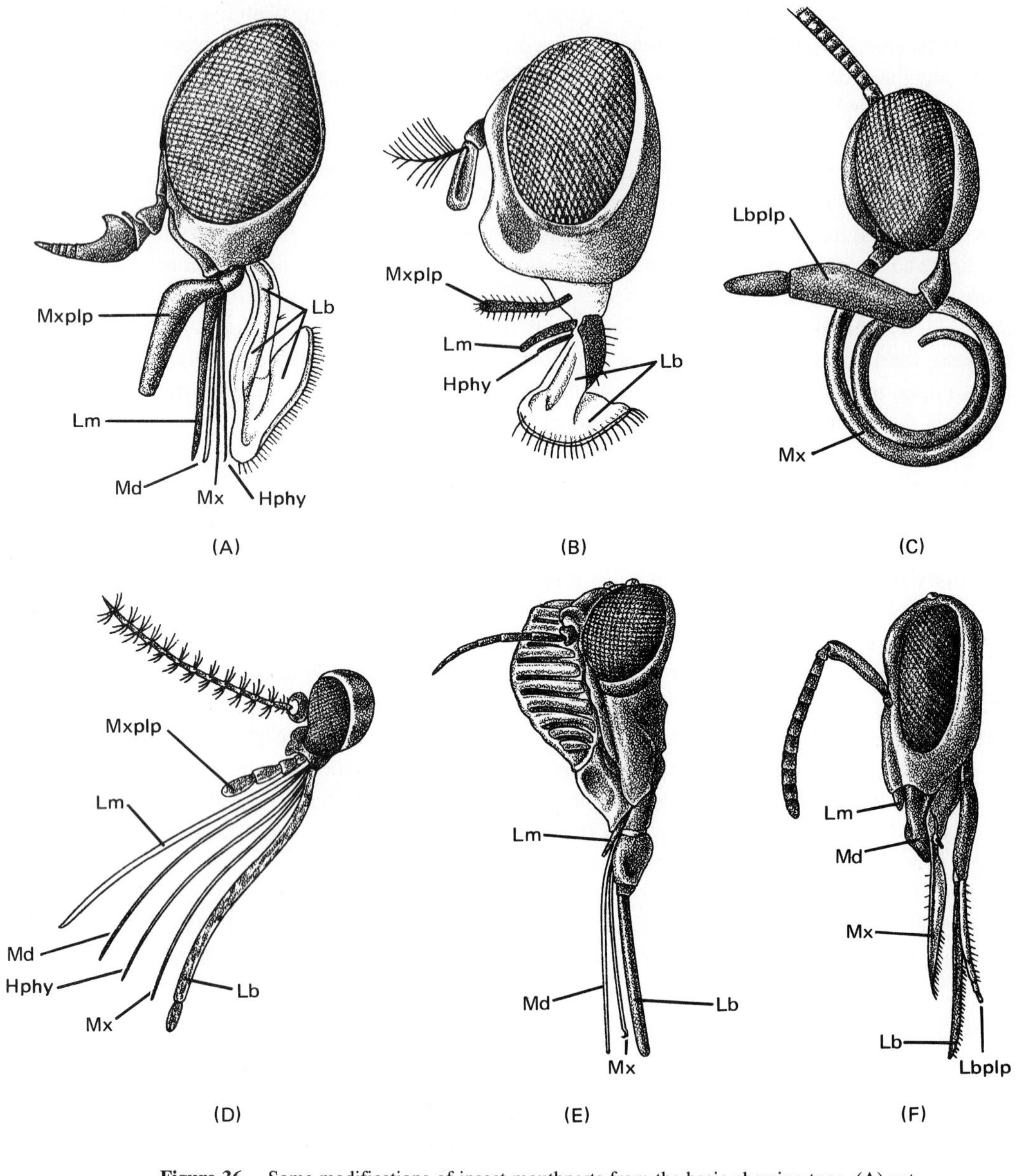

Figure 36. Some modifications of insect mouthparts from the basic chewing type. (A) cutting-sponging of a horse fly; (B) sponging of the house fly; (C) siphoning of a butterfly; (D) piercing-sucking of a mosquito; (E) piercing-sucking of a cicada; (F) chewing-lapping of the honeybee. *Hyphy*, hypopharynx; *Lb*, labium; *Lbplp*, labial palp; *Lm*, labrum; *Md*, mandible; *Mx*, maxilla; *Mxplp*, maxillary palp.

Figure 37. Scanning electron microscope photograph of the labellum or "sponge" portion of the face fly proboscis. Note the canals or pseudotracheae for channeling fluids and the prestomal teeth (*arrow*) surrounding the central "mouth" or prestomum (*P*).

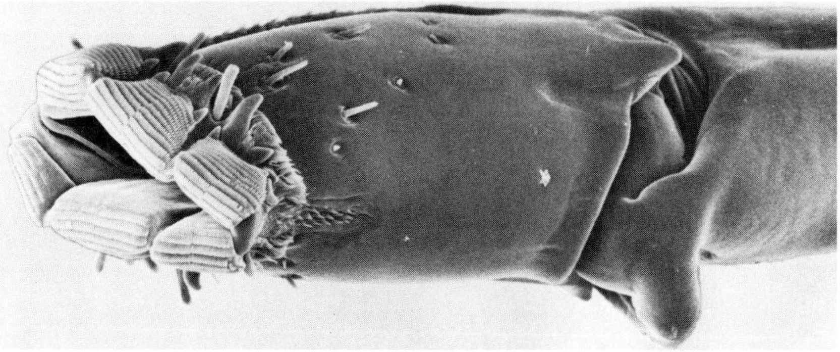

Figure 38. The terminal portion of the proboscis of a tsetse fly, *Glossina morsitans.* Note the rasps that are being everted outward in a cutting action, a motion that ruptures blood from the mammal host.

such as fleas and mosquitoes, and carnivores such as assassin bugs are but a few examples of modifications for different diets. There are three basic subtypes: one has an elongate beak, a second has a short cone-shaped beak (thrips), and the third has no external beak but the parts are retracted into the body (sucking lice). The first type is the most common and consists of a series of elongated and toothed stylets (mandibles, maxillae, and sometimes the hypopharynx and labrum) enclosed by a sheath or labium to hold the stylets in position (Fig. 40). The number of stylets varies up to six in the mosquito. These move up and down, piercing tissues. Molecular cohesion of the water

Figure 39. Siphoning mouthparts of a skipper. The proboscis has uncoiled and has been inserted into the flower nectaries to ingest nectar.

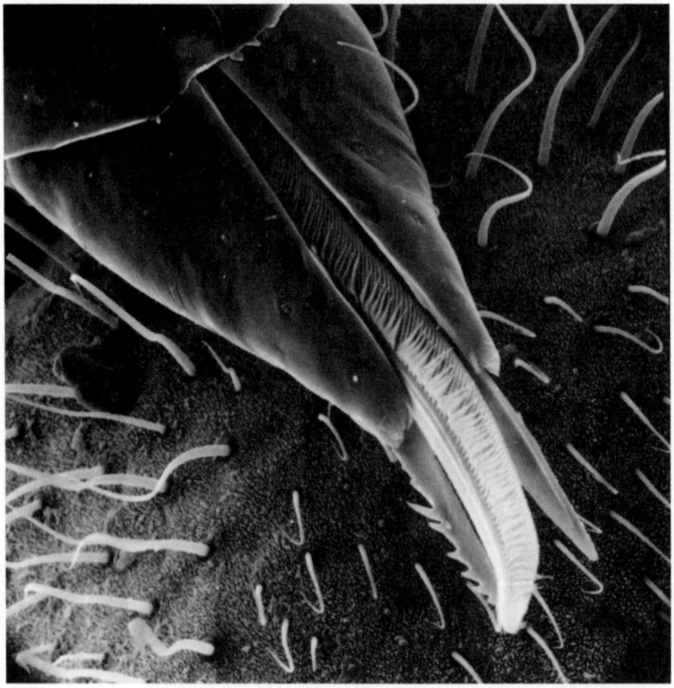

Figure 40. The tip of a giant water bug (*Belastoma* sp.) beak with the toothed stylets visibly exposed.

Figure 41. *Aedes aegypti* (L.) probing into the tissue of a frog's foot at the left and penetrating a capillary to ingest blood at the right. [Redrawn from Gordon and Lumsden, *Ann, Trop, Med. Parasit.,* 33, (1939), 259–78. By permission of the Liverpool School of Tropical Medicine.]

molecules surrounding the stylets aid in holding the stylets together, although they are dovetailed by longitudinal ridges and grooves in some species in which up and down movements are permitted while preventing the stylets from parting. As the stylets penetrate deeper and deeper, the labium is often bent elbow-shaped out of the way and does not enter the wound (Fig. 41). Precise directional probes are possible by varying the tip position.

The second type of piercing-sucking mouthpart is found only in thrips [Fig. 303(E)] and appears to be intermediate between the chewing and piercing-sucking types. Only the left mandible is present, and it and the two maxillae function as stylets in piercing plant tissues. The resulting shallow wound exudes sap and cellular fluids that are sucked into the body.

The third type, found in sucking lice, is composed of three stylets that are retracted into the body when not being used. The origin of the stylets remains in dispute although they are probably from the hypopharynx and labium. The stylets are everted and penetrate the skin of the host and pierce blood vessels, from which the blood is sucked into the body by a sucking pump as in the previous types.

In summary, piercing-sucking mouthparts are modified for penetrating the outer barrier of the host organism, whether it be plant or animal, and fluids are taken into the body, often in a predigested form. This type of mouthpart is commonly associated with disease transmission.

Chewing-lapping. Adult honeybees [Fig. 36(F)] and bumblebees have mouthparts that are modified in still another fashion to utilize liquid food, in this case nectar and honey. The major feeding apparatus consists of a maxillo-labial complex. Surrounding the central "tongue," the glossae of the labium, is a tube formed from the galeal part of the maxillae. With the combined action of both the sucking pump and "tongue" moving up and down, nectar is drawn up into the body. The mandibles usually do not function directly in feeding but may be used for cutting flowers that have long corolla to gain access to the nectar, for defense, and for molding wax into combs for storing honey in the hive.

Other specialized larval mouthparts. Immature terrestrial Neuroptera have sickle-shaped mouthparts [Fig. 305(B)] formed as each mandible becomes appressed to a maxilla. The resulting hollow tubes inject salivary fluids and suck hemolymph from the prey.

Dipterous larvae have numerous modifications, only two of which will be mentioned here. One type has filtering setae along the mouthparts [Fig. 32(C)], which strain plankton from water. A more common second type has tusklike mouthhooks [Fig. 389(D)], a pair of sclerotized mouthparts of uncertain origin, which are often withdrawn into the body when not in use. During feeding, the hooks are protruded and function in tearing or rasping the food.

In summary, insect mouthparts are formed primarily from three pairs of appendages. They are highly adapted to permit specific diets. Immature insects have mainly chewing mouthparts, although some may be highly specialized or piercing-sucking. Those insects undergoing holometabolous development (discussed in Chapter 4) often possess other types (Fig. 33) when they become adults to permit the use of ephemeral resources.

Sensory Role

Compound eyes. As with most animals, photoreceptors are usually located near the mouth at the anterior region of the body. The most conspicuous of these photoreceptors are the compound eyes of adults and of many immatures called *nymphs* (Fig. 110). Each photoreceptor consists of a number of separate receptors or *ommatidia* (Figs. 42, 43, 44); the number varies from a single ommatidium in some ants (Fig. 45) to more than 30,000 in some dragonflies.

The types of compound eyes number at least seven, but these are usually categorized as either *apposition* or *superposition.* The apposition type, common in many day-flying insects, has a light-gathering apparatus, the *corneal lens* and *crystalline cone,* and a light-sensing apparatus, the *rhabdom,* and each ommatidium is isolated

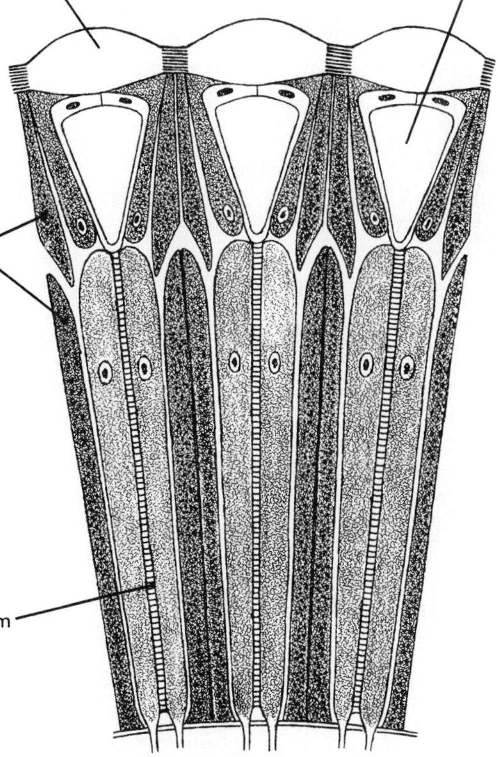

Figure 42. Diagrammatic representation of three ommatidia (apposition type) of the compound eye. (Redrawn with slight modifications from Snodgrass, 1935.)

from one another by pigment cells (Fig. 42). Direct bright light is focused by each lens system only onto its own rhabdom, which contains visual pigments, and initiates a discharge of a nerve. Interpretation of these messages by association centers in the brain is made, and the insect accomplishes vision.

Superposition eyes, in contrast, have a clear zone between the light-gathering and light-sensing apparatus (when pigments are retracted). This permits light from as many as 30 lenses to be focused on a single rhabdom (Fig. 43). Such a recruitment of light results in increased sensitivity, thereby permitting an insect to operate in dim light such as at night. Superposition ommatidia can be converted to daylight use by moving the pigment cells to block the spread of light to other units.

The number of ommatidia varies greatly depending on the biology of the insect. Some, such as most ants, lice, and fleas, have a single ommatidium per side of the head and obviously have poor resolution of images. Others, such as predators and fast-flying species that seek flowers or mates during flight, have the greatest number of ommatidia, and an increased visual acuity. Another reason for the large number of ommatidia in predators and fast-flying species seems to be the advantage of increased

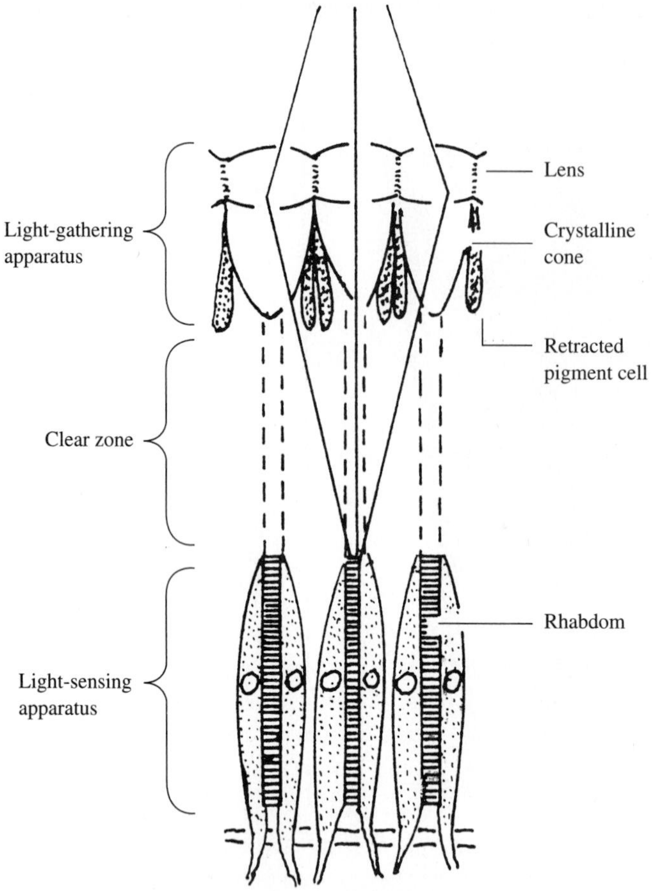

Figure 43. Diagrammatic representation of three ommatidia (superposition type) of the compound eye. With pigments retracted, light can be focused from many lenses through the clear zone to a single rhabdom.

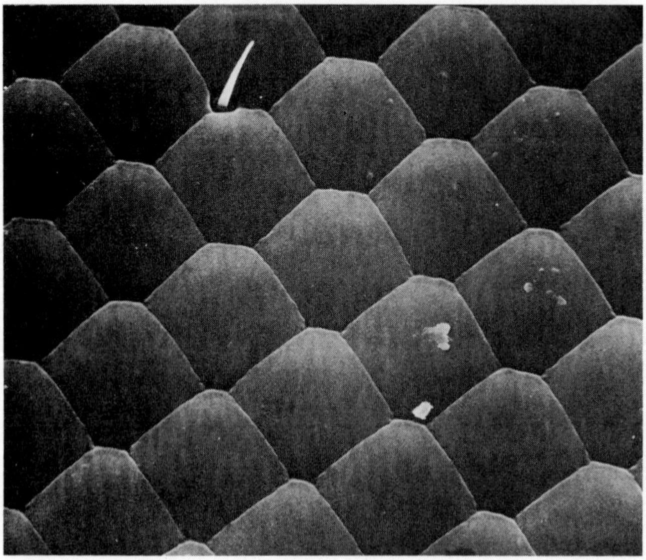

Figure 44. Scanning electron microscope photograph of the compound eye of a wasp (*Vespula maculata*) showing the individual ommatidial lenses.

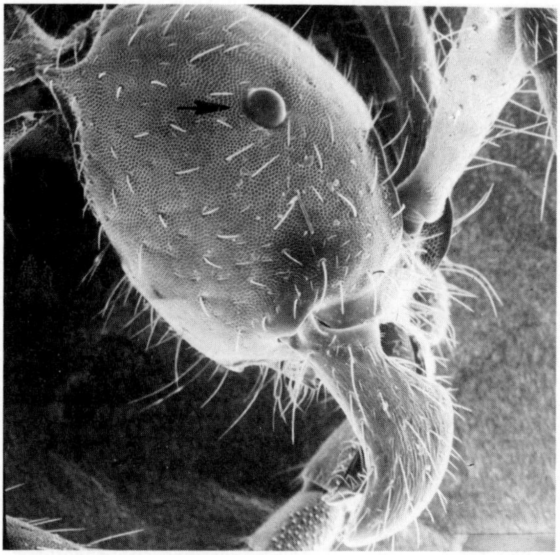

Figure 45. The head of an army ant (*Eciton* sp.) as seen through the scanning microscope. Only a single ommatidium is present, indicating an essentially blind situation as far as image formation is concerned.

depth perception. If an insect faces an object, the intervening distance is calculated by the angle between the eyes. As the two come closer, the ommatidia nearer and nearer the meson are used, thereby providing a measure of distance. The greater the number of ommatidia, the more precise will be this measure. Objects laterally positioned cannot be seen by both eyes, and hence distance cannot be adequately measured.

Complexity and variation of the compound eye cannot be overstated. Variations from the above simplified explanation include:

1. Lens size of eyes of fast-flying insects differs from the lens size of eyes of slow-flying insects.
2. Eyes of different insects vary in their capability to perceive different wavelengths, although most insects' eyes are able to see ultraviolet, and a few can distinguish red, blue, or green wavelengths and polarized light. Ultraviolet receptors are most frequent in the dorsal part of an eye.

Other photoreceptors. Located on the frons or vertex of many adults and nymphs are the *ocelli.* A maximal number of three per individual are known. Ocelli probably do not perceive form because light is not focused on the sense cells. Their function is believed to be one of increasing body tonus, in preparation to jump or fly, in response to drastic changes of illumination, such as would be produced by a shadow of an approaching predator. In addition, daily rhythms are affected when the ocelli are experimentally covered with paint or when light reception is interfered with in other ways. Some evidence exists that the insects can receive polarized light and perceive the horizon using these ocelli for orientation.

Larval insects, such as caterpillars, have photoreceptors structurally intermediate between typical ocelli and ommatidia. These *stemmata* (Fig. 46) are usually in two

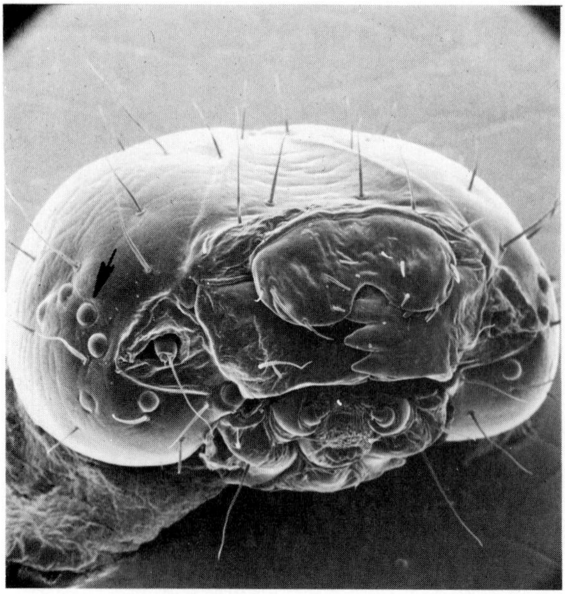

Figure 46. A scanning electron microscope photograph of a noctuid caterpillar head illustrating the relative position of the 12 stemmata.

scattered groups on the gena. They probably have weak powers of form perception because light can be focused in some types, but visual acuity is probably low because there are usually fewer than 12 stemmata. During metamorphosis to adulthood, stemmata are lost and are replaced by the adult compound eye.

Antennae. Few people have picked up an insect without noticing the conspicuous "feelers" projecting from the head. Although reduced in many immature forms, these *antennae* (Fig. 47) are frequently large in adults in order to aid in the increased sensory activities necessary for the specialized food and mate location at this stage in the life cycle. These are appendages of the second head segment.

Only the two basal segments, the *scape* and the *pedicel,* are operated by direct musculature. The remaining segments, often collectively termed the *flagellum,* maintain their position by blood pressure and normal rigidity and move only when environmental factors contact and move them physically. The animal then interprets this contact and reacts accordingly.

In addition, more delicate tactility, odor, humidity, hearing, and various other stimuli are detected by appropriate receptors found on these appendages. Tactile setae are solid and have a sense cell at the base to detect movement. Other sensory setae are hollow and have sensory cells extending to openings for chemoreception. Some sensory structures are not located in setae but are in membrane or beneath the exoskeleton for determining pressure.

Many shapes of antennae may be seen, varying from the long filiform antennae to the short and tapering setaceous type. Major variations are illustrated in Figures 47 and 48. A knowledge of the types of antennae is helpful in identifying insects. Many types of receptors are found on these structures (Fig. 49).

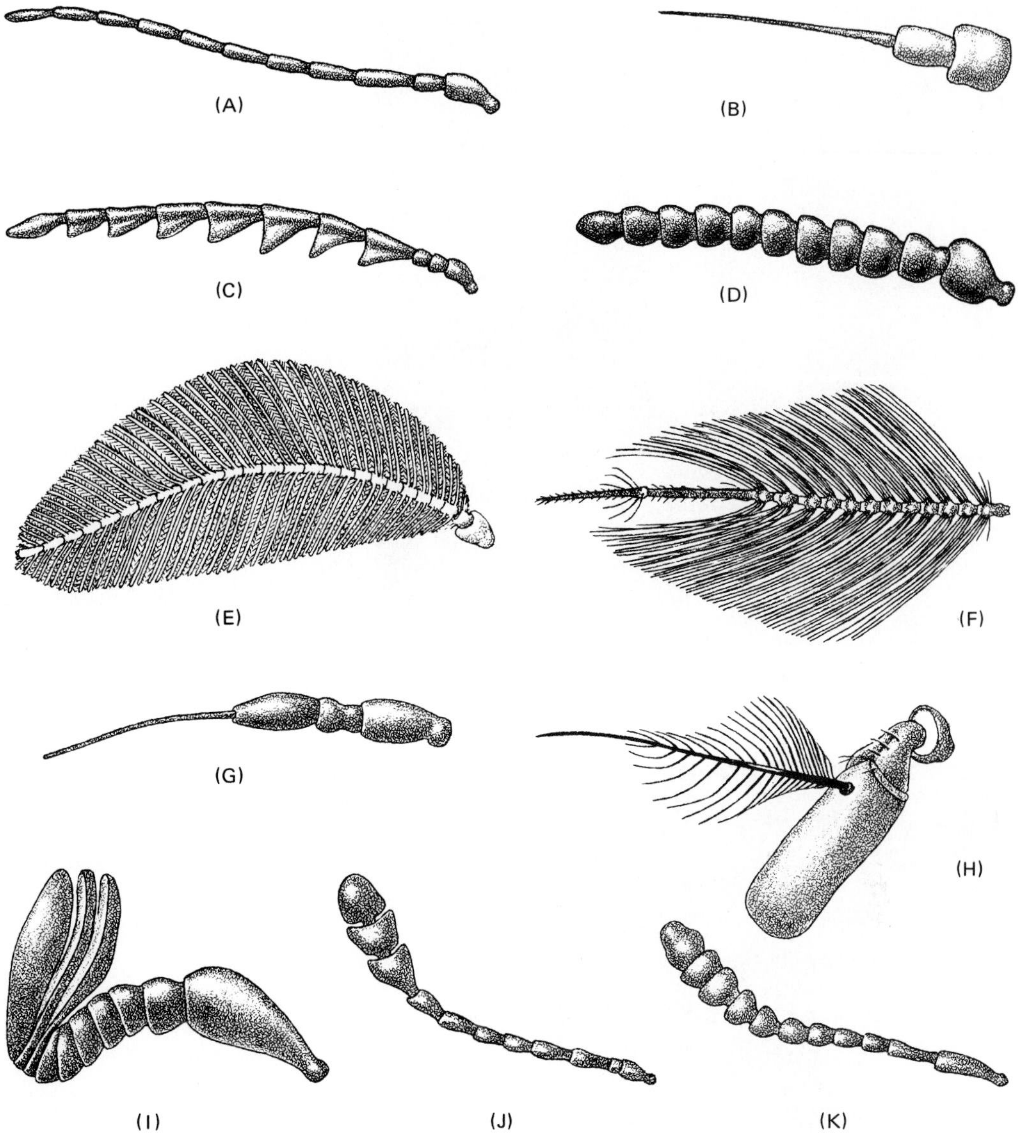

Figure 47. Types of insect antennae. (A) filiform; (B) setaceous; (C) serrate; (D) moniliform; (E) pectinate; (F) plumose; (G) stylate; (H) aristate; (I) lamellate; (J) capitate; (K) clavate.

THE THORAX

The function of locomotion in adult and nymphal insects has been taken over almost exclusively by three body segments, collectively referred to as the *thorax* (Fig. 50). For ease in discussion, the seventh body segment will be referred to as the *prothorax,*

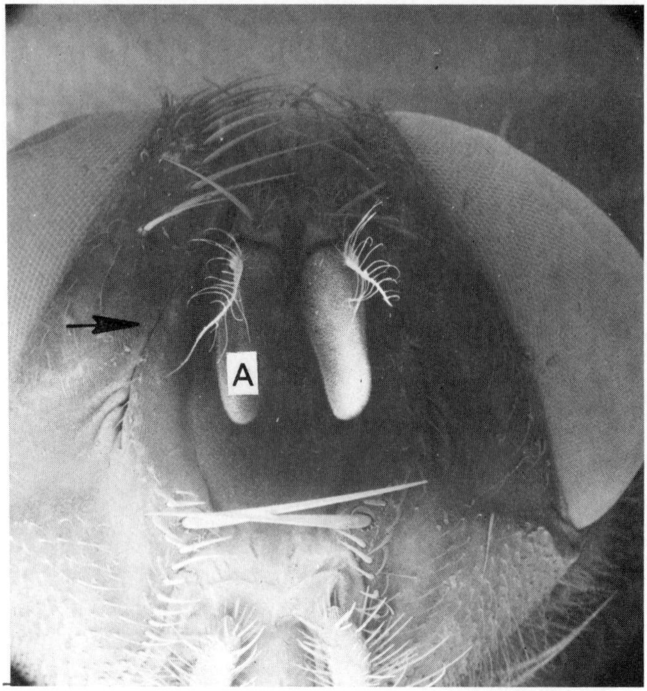

Figure 48. Antennae (*A*) of a face fly (*Musa autumnalis*). Located to the side of the antennae is the ptilinal groove (*arrow*) through which the ptilinum was everted to push off the cap of the puparium (see Fig. 115).

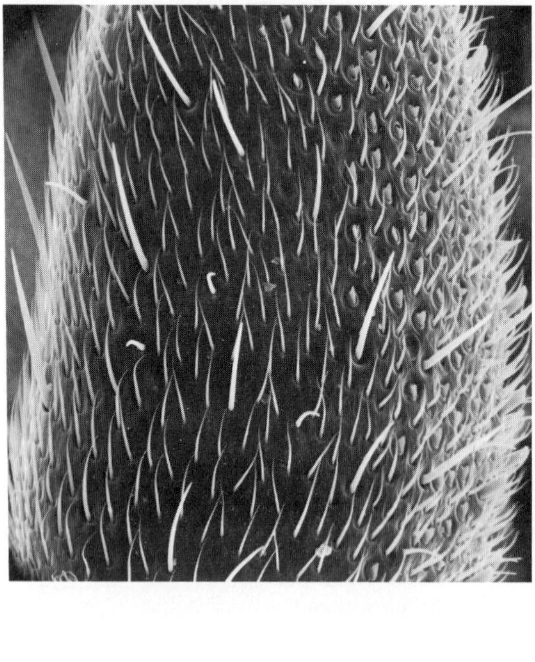

Figure 49. The terminal segment of an army ant (*Eciton* sp.) antenna. Note the disproportionate position of the chemoreceptors (peglike and many to the right). Because the ant is essentially blind and utilizes its antennae to follow chemical trails, the majority of the sensory setae are located ventrally where they constantly touch the substrate as the insect moves about.

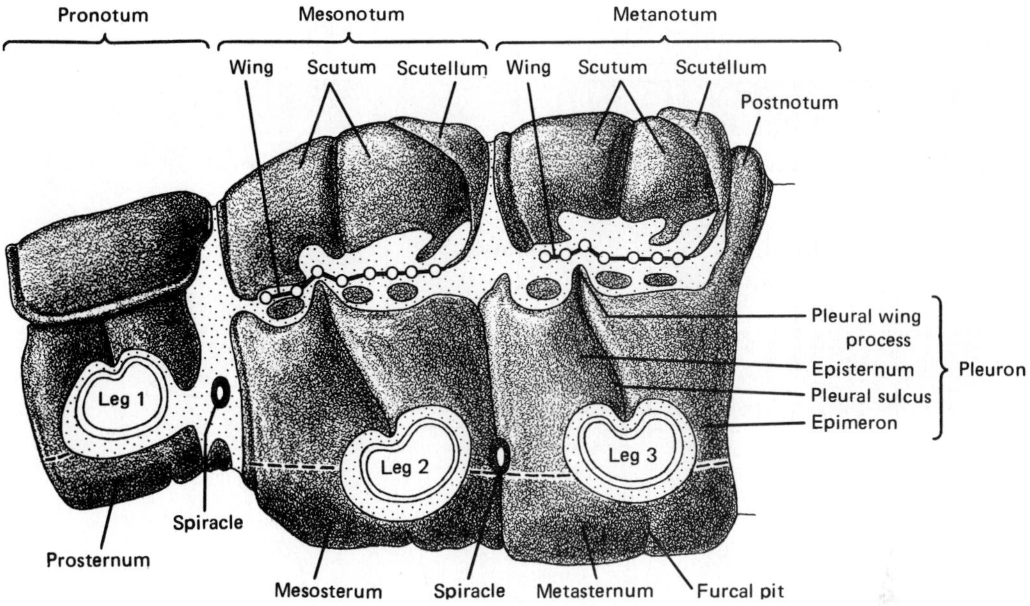

Figure 50. Diagrammatic representation of the insect thorax indicating the three segments involved, as well as showing sclerites and position of legs and wings.

the eighth as the *mesothorax,* and the ninth as the *metathorax.* Each segment is usually sclerotized to maintain a more or less rigid position and to prevent the body wall from flexing during movement of the appendages. Each sclerite is named by its position, and its segmental affinity is indicated by a prefix. The prothoracic sclerites are prefixed by *pro-*, the mesothorax by *meso-*, and the metathorax by *meta-*.

The basic plan of each of the thoracic segments consists of a dorsal *tergum* or *notum,* a ventral *sternum,* and two lateral *pleura* (singular, *pleuron*). Each of these basic sclerites may be subdivided by sulci or membranes. The tergum or notum is subdivided into the *prescutum, scutum, scutellum,* and *postnotum.*

The pleura are the result of incorporating a leg segment into the body wall, and each is braced by a linear invaginated ridge, the *pleural sulcus.* The basal part of the sulcus serves as a point of articulation for the legs and braces the pleural plates against the tension of muscles that originate on the tergum and sternum and move the legs. For descriptive purposes, the pleural plate anterior to the sulcus is called the *episternum,* the posterior one the *epimeron.* The sternum is the least likely to be subdivided; however, it is sometimes subdivided into a presternum, basisternum, and sternellum by sulci, and the lateral sides often fuse with elements of the pleuron. A pair of *spiracles,* openings into the tracheal system, are found between the prothorax and mesothorax and between the mesothorax and metathorax.

In addition to movement by appendages, most advanced adult insects are also able to fly. Two of the thoracic segments, the mesothorax and metathorax, have wings. They contain no internal muscles or tendons; flight usually comes from movement of the notum, which in turn causes the wings to move. Because the notum or tergum is so

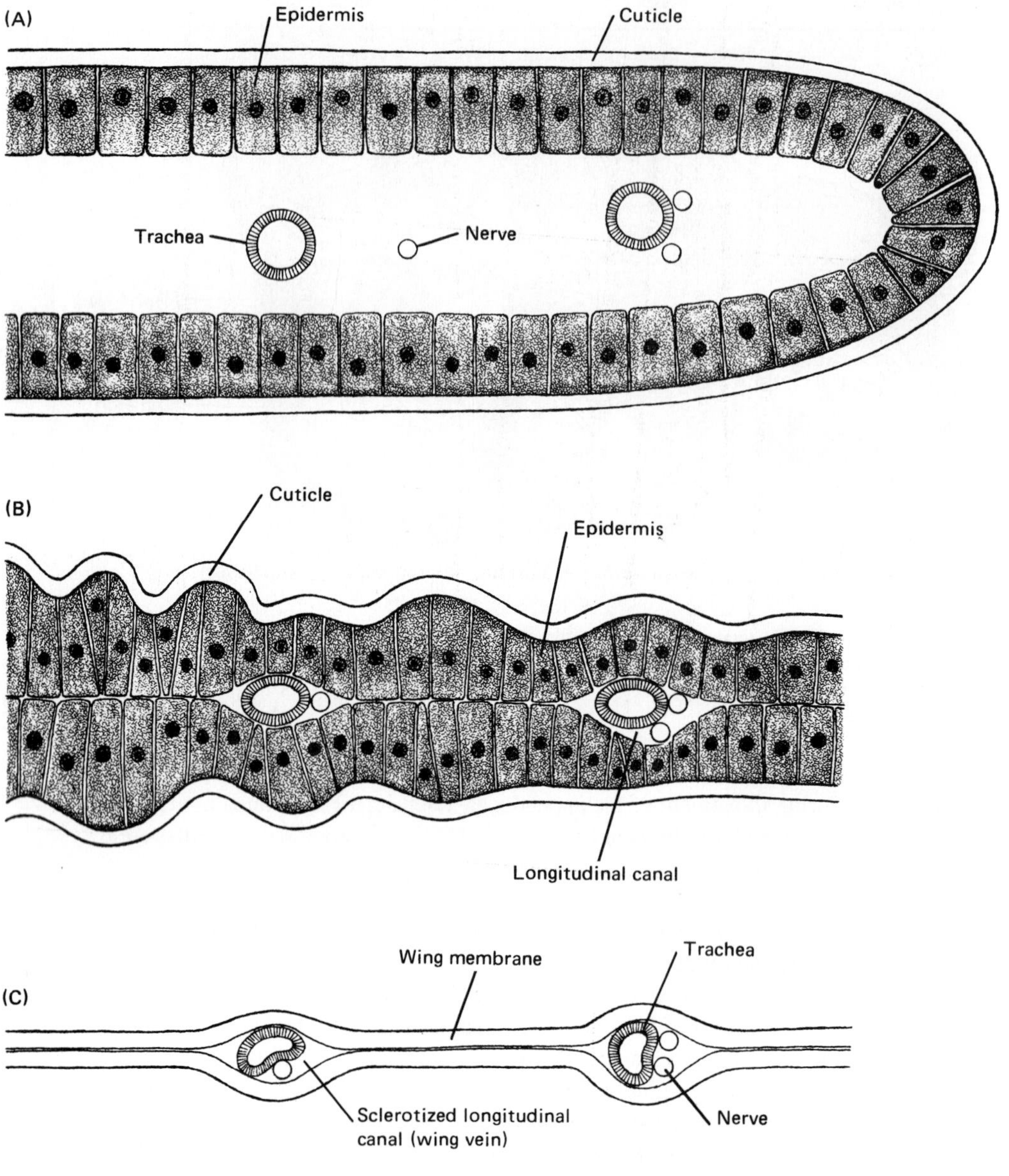

Figure 51. Development of the insect wing as seen in cross section. (A) in the early pupa; (B) in the late pupa; (C) in the adult phase after wing expansion.

intimately involved with flight (see Chapter 3), it has become highly modified in winged species. The other sclerites of the mesothoracic and metathoracic segments also are altered but to a lesser degree. Both the pleura and sterna tend to fuse solidly with one another and between the wing-bearing segments to form a U-shaped brace beneath the wings. The dorsal extension of each pleural sulcus, the *pleural wing process,*

Figure 52. The immature long-horned grasshopper nymph possesses developing wings in the form of wing pads. Note the shape and size. During subsequent molts the wing pads will enlarge until they form functional wings after the final shedding of the exoskeleton.

functions as a fulcrum for wing movement. Often the sternum has been invaginated into the body to form a *furca* for more efficient bracing and muscle attachment as well as to enable the legs to be placed directly under the body for better movement.

Wings

Wings are first evident as *wing pads* in immature insects. In wing pads that develop externally (in nymphs), a lateral evagination occurs that has both upper and lower cuticula and epidermis (Fig. 51). Between these two layers may be seen tracheae, nerves, and an extension of the body cavity. With each successive molt, the pads enlarge until they appear as miniature wings (Fig. 52). After the last molt, blood pressure enlarges these structures, and fully developed wings of the adult individual result. During expansion, the epidermis is stretched and disorganizes, thereby allowing the two cuticular layers to fuse and sclerotize around the tracheae and nerves to form longitudinal *veins* and *crossveins* that serve as braces for the wings and prevent ribs in the wing from spreading. Veins also serve as conduits for the movement of oxygen (through tracheae), sensory information (through nerves), and hemolymph or blood

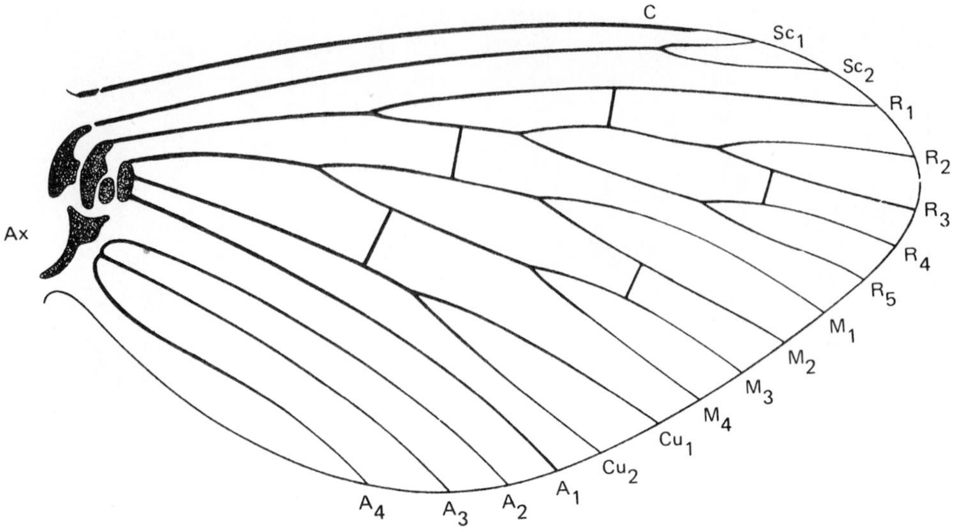

Figure 53. Generalized wing illustrating venation. *A*, anal veins; *Ax*, axillary sclerites; *C*, costa vein; *Cu*, cubitus veins; *M*, median veins; *R*, radius veins; *Sc*, subcosta veins.

(between abutting cuticle layers). Veins are valuable in demonstrating relationships and are named to permit studies on classification (Fig. 53).

What is the evolutionary origin of wings? One would think that the answer would come easily by studying fossils and current insect structure, but this is not the case. Until recently wings were believed to have evolved from short winglike lateral expansions of the tergum, the *paranota.* Why and how they evolved from such paranota has been the subject of intense debate. Some morphologists believed they represented parachute-like or gliderlike adaptations, while others saw them used as camouflage structures, as sexual expressions, as heat-regulating surfaces, or in combinations for these purposes. Modifications or just plain disagreement with this paranotal theory have been published by Kukalova-Peck (1978), Matsuda (1981), Rasnitsyn (1981), Robertson and Reichert (1982), Birket-Smith (1984), and Marden and Kramer (1994). Each of the authors has rejected the paranotal theory and provided evidence for a pleural and/or appendicular origin for insect wings, but they differ on some of the specifics involved. Appendicular gill exites are now believed to have been the origin of insect wings (Kukalova-Peck, 1978). This revives an old discarded theory widely held earlier. See Kukalova-Peck (1978, 1983, 1986), Marden and Kramer (1994), and Kingsolver and Kohl (1994) for further details and interesting reading on this subject.

Viewing wings from functional entities seen in modern insects can be fruitful in understanding change and function (Fig. 54). Wings are not solidly attached to the body as in airplanes or in paranota. They consist of two major types depending upon their attachment mechanisms. In the more primitive insects, wing movement is restricted to essentially one plane; dragonflies serve as an example of this *paleopterous* condition. Although primitive, this wing type is very efficient. In most insects, however, wings are capable of being *flexed* or folded posteriorly over the body when not

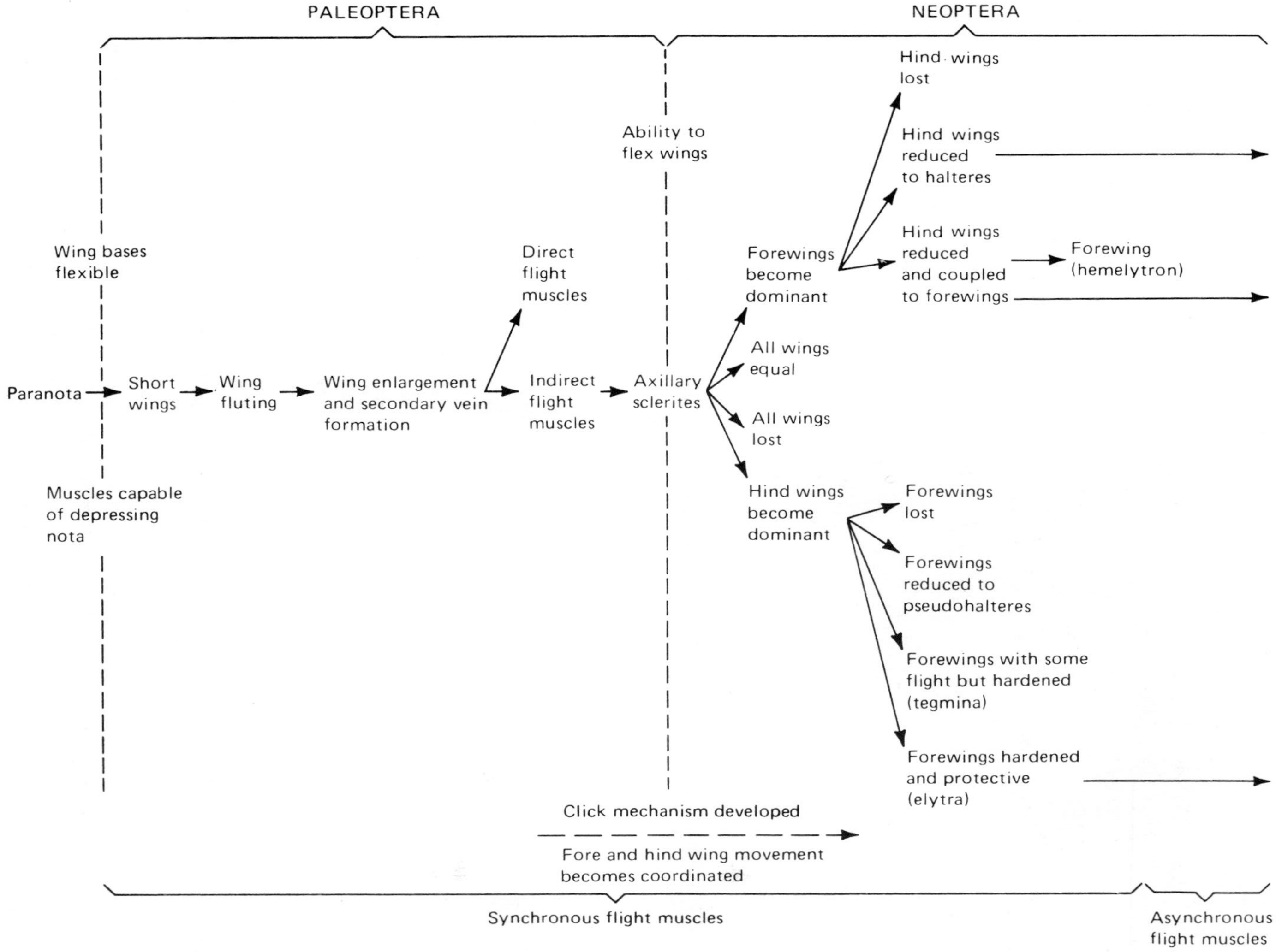

Figure 54. Structure-function evolutional diagram of insect wings. Because of simplification, phylogeny is not necessarily correlated.

Figure 55. Wings of a grasshopper (*Dissosteira carolina*). The forewing or tegmen is mainly protective during periods of rest. The hind or flight wings are large and fluted to give strength to the large surface.

in use and hence can be moved on more than one plane. The flexing mechanism results from the development of *axillary sclerites* in the membrane at the wing base. These *neopterous* wings give an insect the advantages of flight while permitting these structures to be folded to enable the individual to crawl under bark or rocks and to hide from predators or extremes in weather without damaging the wings.

Early in wing evolution, wings had extensive *fluting* (Fig. 55) or longitudinal creases with many veins for support (Edmunds and Traver, 1954). Many crossveins further brace this fluting as seen in the primitive wings of dragonflies, mayflies, and grasshoppers. The rigidity obtained from fluting permitted flight, but sculling action, a common figure-eight movement of the wings, is interfered with, and drag results. Today most highly developed wings, as seen in flies, are slender with a few strong anterior veins for strength and permitting sculling. The membranous wings of insects are in sharp contrast to the wings of birds, yet, when wing movement starts, the flattened form takes on a slightly cambered shape to obtain lift as the air moves across the wing in both groups.

Usually one pair of wings becomes dominant. In some insects the hind wings carry much of the burden of flight, whereas the forewings become increasingly sclerotized to protect the wings when they are flexed, or folded back over the body. Forewings so modified are called *tegmina* (as in grasshoppers) when they are leathery in texture (Fig. 55), *elytra* (as in beetles) when they are heavily sclerotized (Fig. 56) and cover the membranous flight wings, or *hemelytra* (as in true bugs) when the wing base is sclerotized but the distal portion remains membranous. Elytra are usually displaced dorsally and are not moved substantially during flight, whereas tegmina move with

Figure 56. A scarab beetle (*Melionota punctata*) with its heavily sclerotized elytra separated and the hind membranous wings unfolded into the flight position.

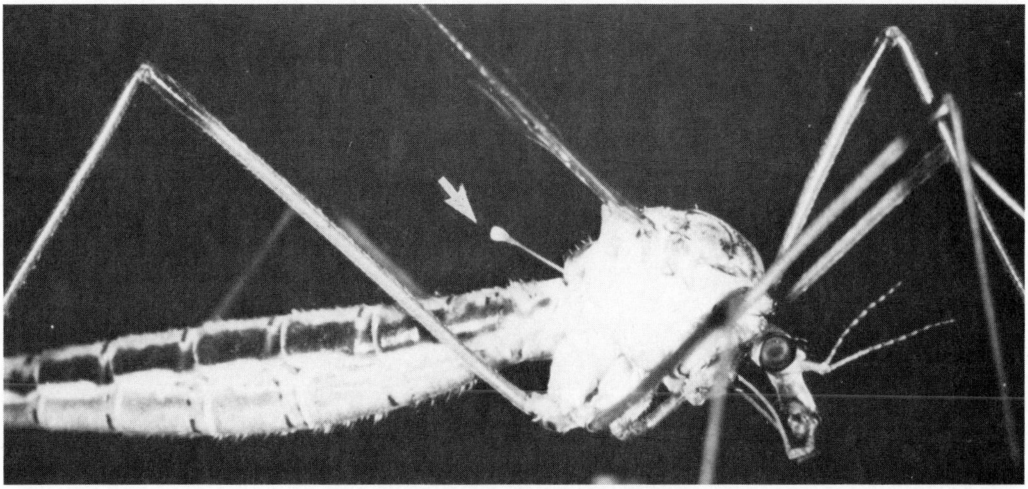

Figure 57. Lateral view of a crane fly (*Tipula dorsimacula*). Note the presence of the clubbed halteres.

the hind wings during flight but probably have little locomotory effects. The hemelytra, however, are used extensively for flight.

In insects in which tegmina and elytra are not developed, the forewings become the primary organs of flight with the hind wings being reduced, lost, or highly modified (Figs. 57, 58). When reduced but functional, hind wings usually are coupled to the forewings to act as a single unit for more efficient flight. The coupling may be by folds,

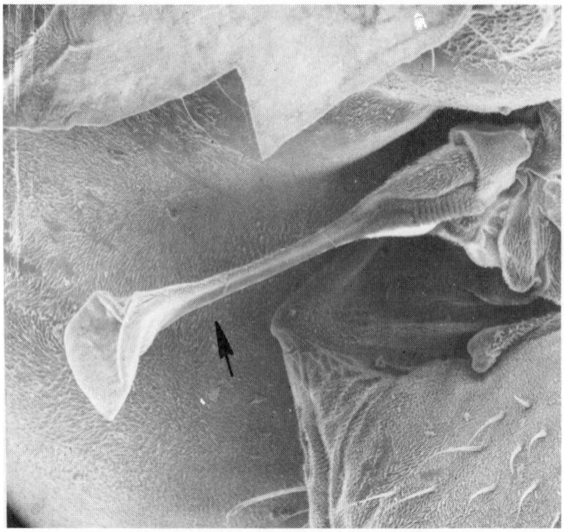

Figure 58. Scanning electron microscope photograph of the flesh fly (*Sarcophaga bullata*) haltere.

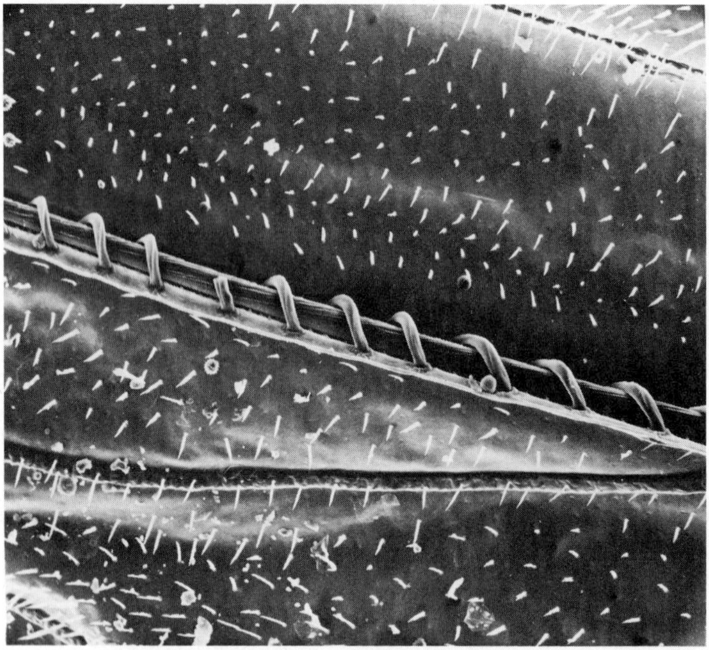

Figure 59. Scanning electron microscope photograph of the hooks from the hind wing, which connect to the forewings in a social wasp (*Polistes* sp.), producing a single functional unit for flight.

hooks (Fig. 59), bristles, or enlarged lobes from either wing. Extreme modification of the metathoracic wing is seen in the fly in which it becomes a vibrating gyroscope-like structure or *haltere,* a vital role for an insect with a stout body and narrow wings. These halteres oscillate at the same frequency as the wings but out of phase with them. Deviations in flight course are detected as the halteres twist due to inertia.

In some species, wing length can vary greatly within each population. In these species, normal wing length is termed *macropterous,* shortened but functional wings are referred to as *brachypterous,* vestigial wings are termed *micropterous,* and those individuals lacking wings are named *apterous.*

In addition to being used in disperal and locating food, wings also have other roles. One is protection against physical damage. Tegmina and elytra not only cover the hind flight wings, but also afford protection for the dorsum of the abdomen, the resultant cavity also serving as a reservoir for air in the case of beetles. In some beetles and butterflies, wings shield the body from excess solar radiation, whereas in others they serve to capture this solar heat to warm the body during cool or cold temperatures.

Wings may be colored by pigments, refractory surfaces, or colored scales. These colored patterns may serve for attracting and recognizing mates, but they normally assist the insect in either hiding, when the color matches the background, or in warning predators of the insect's unpalatability, if brightly colored (see Chapter 6). Attractant sounds are sometimes produced by wing vibrations, such as in mosquitoes, or by vibrations caused by rubbing the wings against one another or against a leg, the latter adaptations common in crickets and grasshoppers. In many social insects, such as honeybees, wings may be used to cool nests. We will elaborate on this summary of wing functions in future chapters.

Major Types of Legs

The insect leg consists of six segments (Fig. 60): the *coxa, trochanter, femur, tibia, tarsus,* and *pretarsus.* The femur and tibia are longer than the other segments and have a conspicuous "knee" between them that permits the insect to be slung low to the ground for stability. Although the tarsus appears to have segments, these segments are actually pseudosegments or *tarsomeres* because each lacks independent musculature. The pretarsus consists of only the claws or *ungues* and often either a single lobelike *arolium* or the two lobed *pulvilli.* Claws enable insects to move on rough surfaces, whereas the arolium and pulvilli or other types of adhesive pads and hair on the tarsi assist in movement across smooth surfaces.

The entire leg is moved by muscles originating on the tergum or sternum and inserted on the coxa. In addition, movement of individual segments is accomplished by muscles within the leg segments (Fig. 60). These muscles either extend or flex, thereby pushing or pulling as the entire appendage is moved basally. In some insects, these muscles also suspend the body above the substrate, whereas other insects permit their bodies to contact the surface when they are not active.

In walking or running, the front and hind legs on one side and the middle leg of the opposite side are raised and moved forward together. Once these legs have completed their movement, the opposite three legs are moved. This alternating movement of two tripods coupled with the low slung body results in extremely well-balanced locomotion. Exceptions include insects that have highly modified front legs, such as the mantids during slow walking, and insects that have lost legs; in such examples, a tripod is maintained by moving only one leg at a time, or else a modified behavior is evoked in which a staggering movement occurs during rapid locomotion.

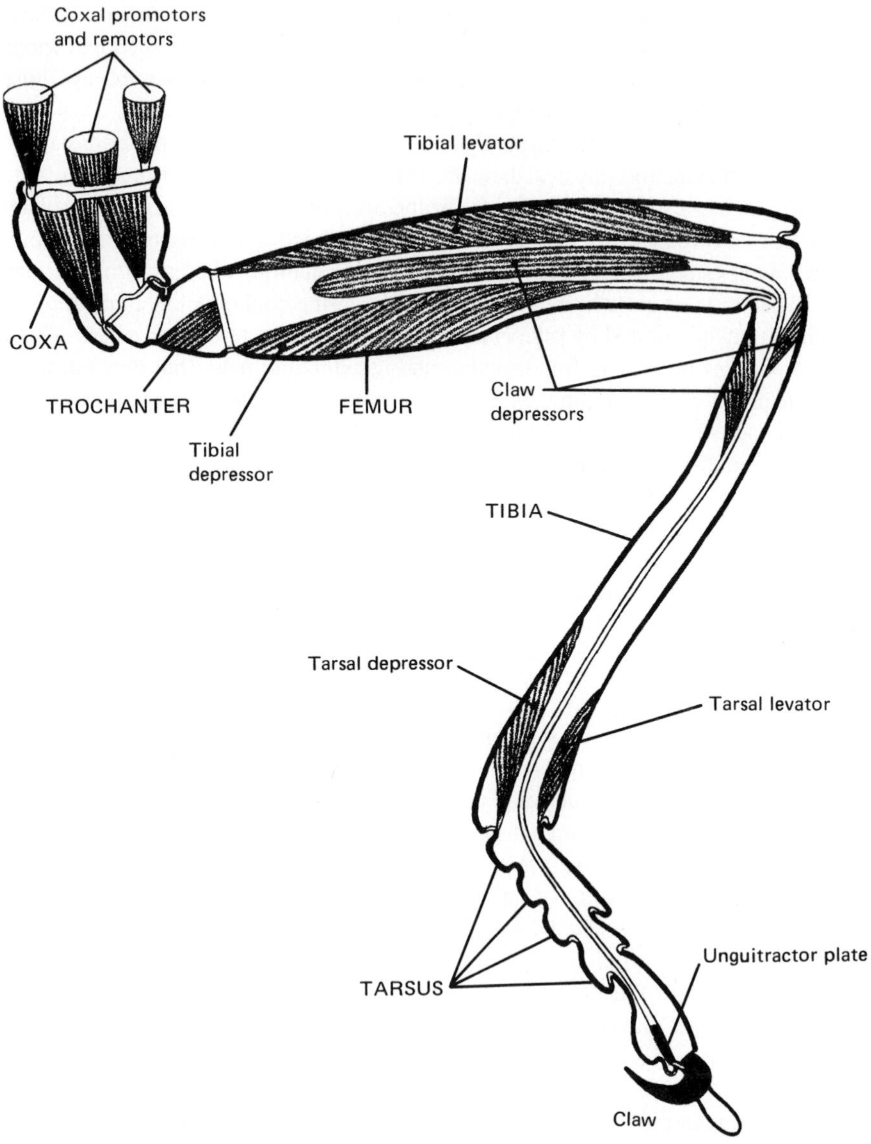

Figure 60. Segmentation and major musculature of a walking leg. (Redrawn with slight modifications from Snodgrass, 1935.)

Ambulatorial and cursorial. Ambulatorial and cursorial legs are the basic type of insect legs. It is difficult to separate these two because walking (ambulatorial) and running (cursorial) are so similar; the former are usually thicker and move slowly, such as the first two legs in grasshoppers, whereas the running legs are usually slimmer and more elongate. Tiger beetles and cockroaches are examples of rapidly moving insects with cursorial legs.

Saltatorial. To *saltate* means "to jump or vault." Legs modified for this function [Fig. 61(A)] commonly have greatly enlarged femora to accommodate the enlarged extensor muscles that extend to the tibia. When the legs are well anchored by large tarsal pads, claws, and often spines, a rapid contraction results in the entire body being propelled (Fig. 62). Most legs of this type are located on the metathorax, as seen in the grasshopper, so that the direction of jumping is forward, where the major sensory structures can perceive the upcoming environment.

Raptorial. The front pair of legs is often modified to grasp and hold prey for feeding [Fig. 61(B)]. The large muscles here are depressors, and the tibia is pulled back against the femur when the muscles contract. Spines also may be present on the femur and tibia, as in the mantid, to impale the prey and decrease the likelihood of escape by the prey.

In a few parasitic wasps the hind legs are raptorial. These legs hold the host near the ovipositor so the egg can be deposited. Modification of all legs to grasp the host's hair may be seen in sucking lice (Fig. 233).

Fossorial. The forelegs of the mole cricket and cicada nymph [Fig. 61(C)] are shortened and heavily sclerotized. Large-toothed projections from the femur or tibia are used to "rake" through the soil to dislodge soil particles. The tarsi, as in raptorial legs, are reduced and usually fold back out of the way during excavation activity.

Natatorial. Most of us are familiar with oars and swimmers' arm movements. These principles in their use are similar to the activities and modifications seen in the swimming legs of insects. Diving beetles, for example, have the middle and hind legs flattened, and the segments often are approximately the same size. When these legs are straightened and are rapidly moved posteriorly, the maximal surface area is exposed to the water. Friction is further increased by the rows of strong setae on the legs [Fig. 61(D)]. The legs are returned more slowly and are rotated so that the thin sides are exposed during the recovery stroke to reduce friction. Swimming hairs also fold back during this maneuver. The net effect is to propel the organism forward.

Clasping. The forelegs of certain aquatic beetles [Fig. 61(E)] are modified for holding the female during copulation. Several tarsomeres are usually enlarged with suckers and large claws to produce effective holdfast organs.

Sensory Structures

Although structures functioning as ears may occur in the head and abdomen, many insects have them in the thorax; therefore, the subject will be briefly mentioned at this time. Ears have evolved at least a dozen times, and they have been refined further in many additional ways. They vary from a single receptor cell in some moths to thousands in cicadas and mosquitoes. Most detect sound by means of a membrane stretched over an internal air sac. They may be located on the front legs of some Orthoptera (Fig. 169), metathorax to the first abdominal segment in some Hemiptera, metathorax in some Lepidoptera, forewing in lacewings, first abdominal segment in short-horned

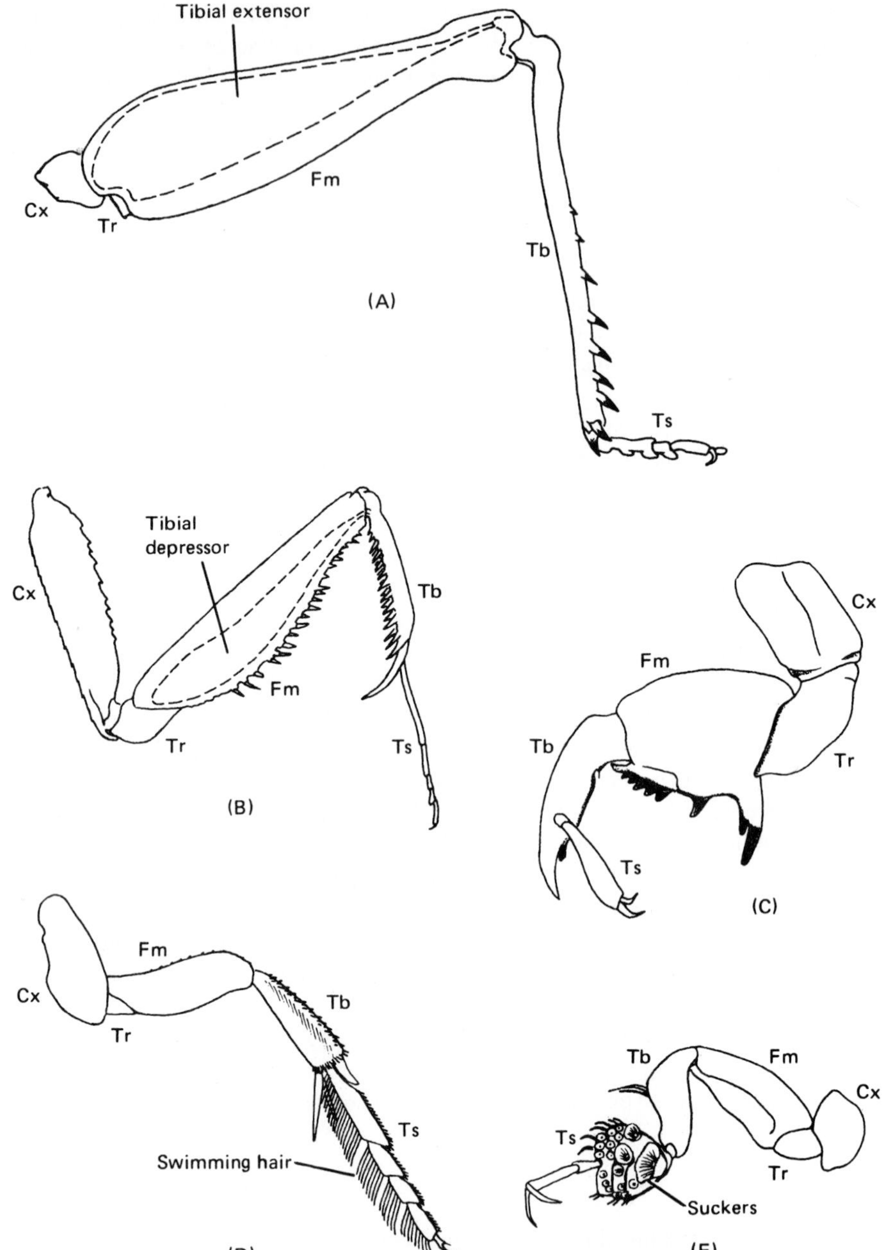

Figure 61. Modifications of the insect leg. (A) saltatorial (hind leg of grasshopper); (B) raptorial (front leg of mantid); (C) fossorial (front leg of cicada nymph); (D) natatorial (middle and hind legs of water scavenger beetles); (E) clasping (front leg of male predaceous diving beetle). *Cx*, coxa; *Fm*, femur; *Tb*, tibia; *Tr*, trochanter; *Ts*, tarsus.

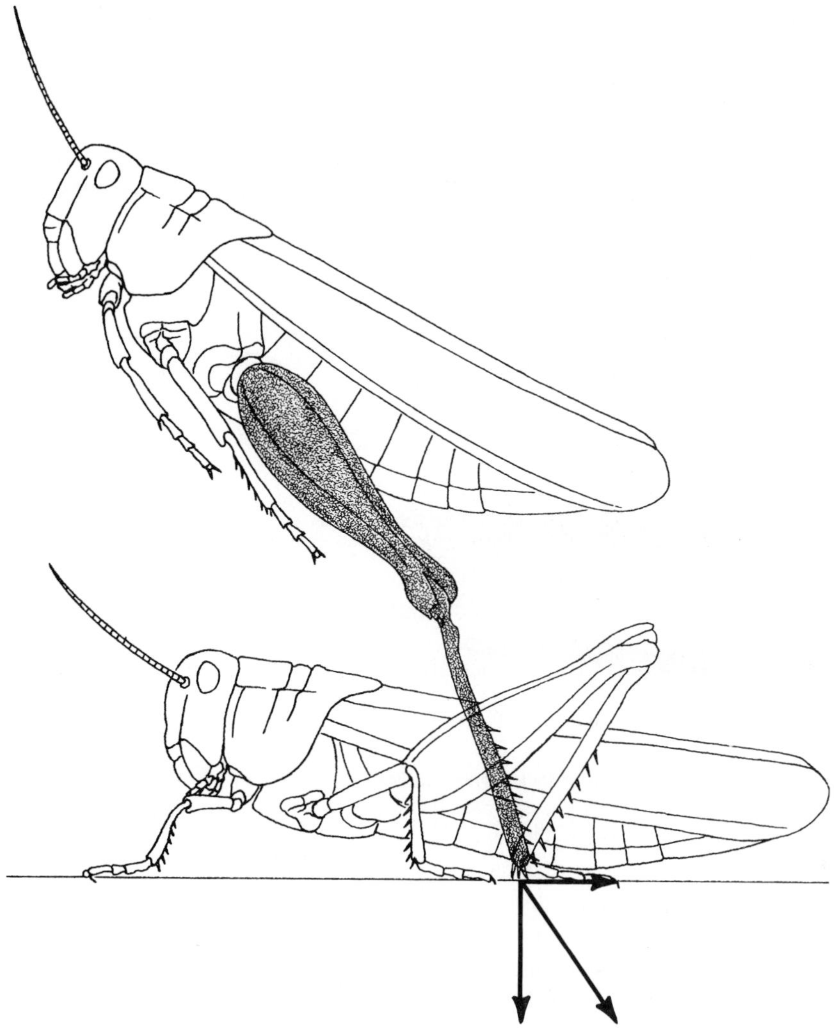

Figure 62. Diagram of grasshopper jumping, showing thrust exerted by hind leg and its vertical and horizontal components. (Redrawn from Chapman, 1969.)

grasshoppers, second abdominal segment in cicadas, palps in some hawkmoths, and the antennae in moquitoes. The latter have no membrane but are entirely setal.

THE ABDOMEN

The posterior body region in insects is called the *abdomen.* The number of segments usually varies from 9 to 11 except in Collembola. The first segment may fuse with the thorax and appear to be part of the thorax, as in ants. The remaining segments, however,

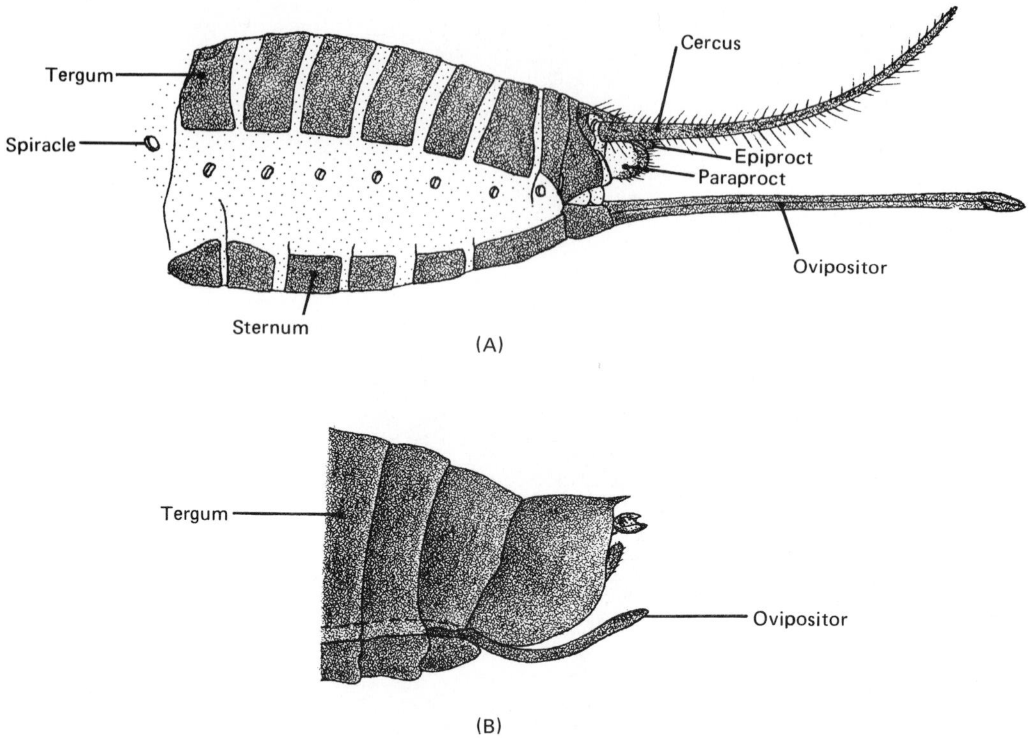

Figure 63. Basic structure of the abdomen with modifications for oviposition. (A) cricket; (B) cicada.

are very similar and consist of a dorsal *tergum* and a ventral *sternum* (Fig. 63). The anterior eight segments usually have a pair of spiracles. Spiracles are usually absent in the terminal segments, and these segments are often fused together or are reduced in size. Variation from the adult plan may be seen in some immatures in which sclerites may be lost or modified into many small plates and the number of spiracles vary greatly.

The functions of this region are vital to the organism because it is in this region that the major viscera, heart, and reproductive organs are located. The reproductive openings and genitalia are found on the ninth abdominal segment in males and on the eighth and ninth abdominal segments in females. Also, in contrast to adults and nymphs, the abdomen serves a major role in locomotion for many larval insects.

External Genitalia

Male. In most terrestrial organisms, intromittent structures are present for penetrating the body of the female and depositing sperm. In insects, the *aedeagus* is often sclerotized in varying degrees and modified sufficiently to prevent, in many species, interspecies mating. The aedeagus is usually a single phallus but it may be paired (mayflies) or many-lobed (cockroaches). In addition, a pair of *claspers* may be pre-

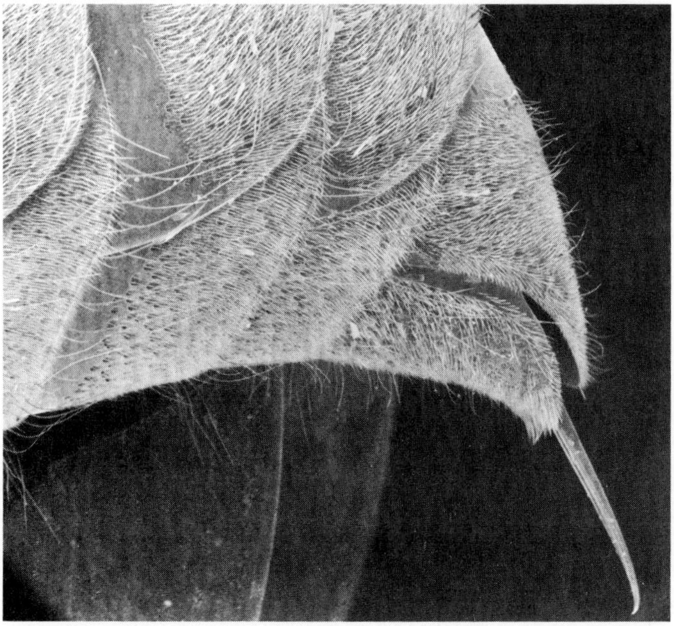

Figure 64. Abdomen of a vespid wasp (*Polistes* sp.) with the terminal sclerites parted and sting thrust into stinging position as seen through the scanning electron microscope.

sent to maintain the correct positioning of the female during copulation. In most insects the aedeagus and claspers are usually considered to be structures of the ninth abdominal segment, but there is some confusion as to their origin and whether they represent modified appendages or secondary development from sternal lobes.

Female. Correlated with copulation (to be discussed under reproduction) is the need to deposit eggs. In primitive insects, eggs were undoubtedly just "dropped," but most present-day insects have an ovipositor of some sort for placing and positioning the eggs in an appropriate microhabitat. A primitive type, seen in crickets [Fig. 63(A)] and cicadas [Fig. 63(B)], consists of a sclerotized tube formed from the paired appendages of both the eighth and ninth abdominal segments. The shape, degree of hardness, and length of this structure usually determine where the eggs are to be placed, in the ground, for instance, or under bark. In a few insects, such as bees and wasps, this ovipositor has been further modified into a sting (Fig. 64), and the primary role of egg deposition has been lost.

The most advanced insects, such as flies, have lost their appendicular structure, and another type of ovipositor consisting of a tube formed from the terminal abdominal segments has evolved (Fig. 65). Under normal circumstances the segmented ovipositor remains telescoped within the anterior segments of the abdomen, but when blood pressure is increased, it is extruded and capable of depositing eggs. Various sense receptors are located at the terminus (Fig. 65), and each segment has retained muscles for movement to permit very exact positioning of the ovipositor in cracks of suitable materials. In others, either the chewing mouthparts open up the substrate or sharp elements of the tubular ovipositor (Fig. 66) itself pierces the site to permit internal oviposition.

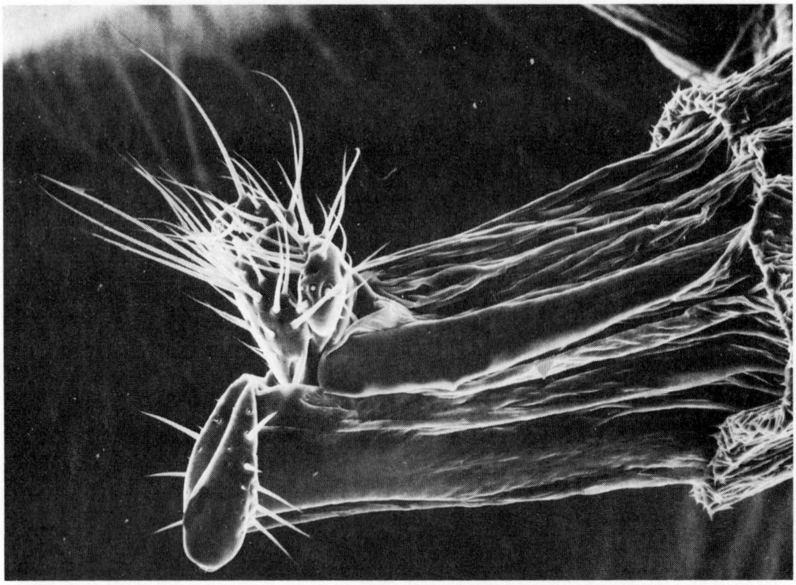

Figure 65. Scanning electron microscope photograph of the terminal end of the tubular ovipositor everted from the face fly (*Musca autumnalis*). Note the numerous sensory setae, particularly at the terminus.

Figure 66. Sclerotized terminus of the tubular ovipositor in the Bogus yucca moth, *Prodoxus* sp.

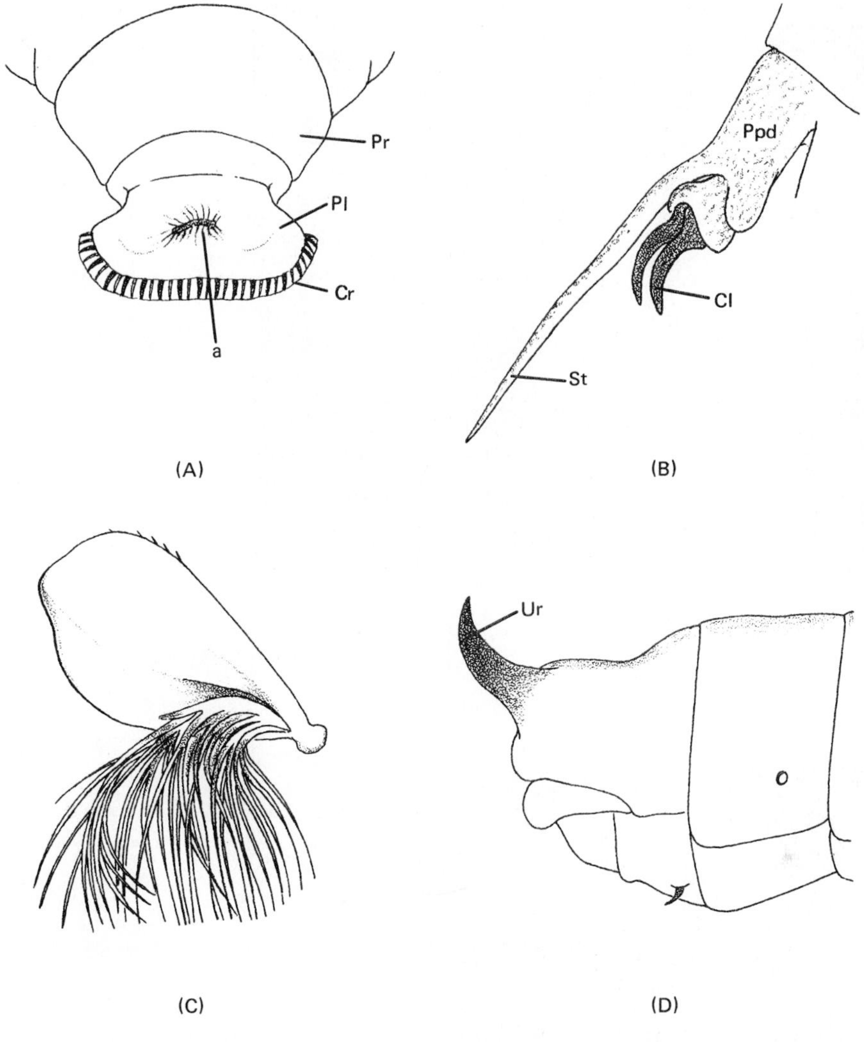

Figure 67. Some abdominal appendages in immature insects. (A) proleg of lepidopteran caterpillar with numerous hooks or crochets for grasping rough surfaces and a suckerlike area (*a*) for smooth surfaces; (B) pygopod (appendage of the tenth abdominal segment) of dobsonfly larva for grasping rocks and maintaining position in swift water; (C) gill of mayfly nymph; (D) urogomphus (appendage of ninth abdominal segment in larvae) of ground beetle. *a*, sucker depression; *Cl*, claw; *Cr*, crochet; *Pl*, planta; *Ppd*, pygopod; *Pr*, proleg; *St*, stylus; *Ur*, urogomphus.

Cerci

Embryologically, each abdominal segment bears a pair of appendages, but these are usually lost in the abdomen prior to hatching from the egg. The eleventh abdominal segment, however, often retains its appendages, the *cerci*. In the most primitive species the cerci are long and multisegmented and have numerous tactile sensory setae. These

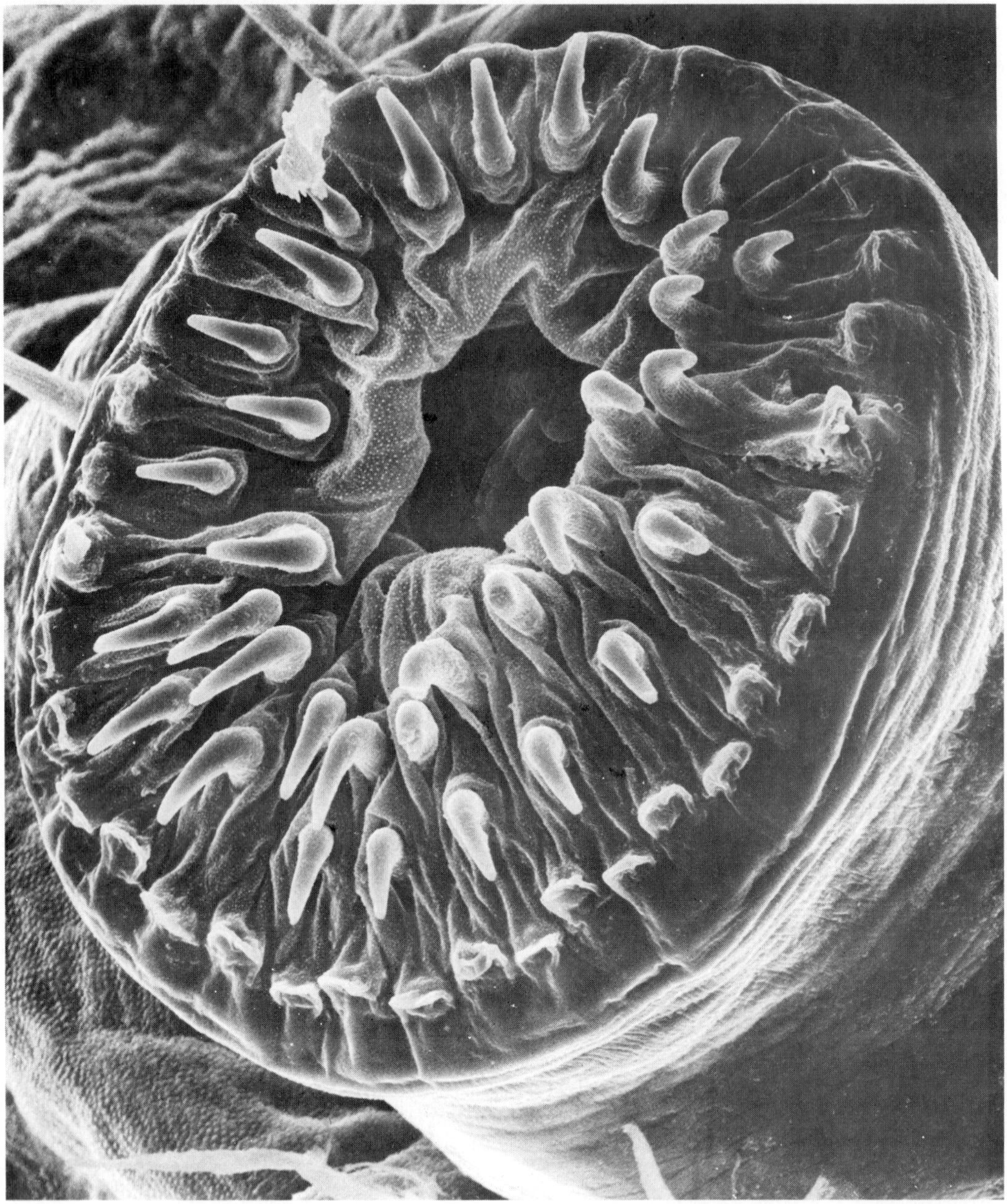

Figure 68. Crochets on the proleg of a lepidopteran caterpillar.

are especially advantageous in soil dwellers because they permit sensory pickup when the insects move backward. Cockroaches and crickets have special wind receptors on their cerci that detect lunging motions toward them by such predators as toads and tarantulas, and these receptors elicit immediate running and evasive maneuvers. In the more advanced insects, these structures tend to become reduced or lost.

Larvapods

Retention and development of abdominal appendages in immatures may occur. For convenience, these appendages are collectively termed *larvapods.* These adaptations are lost when metamorphosis occurs, but their importance in the life cycle should never be minimized.

Some larvapods, such as those found in mayfly nymphs, are modified into tracheal gills [Fig. 67(C)] and are present on most abdominal segments. It is an interesting experience to sit and watch as these gills are vibrated, thereby increasing the water flow across these respiratory organs.

Other aquatic organisms, such as the caddisfly and dobsonfly larvae, have hooked holdfasts [Fig. 67(B)] at the posterior end of their bodies that enable them to retain their position on rocks or to hang onto a portable "house" or case (Fig. 369). These appendages are most often noted in species found in moving water.

There are few individuals who have not observed creeping caterpillars and noticed the lobelike *prolegs* that assist the caterpillars in locomotion. In lepidopteran species, little hooks or *crochets* [Figs. 67(A), 68] are present to permit tenacious gripping on rough substrates. When smooth surfaces are encountered, the region with hooks is rotated dorsally to expose a suckerlike area for increased efficiency in locomotion.

Some larvae possess appendages on the ninth segment, the *urogomphi* [Fig. 67(D)]. The development of these varies from cercilike to fixed horny outgrowths.

QUESTIONS

1. What are the major functions of an insect head? How do these compare with those of other animal heads?
2. What insects would be expected to have the most sclerotized heads? Why?
3. What appendages are associated with the head? What are their functions?
4. What are mouthparts? What types are modified for fluid uptake? Are these found in the adult, larval, or nymphal stages? What are the modifications and functions of the mandibles, maxillae, and labium in each type?
5. Which of the mouthpart types are associated with the greatest range of diets?
6. Is there a correlation between diet and the number of ommatidia present in the compound eye? Elaborate.
7. How many segments make up the thoracic region? Which of these is or are the most highly modified? Does any change in the thorax occur in progressing from an immature to an adult?
8. Are wings found in all insects and all stages? How do wings develop? What is the function of wing veins?
9. What is fluting? How is it advantageous?

10. Do we know how wings evolved? Explain your answer.
11. What is the difference between neopterous wings and paleopterous wings? Which is the most common type?
12. What advantages are gained by the presence of wings? What are some disadvantages?
13. Why is it that only the forewings and not the hind wings become protective structures? What are the names of these protective wings?
14. What are some of the modifications of insect legs? How does each modification better equip the organism to survive?
15. How many segments are in the abdomen? Are they the same structure as the head and thoracic segments?
16. Are any appendages associated with the abdomen? What are their functions?
17. What is an *ovipositor?* What selective advantage does it give the possessor?

3 The Insect Internally

DIGESTIVE SYSTEM

Single-celled and some small multicellular animals obtain nutrients either by diffusion or by engulfing food particles. The exchange rate is proportional to the amount of surface area. Once inside the cell, the molecules provide building blocks necessary for growth (assimilation) or are directed into an energy cycle (respiration). In contrast to food uptake, the rate of metabolism is proportional to cellular volume. With continued growth, size eventually becomes limiting, because the volume, which determines metabolism, increases at a more rapid rate than the surface, which supplies nutrients and oxygen. Various internal chambers and tubes (i.e., respiratory and digestive systems) have been incorporated in most animals to provide the necessary increase in surface area (Fig. 69). Those involved with digestion in insects will now be presented.

Insects possess a complete tube or *alimentary canal* that takes in food through an anterior mouth and breaks down this food by enzymatic hydrolysis. A summary of this reaction is as follows:

$$H_2O + \text{food} \xrightarrow{\text{enzymes}} \text{absorbable molecules} + \text{energy} + \text{wastes}$$

The digested food is absorbed into the body, and the remaining waste material is evacuated through a posterior anus. Enzymes secreted are specific to the diet of the individual. Various glands to increase enzyme production have also evolved.

It is vital to understand the embryonic origin of the insect alimentary canal to comprehend how the system functions. The alimentary canal develops from two invaginations, one anterior and one posterior, of the embryonic tissues termed ectoderm; each grows toward and unites with a hollow sac (Fig. 70) formed from what most believe is endoderm. The boundaries between these three sections break down to form

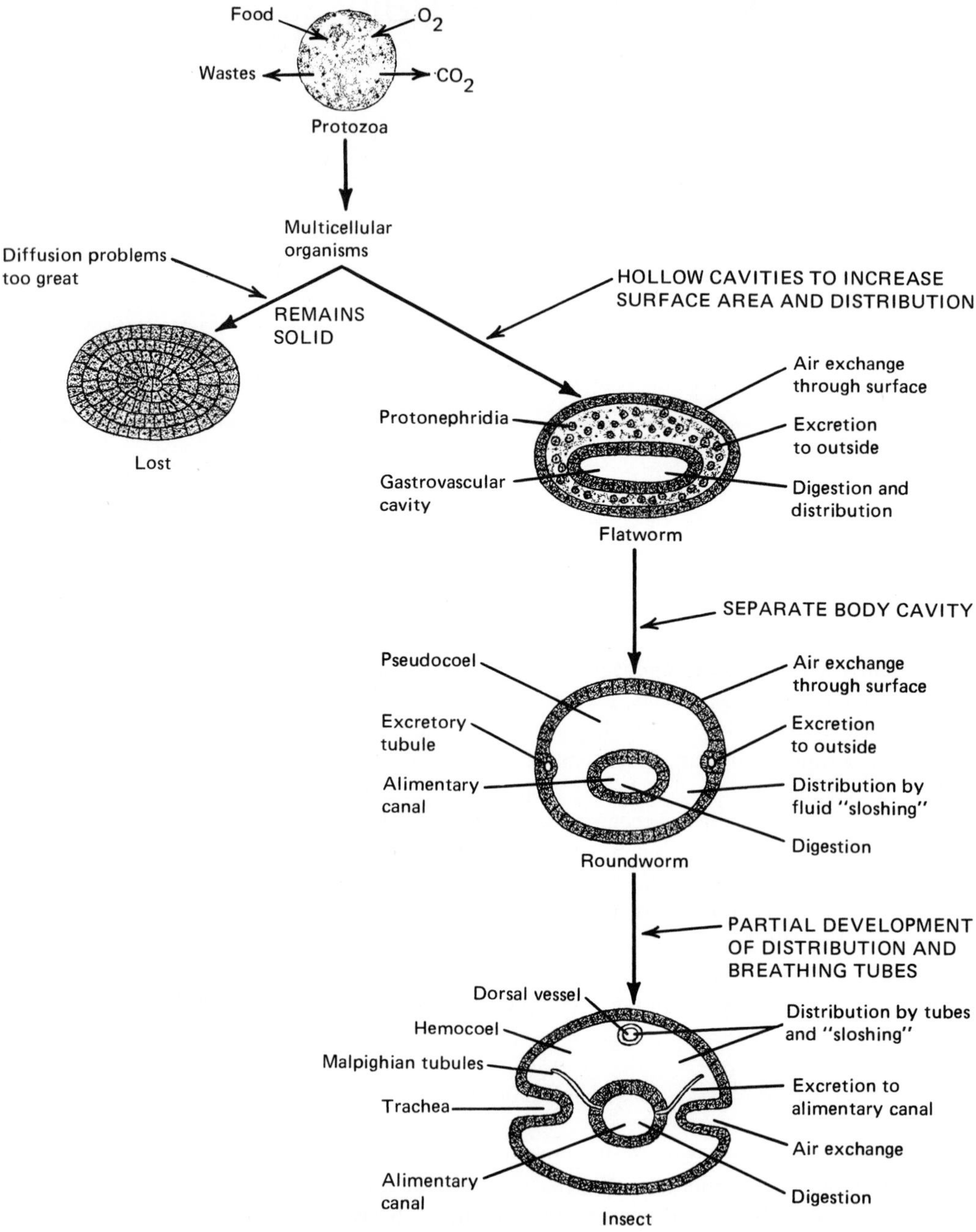

Figure 69. Diagram of means of increasing surface area to carry on the biological needs as organisms get larger and decrease the surface to volume ratio and hence the movement of vital substances.

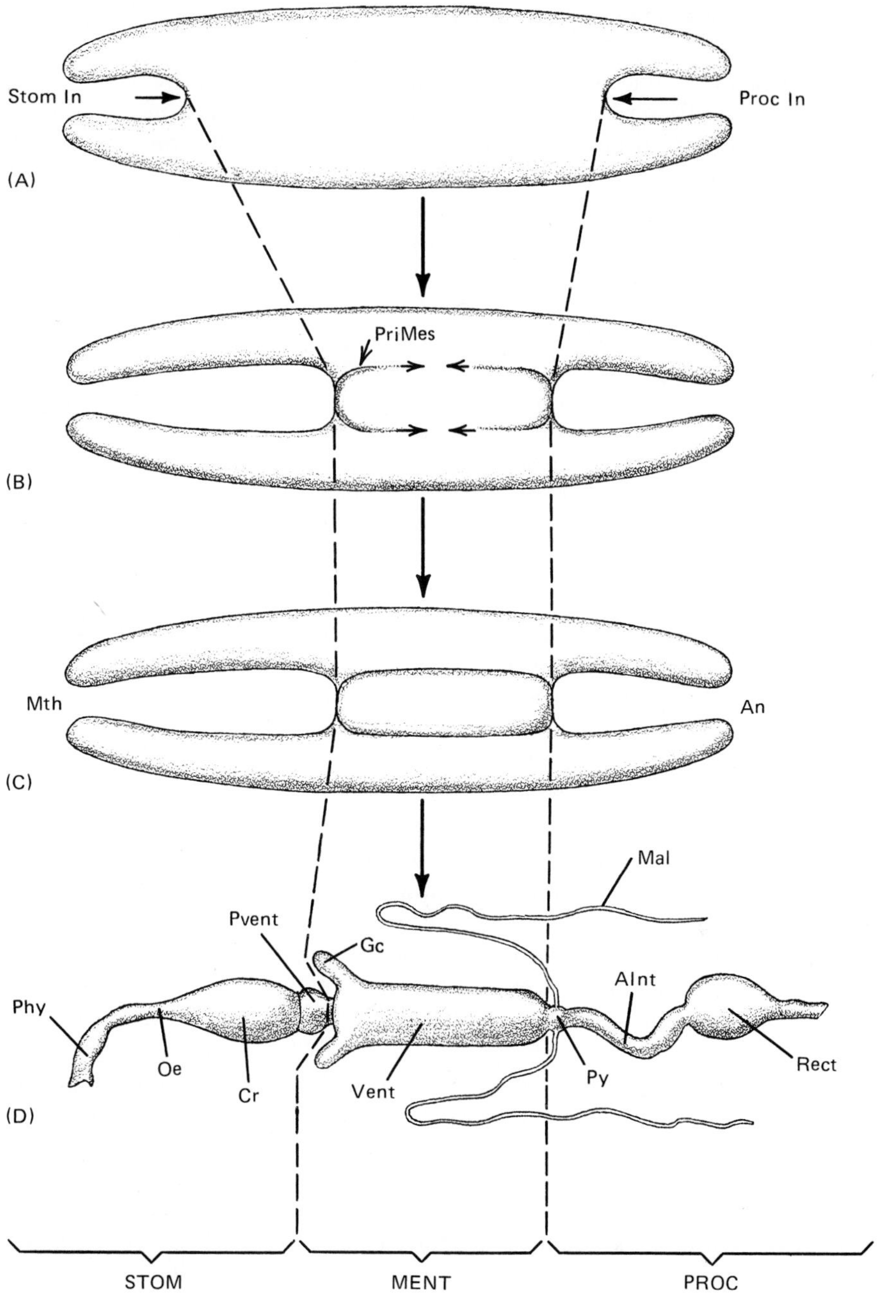

Figure 70. Embryonic development and specialization of the alimentary canal. (A) stomodeal inpouching occurs at the anterior invagination, and proctodeal inpouching at the posterior invagination; (B) primordial mesenteron forms at ends of inpouchings; (C) mesenteral elements unite to form blind sac; (D) regions between stomodeum, mesenteron, and proctodeum split to form continuous tube or alimentary canal. Specialization occurs to produce specific structures. *AInt*, anterior intestine; *An*, anus; *Cr*, crop; *Gc*, gastric caecum; *Mal*, Malpighian tubule; *Ment*, mesenteron; *Mth*, mouth; *Oe*, esophagus; *Phy*, pharynx; *Pri Mes*, primordial mesenteron; *Proc*, proctodeum; *Pvent*, proventriculus; *Py*, pylorus; *Rect*, rectum; *Stom*, stomodeum; *Vent*, ventriculus. [(D) redrawn with slight modifications from Snodgrass, 1935.]

a single continuous tube. The anterior section or invagination is called the *stomodeum,* the posterior section or invagination is the *proctodeum,* and the central endodermal section is the *mesenteron.* Because of their ectodermal origin, both the stomodeum and proctodeum have a lining of cuticle, the *intima.* We will see later that the presence of this intima has a great restrictive impact on the movement of molecules both into and out of the system and also protect the cells lining the gut (Fig. 71). This intima is not found in the endodermal mesenteron; therefore, molecules can enter or leave the gut cells there with relative ease. Muscles of all three sections of the gut have a common origin, which is the migration of mesoderm into the regions.

Variation within each of the three sections of the alimentary canal depends on diet (Fig. 72). Also, the digestive tract may change in insects that undergo complete metamorphosis to the adult stage, with the result that an immature equipped to handle solid food may emerge as an adult feeding on nectar or blood (Fig. 73). This metamorphosis certainly must be classified as one of the marvels of nature.

Solid Food

Solid food requires grinding prior to ingestion. Grinding is normally carried out in a chamber as it is in most higher animals. This chamber, however, is not located within the insect body but outside—thus the name *preoral cavity.* The cavity is formed by the anterior labrum, the lateral mandibles and maxillae, and the posterior labium (Fig. 35). Food is chewed into small pieces, mixed with salivary secretions from *labial glands* for lubrication and very limited carbohydrate digestion, pushed into the mouth by the maxillae, and swallowed with the assistance of a muscular *pharynx.* Since the food often contains sharp materials such as spines on leaves [Fig. 71(A)], the foregut cells are protected by the intima [Fig. 71(B)]. Food is moved down the tubular *esophagus* by waves of muscular contractions termed *peristalsis.* In primitive insects, such as springtails, the food bolus now enters the mesenteron; in more specialized insects, such as the grasshopper [Fig. 72(A)] and the ground beetle [Fig. 72(B)], the posterior part of the stomodeum has been modified into an expansible *crop* for temporary food storage and a muscular *proventriculus* that acts primarily as a valve (some insects also have teeth in this region to shred food as it passes through). Because of the presence of the intima, little to no food is digested or absorbed except in species such as cockroaches and some beetles with large crops where enzymes may be regurgitated forward into the crop from the midgut to effect varying degrees of digestion.

When the proventriculus relaxes, food is moved posteriorly into the mesenteron. The tube here is called the *ventriculus.* Lateral diverticula or *gastric caecae* may expand outward from the ventriculus to increase surface area. It is in this region (mesenteron) that the major enzymes for digestion are secreted. This section lacks an intima so cells can easily secrete these enzymes directly into the cavity. Enzymes are liberated from the midgut epithelial cells by releasing vacuoles or by a complete breakdown of cells containing them. Once liberated, enzymes are activated and proteins are broken down to amino acids by a group of enzymes termed *proteinases* and *peptidases,* the former initiating the breakdown and the latter completing the process. Carbohydrates are acted upon by various enzymes collectively termed either *amylases* (which break down starch and glycogen) or *carbohydrases* (which complete the

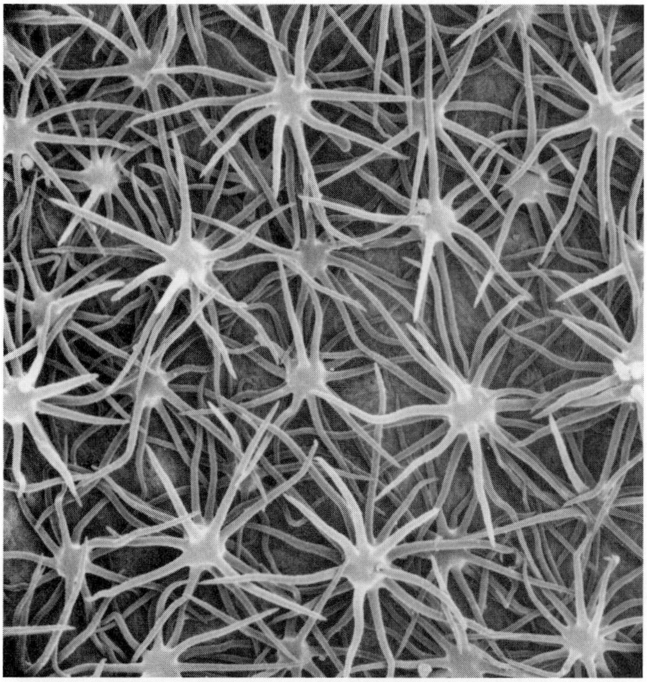

(A)

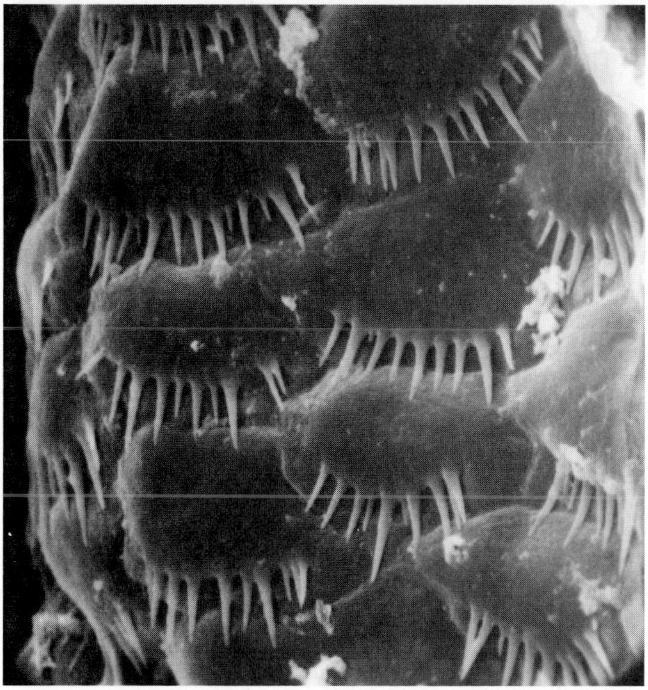

(B)

Figure 71. Importance of stomodeal intima. (A) spines on the surface of a leaf that could lacerate cells; (B) pharyngeal intima with protective overlapping microspines in the cockroach, *Supella longipalpa.*

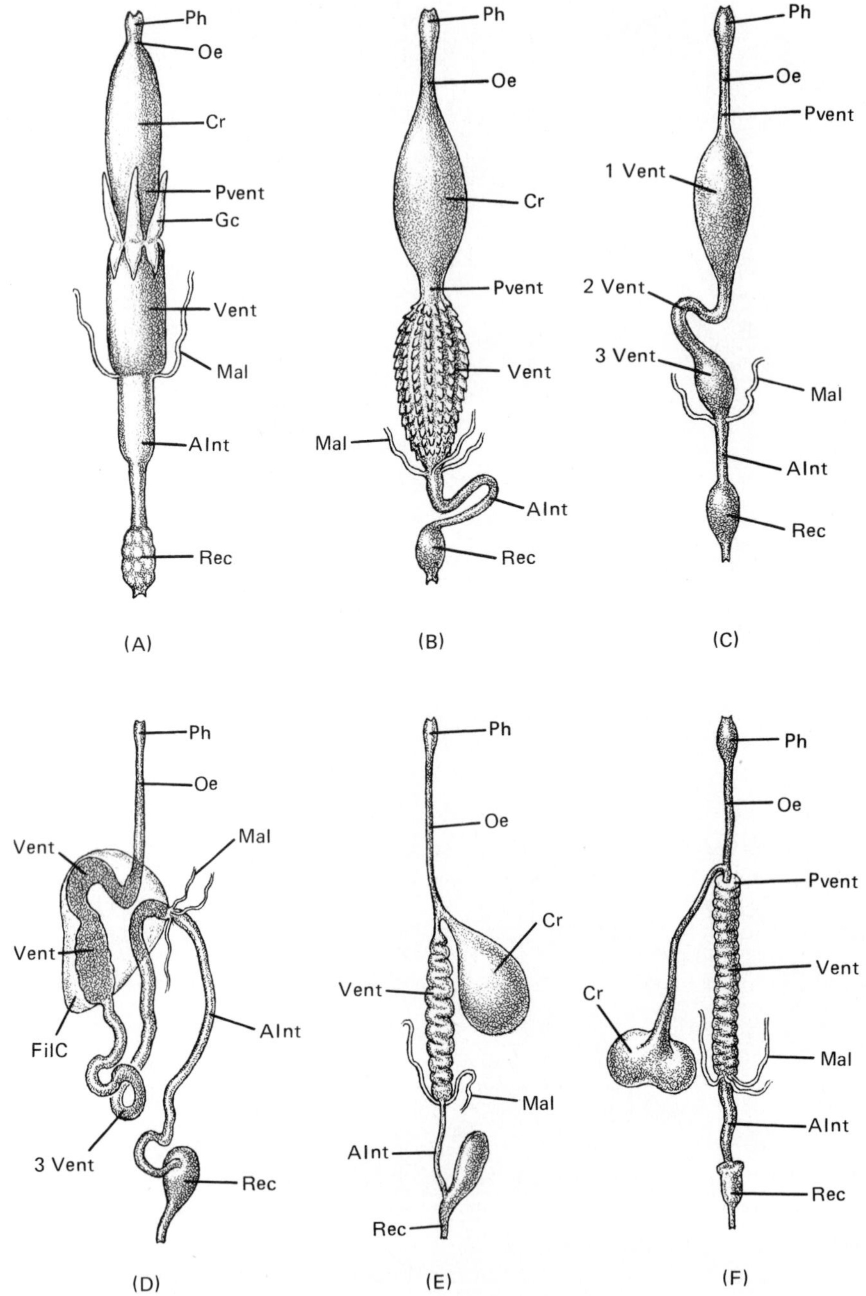

Figure 72. Some variations in the alimentary canal in insects. (A) grasshopper; (B) ground beetle; (C) water strider; (D) cicada; (E) moth; (F) house fly. *AInt*, anterior intestine; *Cr*, crop; *FilC*, filter chamber; *Gc*, gastric caecum; *Mal*, Malpighian tubules; *Oe*, esophagus; *Ph*, pharynx; *Pvent*, proventriculus; *Rec*, rectum; *Vent*, ventriculus.

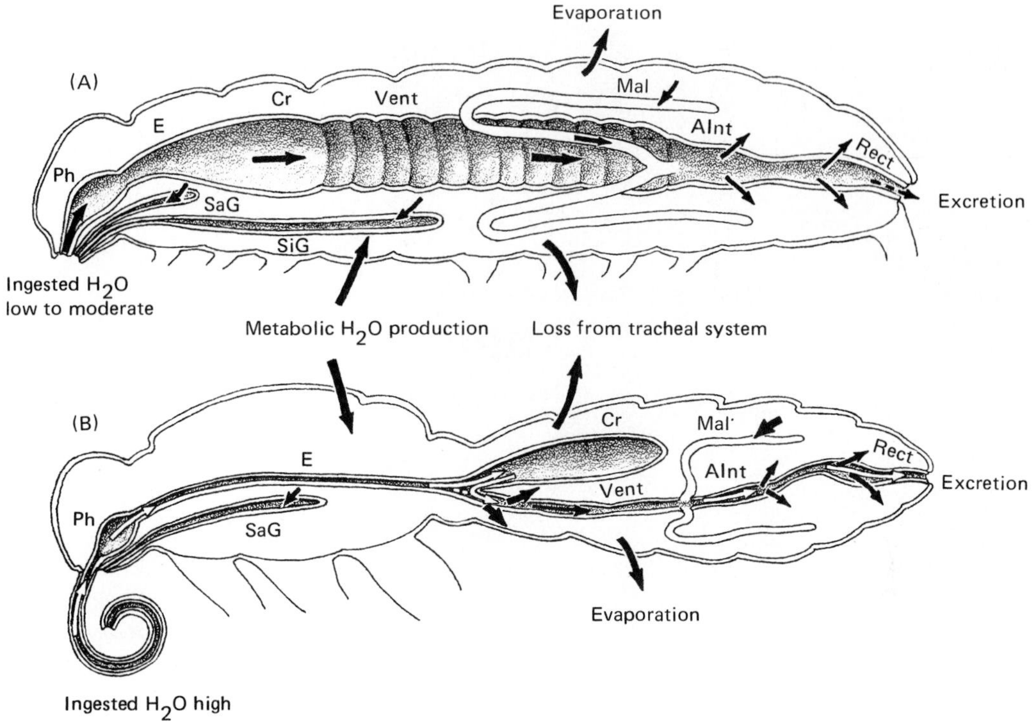

Figure 73. Movement of water into, around, and out of the body. (A) caterpillar ingesting solid food; (B) adult butterfly of the same species ingesting liquid food. Open arrows indicate high water content, black arrows moderate water content, and broken arrows low water content. *AInt*, anterior intestine; *Cr*, crop; *E*, esophagus; *Mal*, Malpighian tubules; *Ph*, pharynx; *Rect*, rectum; *SaG*, salivary gland; *SiG*, silk gland; *Vent*, ventriculus.

process of digestion of intermediate compounds to simple sugars or monosaccharides). Fats and oils are broken down by *lipases* into fatty acids and glycerol. Specialized enzymes such as chitinase, keratinase, and cellulase may be present in species with specific diets.

Correlated with the absence of the intima in this region is the absorption of food. Both the ventriculus and gastric caecae generally absorb the products of digestion. The uptake may be either passive (diffusion) or require energy (active transport). Food is retained in the midgut until the digestive and absorptive periods are complete, whereupon the pyloric valve of the hindgut relaxes and the remaining material is moved posteriorly.

The hindgut is divided into at least three regions, the muscular *pylorus,* the *anterior intestine,* and the *rectum.* The anterior intestine also may be subdivided into an *ileum* and a posterior *colon.* Wherever visual separations are found, an internal valve usually exists to regulate fecal movement between gut sections.

The intima in the proctodeum, especially the rectum, is sufficiently permeable to permit absorption and reabsorption of water (Fig. 73), salts, and some sugars and amino acids. Many insects have inwardly directed papillae or pads that increase the

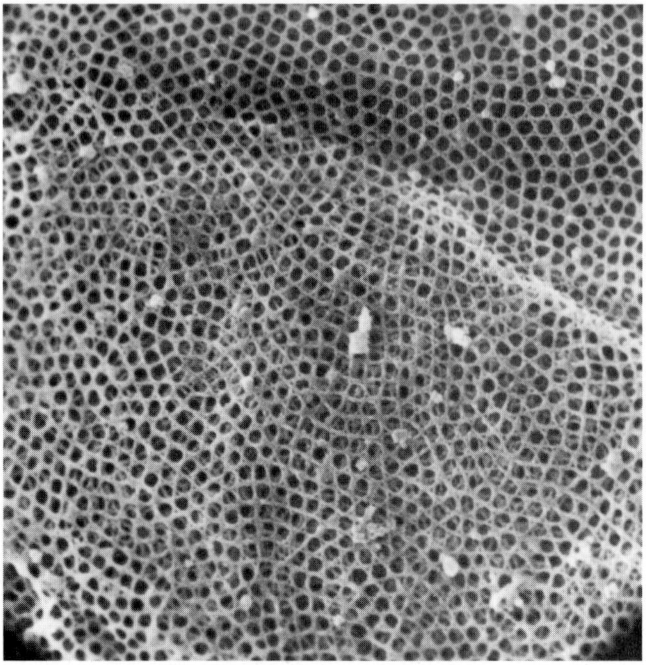

Figure 74. Peritrophic membrane in the hissing cockroach, *Gromphadorhina portentosa.*

effectiveness of this rectal region. As the fecal material is moved posteriorly, it usually becomes progressively dehydrated until it is formed into a pellet in the rectum and subsequently evacuated.

The cellular lining of the alimentary canal is protected from plant spines and other ingested roughage as well as most microorganisms by the intima of the stomodeum and proctodeum, and by the *peritrophic membrane* (Fig. 74) in the ventriculus in many insects. The peritrophic membrane is formed either by cellular secretions or by the loosening of cellular borders of the mesenteron; it encloses the food, but it remains permeable to enzymes and permits digestive products to leave and be further digested or absorbed by the midgut cells, especially in the gastric caecae. After completion of the digestive process, the membrane is moved down the canal and is often evacuated with the feces.

Liquid Food

Modifications of the digestive system for liquid diets are many [Fig. 72(C), (D), (E), (F)] and are usually correlated with a corresponding change in mouthparts. The most common structural modifications are the narrowing of the tube, a modification that increases fluid movement through the system, and the presence of a diverticulum or crop off the esophagus for storage of liquids [Fig. 72(E)] high in diluted carbohydrates such as nectar. Excess water in insects that ingest liquids often becomes a problem, especially in immatures in which large amounts of food must be taken in to facilitate growth. Excess water also makes the enzyme-substrate encounter more difficult because of the increased distance between food particles. This water is usually absorbed at the anterior part of the ventriculus to concentrate the food. This physiological bypass produces a temporary osmotic problem as the blood becomes temporarily diluted prior to its removal by the Malpighian tubules (Fig. 73).

Another modification, as seen in Figure 72(C), results in sections of the ventriculus becoming morphologically distinct; these are named ventriculus 1, 2, 3, and sometimes 4. Ventriculus 1 serves in carbohydrate and lipid digestion and in water removal for concentrating the food, but not in protein digestion, which occurs in the other sections.

Some insects often ingest excessive liquids to obtain minimal amounts of nutrients. This is particularly true of sap feeders. One such group, the Homoptera [Fig. 72 (D)], has solved this by concentrating the food in a *filter chamber.* An anterior section and a posterior section of the alimentary canal touch one another in an enclosed sac in which much water is short-circuited between these sections and is evacuated to the exterior. The net effect is food concentration for efficient digestion in the ventriculus without disrupting the osmoregulation system of the body. In aphids, more than 90 percent of ingested sap ends up in the feces.

Of special importance are the enzymatic changes that occur in the evolution of liquid feeding. The labial glands and the regurgitation of ventricular fluids become exceedingly important and produce anticoagulants in most blood feeders, toxins in many carnivores, and histolyzing enzymes in many herbivores and carnivores. The change in diet also results in varied enzymatic changes in the digestive materials secreted by the mesenteron and by the microorganisms that may exist symbiotically in the gut or gut lining.

Metamorphosis can have critical impact on this system. Although few changes in the type and enzymic action are seen in the Ametabola and Hemimetabola, an extensive change occurs in those insects undergoing Holometabolous development. Here we often see larvae adapted for ingesting solid food transformed into adults with a digestive system for handling liquid food. This marvelous change occurs during the nonfeeding pupal stage and will be discussed further in Chapter 4.

In summary, the digestive system is responsible for the breakdown of large nonabsorbable organic compounds by enzymic hydrolysis into small absorbable-sized molecules and then absorbing them into the body. The structures in the system vary with diet and the type of metamorphosis. Materials not digested, along with excretory materials, are then defecated.

Symbionts

Many insects have cultures of microorganisms within their bodies that provide their host with vital nutrients and the capacity to utilize many nutrient-deficient foods. Many symbionts are found intracellularly throughout the body, but we will restrict our discussion to those found in the gut.

Although many insects ingest cellulose, few have the enzyme *cellulase* to digest this complex polysaccharide. Digestion of this carbohydrate is often accomplished by bacteria and protozoa located in the alimentary canal of the host. Some species of termites, for example, feed on fecal material and oral feedings from older members of the colony after hatching from the egg and are thereby inoculated with the microorganisms necessary to digest the very food they feed upon, although other termites are capable of producing their own cellulase (O'Brien and Slaytor, 1982).

Some insects may have specialized chambers to "house" these organisms. If these chambers are part of the hindgut, as in termites, the lining of this region along with the symbionts is shed at each molt. Reinoculation must then occur if these symbionts are needed for survival.

CIRCULATORY SYSTEM

With multicellular organisms and their increased size, distribution of metabolic materials becomes a problem to be solved first by a body cavity filled with "sloshing" fluids (Fig. 69) and later by incorporating distribution tubes and a pump into this system. When fluids are completely confined within vessels, the system is said to be *closed.* However, when tubes or vessels are absent, or when they are present but the fluid is not completely restricted within them, as in insects, the system is referred to as *open* (Fig. 75).

Insect blood, or *hemolymph,* comprising from 5 to 40 percent of body weight, varies from vertebrate blood in that red blood cells are absent; therefore, it is more similar to lymph. This fluid is enclosed within the body cavity or *hemocoel* [Fig. 75 (A)]. Within the hemocoel is a longitudinal *dorsal vessel* that is divided into a series of pumping chambers (*heart*) in the abdomen [Fig. 75(C)] and the *aorta,* which extends forward to the brain region [Fig. 75(A)]. Some insects possess a large aortal sac in the head and, an accessory pulsating organ in the thorax [Fig. 75(B)], and segmental vessels may be seen in a few orthopteran taxa such as cockroaches and mantids. Changes in overall aortal shape and position, in the number of heart chambers, and in pulsating rhythms may occur during metamorphosis from a larva to an adult.

How does hemolymph move throughout the body? Hemolymph enters each chamber of the heart through openings, the *ostia,* and is pumped forward, either by peristaltic contractions or by a single uniform contraction of the entire heart, through the aorta to beneath the brain where it flows out into the hemocoel. The brain, therefore, is the immediate recipient of the blood as it leaves the heart (as in many animals). The hemolymph then generally flows posteriorly to the legs and wings and along the alimentary canal, where the absorbed nutrients are picked up from the ventriculus for distribution to the body. The hemolymph now flows dorsally through openings in the *dorsal diaphragm,* which with the *alary muscles* suspends the dorsal vessel, and the hemolymph then reenters the heart [Fig. 75(A)], thereby completing the cycle. The amount of hemolymph varies with the species, size, stage of development, and physiological state of the insect.

The previous description applies only to small resting insects. Larger resting insects often require abdominal segment contractions, as occur during breathing, to assist in fluid distribution. Active insects "slosh" the hemolymph about during locomotion movements, especially flight, and this aids the mixing action and hemolymph movement. Efficiency by exploiting contractions by locomotory muscles is some-

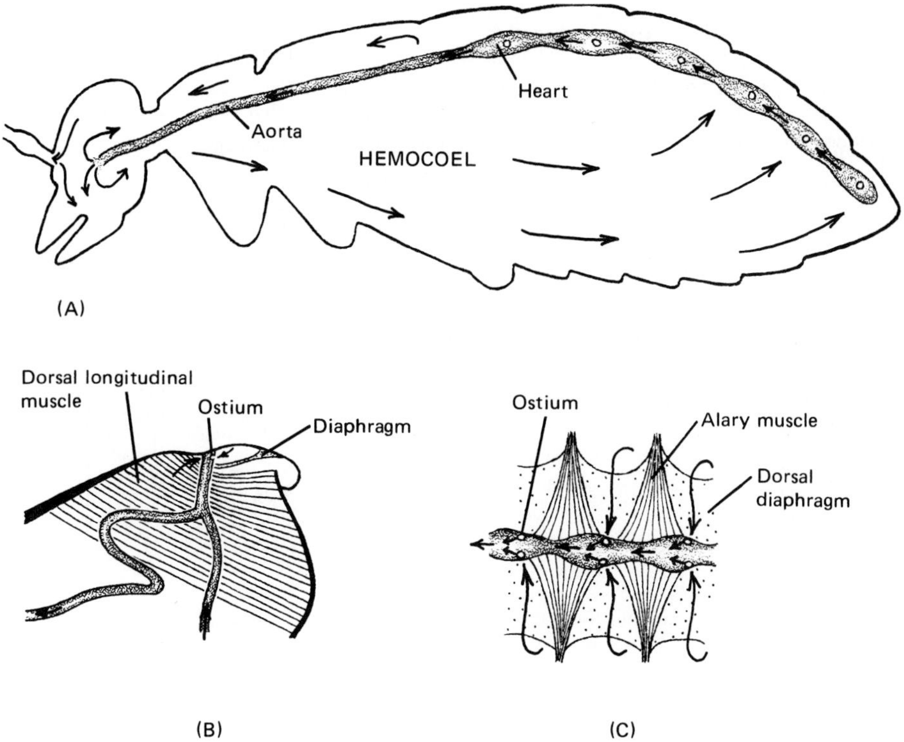

Figure 75. Circulatory system structure. (A) lateral view of hypothetical insect indicating directional blood flow; (B) accessory pulsating organ; (C) dorsal view of heart section. [(C) redrawn with slight modifications from Snodgrass, 1935.]

times enhanced by specialized *accessory pulsating organs* [Fig. 75(B)] near the antennae, legs, and wing muscles.

Hatching from eggs, wing enlargement, and molting all require increased blood pressure. Air is first taken into the body and then the skeletal muscles contract, thereby increasing blood pressure to rupture weak spots in either the eggshell or in the insect exoskeleton. For example, a sac in the head of certain flies, the *ptilinum* (Fig. 115), is everted by blood pressure and fractures both the last larval instar exoskeleton (puparium) and the pupal exuvium. After emerging, the ptilinum deflates and is retracted into the head and is lost (Fig. 48) as blood pressure is reduced.

Locomotion, especially by larvae, also may involve increased blood pressure and localized muscle contractions working against this pressure. Such hydraulic pressure is possible only by having a soft pliable cuticle; however, it is costly since such locomotion requires four to five times the energy as moving a rigid, heavily armored adult exoskeleton where the hardened elements act as a series of levers.

Heat transfer is also a function of the circulatory system, particularly during flight when extensive buildup of heat occurs around flight muscles. Blood is circulated posteriorly into the abdomen where temperatures drop as heat is lost through radiation;

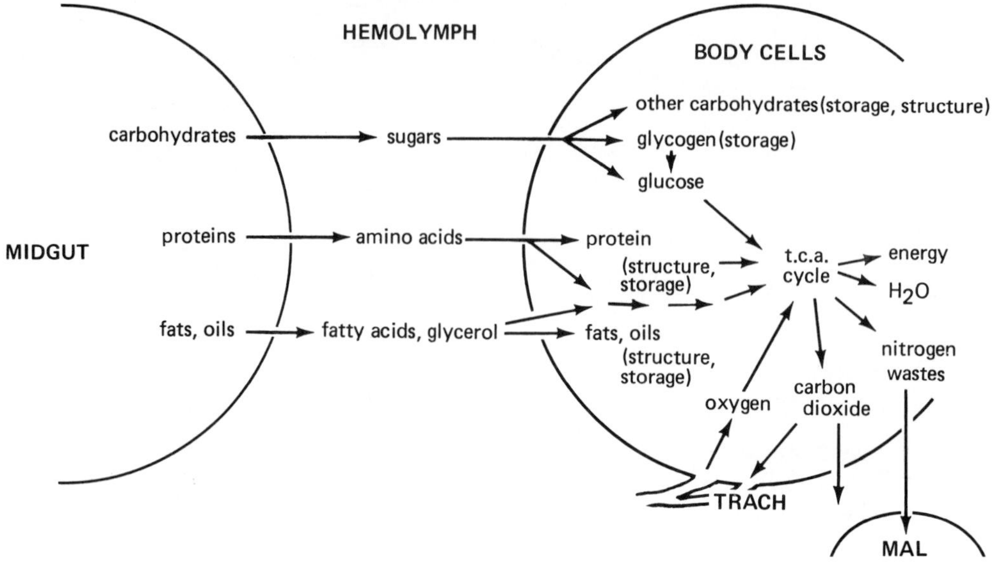

Figure 76. Diagram indicating the interrelationship between the digestive, circulatory, tracheal, and excretory systems. *MAL*, Malpighian tubule; *t.c.a. cycle*, energy conversion; *TRACH*, trachea.

bees, for example, lack insulating hair under the abdomen for this purpose. Cooled blood is then circulated back into the thorax. The reverse movement of heat may also occur—from the environment into the insect with the blood transferring the energy throughout the body. For example, butterflies bask in the sun during cool days and utilize their wings, especially the basal regions which are darkly pigmented, as organs of heat uptake to increase body temperature.

Blood cells, or *hematocytes,* are present in the hemolymph but do not carry oxygen. Their function is usually similar to that of the white cells of humans; they are mainly phagocytic, and they act individually to engulf foreign objects or in concert against large invaders such as nematodes and insect parasites. The former action is termed *endocytosis* and is also used in ridding the body of dead cells and during metamorphosis, whereas the latter is termed *encapsulation* and appears to either starve or asphyxiate the foreign invader by secreting multiple layers of hematocytes. Hematocytes also are involved in coagulation and wound healing. More than seven types of hematocytes have been classified, although not all are found in the same species or at the same time in an insect. Many of these cells circulate at specific times but localize to form phagocytic organs at other periods, especially near the heart.

In summary, the circulatory system functions in the transportation of nutrients (Fig. 76), hormones, etc., about the body but has little to do in oxygen transfer because red blood cells with their hemoglobin are absent. Of less conspicuous but perhaps equal importance is its role in keeping cells moist and maintaining osmotic pressures, storing reserves of certain metabolites and water, regulating heat within the body, buffering or detoxifying reactive molecules, healing wounds, protecting against foreign invaders, maintaining blood pressure to assist in molting and wing inflation, and aiding in locomotion, particularly for legless forms.

TRACHEAL SYSTEM AND RESPIRATION

The word *respiration* has acquired several different meanings; to some it indicates the exchange of air or ventilation, and to others it refers to the breakdown of food to obtain energy. The word is best used to indicate a series of biochemical reactions occurring in cells that liberate energy. In the presence of oxygen, the reactions may be summarized as:

$$6O_2 + 6H_2O + C_6H_{12}O_6 \rightarrow 6CO_2 + 12H_2O + \text{energy} \quad (1)$$

The above equation is an extremely simplified one in which the initial compound is a simple sugar or monosaccharide. In insects, monosaccharides come from the breakdown of cellular glycogen and the blood disaccharide sugar trehalose. Other substances, such as amino acids and fatty acids, also may be shunted into this energy cycle at various times. The sequence and its tie-in with the digestive and circulatory systems are seen in Figure 76.

Carbohydrates are the major source of energy reserves drawn from during and shortly after feeding, whereas fats and oils and protein are used at other times, especially during extended activity periods. The liberated energy from the tricarboxylic acid (TCA or Krebs citric acid) cycle is normally used to convert adenosine diphosphate (ADP) to adenosine triphosphate (ATP), which in turn serves as a ready source of high energy.

$$\text{Energy} + \text{ADP} + \text{inorganic phosphate} \rightarrow \text{ATP} \quad (2)$$

Approximately 55 percent, or 380,000 calories, of the total energy of the sugar molecule is released and stored in the resulting 38 molecules of ATP when going through the oxidative cycle (*aerobic respiration*). Carbohydrates and protein respiration result in similar amounts of energy (4 calories/g), but fats are more productive (9 calories/g) and also yield more metabolic water. If oxygen is not available, the food is shunted into another metabolic pathway (*anaerobic respiration*), yielding only a small fraction of the potential energy. Because only two molecules of ATP result from anaerobic means, nearly 20 times as much glucose must be metabolized to obtain as much energy as through the aerobic form. ATP is used to synthesize new structure and maintain current structure and processes for specialized activities such as muscle contraction or reproduction.

Insects are ectothermic, and the respiratory rate is usually proportional to the temperature of the external environment. Species in the tropics have little difficulty in maintaining an active metabolism. For those surviving in cold regions of the world, however, nearly total inactivity occurs for significant periods of their life. This inactivity is advantageous especially during the winter for herbivorous species because their food source is not available and energy would be wasted by fruitless searching and activity.

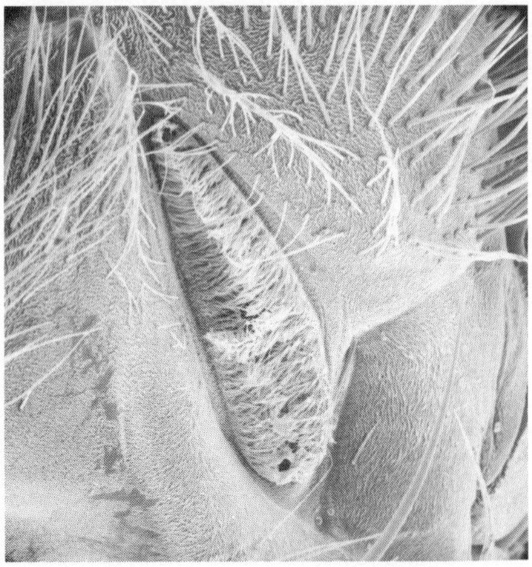

Figure 77. Prothoracic spiracles of an adult flesh fly (*Sacophaga bullata*) showing filtering setae, which prevent dust and foreign particles from entering the tracheal system and slow down water and heat loss from the tracheal system.

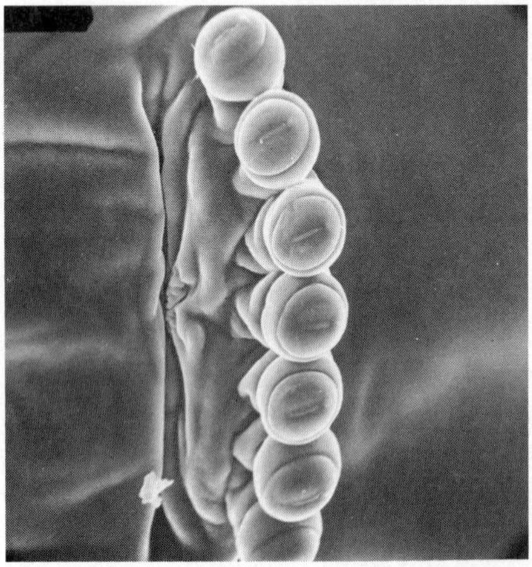

Figure 78. The prothoracic spiracle of a house fly (*Musca domestica*) larva. This spiracle, as seen through the scanning electron microscope, has multiple openings into a common internal atrium.

How does oxygen get into the body and reach tissues? We have previously ascertained that hemolymph and hematocytes play no role in this transfer. Obviously oxygen must enter the insect through some surface. Because the oxygen molecule is larger than the water molecule, a paradox exists: if there is sufficient surface for effective uptake of oxygen, this also results in an excessive loss of water. Throughout geologic time, a water-proofed exoskeleton has been selected externally, but the surface area needed for oxygen uptake has come from an extensive invagination of tubes (*tracheae*) into the body whereby water loss can be reduced, because of this internal location, to periods when air is exhaled. Atmospheric oxygen is carried directly to tissues by these tubes (Fig. 76).

Most insects have openings or *spiracles* into the tracheae (Figs. 77, 78, 79, 80, 81). This system is referred to as *open* [Fig. 80(A), (B)]. Spiracles vary from simple openings to highly modified structures with filters and valves for regulating the openings (Figs. 78, 79). Most species have two pairs of thoracic and eight pairs of abdominal

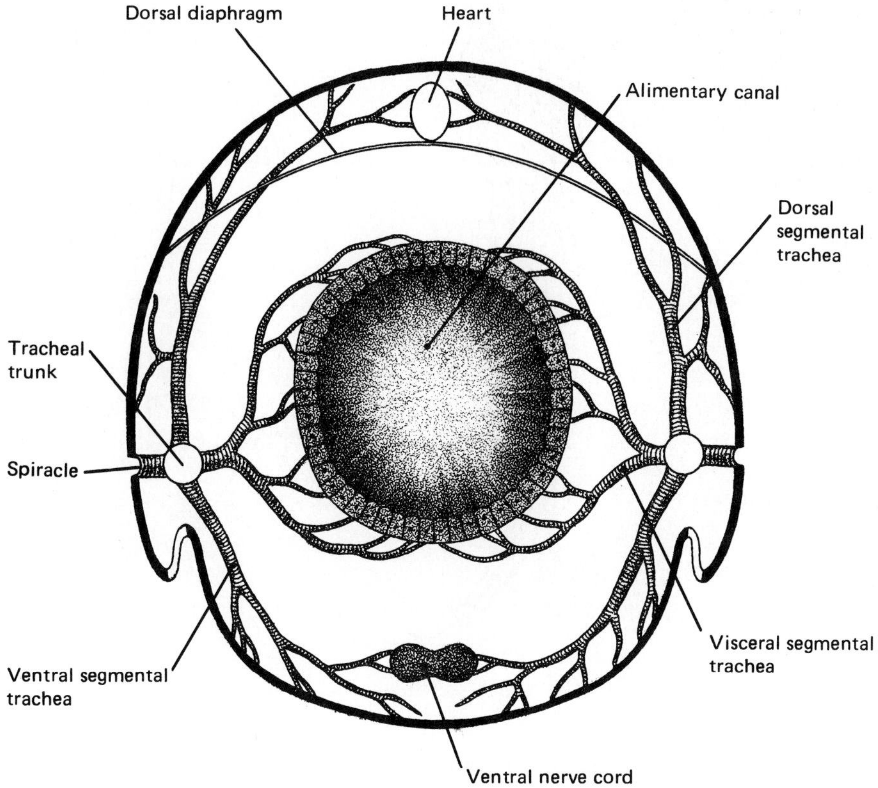

Figure 79. Cross section through the abdomen illustrating some of the tracheation. (Redrawn with slight modifications from Snodgrass, 1935.)

spiracles [Fig. 80(A)]. However, the trend is for a loss of spiracles in advanced Holometabola where the terminal segments become involved into a tubular ovipositor. Reduction in numbers is also common in larvae. A loss of the metathoracic pair reduces the *holopneustic* state of two pairs of thoracic and eight pairs of abdominal spiracles to the *peripneustic* state. When only the metathoracic and last abdominal pairs of spiracles are functional, the insect is *amphipneustic.* If only one pair of spiracles remain open, the condition is *propneustic* if they are on the prothorax (Fig. 132) or *metapneustic* [Fig. 80(B)] if on the terminal abdominal segment.

Each spiracle opens into *segmental tracheae* that send tracheal branches to any structure located within that segment (Fig. 79). A dorsal branch carries oxygen dorsally to the muscles located there and the dorsal vessel. A medial branch goes to the alimentary canal and gonads, while a ventral branch carries air to the ventral muscles and the nerve cord. Tracheae originate as invaginations of the epidermis and often have a waxy coating or epicuticle that prevents most oxygen uptake and reduces water loss. They also have spiral sclerotized braces (*taenidia*) in the walls that prevent collapse, and each tube progressively branches into smaller tracheae until minute *tracheoles* (nontapering and thin-walled tubes) are reached. The radiating tracheole network

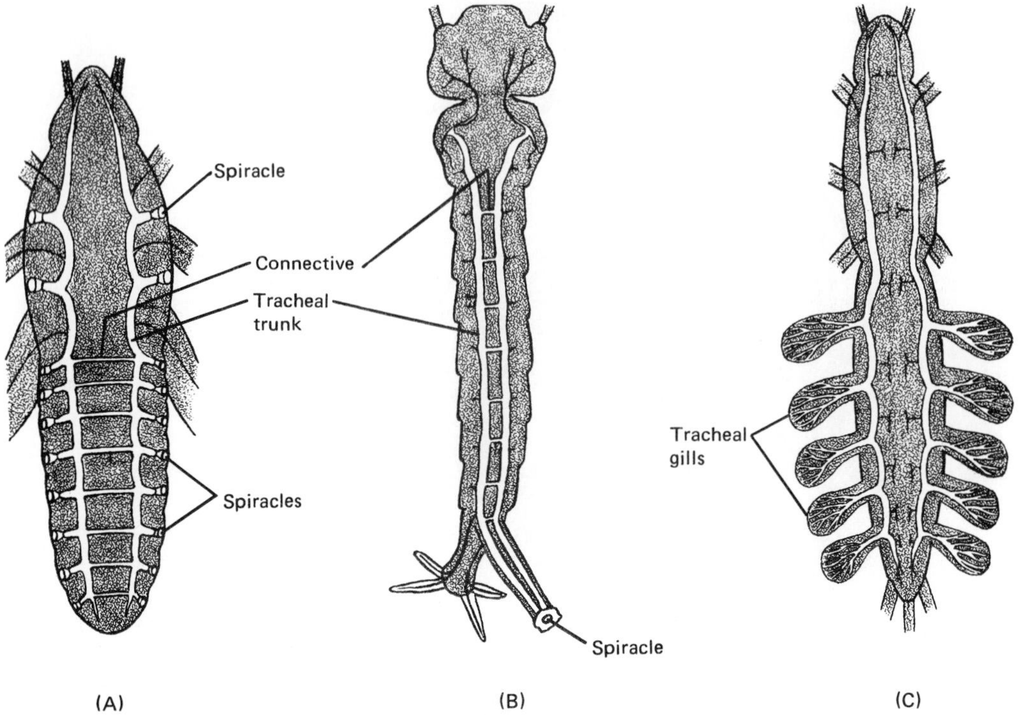

Figure 80. Some variations on the tracheal system of insects. (A) typical type of open system; (B) open system with reduction in number of spiracles; (C) closed system with cutaneous and tracheal gill gas exchange.

resembles the capillary network in humans, and it is through such tracheoles with their extensive surface area that most gaseous transfer occurs. Tracheoles are permeable to oxygen uptake and carbon dioxide loss. These tracheoles are the only part of the system that are not shed at a molt.

What about the body segments that do not have spiracles, such as the head? What happens when a spiracle becomes covered or blocked? The answer is intercommunication. All segments are interconnected by longitudinal *tracheal trunks* as seen in Figure 81. Oxygen enters the segments that have spiracles and moves throughout the body to segments lacking spiracles by diffusion and muscular breathing contractions of the abdomen. Air normally enters the anterior spiracles and is directed anteriorly into the head, or posteriorly and out through the abdominal spiracles. Tracheal trunks and segmental trachea also help suspend the gut within the hemocoel.

Air sacs of various sizes may be present and are recognized by their reduction or lack of braces. Their functions are (1) to serve as reservoirs of oxygen, (2) to serve as bellows in distributing air and cooling the body, particularly during flight, (3) to decrease weight in fast-flying species, (4) to increase body pressure during certain periods such as molting, and (5) to provide space storage during molting into which future growth may occur (they occupy more than 40 percent of body volume at the beginning of an instar but only 4 percent just prior to molting in grasshoppers).

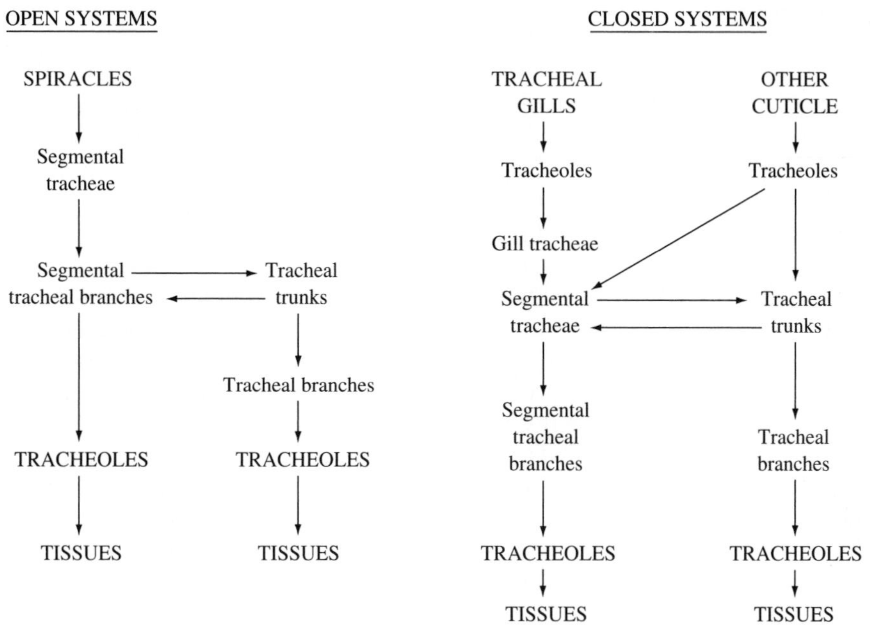

Figure 81. Directional flow of air to internal body tissues in open and closed tracheal systems.

Lack of oxygen, excess of carbon dioxide, or both are the normal initiators of spiracular and ventilatory activity. Having the spiracles closed except for brief periods of breathing results in the capability to control water loss more efficiently. The ability to close the spiracles and to have a hard exoskeleton and air reservoirs internally are beneficial during certain periods of environmental stress. For example, attempts to control grain beetles by fumigation are successful only after long periods of exposure, which result in oxygen depletion, the system opening, and the active agent being inhaled.

A few tracheal systems are *closed,* and gaseous movement into the tracheae for distribution must first pass through the exoskeleton proper. Some aquatic insects have increased surface area in the form of tracheal gills, evaginations of the exoskeleton with extensive tracheole networks [Fig. 80(C)]. A detailed discussion on aquatic adaptations is found in Chapter 5.

Because the tracheal system originates from invaginations of the epidermis and, therefore, has cuticular linings, what happens to this system at molting? During ecdysis all tracheal intima or linings are shed, but the old tracheoles remain along with new ones formed by tracheoblast cells located near the tip of the new tracheal intima.

In summary, in open systems air extends from the external environment directly to the individual cell through a series of tracheae and tracheoles. Cells receive oxygen through the tracheoles. In aquatic forms, oxygen enters the body through terminal spiracles or from the water through tracheal gills or body surface. The directional flow differences between the open and closed systems may be seen in Figure 81. Once within tissues, oxygen can be used as necessary for metabolism.

MUSCULAR SYSTEM

Although most cells are capable of limited contraction, certain cells have great ability to contract. The degree of contractibility and the rate are determined by intracellular bands or *striae* that carry out the actual process. Those muscle cells with relatively few striae are characterized by a slow rhythmic contraction, whereas vigorous ones have extensive networks. Unlike the muscle cells in mammals, all insect muscle cells have striae, but the number varies greatly.

Muscle cells are aggregated into muscle fibers and in turn into functional units called *muscles.* The amount of muscle varies up to a maximum of 60 percent of the body mass (in dragonflies). The number of individual muscles in an insect also varies markedly. The adult grasshopper (Fig. 82) has approximately 900 muscles, and the butterfly caterpillars have approximately 4,000 (humans have 800).

There are four basic morphological types of muscle: (1) minute fibrillae, with little differentiation and with the muscle encircled by a thick layer of sarcoplasm with nuclei (larval Diptera), (2) large in cross section, with a tough covering membrane, little sarcoplasm, and nuclei and sarcoplasm between the muscle fibrillae (most common type), (3) well-developed fibrillae surrounding an axial core of nuclei and sarcoplasm (most appendicular muscles in adult Diptera, Hymenoptera, and Odonata), and (4) bundles of loosely bound large fibrillae with no covering membrane, and nuclei and sarcoplasm scattered between and throughout fibrillae (see indirect flight muscles below). More details on these muscles may be found in Wigglesworth (1984).

The power of a muscle is proportional to the area of its cross section. Because the mass that must be moved is proportional to its volume and because of somewhat better leverage systems than humans, small insects can carry out what appears to be miraculous performances. Fleas, aided by special elasticity in the pleuron, are capable of leaping distances that would approximate a human's leaping more than 1,000 ft (305 m). Ants are capable of moving weights that would approximate several tons when their size and weight are proportionately increased to compare with those of a human. These examples, however, are only valid for comparisons and cannot be construed to indicate potentials, because size is such a limiting factor.

Muscles in insects are often artificially segregated into four types: *visceral, segmental, appendicular,* and *flight.* Visceral muscles surround the ducts and tubes and produce directional waves or peristalsis to move products from one region to another. Segmental muscles cause telescoping of segments necessary in molting, inhalation and exhalation, increasing body pressure, and locomotion in legless individuals. Appendages are moved as a unit by muscles originating on either the tergum or sternum and inserted on the coxae. In addition, as previously discussed, each segment is operated by muscles originating in the previous segments (Fig. 60). Movement of segments, either body or of the appendage, is the result of coordinated contractions of antagonistic muscles working against blood pressure within the hemocoel.

It is in the flight muscles that one sees the greatest specialization. One type of muscle is termed *synchronous* and produces one contraction per nerve impulse. Because of this limit, the rate of wing movement is restricted to a maximum of approximately from 30 to 40 beats per second with most species in the 5 to 15 beats per second range.

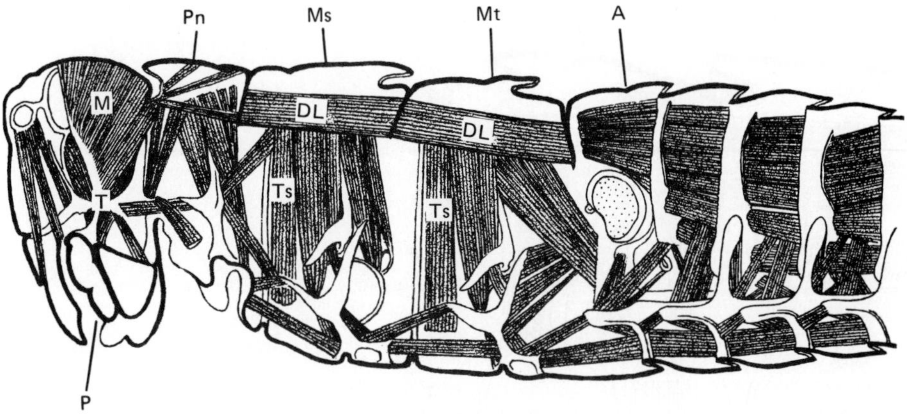

Figure 82. Musculature of a grasshopper (*Melanoplus differentialis*). *A*, abdominal segment; *DL*, dorsal longitudinal muscle; *M*, mandibular muscle; *Ms*, mesonotum; *Mt*, metanotum; *P*, preoral cavity; *Pn*, pronotum; *T*, tentorium; *Ts*, tergo-sternal muscle.

Another type is *asynchronous* (sometimes termed fibrillar), which is found in highly specialized insects such as many beetles, wasps, and flies. This type has an innate rhythm much faster than the nerve impulse rate, because each impulse initiates a series of contractions of the muscle maintained by stretching antagonistic muscles as the thoracic exoskeleton is deformed. This enables some gnats and mosquitoes to move their wings from 500 to 1,000 beats per second, which is fast enough to produce air vibrations that insects and humans can hear.

Such accelerated contractions obviously consume great amounts of energy, approximately 1,000 calories per gram of muscle per hour. This is from 10 to 20 times greater than the amount of energy used by leg or heart muscle during vigorous exercise in humans. Although flies and bees use trehalose and glycogen as energy reserves, many other insects obtain energy from proline or fats, the latter being most efficient (eight times) and common, especially in those insects that fly great distances. These food sources provide the energy necessary to convert ADP to ATP in the numerous mitochondria, up to 40 percent o e volume, found in insect muscle cells. With such energy consumption, one would expect flight muscles to become fatigued during extended flight. This does not occur, however, because these muscles lack lactic dehydrogenase (an enzyme that produces lactic acid during anaerobic glycolysis) and because they receive such a great amount of oxygen through the tracheal system.

Because insects are ectothermic, their temperature varies directly with the environment. This variation is beneficial to the small insect, because great amounts of energy loss (more than 90 percent) would be required to metabolically burn sufficient food to maintain a uniform body temperature with such a high surface/volume ratio. However, the temperature of flight muscles must normally approach 30°C to 35°C for flight to occur. How is this possible? In some insects this temperature is not maintained, and flight becomes greatly restricted. Some insects expose themselves to maximal sunlight whereby the body heats up sufficiently for flights. However, in some of the larger species whose surface/volume ratio is too low or when the insects are nocturnal,

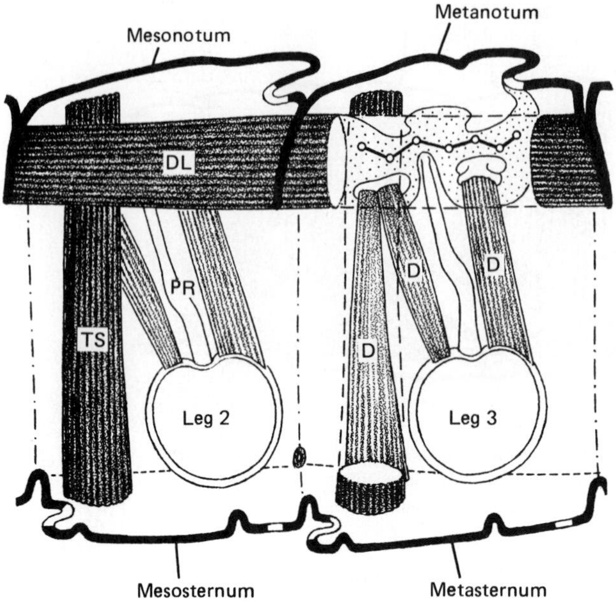

Figure 83. Meso- and metathorax of a grasshopper with all but the flight muscles removed. *D*, direct flight muscles; *DL*, dorsal longitudinal muscles; *PR*, pleural ridge; *TS*, tergo-sternal muscles.

sufficient warming cannot occur by solar radiation. These species require a warmup period of wing vibrations to generate sufficient metabolic heat to raise the muscle temperatures into the flight range and usually have insulating scales and hair to reduce heat loss. In cool weather this may require 10 to 15 minutes. To prevent excessive temperature buildup during flight, particularly during warm weather, heat is transferred from the thorax to the abdomen and/or head by the hemolymph and is lost through radiation to the air, either through the air sacs during breathing or from the external exoskeleton. Those insects with insulating hair over most of their body, such as bumblebees, have bald abdominal sterna that act as cooling coils.

The mechanics of flight differ from that of other animals capable of flight (bats and birds) because the wings are not modified appendages. Except in a few insects such as dragonflies, movement of the wings results *indirectly* from the distortion of the thorax, especially the tergum. Because the pleural wing process serves as a fulcrum, the contraction of the *tergo-sternal muscles* (Fig. 83) lowers the tergum, which in turn raises the wing tips. The wing tips are lowered when the tergum is vaulted upward as a result of the contraction of the *dorsal longitudinal muscles.* A complex click mechanism (Fig. 84) greatly amplifies the power of each stroke because of the sudden release of tension built up in the thoracic plates as the wing clicks into a new position of stability. The shape and rigidity of the thorax play an important role in determining how this system functions.

The simple movement of wings up and down, however, does not produce the flight as seen in insects. Directional movement is accomplished by contraction of *direct flight muscles* (Fig. 83), which twist the wings by raising and lowering the front and back edges of the wings. The combined action of the indirect and direct flight muscles is, therefore, to push air down and backward, and the insect is drawn forward into

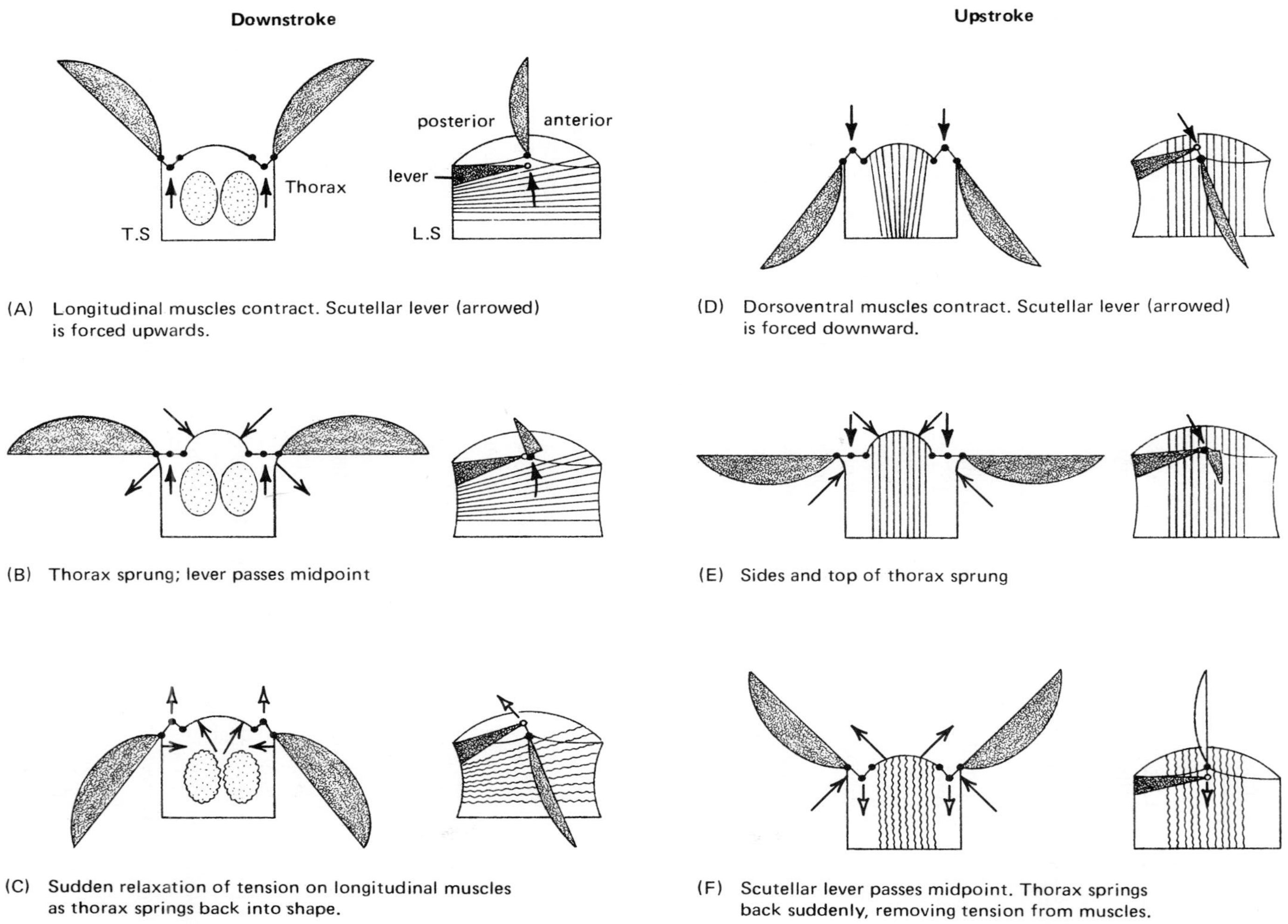

Figure 84. The click mechanism of a fly's wing. Wing movement is driven by muscles that distort the thorax. (Redrawn from Wells, 1968.)

the resulting region of low pressure above the wings. Further changes in wing pitch determine directional changes, elevation, and backward movements.

Although insects can move their wings rapidly, they are nevertheless slow fliers. The maximal speed known, based on experimentation rather than speculation, is approximately 36 mph (58 km/h) by a dragonfly. Average speed, however, is probably in the 5 to 10 mph (8 to 16 km/h) range.

EXCRETORY SYSTEM

As the direct result of metabolism within the cell, certain biproduct molecules, such as protein wastes, must be eliminated. Single-celled organisms excrete these wastes directly into the external aquatic environment with little or no change. However, multicellular organisms usually must detoxify wastes to prevent bodily damage prior to evacuation by combining them with other substances unless copious amounts of water are present to dilute them. If these products cannot be used in the synthesis of organic compounds, such as amino acids, specialized organs and ducts are present in most animals to remove them from the blood and eliminate them from the body (Fig. 76).

The major nitrogenous waste product in terrestrial insects, about 80 to 90 percent, consists of *uric acid,* a substance that is relatively nontoxic and highly insoluble. These insoluble molecules become vital to a small animal that must economize on water. Why? Because uric acid is removed from the blood by the *Malpighian tubules* (Table 1) and deposited into the hindgut where it crystallizes as the water is resorbed. Therefore, water does not need to be used as a carrier for elimination of wastes because this takes place through defecation with the feces. Some uric acid, however, is incorporated into tissues and the exoskeleton. In some cockroaches, up to 10 percent of the total dry body weight consists of uric acid, which can be utilized later in metabolism during periods when diet is deficient in nitrogen. Another waste product, *ammonia,* is found in freshwater insects and in a few terrestrial insects, such as blowfly larvae, that live in very moist habitats. Urea, although isolated from some insects, is inconsequential.

As in other animals, excesses of any type are eliminated to prevent osmotic imbalances from occurring. Insects that ingest much liquid food (Fig. 73) and some

TABLE 1. Representative Numbers of Malpighian Tubules as Related to Metamorphosis

No Metamorphosis		Incomplete Metamorphosis		Complete Metamorphosis	
Order	*Number*	*Order*	*Number*	*Order*	*Number*
Collembola	0	Ephemeroptera	40–100	Coleoptera	4–6
				Neuroptera	6–8
Thysanura	6	Odonata	50–60	Hymenoptera	12–150
Protura	6	Orthoptera	2–12	Lepidoptera	6
Archeognatha	20	Plecoptera	50–60	Mecoptera	6
Diplura	16	Hemiptera & Homoptera	0–4	Trichoptera	6
		Isoptera	2–8	Diptera	4

freshwater-inhabiting insects (Fig. 134) must eliminate water to maintain proper osmotic balance. During these periods of excesses, water is removed from the blood by the Malpighian tubules, and little or no absorption occurs in the proctodeum.

In addition to the Malpighian tubules, various phagocytic cells and the fat body play an important role in the elimination or detoxification of waste products from the circulating blood.

FAT BODY

The fat body is a loose network of cells associated with the connective tissues of the body. It is located around the gut and other organs and may account for up to 65 percent of the total body weight in some larvae. The primary functions of the fat body appear to be the interrelated activities of storage and intermediary metabolism.

As the name implies, fats are synthesized and stored in these tissues. In fact, the majority of all lipids found in an insect are localized here. The majority of the lipids are triglycerides. Lipids become particularly high in reproducing female insects, in insects entering diapause, and in some insects readying to migrate.

The fat body is also the major site of glycogen deposition and storage. Sugars are removed from the hemolymph (Fig. 76) and are converted to this storage molecule, although some sugars are converted to the disaccharide trehalose. Trehalose, the blood sugar of insects, is constantly released from the fat body for distribution throughout the body. Although this tissue is the main source of hemolymph proteins, adult insects store little protein in the fat body. Immatures, however, can and often do store these and other nitrogenous organic compounds.

NERVOUS SYSTEM

All cells are *irritable,* that is, capable of response to stimuli. In multicellular organisms certain cells have become specialized to carry these responses and to coordinate the incoming information into behavioral action. The potential number of these *neurons* is often influenced by the size and specialization of the organism. Insects are relatively small and have a restricted number of neurons. Nevertheless, these few neurons are used very efficiently through a series of "built-in" or innate behavioral patterns.

Neurons are of two basic types. The first type, the *sensory neuron,* has one to many dendrites. The second type, the *motor and interneuron* (or association neuron) lacks dendrites. Impulses in motor and association neurons travel only along the axon originating on a *collateral branch* and normally do not pass through the cell body (Fig. 85). In sensory neurons the impulse is passed along the dendrite to the cell body before reaching the axon.

The nerve impulse starts at some sensory structure and represents an ionic change as depolarization of the membrane passes progressively along the cell. Because of the long length of neurons, impulses are carried more quickly and efficiently than if the message had to be passed through a series of normal-sized cells. The impulse or action

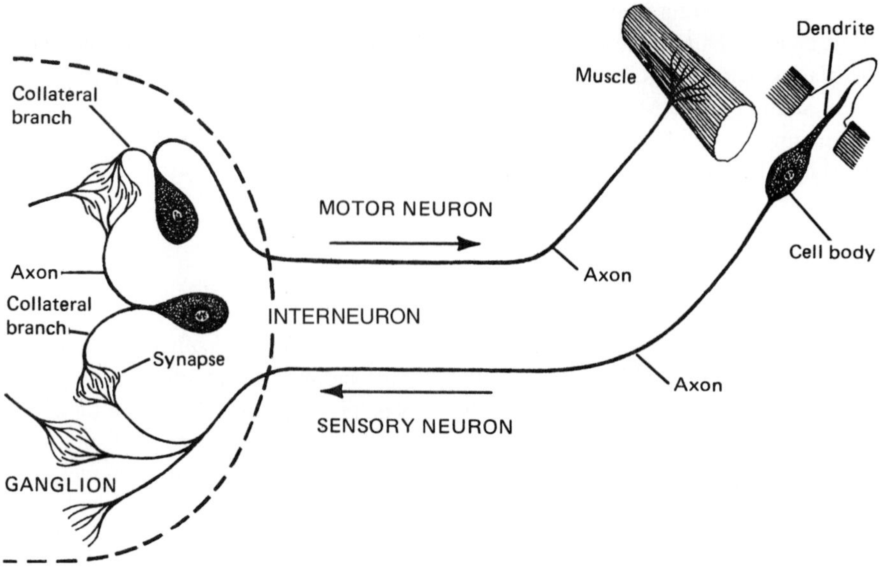

Figure 85. Diagram of the reflex mechanism of the insect nervous system. Note that it is the opposite of the chordate type in vertebrates where the sensory neurons enter the dorsal part of the ganglion and the motor neuron ventrally. (Redrawn with slight modifications from Richards and Davies, 1977.)

potential triggers the release of specific chemicals or *neurotransmitters* from one cell (presynaptic) into an area between two nerve cells (*synapse*), which then initiates an impulse in the second cell (postsynaptic).

Nervous tissue arises early during embryonic development from *neuroblasts* and becomes segmented as the individual metameres are formed (Fig. 86). These neural tissues form paired ganglia in each segment and are the bases of the *central nervous system.* Ganglia become interconnected as the neuron fibers grow from one ganglion to another, giving the central nervous system a ladderlike appearance. The position of most ganglia is opposite to that of humans, being located below the digestive system; each pair of ganglia coordinates the activities of the structures of the segment in which it formed. Some of the cells in ganglia secrete hormones and are called *neurosecretory* (see Chapter 4).

The ladderlike appearance is only relative, however. The anterior three pairs of ganglia fuse to form the brain or *supraesophageal ganglion,* and the fourth to sixth pairs unite into the *subesophageal ganglion.* The remainder of the central nervous system is termed the *ventral nerve cord,* and these ganglia also tend to fuse, especially each segmental pair and the last three to four pairs in the abdomen. In some specialized insects, all ganglia fuse to form a single ganglionic mass in the head and prothorax (Fig. 87). The primitive origin of each segmental pair of ganglion in this cephalized mass can be determined by tracing the nerves to the segments they innervate.

The supraesophageal ganglion, as the name indicates, is located dorsal to the esophagus (Fig. 88). The first pair of lobes, the *protocerebrum,* receives nerves from

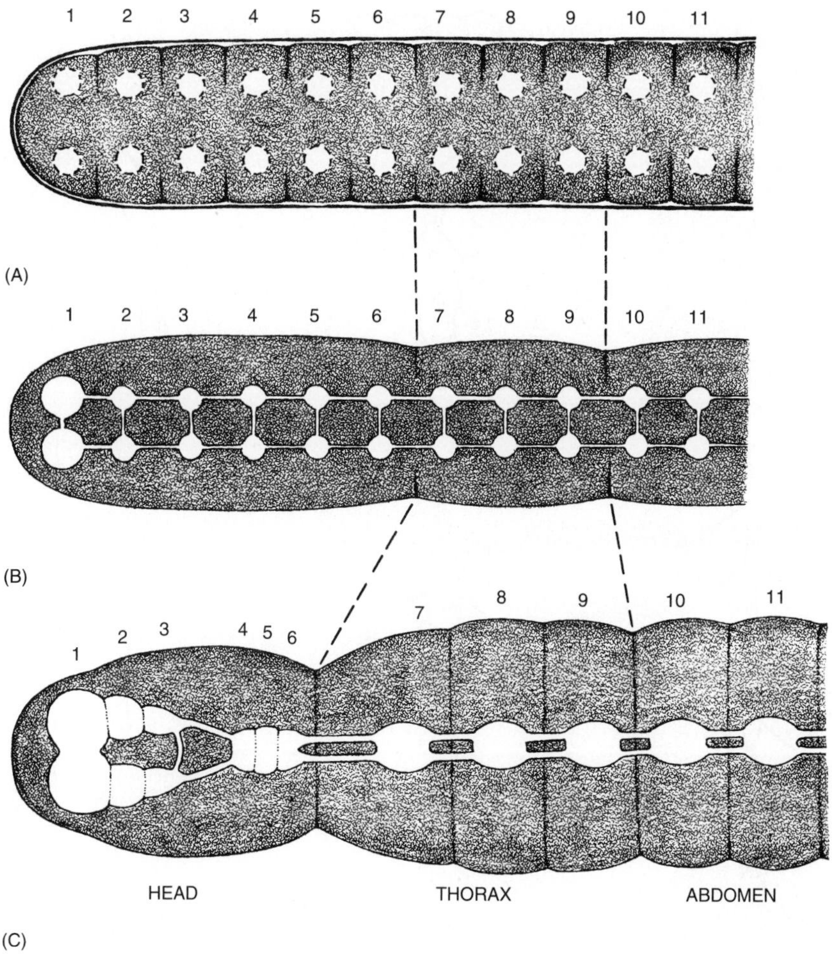

Figure 86. Diagrammatic representation of the development of the central nervous system. Nerves to the periphery have not been included. (A) neuroblast formation during segmentation; (B) interconnection of ganglia; (C) differentiation of ganglia as tagmosis occurs.

the compound eyes and ocelli. The protocerebrum is the major association region in the central nervous system (about 100,000+ neurons), and its direct connection to the photoreceptors indicates the great effect light stimuli have upon most insects. Also, experiments on social insects have shown that the comparative size of this area (850,000 neurons in the honeybee) seems to be correlated to the increased ability to learn. The *deutocerebrum* or second pair of ganglionic lobes receives impulses from the antennae (the appendages of the second body segment), coordinates this sensory input with the brain, and controls the movement of the antennae. The *tritocerebrum,* unlike the other sections of the brain, remains separated into two lobes and receives nerves from the frontal ganglion, labrum, and subesophageal ganglion. All lobes of the brain are interconnected through nerve fiber tracts.

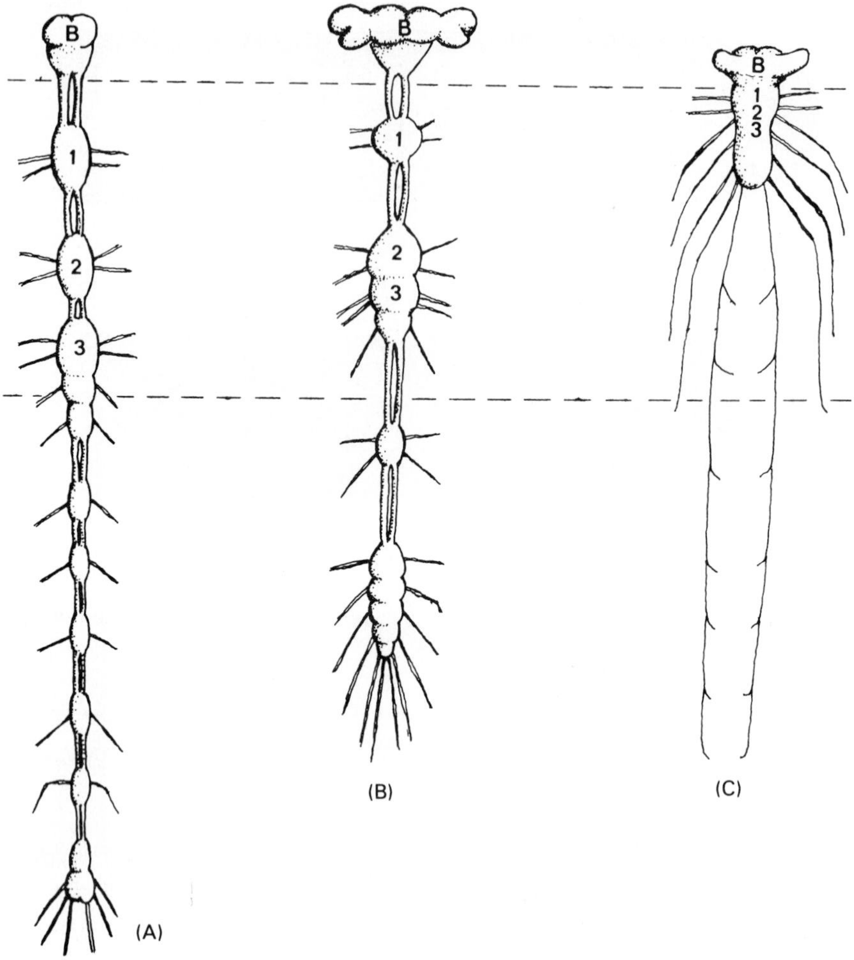

Figure 87. Diagram indicating cephalization (fusion and shortening anteriorly) of the insect's central nervous system. (A) silverfish; (B) whirligig beetle; (C) waterstrider. *B*, brain, *1*, prothoracic ganglia; *2*, mesothoracic ganglia; *3*, metathoracic ganglia.

The subesophageal ganglion is located below the esophagus and coordinates the sensory and motor activities of the appendages of the fourth, fifth, and sixth segments (the mandibles, maxillae, and labium, respectively). Nerves also supply the hypopharynx and salivary glands.

The *stomodeal sympathetic system* is connected to the central nervous system through the tritocerebrum. It commonly consists of a single *frontal ganglion* plus one or two recurrent nerves along the dorsal surface of the stomodeum. In addition to controlling the peristaltic movements of the anterior part of the alimentary canal, this system also sends nerves to the dorsal vessel and two types of endocrine glands, the paired *corpora allata* and the single *corpus cardiacum* (Fig. 88).

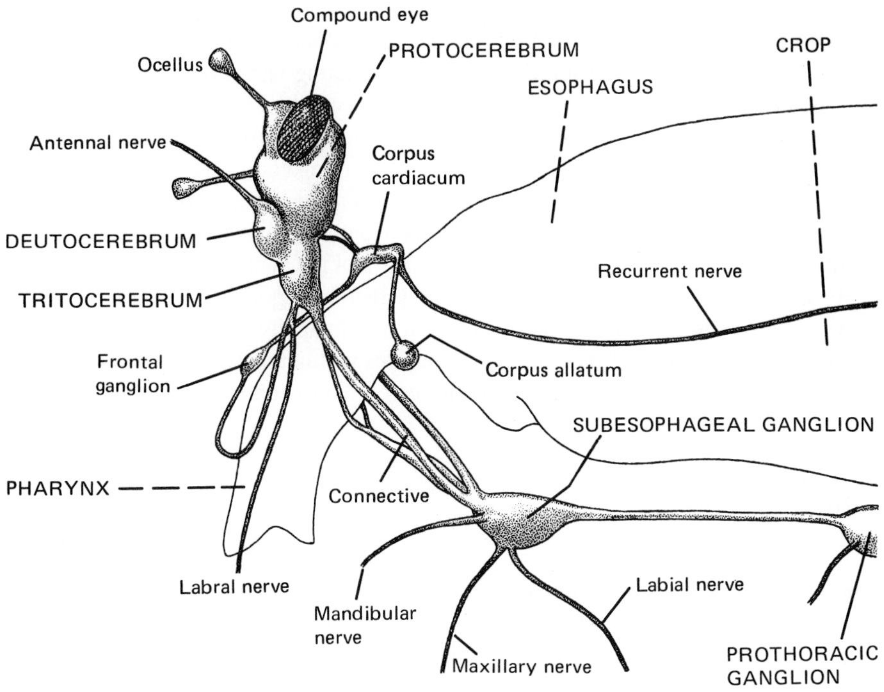

Figure 88. Insect brain and associated structures. (Redrawn with slight modifications from Snodgrass, 1935.)

Other parts of the nervous system include an *unpaired ventral nerve* that controls the spiracles and a *caudal sympathetic system.* The latter system arises from the last compound ganglion of the ventral nerve cord and sends nerves to the reproductive system and posterior part of the alimentary canal.

The nerves radiating from the ganglia, especially in the abdomen, to various structures are simple when compared to vertebrates and contain few neurons. Each neuron can transmit impulses of only one magnitude. Sensory information is coded into the system according to the number of impulses, their frequency, and the number of neurons involved. Low-level stimuli may only produce a simple reflex through the segmental ganglion. Strong stimuli, however, produce a great number of impulses along many sensory neurons, which "overflow" at synapses into additional neurons and result in integrated motor patterns. In addition, two nerves often run to each skeletal muscle, one a "slow fiber" and the other a "fast fiber." Suitable combinations of impulses between the two fiber tracts produce the necessary responses. Seven pairs of long giant neurons connect the leg with special setal wind receptors on the cockroach cerci, *Periplaneta americana,* and wind stimuli received by the roach, such as produced by toads as they lunge toward the insect, activate leg movements directly (Camhi, 1980). The rapidity of the roach response is within 11 milliseconds of cercal setae activation.

Unlike vertebrates, much of the coordination of body functions and behavior has been decentralized. For example, the abdomen of a female *cecropia* moth is capable of maintaining the various life-support activities, attracting and copulating with males, and oviposition, even when separated (but experimentally sealed to prevent bleeding, and healing of wounds has taken place) from the remainder of the body. The heads of some male mantids often may be eaten by females during courtship, yet the male often continues copulation and walking activities. In this latter instance, the subesophageal ganglion is the inhibitory center for this behavior, and its removal permits these movements to continue with coordination by the ganglia of the ventral nerve cord.

In summary, the immediate control of the insect in response to its environment is carried out by the central nervous system. Embryologically, each segment develops a pair of ganglia, and the functioning of structures within that segment is coordinated by these ganglia. Metamorphosis alters these ganglia to control the new adult adaptations and behaviors.

REPRODUCTIVE SYSTEM

The reproductive system is unusual in that it is often not essential for daily self-maintenance; that is, it usually requires involvement with other individuals and functions to produce offspring. The reproductive process in sexual individuals of most animals (some modifications exist in insects and these will be discussed subsequently) is summarized as follows with *n* representing one set of chromosomes:

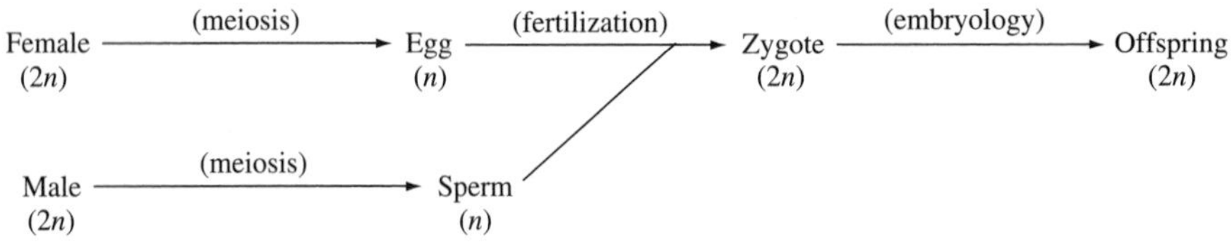

As can be seen, this type of reproduction (sexual) has obvious disadvantages in that it takes two individuals to produce one, but advantages include new genetic variability in the zygote and masking deleterious affects of single genes. In the alternative types of reproduction (asexual) only a single individual is needed to produce offspring; therefore, reproductive potential is theoretically increased but at the sacrifice of lower genotypic variability. In insects, sexual reproduction is the most common type, but asexual types are found in most orders.

Insect sperm usually have either a short end piece [Fig. 89(A)] or are aflagellate. In the least specialized insects, sperm are protected by gelatinous membranes in specialized sacs or *spermatophores.* These spermatophores are either deposited on the soil surface for the female to locate, as in some Collembola, or are transferred into the body of the female during copulation. Females of some insects may utilize nutrients from these spermatophores, either in egg production or for protection of themselves and their eggs. The amount of nutrients obtained may be significant, for instance, about 20 percent of the body weight of male Mormon crickets may be used in

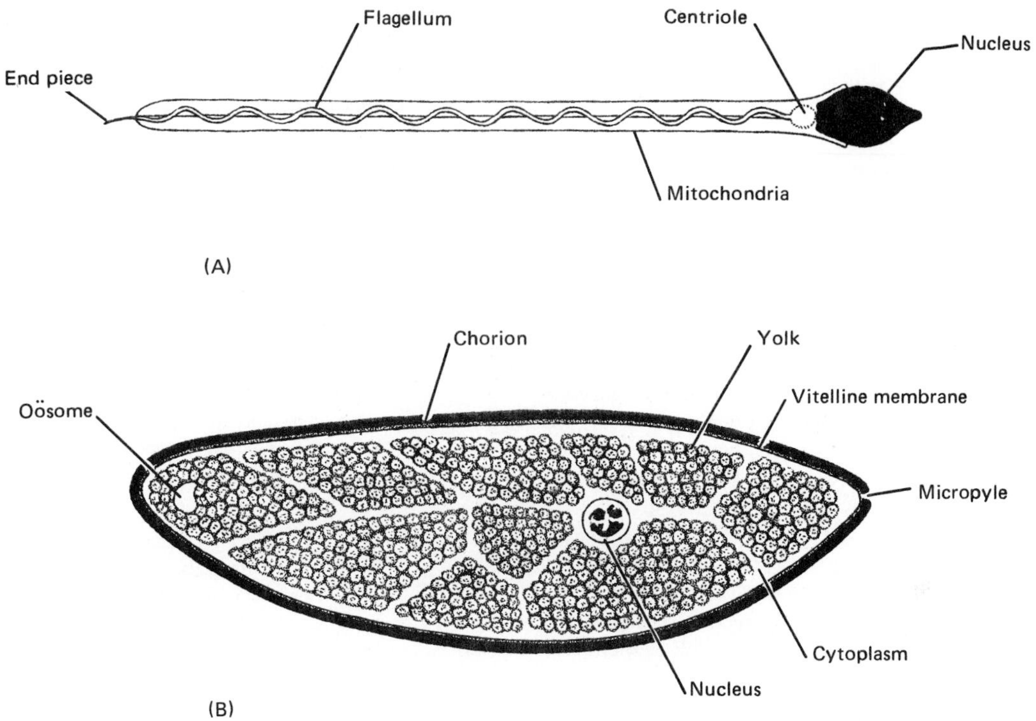

Figure 89. Insect gametes. (A) sperm; (B) egg.

their production. These compounds are either eaten directly (if the spermatophore extends out of the genital tract, as with certain grasshoppers and crickets) or is absorbed into the system directly (most insects). However, spermatophores have been lost in various insects or are represented only by vestiges. A summary of the evolution of sperm transfer is given in Figure 90.

As in sperm, insect eggs are also subject to dessication and are more subject to predators than sperm because they eventually leave and may spend considerable time outside a parent's body. For protection, they are covered by an exoskeleton or *chorion.* Modifications in the chorion include pores such as *micropyles* to permit sperm penetration and *aeropyles* that allow oxygen to penetrate. Some eggs are elongate, while many have peculiar shapes and sculpturing (Fig. 91). Specialized hatching sutures may also be present in eggs whose emerging immatures lack chewing mouthparts for gnawing their way out. Within the eggs is located a nucleus, and most cytoplasm is located peripherally [Fig. 89(B)].

Male

The gonads are normally functional only in the adult; however, precursors are found in early instars. Rapid development to maturity occurs in the last preadult instar or in

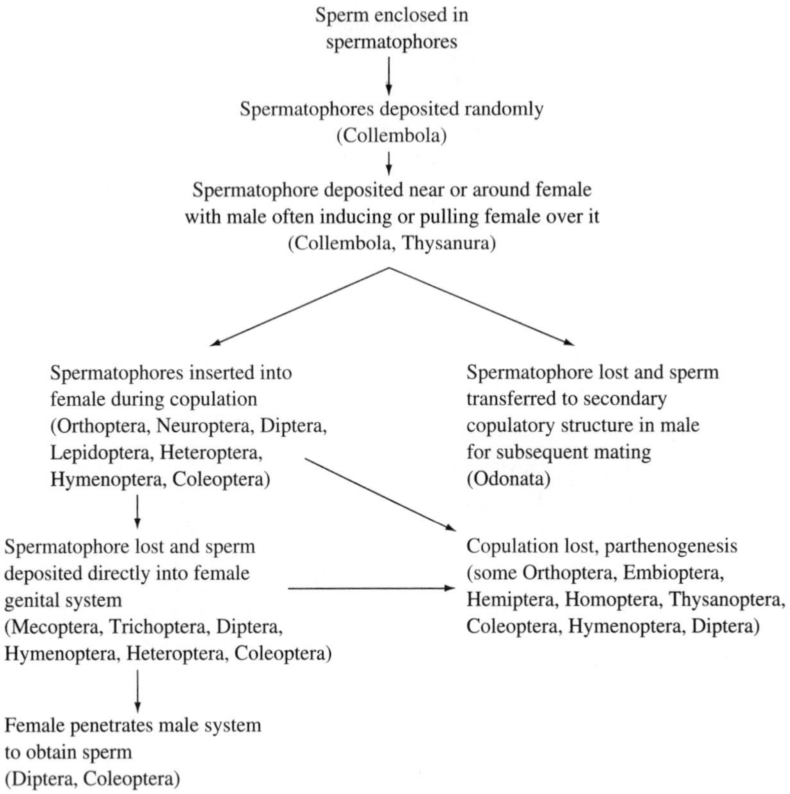

Figure 90. Evolution of sperm transfer in insects.

the early adult stage. Each mature *testis* consists of a definitive number of *sperm tubes* enclosed in a membranous sac. Within these tubes (Fig. 92) unspecialized cells, *spermatogonia,* increase in size within cysts to become *primary spermatocytes,* undergo meiosis to become haploid *secondary spermatocytes* and *spermatids,* and are then transformed into *sperm,* a process called *spermiogenesis.* Cyst walls are later dissolved. Leading from each testis is a *vas deferens* (Fig. 93), which carries these gametes down to the single *ejaculatory duct.* The *accessory glands,* located as outpouchings of the vas deferens and/or ejaculatory duct, add fluids and a covering if spermatophores are used. During insemination, sperm is transferred into the female genital system by the *aedeagus.*

Female

The female system matures late in development and has a dual role: (1) to receive the male aedeagus and sperm and (2) to produce eggs. Copulation occurs when the aedeagus is inserted into either the *genital chamber* (Fig. 93) or into a specialized diverticulum from the genital chamber, the *bursa.* Sperm is transferred either as liquid

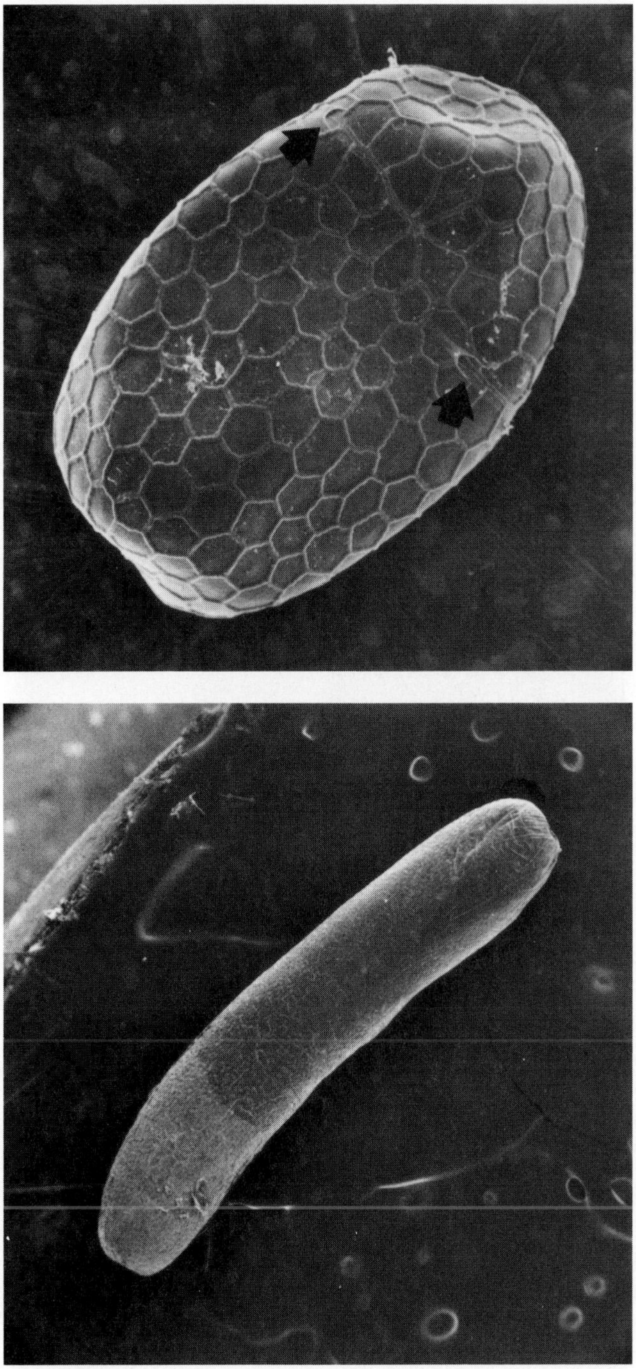

(A)

(B)

Figure 91. Egg variation. (A) oval egg with cuticular sculpturing and several micropyles (*arrows*) (Ephemeroptera); (B) elongate egg with minimal sculpturing (Hydrophilidae).

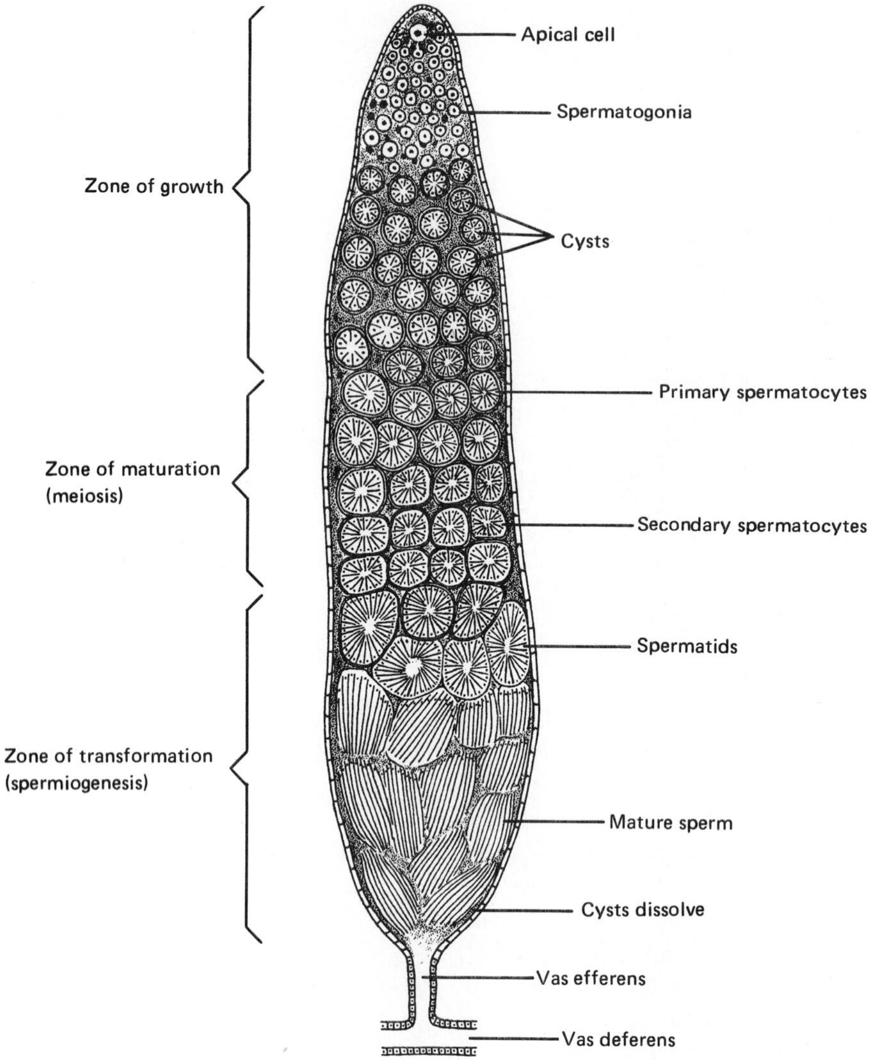

Figure 92. Diagram of sperm tube or follicle indicating sperm development. (Redrawn with slight modifications from Snodgrass, 1935.)

semen or within a spermatophore (Fig. 90); if the spermatophore is received by the female, then the protective coverings are digested or torn by teeth located in the genital tract of the female and the sperm liberated. These gametes migrate or are moved by peristalsis up into the *spermatheca* via the *spermathecal duct,* where they are stored until an egg or eggs are released. In a few insects, sperm is deposited directly into the spermatheca. The length of survival of sperm varies from only several weeks in bedbugs to many years in queen ants and bees. In the latter situation, food of some sort must be supplied and the metabolism of the sperm must be slowed down.

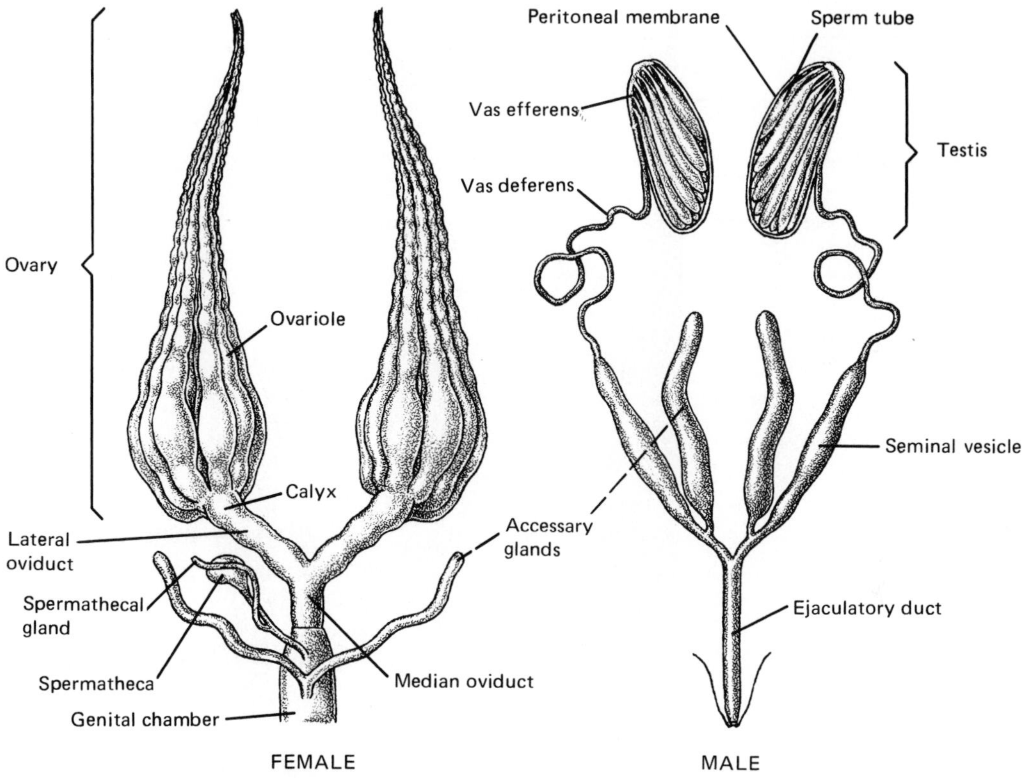

Figure 93. Diagrammatic representations of the female and male reproductive systems. Many variations from these hypothetical models occur. (Redrawn with slight modifications from Snodgrass, 1935.)

The gonads of the female sex are two *ovaries* consisting of a definite number of tubes or *ovarioles* (Fig. 94) linked apically by their collective *terminal filaments. Oögonia* are located in the terminal region and enlarge into *primary oöcytes* as they move down the ovariole and as they receive yolk from the follicular epithelium (*panoistic ovarioles*) or from special nurse or nutritive cells (*meroistic ovarioles*). This enlargement of the descending oöcyte produces a distension of the ovariole into a *follicle.* Finally, a chorion is secreted that gives the characteristic species shape and covering to the egg prior to its release from the ovariole. The egg leaving the ovariole, however, is not yet a gamete as meiosis has not occurred and the diploid state remains.

After leaving the ovariole, eggs are often stored in the *calyx,* particularly in insects that deposit large numbers of eggs at a single laying. The egg, *oöcyte* at this stage, is then moved by peristalsis down the *lateral oviduct* and the *median oviduct* into the *genital chamber* where it passes the *spermathecal duct* pore. Sperm are released at this point and penetrate the micropyle of the egg. The sperm entry initiates oögenesis in which the oöcyte nucleus undergoes meiotic division to produce a haploid gamete nucleus or ovum and polar body nuclei. The polar body nuclei are lost, but the

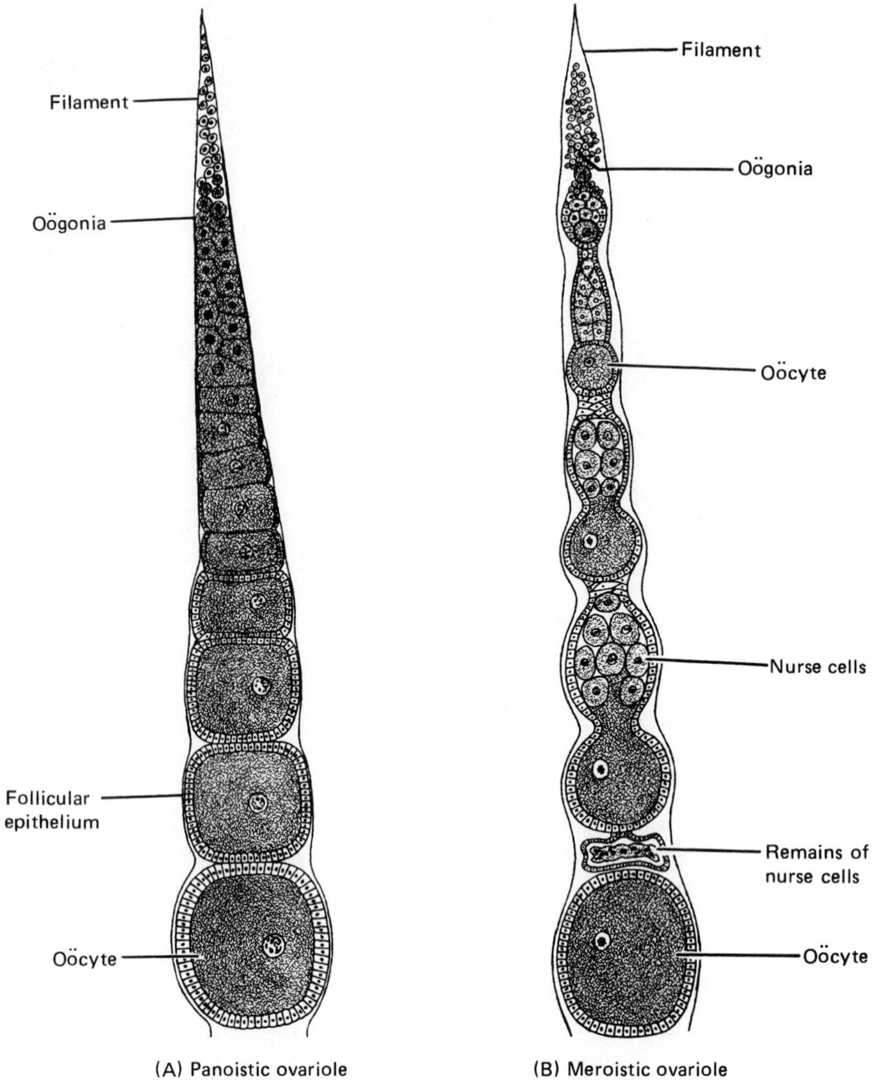

Figure 94. Two types of ovarioles. (A) panoistic from Orthoptera; (B) meroistic from Hymenoptera. (Redrawn with slight modifications from Richards and Davies, 1977.)

ovum and sperm nuclei fuse, and the normal number or diploid state of chromosomes is restored. Once the sperm has penetrated, the egg is moved down the genital tract and is coated with various materials from the *accessory glands* (when present) for sticking the egg to a specific substrate. *Oviposition* then occurs. Eggs may be laid singly, in groups, or glued together into a single structure, the *oötheca* (Fig. 95).

The ability to produce eggs and the number deposited depend on at least four factors. First, proper nutrients must be available. Although many insects utilize food acquired by the immature stages to produce at least the first batch of eggs, others require

Figure 95. An American female cockroach (*Periplaneta americana*) carrying an oötheca (*O*).

additional food or the stimulus of a full gut to manufacture eggs. Second, the number varies as to the number of ovipositions. Some insects deposit the entire batch at once and then die (mayflies), whereas other species may produce many batches of eggs over an extended period of time. Third, the number of eggs varies with the number of ovarioles present. Fourth, the number produced varies with the type of reproduction.

Sexual Reproduction

Oviparity. The majority of insects undergo the type of reproduction described above in which an egg is produced, fertilized, and oviposited by the female. This is *oviparity* [Fig. 96(A)]. Intricate behavior patterns are involved, and the eggs are normally deposited in precise microhabitats, near or on the required food. Sufficient yolk is present to permit embryology to be completed within the egg. Mortality from water loss is slowed down by a waxy layer secreted beneath the chorion by the egg itself.

Ovoviviparity. In ovoviviparity [Fig. 96(B)], eggs are normally developed and fertilized, but they are retained and hatched within the body of the female. Sufficient yolk is present in the eggs for the embryo to complete its development without the female providing further nourishment. The number of eggs produced may be restricted, but the protection offered plus the depositing of hatched immatures that are ready to commence feeding are obviously of selective advantage. Flesh flies are good examples of this variation, and the larvae are deposited on fresh carcasses or hosts rather than eggs as in most flies; this gives them a head start in the competition for food. The sheep bot fly is another good example of this type of reproduction; it deposits larvae into the nostrils of sheep.

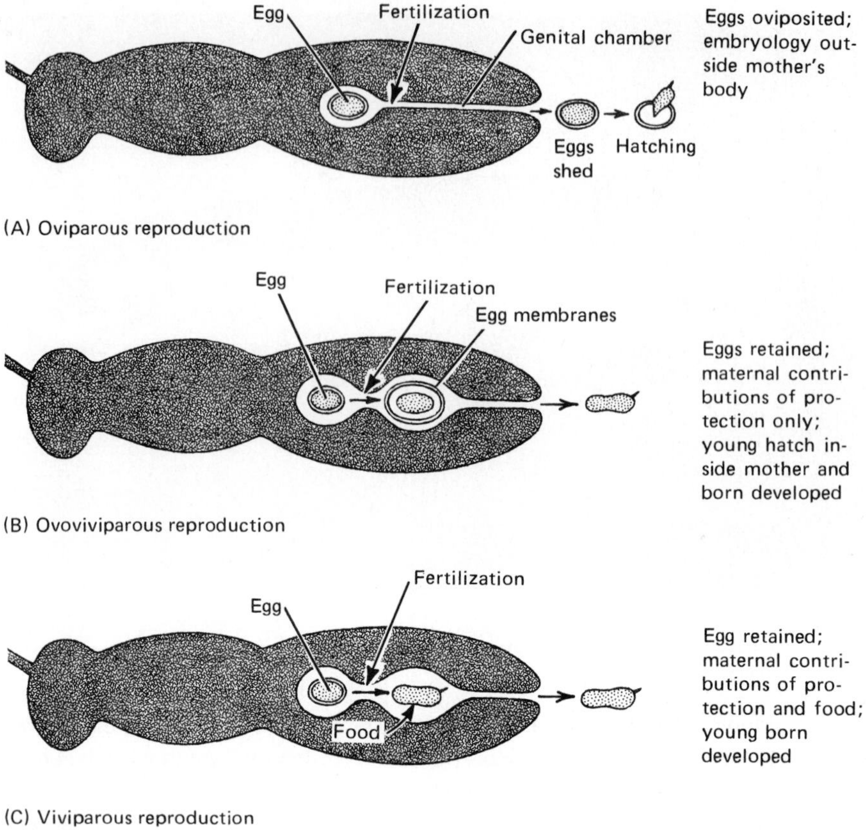

Figure 96. Comparisons of oviparity, ovoviviparity, and viviparity.

Viviparity. In viviparity [Fig. 96(C)], development takes place within the female body. In some variations the eggs do not contain sufficient yolk to permit the embryos to develop fully, and the young (underdeveloped embryos) hatch through their thin chorion-covered eggs to be nourished by the mother. Most examples of this type of reproduction, however, lack a chorion, and the embryos are retained and fed directly by female secretions until embryonic development is complete, at which time they are birthed. Aphids, during the summer months (Fig. 97), and certain flies, such as the tsetse of Africa, are good examples of this type.

Asexual Reproduction

Parthenogenesis. Some adult female insects are capable of reproducing without male involvement and subsequent fertilization (Fig. 98). The resulting individuals may be either diploid ($2n$) or haploid (n), depending upon the cytological mechanisms involved. Various terminologies are used to define the phenomena. *Apomictic* parthenogenesis [Fig. 98(A)] produces females that are diploid, because chromosome reduction

Figure 97. Aphid giving birth to nymph by viviparous reproduction in its typical breech delivery method.

does not occur and the offspring are essentially identical to the mother in genotype. It is a common type, highly stable, and found in aphids, scales, and some beetles, flies, wasps, and psocids. A second type is *automictic* [Fig. 98(B)], in which meiosis occurs but a polar body nucleus unites with the egg nucleus to restore the diploid state of the egg nucleus. Some variability between mother and offspring can result, depending upon the amount of crossover that occurred during meiosis and whether or not the egg nucleus fuses with the first or second polar body nucleus. This latter type is found in some scales, walkingsticks, moths, and flies. A third type is *generative* [Fig. 98(C)] and also involves meiosis, but the resulting individual is haploid so all genes are expressed. These haploid offspring are usually males, whereas the females of the species are produced sexually through oviparity and are diploid. This phenomenon of having one sex diploid and the other haploid is referred to as *haplodiploidy* and is the sex determination mechanism in Hymenoptera and most Thysanoptera.

The literature sometimes defines parthenogenesis by the resulting sex rather than by the previous classification by chromosomal activity. If females are produced, the process is called *thelytokous,* or *arrhenotokous* if males develop. Comparisons between the two systems are easy, because apomictic parthenogenesis produces females and haploid males result from the arrhenotokous process.

Paedogenesis. In a few instances, *immatures* are capable of producing offspring asexually, often as part of a complex life cycle involving sexual reproduction as well. A good example of this form of reproduction may be seen in the dipteran family Cecidomyiidae

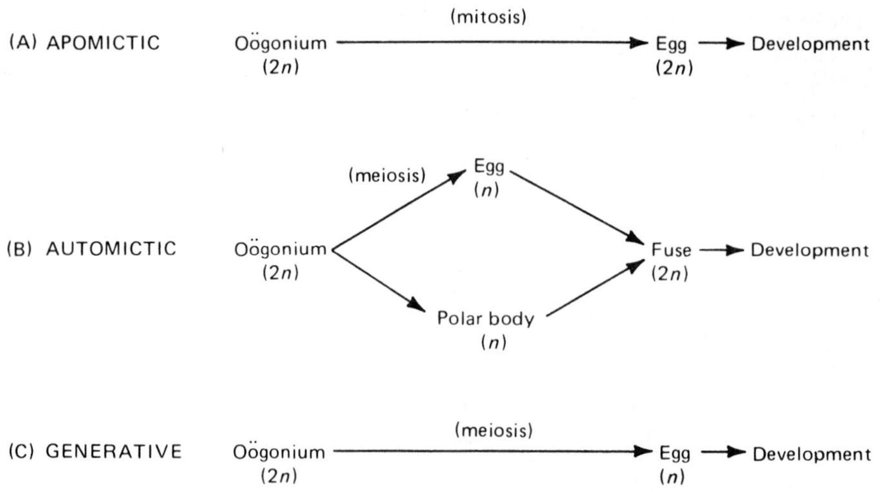

Figure 98. Types of parthenogenesis. Type A has the greatest genetic stability. Type C determines one sex and is haploid, whereas the opposite sex results from fertilization and is diploid.

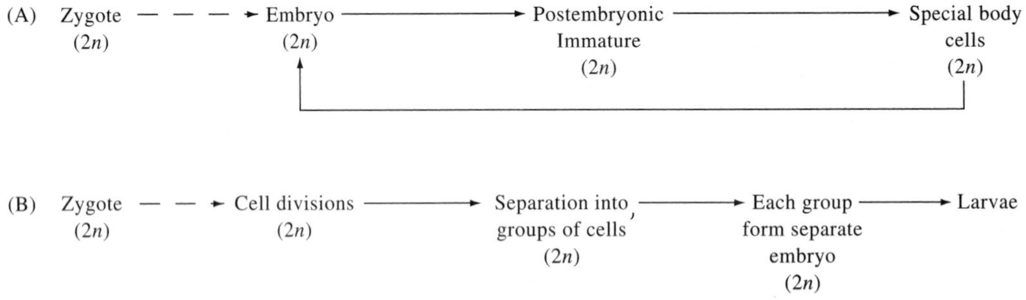

Figure 99. Two types of asexual reproduction. (A) paedogenesis (certain ceccidomyid flies); (B) polyembryology. Solid lines indicate asexual phase in cycle.

[Fig. 99(A)]. Unlike the normal insect route of egg, larva, pupa, and adult, larvae may appear within the body of first instar larvae and devour the "mother's" tissues. In *Mycophila speyeri* where larvae feed on mushrooms, each larva may produce up to 38 offspring, and populations are known to reach 20,000 reproductive larvae per square foot within 5 weeks. As food becomes low and crowding takes place, the life cycle reverts to oviparity until a new food source is located and the asexual phase of the cycle again becomes the norm.

Polyembryony. In polyembryony, reproduction is associated with oviparity or parthenogenesis. There are two variations of the polyembryony type of reproduction. In the first variation, the dividing cells separate during the initial mitotic divisions, and each cell subsequently gives rise to a separate individual. In the second variation, cleavage occurs many times, after which the embryonic "body" subdivides into separate embryos. The number of embryos arising from either type varies from two to several thousand. Polyembryony is normally restricted to a few species of parasitic Hymenoptera and Strepsiptera and is beneficial because it permits a small parasite to exploit a large host for food with minimal danger to the female as she oviposits. In a restricted sense it represents a type of asexual reproduction, although initially the first cell or zygote results from sexual reproduction [Fig. 99(B)].

QUESTIONS

1. How is the surface/volume ratio important in the development of body systems of an organism? Does an exoskeleton have an impact on this importance?
2. What is *digestion?* How does a knowledge of the embryonic origin of the insect alimentary canal assist one in understanding what occurs in each portion? What is an *intima* and *peritrophic membrane,* and what are their functions?
3. Compare a digestive tract specialized for solid food with one modified for a liquid diet. Does the amount of water in the food have an effect upon digestion and the modifications of the system?
4. Correlate liquid or solid food types of alimentary canals with the types of mouthparts discussed in Chapter 2. Does metamorphosis have any effect on the digestive system? How?
5. How might symbionts be important in insect nutrition?
6. Characterize the insect circulatory system and the directional flow of hemolymph throughout the body. Does the size of the insect have any effect upon the degree of specialization? What are the functions of this system?
7. How does the circulatory system of an insect differ from that of a human?
8. What is ATP? Where does the energy necessary to synthesize ATP come from? What is its function? How is body temperature involved?
9. How does oxygen get to the cells of the body? How does the tracheal system originate?
10. What is the importance of tracheoles? Discuss the role of tracheal trunks and air sacs.
11. What modifications of the tracheal system are often seen in aquatic insects? What advantages are gained from these?
12. What is *muscle?* How does the size of an organism affect the muscular system?
13. What are the advantages and disadvantages of synchronous muscles? Of asynchronous muscles?
14. How does flight occur in insects? How important is the notum or tergum to this phenomenon? What is the "*click mechanism*"?
15. What is *excretion?* What are the excretory substances in insects? What advantage is gained by depositing excretory materials into the feces for evacuation? Is there a difference in excretion between insects that eat solid foods and those that ingest liquids?
16. What is a *neuron* and a *synapse?* How do they interrelate?
17. What is a *ganglion?* How many ganglia are there per segment in insects? What role do they play in the nervous system and functioning of the body?
18. How does terrestrial existence influence reproduction?
19. How does the female reproductive system differ from the male system? What is a gamete? Where does meiosis occur in producing sperm and eggs?
20. Compare the types of reproduction. What are the advantages and disadvantages of each?
21. What advantages come from depositing eggs singly or in groups?
22. What is the *fat body,* and what are its functions?

4 Development and Specialization

EMBRYOLOGY

In most insects, penetration of the egg by a sperm initiates meiotic divisions, resulting in a haploid nucleus that fuses with the sperm nucleus. The zygote nucleus next undergoes a series of mitotic divisions. In *Drosophila* flies these occur every 10 minutes, which is considerably faster than every 12 hours in the most rapid divisions in vertebrates. Some of these nuclei remain behind to become *vitellophages,* cells for metabolizing yolk for embryonic use, but most of the cleavage nuclei migrate out through the radiating cytoplasm to the periphery of the egg (Fig. 100). Here, each produces a cell membrane, and the peripheral cells form the *blastoderm.* Some of this blastoderm will become part of the embryonic coverings (*serosa* and *amnion*), and a thickened area or *ventral plate* will give rise to the embryo (Fig. 100). During the next stages, the developing embryo may migrate through the yolk, a process called *blastokinesis.* Such movement is characteristic of eggs that have much yolk, as is common in insects undergoing incomplete metamorphosis, but eggs that have little yolk have reduced movement, or the process is absent.

Johannsen and Butt (1941) and Hagan (1951) give detailed information on the development of embryos. In brief, the ventral plate forms a double layer of cells, the presumptive *ectoderm* and *mesoderm.* Major structures derived from these layers are diagrammed in Figure 101. The presence of a third layer, the endoderm of most animals, is inconclusive but is assumed to be present. The ventral side, including the nervous system, develops first, then the digestive system, and lastly a phenomenon called *dorsal closure* produces the sides and dorsum of the embryo, including the dorsal vessel. A cavity between the alimentary canal and exoskeleton, the *hemocoel,* develops, but it is not the coelom as in Annelida. Segmentation and tagmosis in the exoskeleton, muscles, and nervous system occur early, although complete partitions between the segments or metameres never develop (Fig. 102).

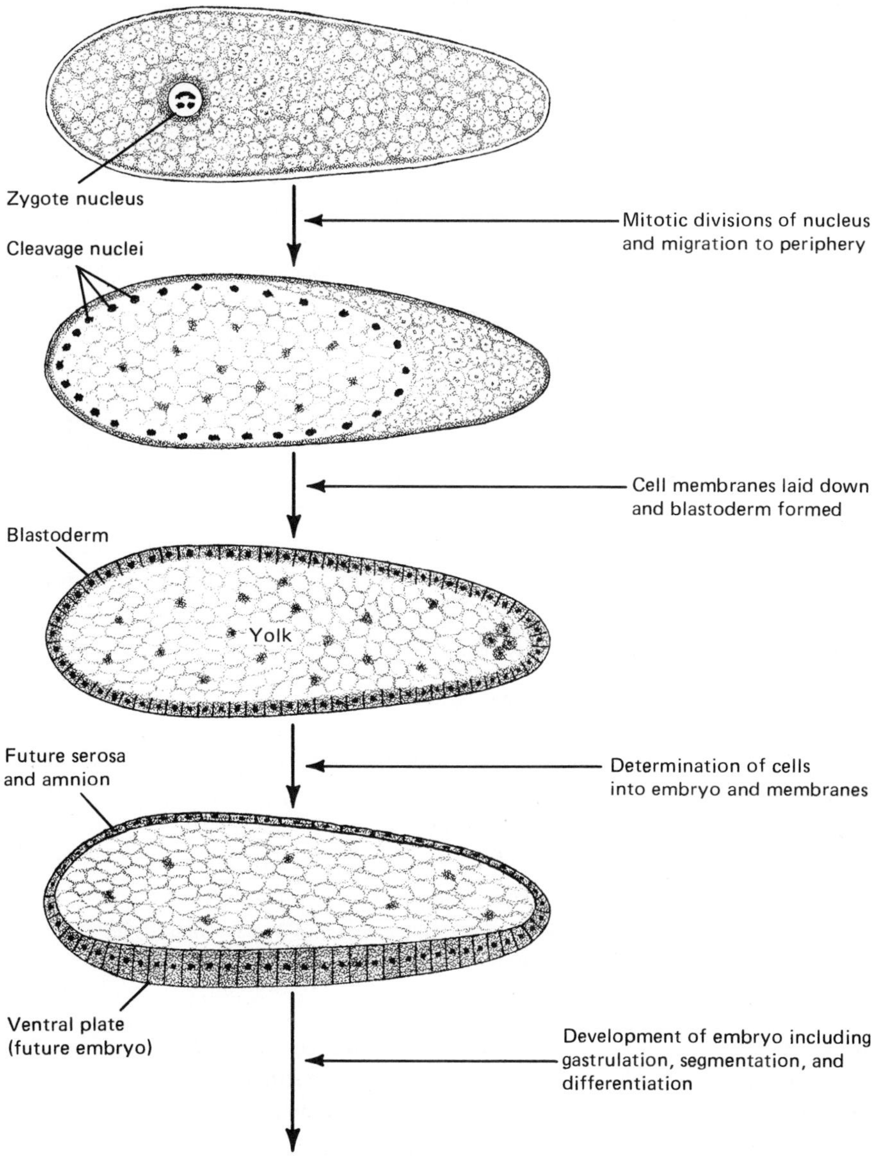

Figure 100. Early stages within the insect egg.

The rate at which development occurs varies mainly with temperature, because insects are ectotherms, and diapause (or suspended development) induced by the environment or obligatory by genetic control. Diapause is especially pronounced where cold winters or long periods of drought occur with regularity.

Emergence from the egg is accomplished in various ways. The immature insect may often chew its way out (Fig. 103) or digest a portion of the chorion. In some

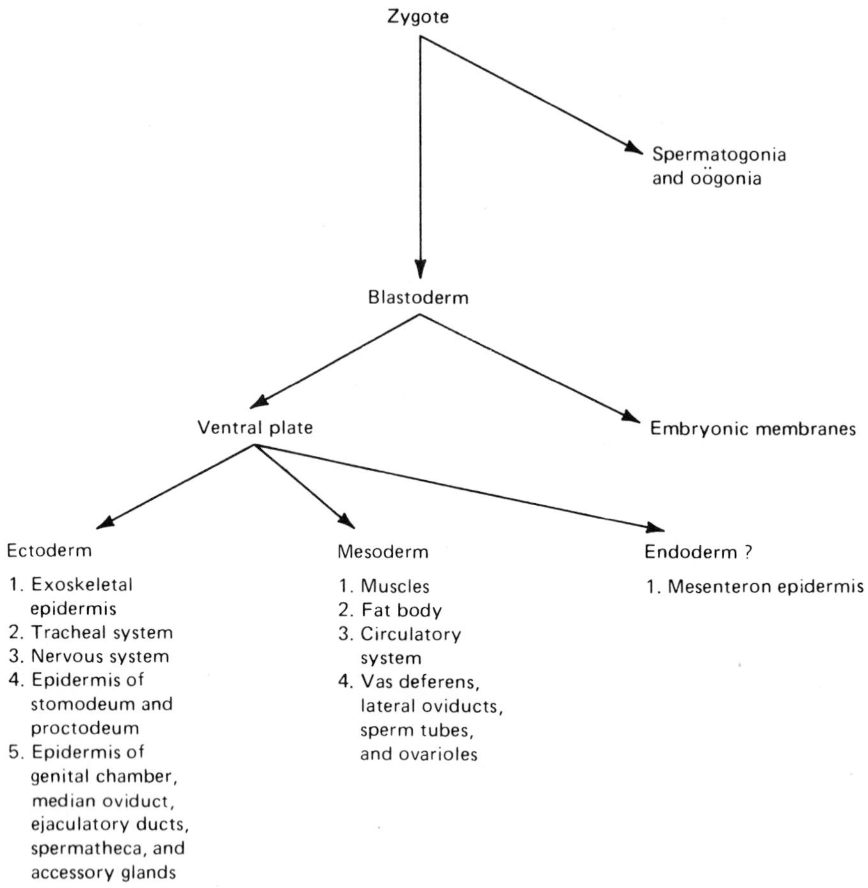

Figure 101. Diagram indicating the development and differentiation from the zygote to the organism with special reference to organ derivation.

species the insect pushes against a preformed weakened area in the egg (Figs. 104, 234) or uses specialized hatching spines or bursters. The stimulus (or stimuli) that initiates emergence is poorly understood except in a few species.

POSTEMBRYONIC DEVELOPMENT

Growth and Ecdysis

Growth in most animals represents the synthesis of new protoplasm and is normally accompanied by cell division and differentiation. Size increases are not good measures of growth in arthropods because of the presence of the integument; this size-restrictive structure must be removed and replaced by a larger one if further growth

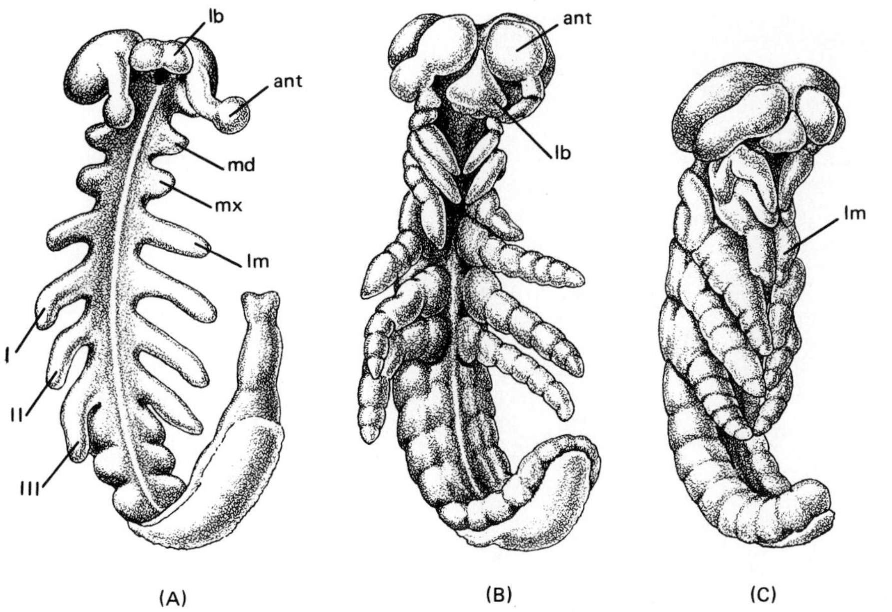

Figure 102. Embryology of the milkweed bug, *Oncopeltus fasciatus.* Envelopes and yolk have been dissected away. (A) 71 hours; (B) 91 hours; (C) 101 hours. *Ant*, antenna; *md*, mandible; *mx*, maxilla; *lb*, labrum; *lm*, labium; *roman numerals*, thoracic legs. (Redrawn after Butt, 1949).

Figure 103. Cecropia moth (*Hyalophora cecropia*) larva chewing its way out of egg.

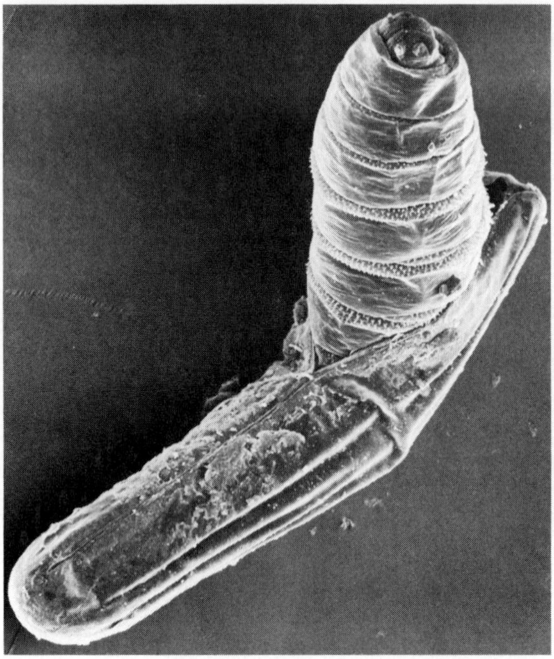

Figure 104. Larva of calliphorid fly, *Cochliomyia macellaria,* hatching. Note the hatching suture along the entire length of the egg, which splits under pressure from the larva.

is to occur. How is this possible if the new integument must be deposited under the old one to be shed or molted (*ecdysis*)? Molting is accomplished as follows. First, the epidermis loosens from the cuticle (*apolysis*), and a new integument is secreted under the old one (Fig. 105) with much of the building blocks or metabolites coming from digesting parts of the original skeleton (up to 85 percent may be digested and absorbed). This new integument is unsclerotized and somewhat folded. The individual insect enclosed within its old exoskeleton is the *pharate stage.* The insect then swallows air and rhythmically contracts its body muscles, thereby increasing its body pressure until the old exoskeleton breaks along certain preformed weak points, or *ecdysial sutures,* in the head and thorax. By continuing peristaltic body movements, the old skeleton, or *exuviae,* is moved off posteriorly (Figs. 106, 112), with any remaining molting fluid acting as a lubricant. Because high body pressure is maintained, the folded new integument is then straightened by stretching (Fig. 107). Once the new integument is expanded, tanning or sclerotization occurs, and the remaining layers of the epicuticle are deposited. After tanning, the muscles relax, and air sacs and hemolymph temporarily occupy this expanded body volume until replaced by normal growth. Such is the basic plan of the more primitive insects, those without complete metamorphosis. However, some poorly sclerotized larval insects, such as lepidopteran caterpillars, are highly specialized for food uptake and rapid growth, and they can stretch and add to the large areas of poorly sclerotized exoskeleton between molts. For example, the tobacco hornworm larva increases its weight tenfold (a 90 percent increase) and its surface area fourfold while in the fifth and last larval instar. This phenomenon results in fewer required molts to adulthood than occurs in those insects with extensively sclerotized exoskeleton; therefore, less energy is lost

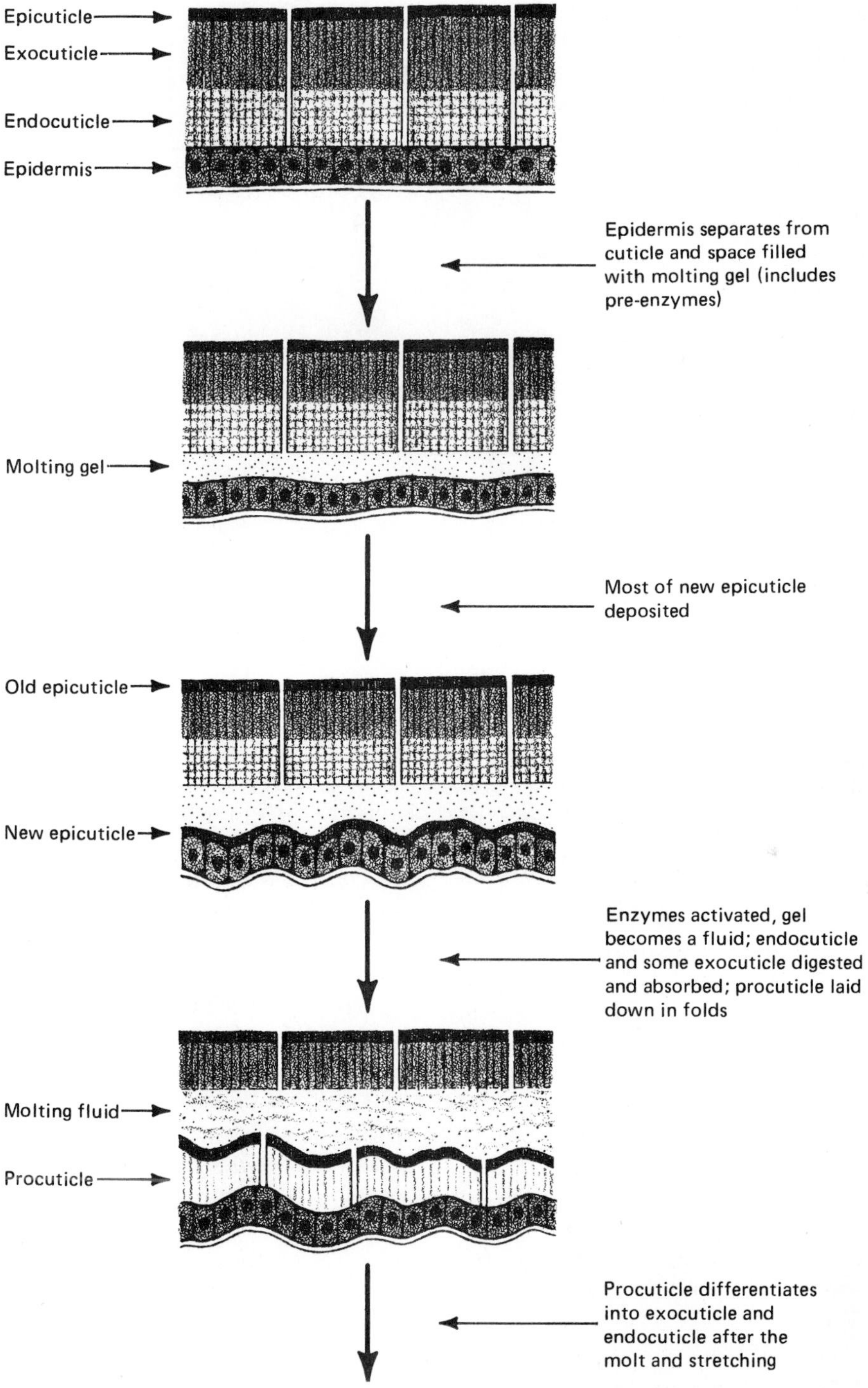

Figure 105. Sequence of events in the molting of the insect exoskeleton.

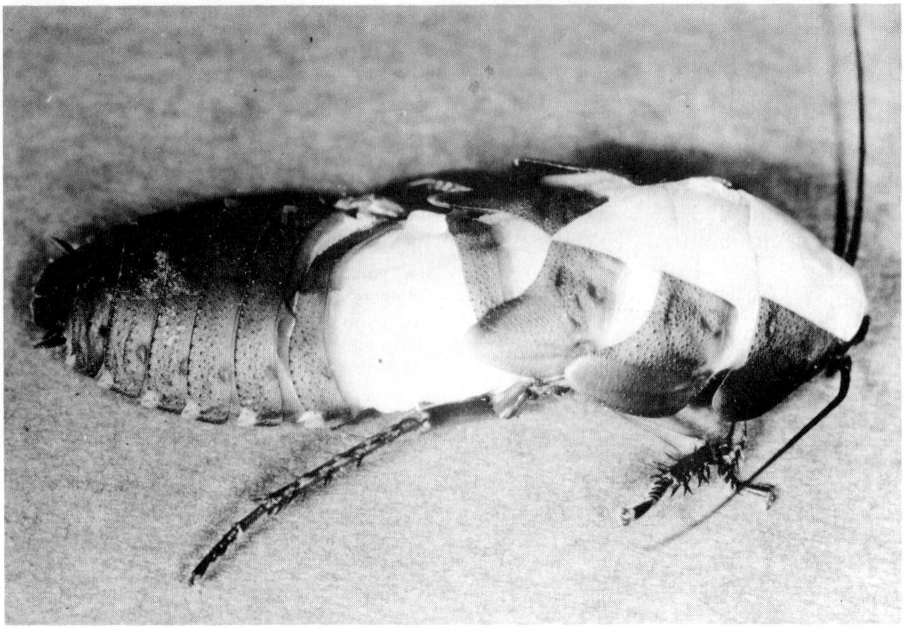

Figure 106. Cockroach (*Leucophaea maderae*) in the process of molting. Note the ecdysial split in the thorax and head that permits the old exoskeleton to be shed.

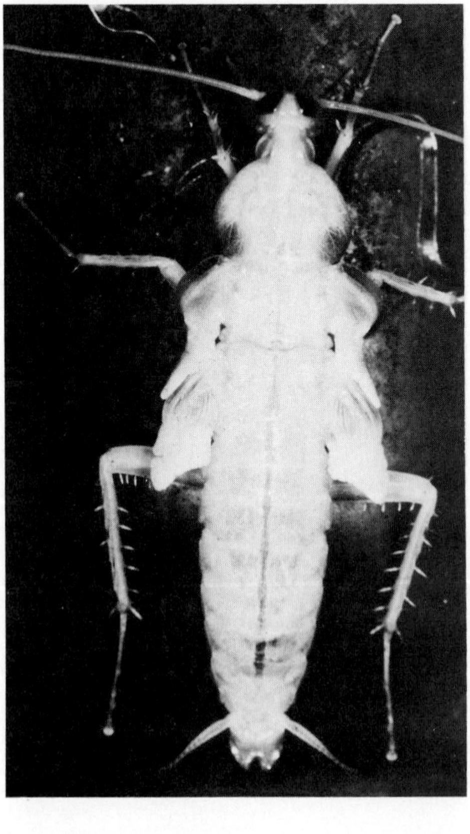

Figure 107. A freshly molted adult American cockroach (*Periplaneta americana*) in the process of expanding its wings. Once expanded, the wings become sclerotized, and the flight muscles mature prior to adult activity. Note the heart and tracheae showing through the unsclerotized integument.

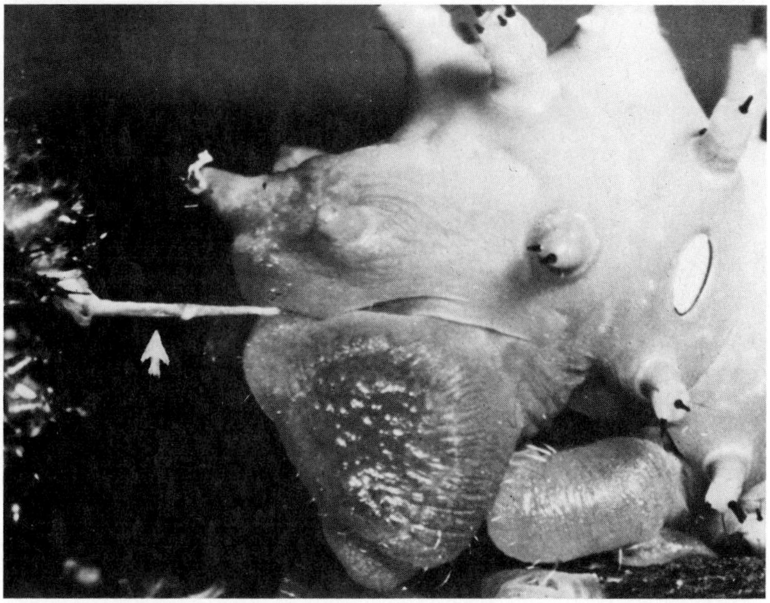

Figure 108. Proctodeal linings being pulled from within the body of a *Hyalophora cecropia* caterpillar during molt.

in molting and new exoskeletal formation, and less hazards are imposed on the individual because of molts.

All parts of the body lined with a cuticle take part in the molt, including the integumental linings of the tracheae (but only the base of the tracheoles), and the stomodeal and proctodeal intima (Fig. 108). It should be obvious from the complexity of this phenomenon that ecdysis is an extremely critical time in the organism's life. The protection once afforded by the exoskeleton is removed, and the insect is highly vulnerable to predation, chemical and physical forces, and water loss (Figs. 5, 142). Water loss is often minimized by limiting the process to periods of high humidity; nevertheless, water loss is critical because the rate is up to six times normal, and the molt must be completed rapidly. Occasionally an insect can be found that has parts of the old exoskeleton still adhering to its body (Fig. 142). If vital areas such as the tracheal system or head capsule are not shed properly, the organism invariably dies.

Although most protoplasm is added between molts, ecdysis, with the rapid visible size increase of the exoskeleton, superficially gives growth a discontinuous appearance and divides the individual developmental cycle into a series of steps or *instars* (Whitten, 1976) (Fig. 109). Although many physiologists today restrict the use of instar to the time period between two *apolyses,* an instar will be defined in this text (for convenience) as the individual between molts. Therefore, the first instar represents the insect from hatching, or *eclosion,* to the first molt, the second instar between the first and second molts, etc. Although variation does exist, the number of molts and instars for a species is relatively constant, with the average insect completing approximately four to six molts. Deviations from the species norm can result from nutritional deficiencies,

Figure 109. Instars of cockroach (*Leucophaea maderae*). The top row are fully sclerotized individuals and the bottom row are newly molted.

temperature, crowding, and the sex of the individual. Duration of development varies from about 1 week in warm weather (mosquitoes, flies, etc.) to as long as 17 years in some cicadas and 32 years in a long-horned beetle. The latter two examples may even be extended an additional year by crowding and intraspecific competition.

Metamorphosis

In the most primitive wingless insects, such as Collembola and Thysanura, body proportions and internal organs remain similar after each molt or ecdysis. Species undergoing such a development are categorized as *Ametabola,* indicating no change in metabolism. Less than 1 percent of all insect species exhibit this growth pattern. The immatures live in the same habitat and feed on the same food as adults, and both immatures and the wingless adults are capable of molting.

Most insects, however, undergo changes in shape or form during their development, a process of *metamorphosis.* Some species have only minor modifications, but a *gradation* may be seen up to and including very drastic alterations. For instance, flies develop from legless, wormlike, immature individuals to adults with legs and wings. Perhaps no concept in entomology has resulted in more polarization of thought than that of categorizing metamorphosis into types and explaining how such differences evolved. One classification includes insects that have immatures with external wing pads, the *exopterygota,* and species that have internal wing pads as larvae, the *endopterygota.* Another classification agrees with the previous demarcations but substitutes *hemimetabola* for most exopterygota and *holometabola* for endopterygota. Still another view separates the exopterygota into *paurometabola* (those whose immatures normally develop terrestrially) and *hemimetabola* (whose immatures inhabit water) and retains the endopterygota as a single unit, the *holometabola.* Fully realizing that no system is completely adequate because of the gradation from one extreme to another, this text will use the terms *hemimetabola* and *holometabola,* because of their more common usage, to illustrate development and metamorphosis.

When a change from the immature to the adult occurs but is not extreme, metamorphosis is *incomplete,* and development is hemimetabolous. Except in the early instars, the immatures or *nymphs* (called *larvae* by some) are normally characterized by possessing external wing pads (Figs. 52, 110, 111), except when wings are secondarily absent as in lice. Most structures of nymphs and adults (Fig. 110) with incomplete metamorphosis are alike, although body proportions differ and changes in the thoracic plates and reproductive system occur in the molt to an adult. Food and ecology are similar in all stages except where nymphs are aquatic and adults terrestrial. Here the aquatic immatures, sometimes called *naiads,* have specialized structures that are vital for survival in water. Odonata nymphs have an enlarged labium [Fig. 305(C)] for capturing prey and gills of two types, the flat tail-like tracheal gills of damselflies (Fig. 316) and the internal rectal gills of dragonflies. Mayfly nymphs have a pair of gills (Fig. 311) on most abdominal segments, and many species have either or both enlarged tusklike mandibles and a front pair of legs for digging. These immature characters, however, are lost in the metamorphosis to adulthood.

The majority of insect species, approximately 85 percent, have a *pupal stage* (Fig. 110) and undergo drastic alterations during development. Metamorphosis is *complete* (holometabolous development), and often the majority of the larval structures are broken down by histolysis. Adult structures within the immatures are built from either small groups of adult tissues, the *imaginal discs,* or from larval cells which transform to adult cells. Wing pads are internal in larvae (Fig. 112) but evert to become external in the pupa (Fig. 113) and enlarge to functional wings in the adults. This type of reconstructive development is found in most advanced species. In such insects, the larva tends to become the dominant feeding stage, the pupa becomes essentially a transformation or second embryonic stage, and the adult is specialized for dispersal and reproduction. Some of the lepidopterous larvae, called caterpillars, have a greater gut capacity and an increased efficiency for converting food to body mass, and less energy goes into cuticular growth than in most other insects. This also reduces the number of molts because the cuticle has great capacity to stretch.

Although most adults do feed, some survive only upon nutrients stored in a fat body built up during the larval stage. In many species, this apparent division of biological labor enables extreme exploitation of an environment, especially where resources are seasonably low or widely distributed in time or space, and reduces intraspecific competition between young and adults.

The pupal stage may be one of three types. *Exarate* pupae (Fig. 139) do not have their appendages appressed to the body, a situation found in Neuroptera, Trichoptera, most Coleoptera, some Lepidoptera, and most Hymenoptera. A second type, the *obtect* (Fig. 114), has its appendages fused to its body, a state common in Lepidoptera, some Coleoptera, and the lower Diptera. The third type, *coarctate,* is enclosed within the last larval exoskeleton, the *puparium* (Fig. 115), and must molt both the larval and pupal integuments to emerge as an adult. Pupae of all three types exhibit only limited

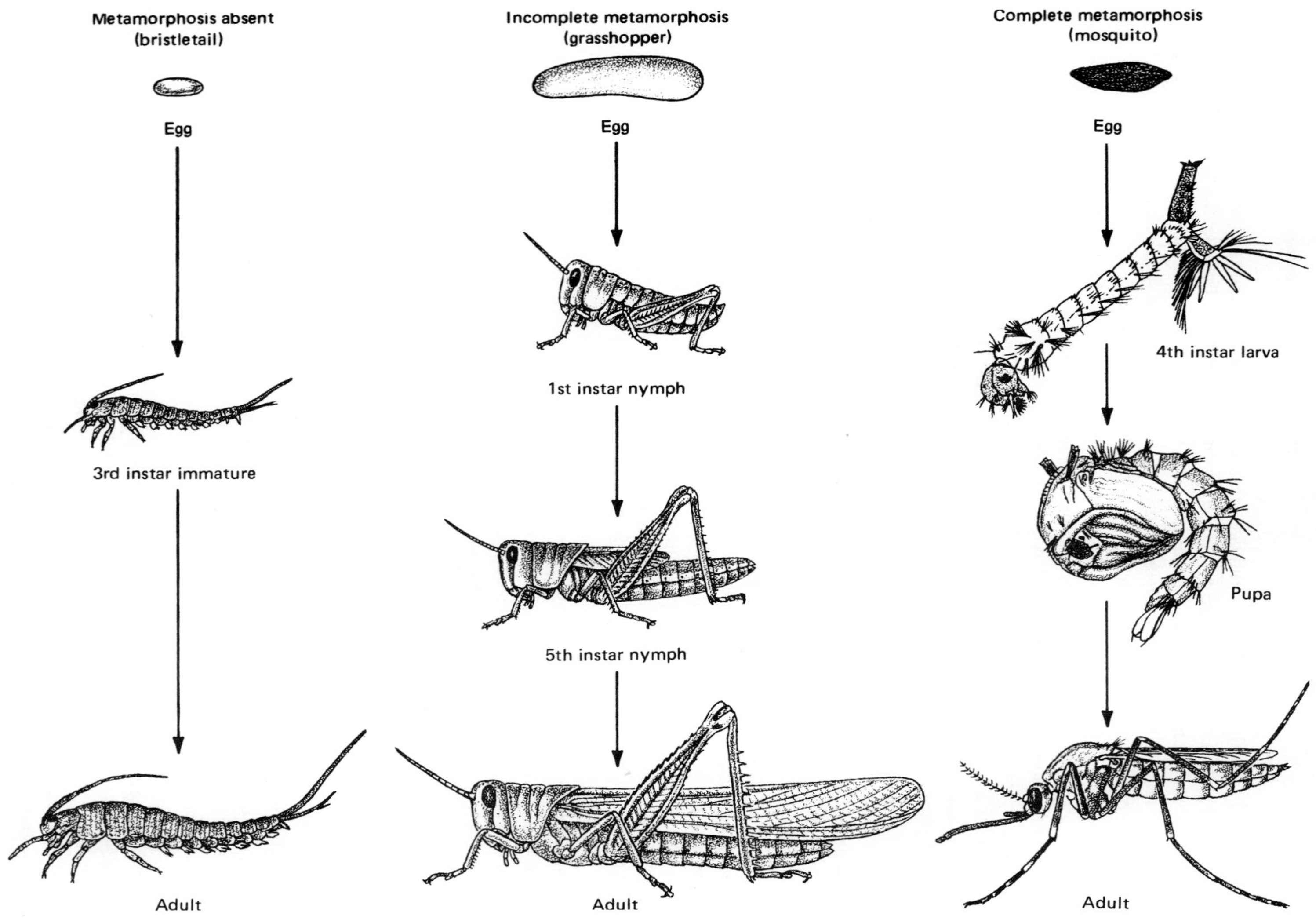

Figure 110. Development from egg to adult, illustrating the degree of metamorphosis that occurs. Most of the instars are omitted to permit enlargement of a few instars for comparative purposes.

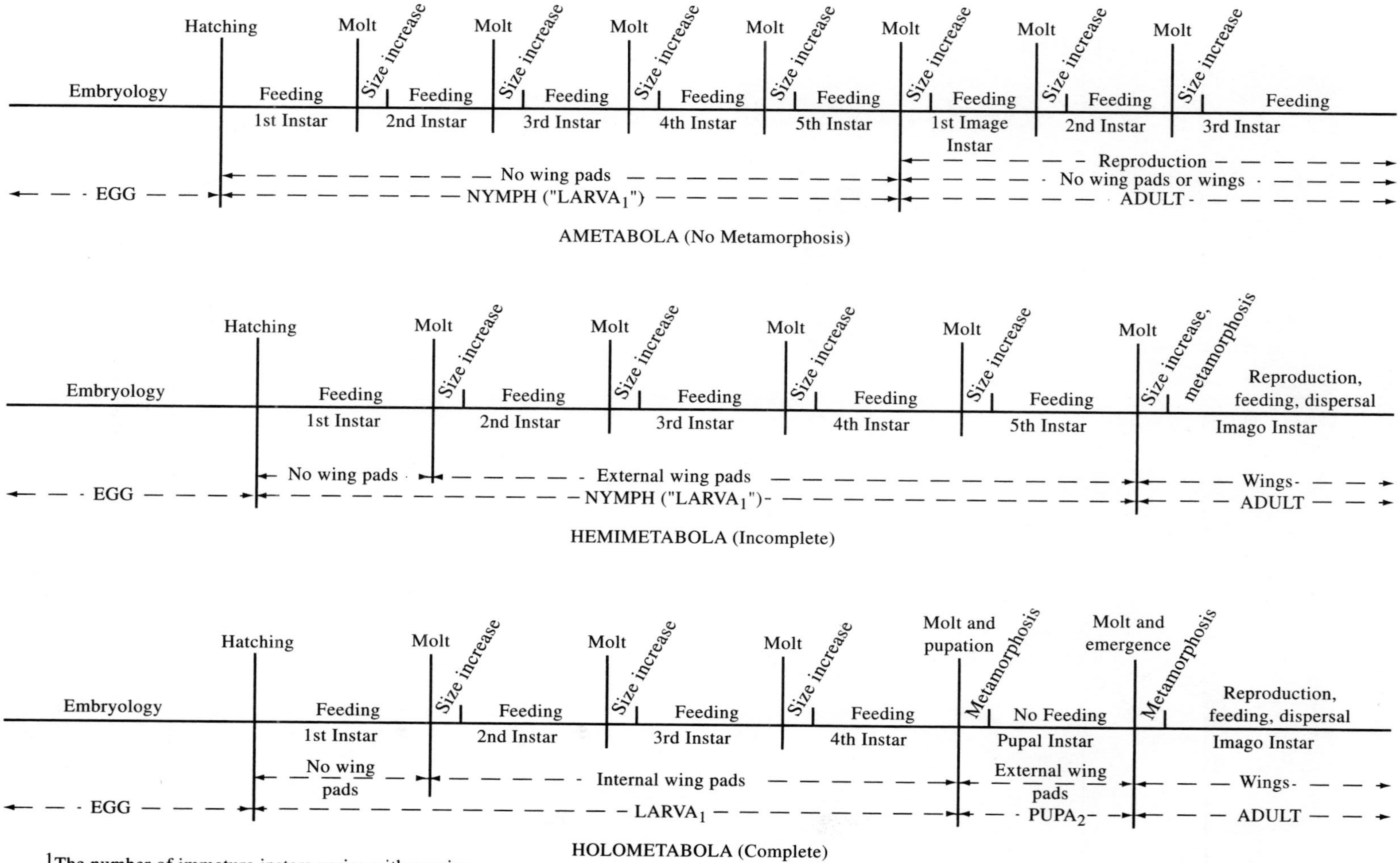

[1]The number of immature instars varies with species.

[2]The pupa is considered the 1st imago or adult instar by some.

Figure 111. Major types of development in insects. (After Fernald and Shepherd, 1942.)

Figure 112. *Hyalophora cecropia* larva molting. The old exoskeleton is thin and has been capable of limited stretching prior to the molt. The white bands along the side represent tracheae being shed, a process necessary because the tracheae are lined with cuticle. Note the size difference between the old head capsule and the new one.

movement, and most pupae are sessile and found in protected environments such as in the ground, under bark, in cases or retreats fashioned by the larvae, or in silken cocoons produced by the last instar larvae.

Both growth and metamorphosis are affected by temperature because insects are ectothermic animals. At 30°C, a house fly becomes a pupa $5\frac{1}{2}$ days after hatching from an egg, but at 10°C, it requires 34 days. Similar examples may be seen in nearly all species.

Sex Determination

With only a few exceptions, insects possess well-differentiated *X* and *Y* sex chromosomes; however, the *Y* chromosome has little to do with sex determination like it does in many mammals. Sex in insects appears to be determined by the balance between the number of *X* chromosomes and the number of autosomes. Having an *XX* combination usually swings the balance to femaleness whereas heterogamety, such as *XY* or *XO* (the Y is missing), induces maleness. Male heterogamety is found in such primitive orders as Collembola, Thysanura, Ephemeroptera, and Odonata, and continues into the advanced insects. A few novel exceptions to this pattern have evolved. In

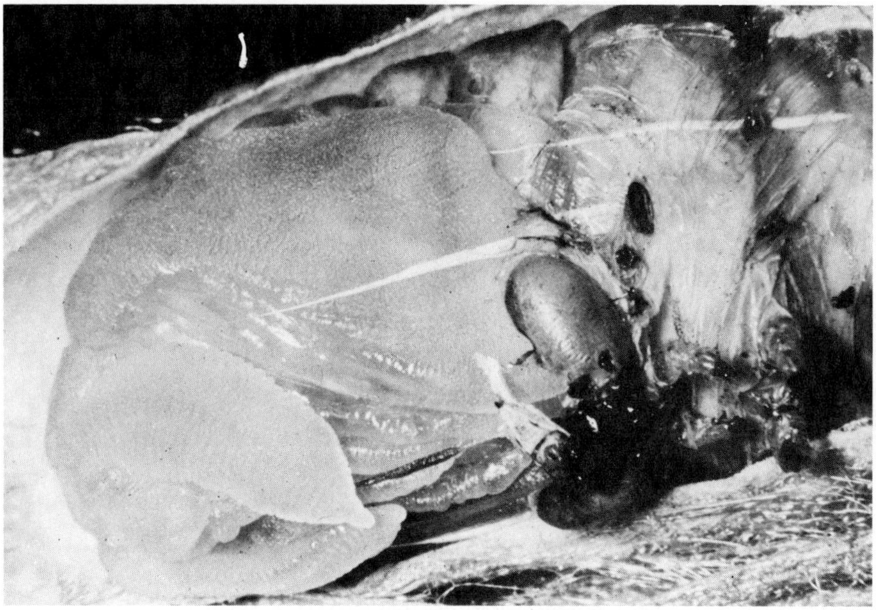

Figure 113. A *Hyalophora cecropia* larva transforming into a pupa within a protective cocoon consisting of three layers, the inner smooth to protect from injury during molt, the middle loosely arranged for insulation, and the outer for proper background color and water-repellency. Note the change in antennal shape, the type of legs, and the tracheae being molted with the larval exoskeleton.

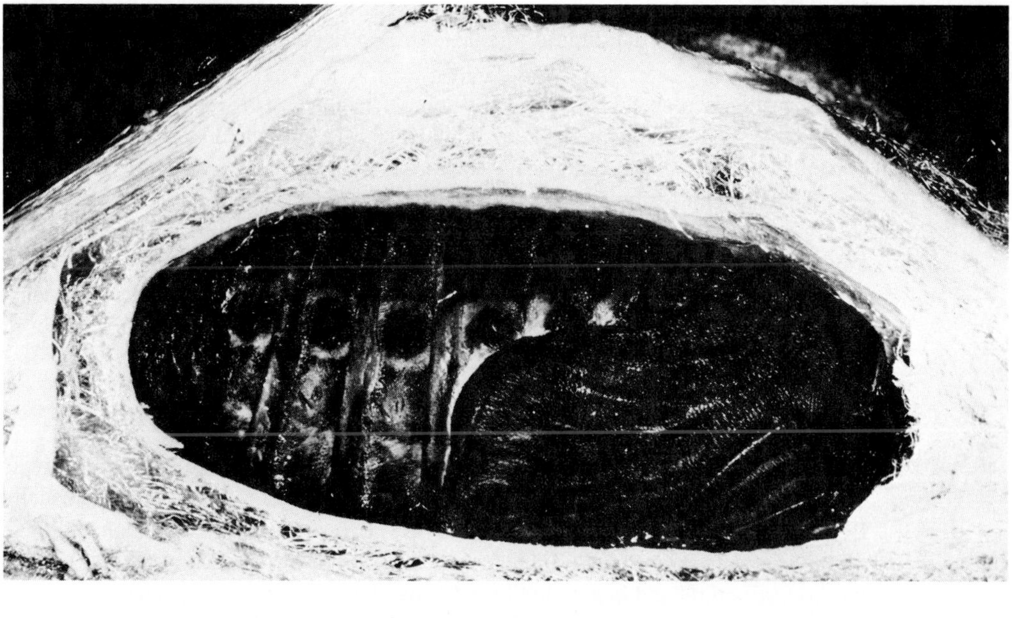

Figure 114. After completing diapause, the *Hyalophora cecropia* pupa completes the development to an adult. Note the adult color patterns under the pupal exoskeleton as the final molt is about to occur.

Figure 115. Flesh fly (*Sarcophaga bullata*) adult emerging from the puparium, a protective sac consisting of the exoskeleton of the last larval instar. Note the ptilinum, an eversible structure at the front of the head that is used to rupture the puparium.

Trichoptera and Lepidoptera the determination is reversed, and the male is *homozygous* for the *X* chromosome.

In Hymenoptera and a few Thysanoptera and Homoptera, sex is determined by *haplodiploidy;* males result from unfertilized eggs (generative parthenogenesis) and are haploid, whereas the diploid females develop from fertilized eggs and have the *XX* condition.

Because each insect cell has a specific number and pattern of sex chromosomes, each cell is either male or female. The total individual, or the summation of cells, is therefore also either male or female because no sex hormones are present in insects to modify the chromosomal expression of sexual characteristics. Occasionally, insects appear that have localized patches of male and female characteristics indicating an abnormal distribution of the sex chromosomes in certain tissues of the body. In some instances each lateral half of the body may be a different sexual expression, but usually the individual is a sexual mosaic. Such abnormal insects are referred to as *gynandromorphs.*

Polymorphism

Variability is extensive within any species and often results in distinctive phenotypic groupings of similar individuals or *morphs.* Evidence indicates both hormonal, usually JH (juvenile hormone), and/or pheromonal involvement in this phenomenon, although the environment often induces phenotypic variation or has at least an indirect effect, and both structural and behavioral morphs may be seen. In many species a particular morph is rarely seen, yet may be of great importance when the environment is at all favorable for its selection. We will concentrate on a few of the more striking examples.

The water strider, *Gerris remigis* (Fig. 126), has two distinct adult morphs or variations. One morph lacks wings (apterous), and the second possesses normal-sized wings (macropterous). The apterous form is usually the most abundant, but up to one third of any population may be macropterous when they reside in temporary habitats. Wing presence is not a selective heritable trait in this species, so the reason why wings appear at all is unknown because the winged morphs fly and disperse only in certain geographic populations.

The migratory locust of Africa, *Shistocerca gregaria,* exists essentially as two different morphs or phases. The *solitary phase* of this short-horned grasshopper has little black pigmentation and normal-sized wings, and the prothoracic glands persist into adulthood, indicating a continued role of juvenile hormone. Adults remain in the same general region where they develop. At the other end of the spectrum is the *gregarious phase,* a morph with much pigmentation, smaller size, elongated wings, a higher metabolic rate, and an ability to store a high amount of fat, that migrates in daylight hours both as a nymph and an adult. In contrast to the juvenilized solitary phase, the prothoracic glands are not retained, and the adult appears to be the result of accelerated growth. Which phase or morph that develops is determined by population density of the immatures; low densities produce the solitary phase, whereas high densities eventually result in the migratory and destructive gregarious phase. Crowding, in turn, is the product of dwindling food resources. The transition from solitary to predominantly gregarious extends over several generations and appears to be the result of a *gregarization pheromone,* a substance produced by all the grasshoppers but whose activity starts only when crowding increases the concentration to a minimal threshold.

Aphids serve as another good example of polymorphism because both winged (macropterous) and wingless (apterous) morphs may be found within the same species. An unused plant that is invaded serves as a superabundant source of food, and populations grow rapidly. Eventually, crowding takes place, and a new generation with wings arises that can disperse to new hosts. Wing development is suppressed by juvenile hormone. High aphid densities, with high frequencies of encounters between individuals, cause a juvenile hormone shutdown resulting in wing formation. Also, winged aphids may be either oviparious or viviparous depending on the season, the former developing late in the season to produce overwintering eggs. In *Macrosiphum,* these overwintering eggs hatch in the spring producing female *fundatrices,* aphids feeding on the primary host plant. This morph is characteristically wingless, parthenogenetic, and viviparous. The fundatrices in turn produce *fundatrigeniae,* which are also wingless, parthenogenetic, and viviparous. After several generations, winged female *migrantes* develop and fly to secondary hosts where they reproduce as parthenogenetic, viviparous individuals. The subsequent generation is composed of wingless females, termed *alienicolae,* which reproduce as previous morphs. Late in the year *sexuparae,* winged females, develop and fly to the primary host plant and reproduce, through parthenogenetic viviparity, winged and wingless males and wingless oviparous females that produce the overwintering ova by sexual means.

In addition to the expected differences in genitalia, variation in form may be evident between sexes, a situation termed *sexual dimorphism.* In most species the morphological differences are minor, consisting mainly of size; however, extreme differences do exist. In some moths the males appear normal, whereas the females are wingless and may even be larviform (bagworms). Male stag beetles have greatly enlarged mandibles, and certain scarab beetles have enlarged horns on their pronotum and head, which are used in duels for mating rights. Females of both groups lack such modifications (Fig. 352).

The mandibles of male dobsonflies are long, curved, and of little use in defense because they often bend when attempting to bite, but the female mandibles are short, toothed, and effective in protection. Colors or color patterns between sexes may be of sufficient variance so as to give the appearance of different species.

Castes in social insects are a type of polymorphism, and immatures in both termites and hymenopteran societies develop into specialized individuals for carrying out specific tasks within the colony. The castes or morphs are the result of hormones, pheromones, and/or diet; the specifics of how they develop and the adaptations of these morphs will be discussed later in Chapter 7.

Coloration

Insect color is of two types, structural and pigmental. Structural colors appear metallic or iridescent and are the result of refraction and interference of light from the integument. In certain leaf, tiger, and ground beetles, the color is produced by a secreted layer that gives an effect much like a film of oil on water. Most structural colors, however, are the result of a series of layers, ridges, or rods deposited at various angles one upon another and separated by material of a different refractive index. Light rays strike these layers and ridges and are refracted differently to interact and produce beautiful interference sheens when viewed at the proper angle. When viewed at inappropriate angles or when movement occurs, the reflectivity diminishes and the insect becomes dark and disappears into the background. These colors also are commonly seen in scales of moths and butterflies, and reflections from the wings of *Morpho* butterflies may be seen for several thousand feet.

Color also results from pigment deposition. *Melanin,* a derivative of tyrosine, is deposited into the cuticle and produces colors from yellow-brown to black. Most colors, however, are from pigments within the gut, blood, or epidermal cells. Some of these pigments are ingested from plants, such as carotenes and carotenoids that produce yellow. Most colors are newly metabolized substances such as the pterines and purines (white, yellow, red) and ommochromes (yellow, brown, red). Greens are normally a mixture of several pigments (not chlorophyll), and the same color may come from different mixtures of pigments. Because most pigments are found in living tissues, death of an insect often results in a loss of coloration, a phenomenon often seen when viewing insect collections.

Some coloration is reversible. The green lacewing, for example, is light green when active and feeding but loses the green pigmentation to become yellow-brown when overwintering. The following spring it regains the green color as activity commences. Walkingsticks and some grasshoppers can vary their color from light to dark by pigment granule movement, the walkingsticks in response to temperature changes and the grasshoppers due to light, humidity, and temperature. Pigment granules within the ectodermal cells move out distally to produce the darkening, whereas retreat and clumping of these granules basally result in a light color. The environment also induces color changes in different generations of the same species. In many areas overwintering tiger swallowtail pupae, for example, become black adults because of excess melanin deposition induced by the cold, whereas the summer morphs have a yellow background because of the warm temperatures.

Crowding also lowers the juvenile hormone concentration and increases melanin deposition in the gregarious phase of the migratory locust. Further discussions on coloration are found in Chapter 6.

Diapause

During embryonic development and in postembryonic stages, growth can be halted, resulting in a dormancy, or *diapause.* This phenomenon permits an insect to survive under seasonal situations when normal metabolism would produce death through lack of food, exhaustion of stored fat, or freezing. It also synchronizes life cycles, and most temperate species and some tropical ones enter into diapause during some specific sensitive stage in the life cycle.

The most common diapause inducing stimuli are photoperiod and temperature. In many instances, this dormancy is genetically programmed, and reversal occurs with difficulty only after unfavorable conditions have passed. The photoperiod of increasingly shorter days, often correlated with cool days, normally initiates diapause in temperate regions, whereas the dry season triggers it in the tropics. Such a response to day length eliminates much of the problems of seasonal climatic variations and permits the insect to make the necessary physiological adjustments prior to winter or hot and dry periods in the summer.

There are two basic types of diapause, *obligatory* and *facultative.* The former always occurs at the same stage and is common in univoltine species, those with only one generation each year (Fig. 116). When diapause is obligatory, it is usually tied to photoperiod, and the environment affects mainly the duration of diapause rather than the time it commences. The insect brain appears to be the site that initiates obligatory diapause, but little conclusive experimental data on the mechanics are available, although it appears that it is either the result of some failure in the neuroendocrine system or the production of a specific hormone, such as in the silkworm moth, *Bombyx mori.* In contrast, facultative diapause occurs when environmental conditions dictate and is probably the most common type. Such stimuli as lowered moisture, temperature, or food often induce this latter type of diapause. Facultative diapause usually is preceded by a preparatory phase in which food reservoirs are accumulated. Once dormancy is initiated, feeding and other normal activities cease, RNA and DNA synthesis stop, and metabolism lowers to preserve resources. Energy comes from stored fats and glycogen, and air intake may be restricted to short inhalation intervals to reduce water loss. Nerve and hormonal rates are markedly lowered. When climatic conditions become favorable, the insect is activated, and the body physiology returns to normal. At this time, diapause is "broken" or terminated.

Diapause is known from every life stage (Fig. 116). It may occur in the egg, when embryological development ceases at some specific point. Larval diapause causes the immature to stop feeding and is a common type. Pupal diapause also exists, except in beetles, and metamorphic changes are halted. Diapause in adults is best known as a reproductive cessation, but feeding also can take place. Not often discussed is the phenomenon of a species undergoing more than one diapause in its life cycle, but a grasshopper is known to have two diapause periods in the same egg, and a butterfly is known with both a pupal and a pharate adult diapause.

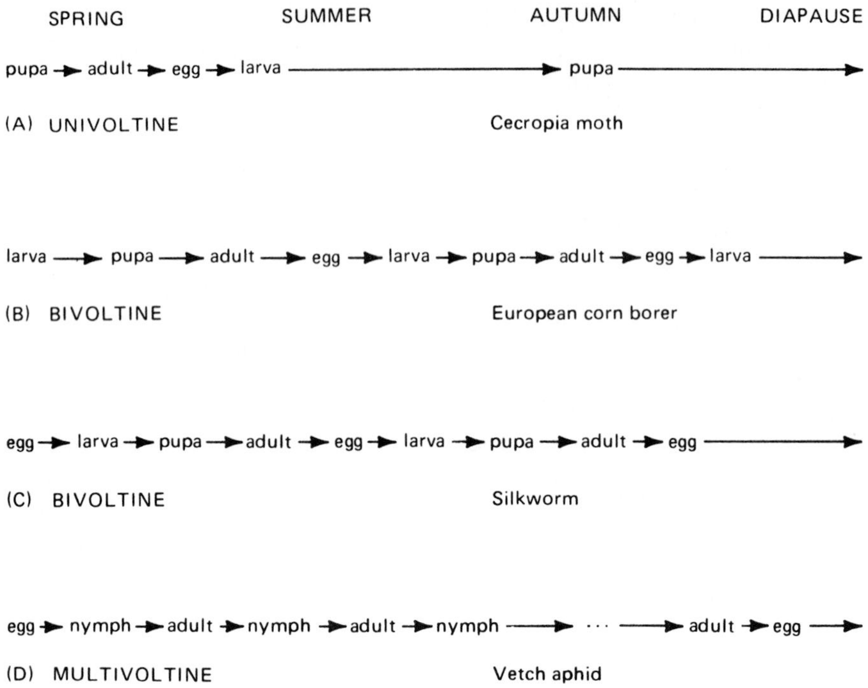

Figure 116. Insect species illustrating stages in which diapause occurs and the number of generations per year that each passes through prior to overwintering. Univoltine = one generation per year; bivoltine = two per year; and multivoltine = many per year.

Regeneration

Insects are capable of limited amounts of regeneration depending on the species, age of the individual, and the extent of injury, but the regenerated parts appear only at ecdysis. In contrast to vertebrates and annelids in which the process is similar to the original embryology, regeneration in insects starts with a division and migration of epidermal cells around the lost structure. Nerves and tracheae then grow into the regenerate area. Muscle formation has not been determined, although this tissue does require the presence of nerves in order to develop (Nuesch, 1968; Goss, 1969). With subsequent molts, part or all the lost structure will develop, depending on the species and the extent of the earlier loss (Fig. 117).

Some insects such as walkingsticks have weak points in their appendages and may lose these structures when they are grasped by a predator. The loss affords the insects a means of escape. The limb often twitches and diverts attention away from the escaping individual. Such a phenomenon is termed *autonomy.* Subsequent molts result in replacement, although the legs may be reduced in size.

Insects that undergo complete metamorphosis are unique in that the loss of larval appendages, unless very extensive, usually does not affect the presence of appendages in the adult. This is because most adult structures arise from imaginal discs that develop into the adult structures at metamorphosis.

Figure 117. These two American cockroach nymphs (*Periplaneta americana*) had their middle two pairs of legs amputated. The nymph's legs on the left were severed early during the previous instar, whereas the other roach was experimented with late. Regeneration occurred on the left individual (one tarsomere was lacking), but no legs appeared at molting on the right roach because of the late amputation.

ENDOCRINE AND EXOCRINE SYSTEMS

An endocrine structure is a ductless gland or tissue that produces *hormones* and secretes them into the blood for distribution and activity within the same individual, whereas *pheromones* are secreted by ducted exocrine glands and are released to the outside environment and usually affect other individuals of the same species. In insects these "glands" are often more appropriately referred to as tissues because of their relative simplicity. The endocrine system will be explored now, whereas the exocrine system with its pheromones will be covered later.

Endocrine tissues very commonly are specialized neurons. Their secretions, hormones, are characterized by their ability to produce specific reactions in minute concentrations upon specific tissues (*target structures*). The reaction can be stimulatory or inhibitory; not all tissues are equally competent or ready to respond to the presence of hormones because the cellular environment in various portions of the body is continually fluctuating.

The presence and effect of hormones are difficult to demonstrate. Once the presence of a hormone is suspected (often suggested accidentally), scientific evidence of cause and effect must be determined and often involves the removal of a tissue with subsequent implantation of the tissue of extract in the same or different individual. Hormones are undoubtedly involved in nearly all physiological activities of the insect, but only a few that have substantial research data as to their activity will be discussed in this chapter.

Growth and Metamorphic Hormones

The first evidence of hormones and the most widely studied ones today are those involved in the phenomena of growth and metamorphosis. The discovery and techniques involved in elucidating their presence and role provide some of the most fascinating experiments in biology. Readers are referred to such works as Williams (1947, 1952, 1958) and Wigglesworth (1984). One experiment (Williams, 1952) is summarized in Figure 118.

Certain neurosecretory cells in the protocerebrum produce the PTTH or *brain hormone.* The number of cells involved in producing this hormone varies from approximately five to several thousand, depending on the insect studied. Once synthesized, the PTTH hormones are moved along the axons of these neural cells from the brain to the *corpus cardiacum* (Fig. 88), where they are released into the hemolymph at specific times (the corpus allatum may be the site of release in at least some insects). The target tissues are the *prothoracic glands,* which become activated.

The prothoracic (= thoracic or ecdysial) "glands" are inconspicuous tissues often associated with the first pair of thoracic spiracles. Although small, their effect is monumental, because they produce a group of hormones, the *ecdysones* (= growth and differentiation or molting) that activate the epidermal cells to produce both a new exoskeleton and molting fluid. When juvenile hormone is absent or in low concentrations, ecdysones may induce metamorphosis of the immature to an adult. The prothoracic glands degenerate at the last molt to adulthood in most insects but remain functional in those Apterygota that continue to molt as adults. The chemical structures of the two ecdysones are as follows:

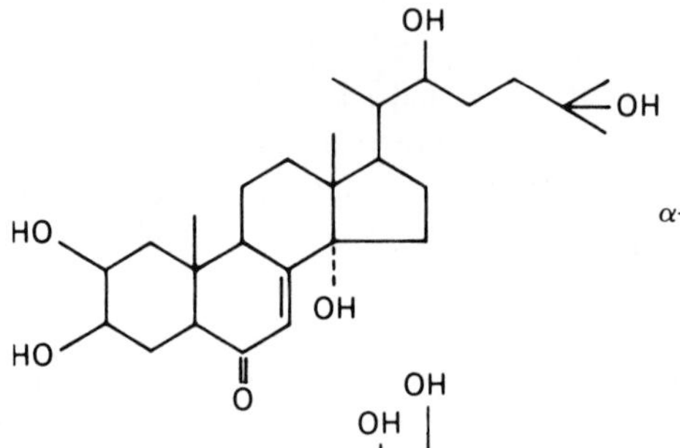

α-ecdysone

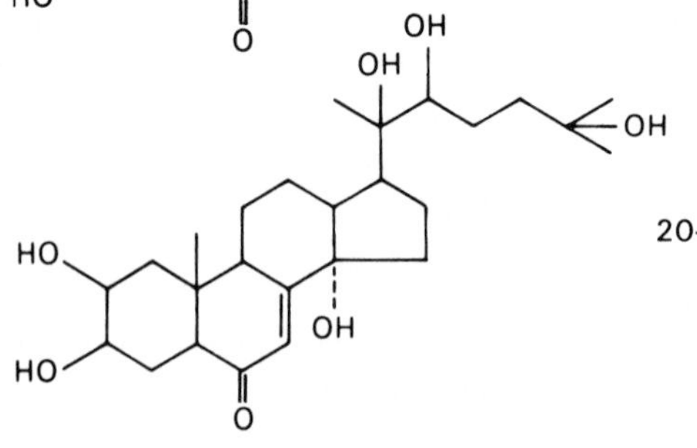

20-hydroxyecdysone

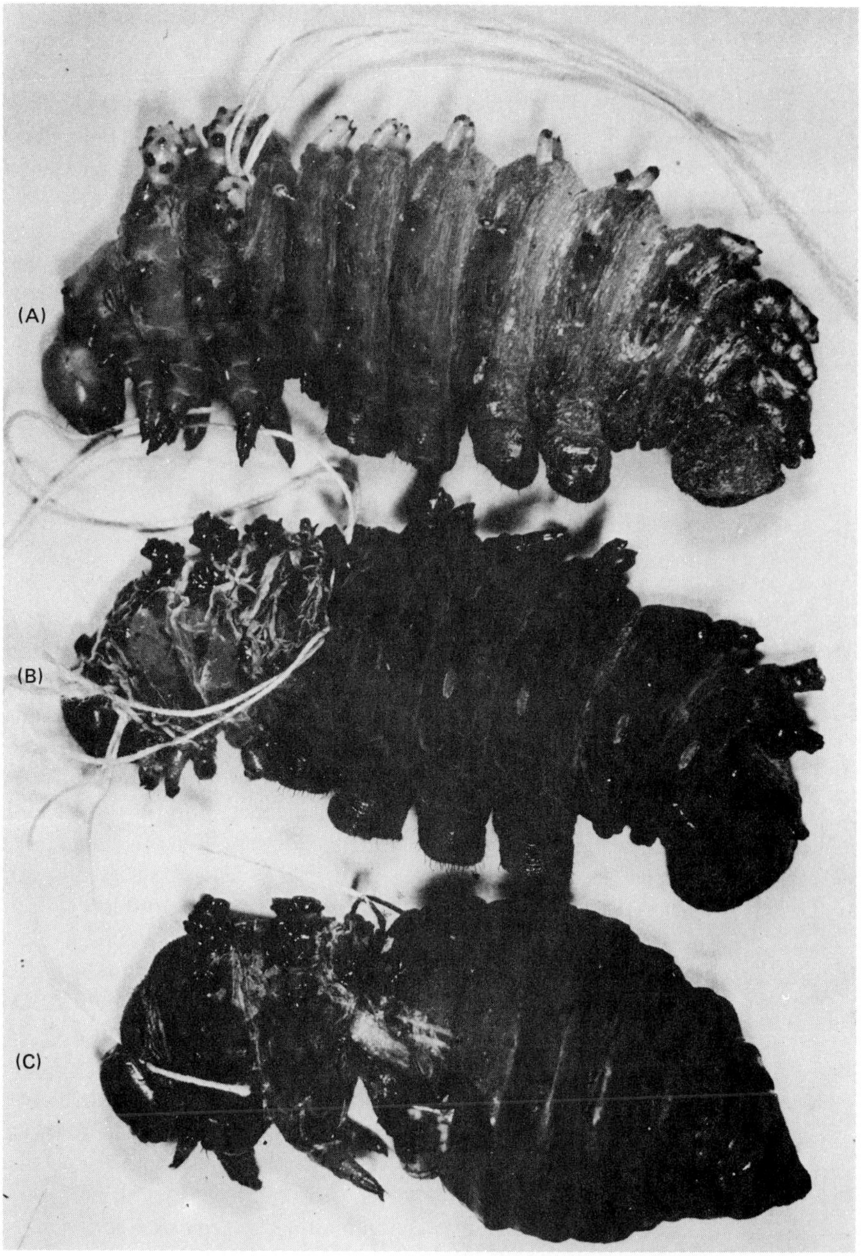

Figure 118. Experiments on *Hyalophora cecropia* larvae illustrating the action of brain hormone and ecdysone by ligation. The juvenile hormone is no longer being produced. The procedures followed were similar to Williams (1952). Ligation by body regions prevented distribution of hormones except within certain areas. The thorax region in (B) and (C) were not able to shed the larval exoskeleton because of the ligations, and the head, deprived of food and oxygen, died. (A) ligation before any hormones were released; no metamorphosis or molting took place; (B) ligation after brain hormone had been released but prior to release of ecdysone from prothoracic glands; only the thorax metamorphosed, the head and abdomen remained in their larval form; (C) ligation after both brain hormone and ecdysone were released; all regions underwent metamorphosis.

A third group of three compounds, often referred to singularly as the *juvenile hormone* or *JH,* is produced in the corpora allata (Fig. 88) and maintains the proper functioning of "larval genes," modulates what ecdysones can do, and prevents degeneration of the prothoracic glands. This group consists of long-chain fatty acids and may be shown structurally as follows:

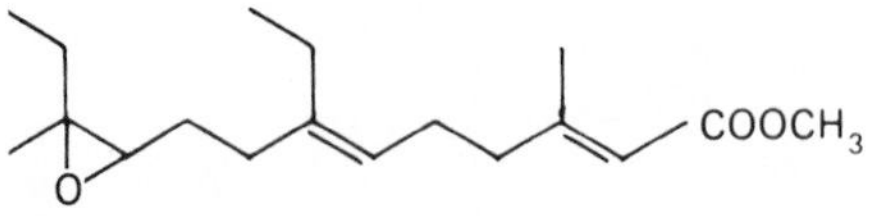

JH I

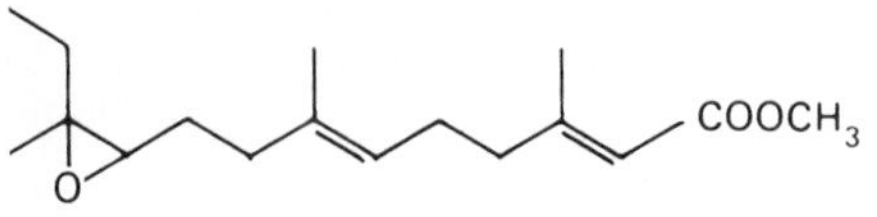

JH II

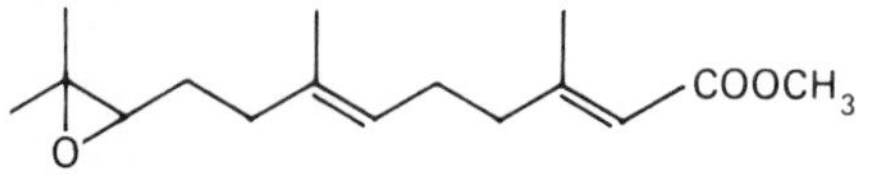

JH III

When both ecdysones and juvenile hormones are present (Fig. 119), growth and molting occur, but the larval characteristics are perpetuated into the next immature instar [Fig. 120(B), (C)]. However, if juvenile synthesis is either shut off or drops precipitously to a low level, the result is inactivation or different rates of activity of these "larval" genes and subsequent differentiation of adult characteristics through the sole action of ecdysone [Fig. 120(B)]. If the immature is similar to the adult (hemimetabolous development), the adult will be produced in the next molt. If, however, drastic alterations must be made (holometabolous development), then two molts are required; the first instar is referred to as the *pupa* and the second instar as the *adult.* Juvenile hormone is also involved in polymorphism, as previously discussed in this chapter, as well as in regulating times of ecdysis and in egg production in adults. Of special interest are the more than 60 compounds or variants found in plants that mimic juvenile hormone, some of which are more active biologically than the insect's own hormones. These variants are mainly found in ferns and gymnosperms and may account, in part, for the few insects that can develop on these plants.

A fourth hormone involved in molting, the *eclosion hormone,* has been found in giant silkworm moths. Produced in the brain of pharate adults, this hormone appears to trigger eclosion from the pupal exuvia. The prevalence of this latter hormone has yet to be determined.

Interesting experiments may be imposed upon the endocrine system. For example, in certain moths, removal of the corpora allata in early instars may result in premature metamorphosis and a dwarf adult. In contrast, implantation of corpora allata or injection of juvenile hormone into a last instar may produce an additional instar and a giant individual (Fig. 121).

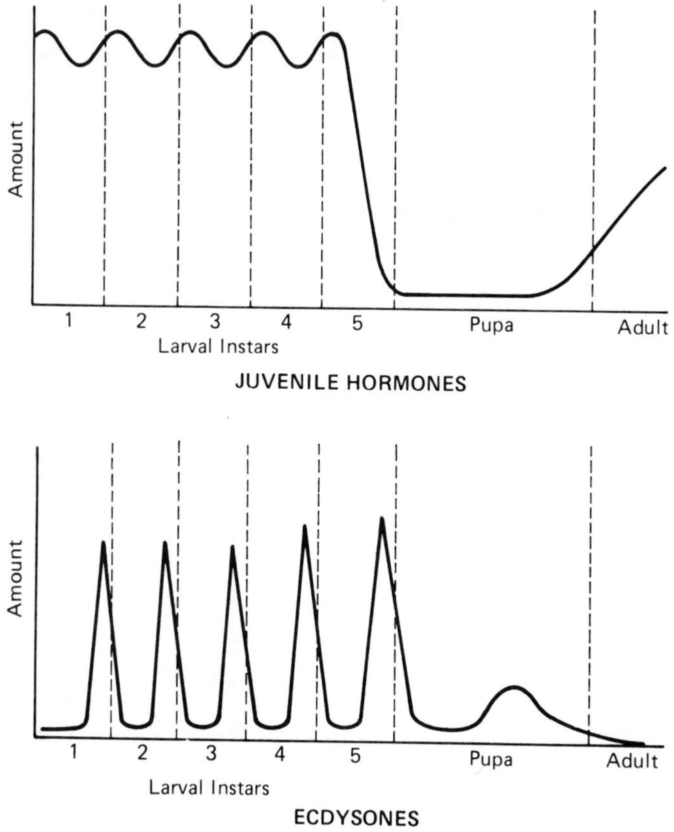

Figure 119. Comparison of hormone levels in developing insects.

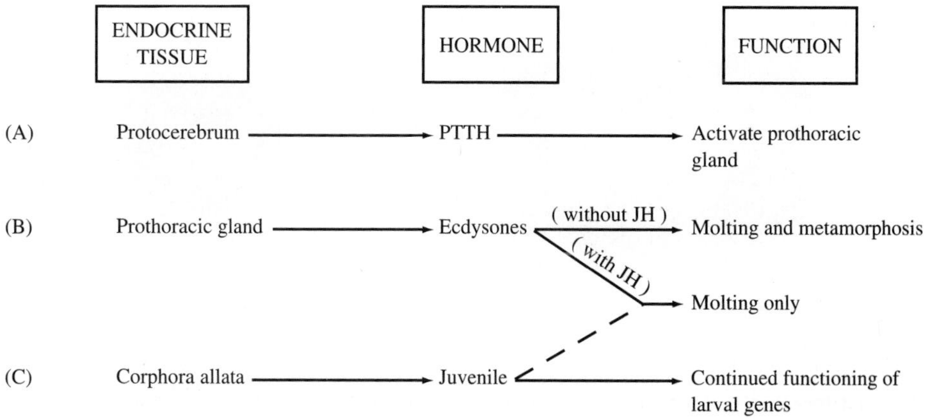

Figure 120. Summary of growth and metamorphic hormones and their functions.

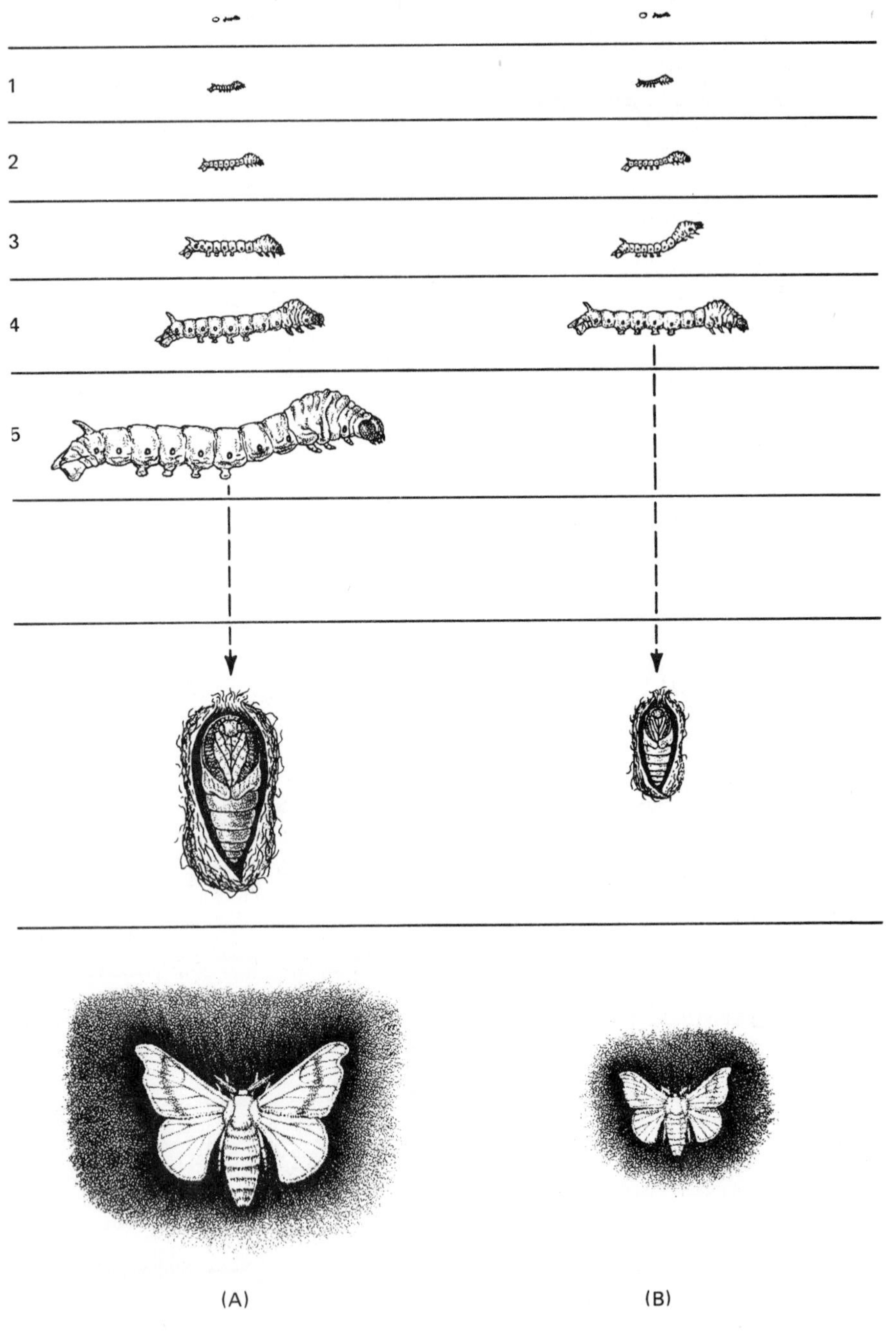

Figure 121. Dwarf and giant moths were made by removing and implanting the corpora allata. The column at left outlines the normal development of the commercial silkworm, *Bombyx mori.* At the top of the column are the egg of the insect and its newly hatched larva. The second column shows what happens when the corpora allata are removed at the fourth instar; the larva immediately changes into a dwarf pupa, and then into a dwarf moth. In the third column

Continued

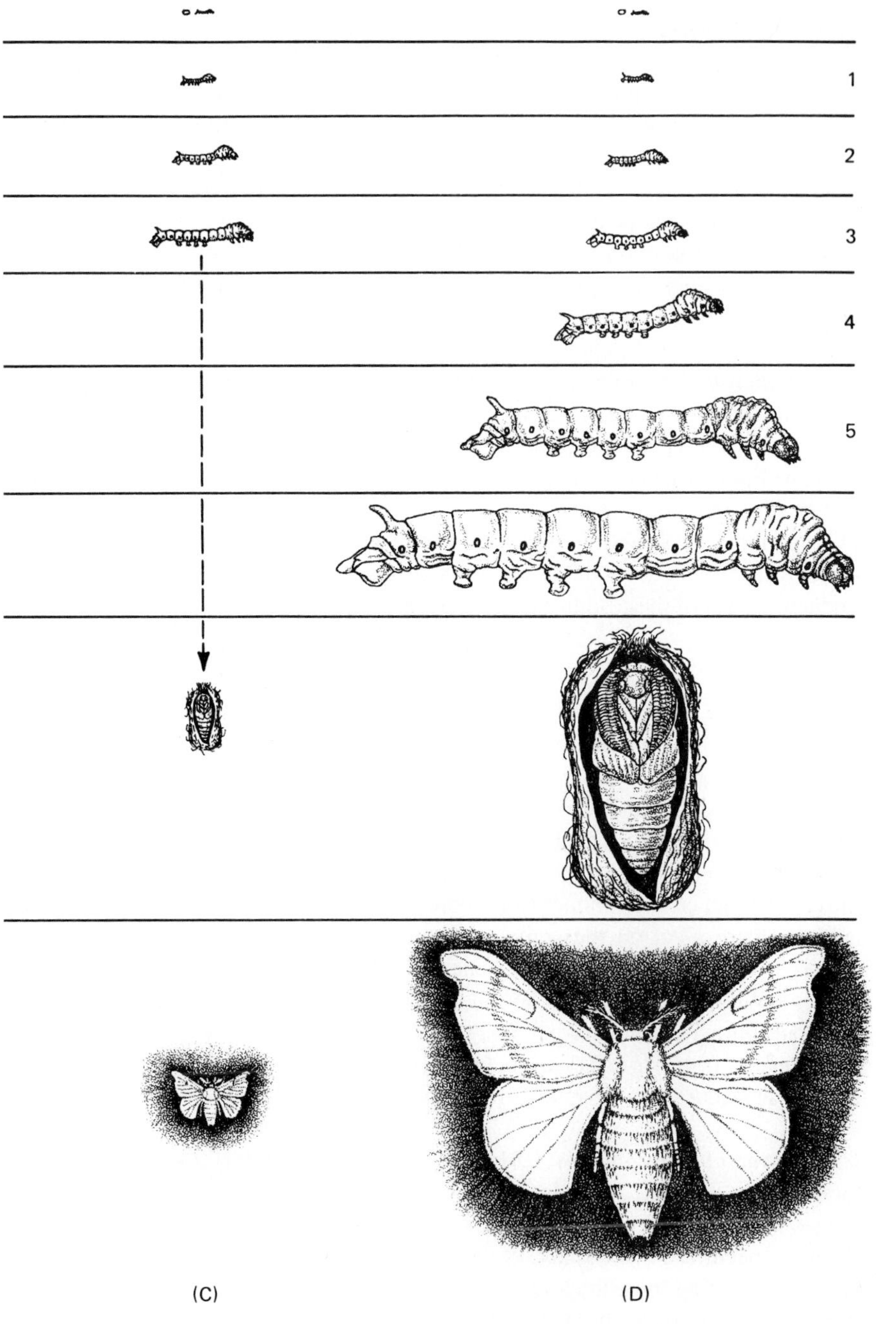

the corpora allata are removed at the third instar, resulting in even smaller pupa and moth. In the fourth column the corpora allata of a young larva are implanted in a fifth instar larva. This larva continues to grow and then changes into a giant pupa and moth. [From C. M. Williams, "The Juvenile Hormone," *Scientific American,* 198, no. 2 (1958), 67–74. Copyright 1958. Used by the permission of Scientific American, Inc. All rights reserved.]

Other Hormones

Bursicon, a hormone produced by neurosecretory cells in ganglia, is necessary for darkening and sclerotizing the exoskeleton. In cockroaches, this hormone first appears after the old cuticle splits, but in other insects such as Lepidoptera and Diptera its release may be delayed until after the adult escapes from either its cocoon or puparium. Bursicon production ceases several hours after molting.

Knowledge of hormones other than those directly involved with growth and metamorphosis is scanty. A *diapause hormone,* originating from the subesophageal ganglion, causes second-generation eggs of the silkworm moth to enter diapause. A *hyperglycemic hormone* released from the corpus cardiacum increases trehalose in the blood. A *diuretic hormone* aids in maintaining the water balance in the body by acting on the Malpighian tubules and rectum. A *wound hormone* prevents molting until after exoskeletal repairs are complete. An *adipokinetic hormone* is involved in regulating the lipid concentration in the hemolymph. Undoubtedly, hormones also are involved in certain color changes and the varied responses of insects to light cycles (photoperiod), but scientific evidence is only suggestive. Most hypotheses include the neurosecretory tissues of the subesophageal ganglion, brain, and corpora allata as production sites of these additional hormones.

Pheromones

Chemicals released externally from an insect that induce interactions with other organisms are termed *semiochemicals.* One group of these substances, the *pheromones,* act on individuals of the same species and induce biological activities at very low concentrations. The word pheromone is coined from Greek, meaning "carrier of excitation." Although some pheromones consist of a single compound, most are mixtures of chemicals. Pheromones from more than 1,000 insects are known. Some have an immediate effect on the central nervous system and behavior, "releaser substances," whereas others trigger a chain of developmental events, "primer substances" (Wilson, 1963a). Releaser pheromones induce such behavior as sex attraction, trail following, and alerting. Only a few of the most common will be discussed subsequently. Primer substances include social pheromones that influence castes in termites as discussed in Chapter 7 on social insects.

Sex or attractant pheromones. More than 100 years ago, J. H. Fabre observed that upon placing a female moth in a cage outside his house, male moths of the same species were attracted in large numbers from nearby woods. No females were drawn in. Fabre speculated that because of their disproportionate size, the large male antennae were detecting something that was attractive. Little follow-up on this and similar observations was made until the 1950s when advanced technology provided a means of isolating, analyzing, and testing compounds that appear to induce these responses. Research is providing a picture of much diversity in the molecular chemistry and action of these attractants, but three generalities may be made. First, it is most often the female (Fig. 122) that releases the pheromones and the male that is attracted. Second,

Figure 122. Lilac borer (*Podosesia syringae*) in the process of releasing attractant pheromone.

sex pheromones are usually produced in abdominal glands, although a notable exception is in the mandibular glands of the queen honeybee. Third, although some pheromones have proven to be single compounds, attractiveness is usually a function of different concentrations or blends of the specific compounds in a pheromone mixture. More specific responses to these pheromones will be covered in Chapter 6 on behavior.

Trail pheromones. Trail pheromones are most common in ants and termites with large colonies. As workers forage and return to the nest, periodic deposits are left by each individual. In some species, such as the fire ant, these pheromones are deposited by periodic dragging of the sting. In others, the substances originate from tarsal glands, abdominal glands, or from the alimentary canal. In most instances, because trail pheromones are highly volatile, continued pheromone deposition is necessary to maintain the trail. Consequently, it is only the routes from the nest to a good source of food that become trails. In a few instances, such as the leafcutter ants (Fig. 123) and eastern tent caterpillars, the trails are persistent and last many days, thereby permitting the colony to continue contact with good sources of food. Because trail pheromones are detected by the antennae, the lower portions of these structures of trail-following species, such as army ants, have a disproportional number of receptors (Fig. 49).

Figure 123. Leafcutter ants (*Atta* sp.) following a chemical trail across a piece of string back to the colony.

Alarm (or alerting) pheromones. These substances are produced by insects that are either social or closely related individuals that aggregate for extended periods of time. Ants and termites produce these pheromones, in addition to other substances, that excite their siblings to activity. The emergency that initiates the release may be a break in the nest wall, as in termites with above-ground nests (Fig. 124), or an attack by predators. As additional individuals arrive on the scene, they also release these highly volatile alarm materials until the necessary number of individuals are recruited to cope with the problem. Soldiers, when present, are especially activated. Different glands are responsible for the alarm secretion. In some ants, it is the mandibular gland; but in other ants, anal glands produce these substances. In termites, it is the frontal glands. The honeybee produces the alarm pheromone from both the mandibular gland and the sting glands. This type of pheromone is usually short-lived to prevent colony disruption beyond the duration of the crisis.

In aphids, the alarm pheromone causes other aphids to disperse. Droplets from the aphids' cornicles apparently initiate this activity, which also causes symbiotic ants to engage any attacking predators.

Aggregating pheromones. Although sex pheromones attract individuals to one another for mating, aggregating pheromones are released in response to other environmental conditions. An example could be bark beetles. Because it is of selective advantage to "attack" a tree while it is in a susceptible stage for larval development, adult bark beetles, in locating a suitable tree, release pheromones that attract other bark beetles to the site. The result is the ability to overcome the tree defenses by mass infestation, thereby enabling the larvae to complete their development prior to tree desiccation and before there is competition with other insect species. Aggregating pheromones also have been discovered in certain desert grasshoppers, cockroaches, and mosquitoes that oviposit in remote areas.

Anti-aggregation pheromones. Anti-aggregation pheromones are produced by insects when high densities will be detrimental because of restricted food supplies. A few hymenopterous parasites of insects have this adaptation.

Figure 124. Termite nasutes (*Nasutitermes* sp.) attracted by alarm pheromones to a hole in the nest.

Social pheromones. Colony cohesion in bees, wasps, and termites is often maintained by pheromones. A queen pheromone in honeybees and in some social wasps inhibits maturation of the ovaries of workers. If the queen dies or leaves, behavior changes drastically within the colony, and a new queen is reared. Similar effects have been noted in termites whose pheromones also influence caste development in reproductives and soldiers. More details on these primer substances can be found in Chapter 7.

QUESTIONS

1. What is *embryology?* What structures come from ectoderm, endoderm, and mesoderm? What factors influence embryology in insects?
2. Which cells in the developing egg do not become part of the embryo? What are the functions of these cells?
3. Why does growth in insects appear to be discontinuous or in sudden spurts? How does the new exoskeleton form?
4. Are growth and metamorphosis synonymous? Explain your position.
5. What are some of the advantages and disadvantages of complete metamorphosis (holometabolous development)? What is a *pupa?*
6. Is metamorphosis restricted to the exoskeleton and its resulting body shape? Explain your answer.
7. What are *hormones?* How do they relate to "target structures"?
8. What is the difference between endocrine and exocrine tissues?
9. What hormones are involved in ecdysis? In metamorphosis? What is the function of each?

10. What are *pheromones?* Discuss their role in the life of the insect.
11. Can insects regenerate lost appendages? If so, how?
12. How is sex determined in insects?
13. What is *polymorphism?* How might it be important in evolution?
14. Define the word *diapause* and describe what selective advantages it gives insects.

5 Ecology

The next logical step in unfolding the life of insects would have to be a discussion of the environment. Although the environment is heterogenous and offers numerous differing habitats, an insect species selects and occupies those areas that fall within certain tolerance limits with the maximal population near the optimal preferences. Many current ecological studies attempt to analyze these habitats, and the concept of the *ecosystem,* or the study of the biotic and abiotic elements of an area and how these affect organisms (Fig. 125), has become dominant. *Abiotic* elements include temperature, soils, water, and gases, to name a few, and are often the first to be considered. *Biotic* elements include the study of organisms and their interactions within and between *populations* (localized breeding groups). Associations of many interdependent populations of different species constitute *communities;* groups of communities having similar dominant plant life forms make up a *biome.* Because of the complexity of the natural environment, this chapter will be restricted to a few of the major factors and concepts.

AQUATIC ENVIRONMENT

Although water may freeze or evaporate, it is usually found in a liquid state. Six characteristics of this substance must be discussed so that the existence of aquatic organisms in this medium can be understood. First, water gets heavier as it cools until its maximal density is reached at 4°C. From this point to 0°C, fresh water becomes lighter and floats to the surface before freezing. As a result, formation of ice progresses downward from the water surface, and whether or not a body of water freezes solidly is determined by its depth and the duration of subfreezing temperatures. Without this characteristic of freezing progressing away from the surface, aquatic biota would not survive freezing weather.

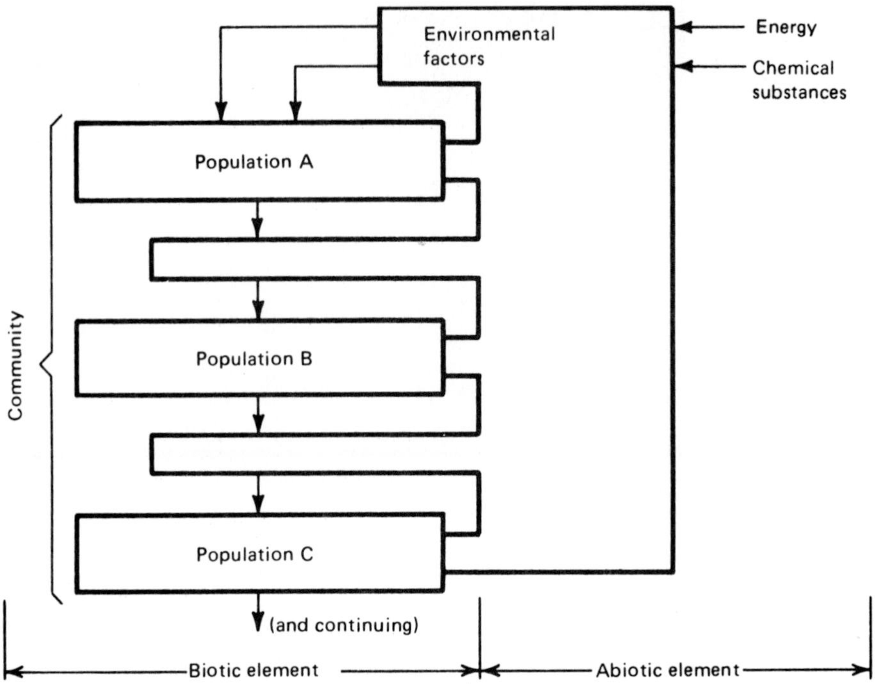

Figure 125. Basic features of an ecosystem and the fundamental relationships within it. Ecosystems are composed of biotic and abiotic elements. The biotic element is formed from groups of similar and related organisms, the *populations.* These are linked together by functional interdependence to form *communities.* The abiotic element is involved in processes of energy absorption and release of chemical substances. Both of these vital activities of an ecosystem are subject to environmental influences, as are the intracommunity and intercommunity biological interactions. (From Boughey, 1972.)

Second, water can absorb large amounts of heat, almost 500 times that of air, without changing its temperature. This heat, along with dissolved substances, diffuses very slowly through the liquid (several thousand times slower than in air). This capacity to store energy and transfer it slowly produces a relatively uniform situation when compared to air, because extremes in temperature within an aquatic environment are greatly moderated.

A third characteristic of water is its relationship to gases. The warmer the water, the lower the amount of oxygen that can remain dissolved. Warm water can be a limiting factor to life because of lowered oxygen levels long before temperature itself causes death. The correlation between temperature and the saturation levels of oxygen may be seen in Figure 130 later in this chapter.

Fourth, water in its liquid state does not expand to fill space as gases do. Molecules that contact the atmosphere arrange themselves to form a distinct film, a membrane-like feature caused by molecular cohesion of the surface water molecules. Some insects such as water striders (Fig. 126) and adult whirligig beetles use the surface film to move about on. Although most structures of water striders and whirligig beetles

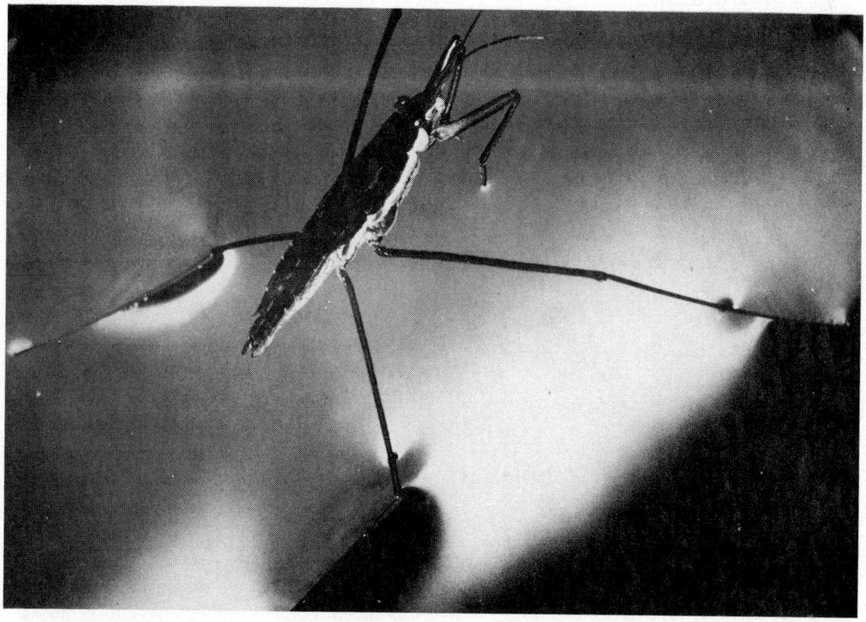

Figure 126. The water strider (*Gerris remigis*) rows its way across the water using its middle legs for the power stroke. Note that the weight of the strider depresses the surface film.

are similar to terrestrial insects, the second and third pairs of legs are significantly modified. In water striders, these legs are greatly lengthened, with the middle pair providing oarlike power strokes for locomotion and the hind pair used for steering. The tarsi of both pairs of legs are hydrophobic and prevent the insect from sinking. Whirligig beetles have their second and third pair of legs shortened and flattened for rapid swimming movements. In addition, the compound eyes of whirligig beetles are divided into dorsal and ventral halves (Fig. 127) with, apparently, the ventral portion functioning below the surface and the dorsal part used for seeing in the atmosphere.

The fifth characteristic of water is the uneven distribution of electrons between oxygen and hydrogen, producing a polar molecule; its charges are not distributed symmetrically but form positive and negative poles. As a result, water molecules arrange themselves into a distinctive pattern that tends to neutralize any weak electrical fields around them. Substances held together by ionic bonding readily disassociate and dissolve. Also, the hydrogen bonding pattern increases the ability of water to serve as a solvent for many other molecules. As a result, water varies in salinity and acidity. Obviously, physiological adaptations to survive under each pH situation must be present.

Last, water is dense and viscous. To be able to move through water, especially in currents, requires some form of streamlining. The entire body may be streamlined, as is usually found in insects that are strong swimmers, or only the dorsum may be rounded and the ventral surface may be flattened. For many years the belief prevailed that streamlining of half the body caused the water passing over the body to press the organism tightly to the substrate, and indeed, it does in a few instances (Fig. 128). In

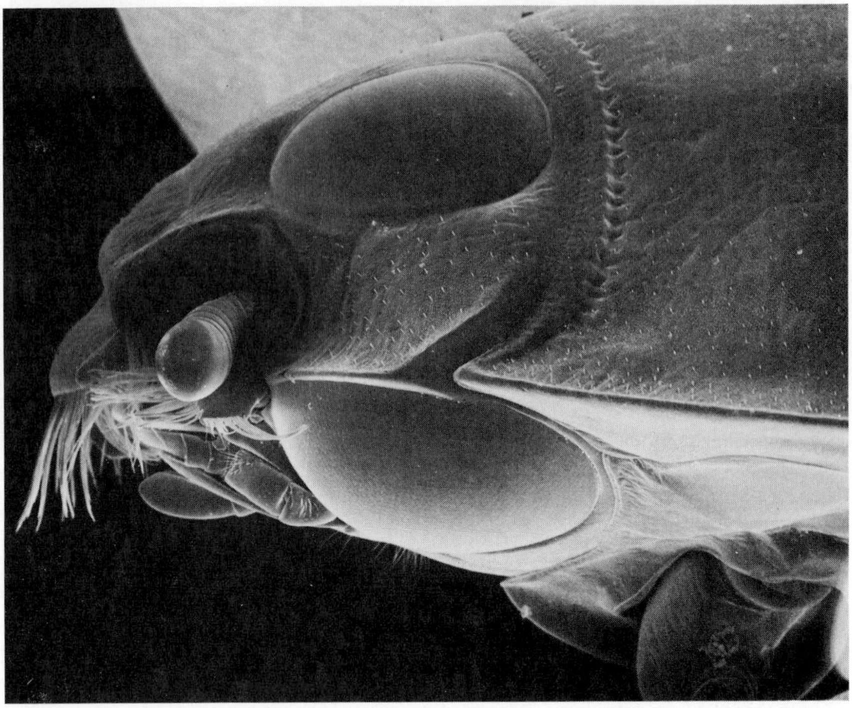

Figure 127. The adult whirligig beetle (*Dineutus americanus*) swims on the water surface. The eyes are separated into upper and lower sections to view above and below the water surface.

most insects, however, flattening is an adaptation that permits the insects to crawl into protected habitats to escape currents except for certain periods of feeding (Fig. 129). Avoiding direct exposure is certainly an efficient means of maintaining position in rapidly moving water.

Oxygen

Although high and uniform in the atmosphere, oxygen is relatively low (measured in parts per million) and varies markedly in aquatic habitats, primarily for five interrelated reasons.

1. Cold water has a higher saturation level than warm water (Fig. 130).
2. The oxygen level is high in the day but drops at night when plants cannot carry out photosynthesis.
3. The greater the distance from the surface, the lower the concentration when diffusion from the air is the main source of replacement.
4. Turbulent water adds oxygen at a more rapid rate than still water does.
5. The presence of high organic matter usually lowers concentrations through decay and respiration.

Figure 128. Ventral view of a water-penny beetle larva (*Psephenus herricki*) found in rapidly moving water. Currents passing dorsally over the body press the organism tightly to the rock substrate. Note the abdominal gills.

Figure 129. Some caddisfly larvae, such as this *Hydropsyche scalaris*, produce a netlike structure that filters organic material from the flowing water. This food is then grazed from the web.

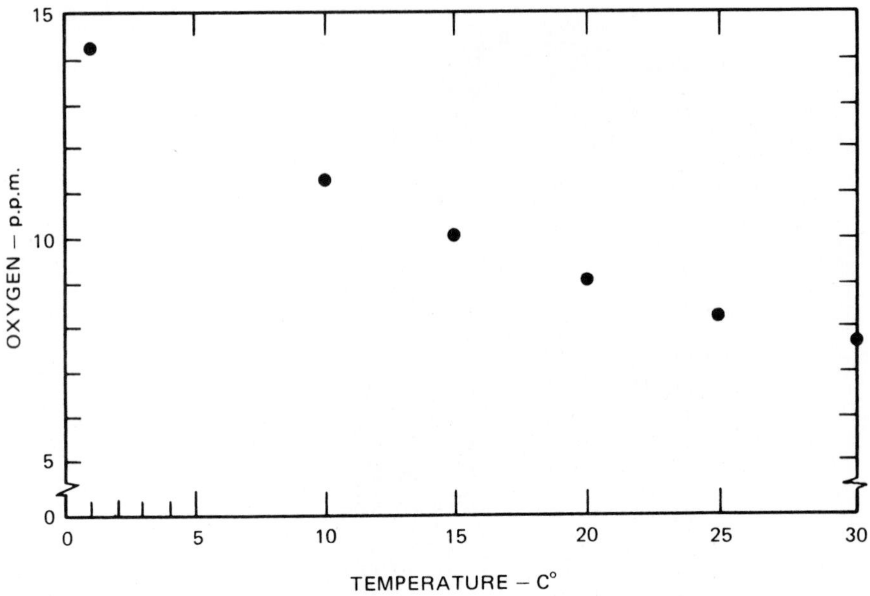

Figure 130. The maximal solubility of oxygen in water at various temperatures at sea level.

What effects do these characteristics of water have on insects? Species that are unable to survive lowered oxygen requirements are restricted in distribution, often to cool turbulent streams and temperate regions of the world. Adaptations for increased surface area (gills) to enhance oxygen exchange are common. Most immature insects are small with a large surface/volume ratio, have a closed tracheal system, and utilize a thin cuticle that has extensive networks of tracheoles and gills. Gills greatly increase the amount of surface area, hence, the insect can be much larger (volume) before the ratio becomes limiting. Gills vary from fingerlike outgrowths on the abdomen (caddisfly larvae) or thorax (stonefly nymphs) to modified appendages as in mayfly nymphs (Fig. 131). The rectal gills of dragonfly nymphs provide an example of internal gills.

Most aquatic adults and some immatures have open tracheal systems and come to the surface for oxygen. This gives the insect the best of both worlds because it can use the atmosphere's high oxygen concentration and yet not be subject to its drying effects. How does this atmospheric oxygen get into the body? A common adaptation, as seen in larval and pupal mosquitoes (Fig. 132), is the location of spiracles at the end of breathing tubes from which a waxy secretion is discharged to part the surface water molecules and admit air. The water-parting mechanism may be on the antenna, as in adult water scavenger beetles, which separates the surface water molecules and causes a funnel of air to extend back to the center of the body and to an air reservoir under the elytra into the abdominal spiracles. A thin film of air over part of the body surface or an air bubble extending through spiracles into the tracheae also may serve as a *plastron*, or artificial gill (Fig. 133). This relieves the insect from repeatedly coming to the surface for air. As long as some nitrogen remains in the bubble (it is slow to

Figure 131. Tracheal gills of the mayfly nymph, *Hexagenia limbata.*

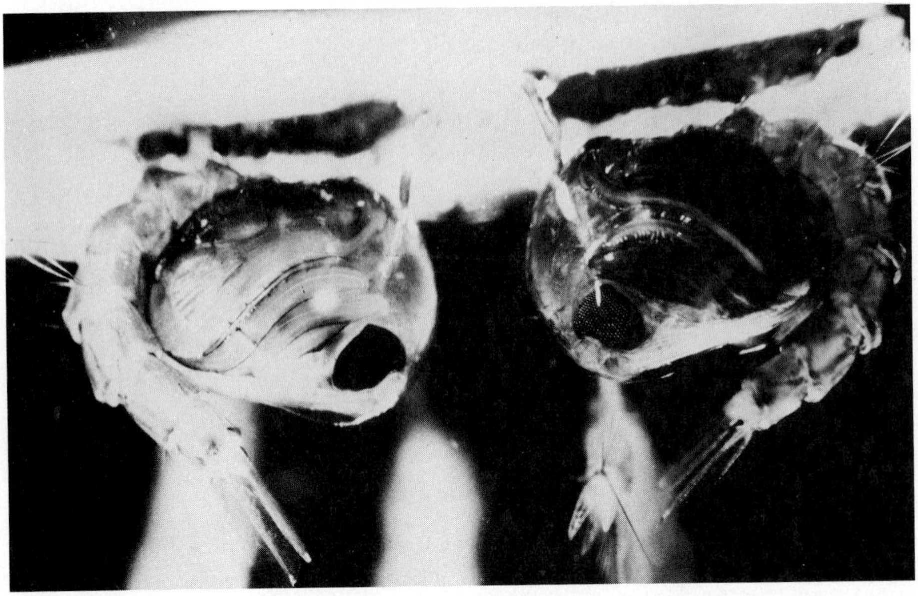

Figure 132. Mosquito pupae taking in atmospheric oxygen at the surface of water through thoracic breathing tubes. Mosquito pupae, contrary to the general pattern, are not sedentary but move about in a tumbling manner by moving the terminal paddles of the abdomen.

dissolve and leave) and if the dissolved oxygen is higher than in the tissues of the insect, oxygen diffuses into the plastron from the water to replace some of that utilized within the insect body. Another interesting example of oxygen-uptake mechanisms is in the mosquito genus *Mansonia.* These mosquito larvae have sharply pointed breathing tubes that pierce submerged plant stems and obtain oxygen from plant tissues to supplement any oxygen obtained through cuticular uptake from the water.

Salts

Dissolved salts influence both chemical equilibria and biological activity. With each increase comes higher water density and reduction in solubility to oxygen. Of far greater influence, however, is the effect these molecules have on osmotic pressures of cells. When salt concentrations are lower outside than within the body, as normally exists in fresh water, the body is *hyperosmotic,* and water and salts diffuse into the body, diluting blood and causing serious osmoregulatory problems. The reverse is true for salty or brackish water in which cellular fluid is *hyposmotic* and water moves from the body. These two situations are contrasted in Figure 134.

Several adaptations are present to handle problems of salt concentration. An insect that has hyperosmotic blood excretes water and ammonia to maintain optimal osmotic pressures. Hyposmotic individuals are similar to terrestrial forms and have cu-

Figure 133. Whirligig beetle (*Dineutus americanus*) with bubble serving as a plastron at the posterior end of the body. Although most of its adult time is spent on the surface film, this beetle can dive and swim beneath the surface using the bubble as an air source.

ticles that restrict salt absorption and water loss and utilize uric acid to decrease water lost through excretion.

Aquatic Representatives

Although there is a vast array of insect species, few are aquatic, and these few are secondary invaders having evolved from terrestrial ancestors. Table 2 summarizes the orders of insects and the stages that are often classified as aquatic. Most insects are aquatic only during their immature stages (Fig. 135); only the Hemiptera and Coleoptera have adults that normally inhabit water, and these return to the surface for oxygen. A study of Table 2, Figures 135 and 157, and Chapter 13 will be useful in understanding these interesting arthropods. An entire spectrum of diet can be found, and preferred habitats vary greatly. Feeding categories include shredders (large particulate detritivores), collectors (small particulate detritus filterers and gatherers), scrapers (periphyton grazers), piercers (herbivores and carnivores), engulfers (carnivores), and parasites. Habit categories include clingers, climbers, sprawlers, burrowers, and swimmers. Some insects select running water such as streams, others stagnant situations, and so forth. The major aquatic habitats where these insects reside will now be discussed.

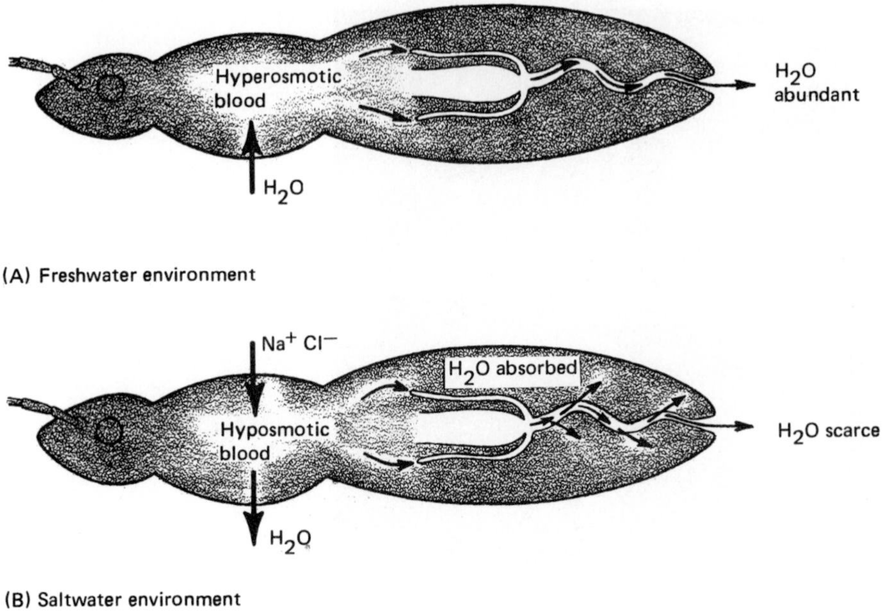

Figure 134. Comparison of osmoregulation strategies. (A) freshwater environment where body water is in excess; (B) saltwater environment where body water must be conserved.

TABLE 2. Orders of Insects with Aquatic Representatives

Order	Percent Species Aquatic	Aquatic Stage	Habitat or Activity	Diet
Collembola	1	All stages	Above surface film	Unknown
Odonata	>99	Nymph	Bottom, climbers, burrowers	Predators
Orthoptera	<1	All stages	Above surface film	Herbivores (emergent plants)
Ephemeroptera	100	Nymph	Bottom, rocks, burrowers, clingers	Herbivores and scavengers, some carnivores
Plecoptera	100	Nymph	Bottom, rocks	Herbivores, scavengers, or carnivores
Hemiptera	10	Nymph, adult	Above or below surface film, climbers	Carnivores, some scavengers
Megaloptera	100	Larva	Rocks, bottom, climbers	Carnivores
Coleoptera	2	Larva, adult	Above or below surface film, bottom, climbers	Carnivores, some herbivores and scavengers
Hymenoptera	<1	Larva	Within insects	Parasites
Lepidoptera	<1	Larva	Rocks, climbers	Herbivores
Diptera	<50	Larva	Below surface film, bottom, burrowers	Herbivores and scavengers, some carnivores
Trichoptera	100	Larva	Rocks, bottom, climbers, few swimmers	Omnivores and herbivores, some carnivores

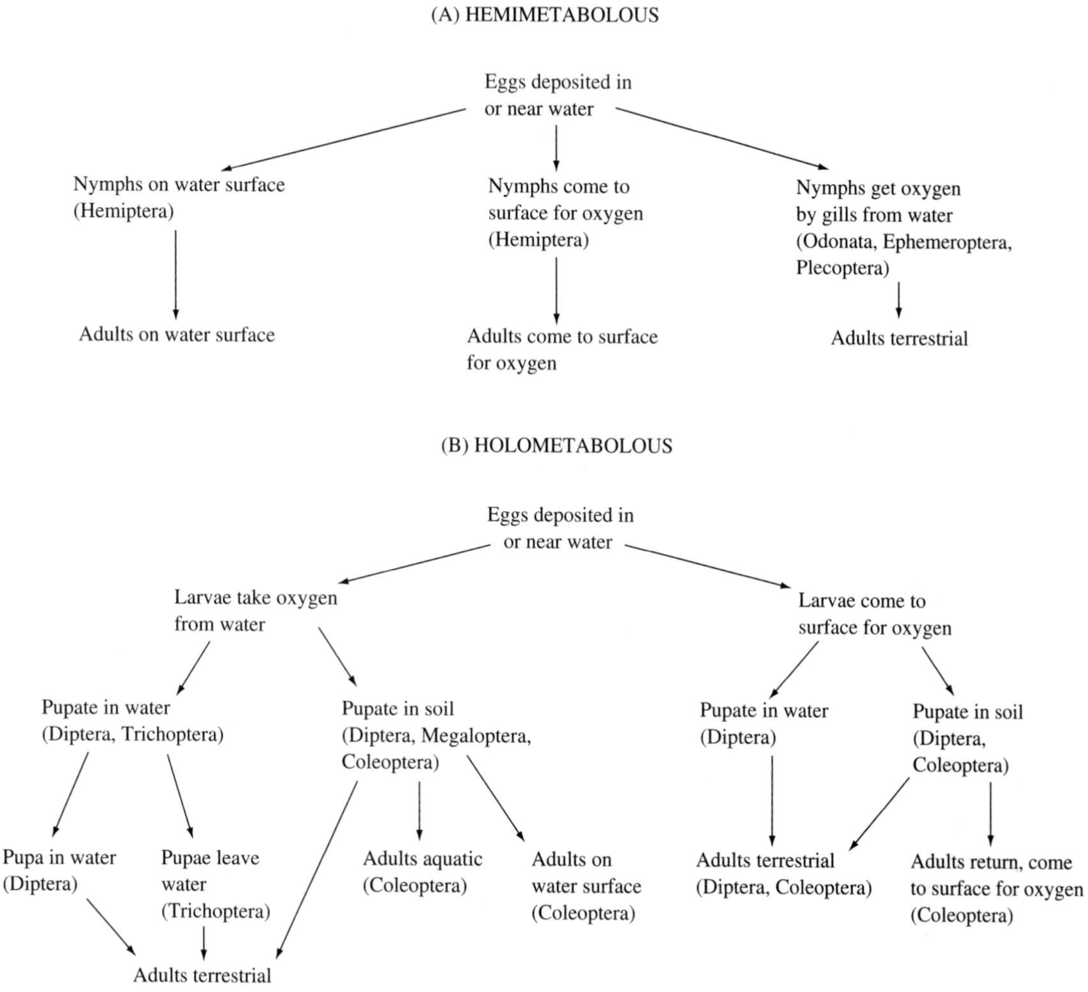

Figure 135. Development of common aquatic insects with emphasis on means of obtaining oxygen.

Aquatic Habitats

Streams and rivers have the following features:

1. Currents that are directionally one-way.
2. Differences in flow rate that are determined by topographic slope.
3. Chemical variations that are usually the result of the substrate over which the water traverses.
4. Physical peculiarities that occur in width, depth, and bottom features.

Zone	No. orders	No. families	No. families abundant	No. restricted to zone
1	8	15	1	5
2	8	11	6	4
3	3	3	3	0
4	6	11	2	2

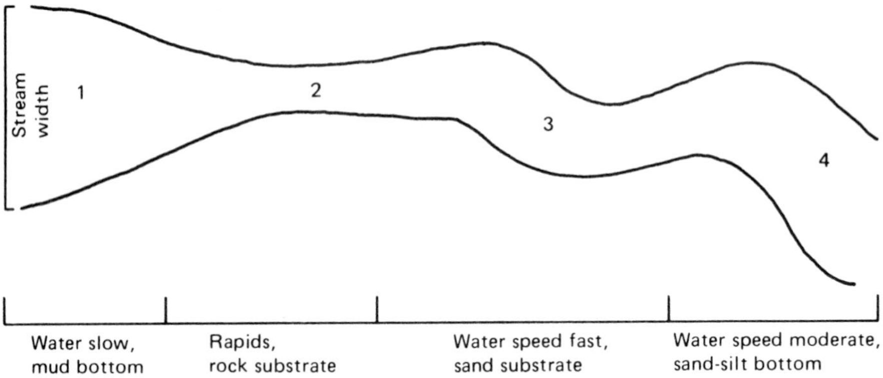

Figure 136. Characteristics of a stream in Kansas between February and April as recorded over a 10-year period. Distance represents about 200 m. Note differences in insect taxa and abundance, and that distribution was often restrictive.

Although oxygen is relatively uniform when minimal stagnation and freezing take place, streams and rivers become hazardous habitats during spring floods. Abrasive action of suspended silt results in high insect mortality, and many of the survivors are washed downstream to less favorable surroundings. Consequently, species living in streams commonly have adults that fly upstream prior to oviposition. These upstream migrations tend to nullify drift by the aquatic immatures. Within a distance of several hundred meters, the bottom, water velocity, stream width, and many other physical differences may be found, each having specific effects on oviposition and on insect distribution and abundance. Figure 136 illustrates such a situation in a small stream in Kansas as seen prior to spring flooding when some of the habitat preferences become obscured. Decided differences are observable, both in species number and where maximal abundance occurs. Some insects, such as black fly larvae and stonefly nymphs, are normally found in rapids except after flood translocation. Others, such as giant water bugs (Fig. 137) and dragonfly nymphs (Fig. 138), prefer slow-moving water. Also, narrow streams (20 m or less in width) support about four times the bottom organisms as streams that are 60 to 70 m wide.

Although some species, such as water scavenger beetles, leave and overwinter in the soil, most aquatic insects remain active during cold temperate winters. Because a decrease of 1°C lowers insect metabolism approximately 10 times, the same amount

Figure 137. Male giant water bug (*Belostoma* sp.) with eggs attached to wings. Such paternal care is rare in insects.

Figure 138. This dragonfly nymph (*Anax* sp.) captures prey by means of its greatly enlarged labium. During most periods the labium is held in repose under the head; however, it can be quickly extended forward to capture its prey as they come within striking distance. Once captured, the prey is drawn back to the other mouthparts for chewing and ingestion.

Figure 139. Dobsonfly (*Corydalus cornutus*) pupa in earthen cell under rock near stream bank. Larva migrated from the aquatic environment prior to metamorphosis.

of food at cool temperatures will sustain a higher population. Therefore, supplied by ample food from autumn leaves, herbivores and scavengers, such as mayflies and caddisflies, accomplish major growth during the winter when cold makes many fish predators sluggish or inactive. Some aquatic immatures require several years to complete development (Fig. 139).

Ponds are small bodies of water characterized by:

1. Shallow depth.
2. Great variation in temperatures.
3. Early freezing and thawing.
4. Much diversity in dissolved gases.
5. Common stagnation.

Insects play vital roles in ponds, and large populations of many diverse taxonomic groups may be seen there. Numerous Diptera larvae, Trichoptera larvae, and Ephemeroptera nymphs are herbivores and scavengers, and carnivores include Odonata and Hemiptera nymphs and Coleoptera larvae. Most pond insects have a short development and pass, or are able to pass, significant portions of their life cycle either out of water or in the basin if the pond dries up. As ponds fill up with vegetation, they become bogs. The dominant herbivores in such bogs are often insects; however, insect detritivores have little impact in this ecosystem.

Lakes that are relatively large in volume have unique characteristics. As in streams, substrate determines much of the water chemistry, but the water depth tends to restrict both plant and insect population distribution; few insect species, mainly Diptera (sometimes more than 90 percent of total insect numbers), can survive deep water, and hence most insects are found in the shallow areas of the shore line.

Oceans are the largest aquatic habitat, but few insects have successfully utilized this resource, and of those that have, nearly all are surface dwellers. Many theories have been proposed to explain this failure, such as competition with Crustacea, dependence on air by the adult stage, and many chemical characteristics of marine water, but none of these are satisfactory explanations. Only five species of water striders, mainly in the genus *Halobates,* live in open water; these wingless gerrids have been found great distances from land and apparently feed on windblown insects. Other truly marine insects include mainly hemipteran species in the families Veliidae, Mesoveliidae, and Hermatobatidae, all of which are also surface inhabitants but are located near shore in estuaries or rock pools (Cheng, 1976).

Numerous habitats exist in brackish water with the number of species increasing with higher percentages of fresh water. Here changes in salinity and temperature must be tolerated, and the insects associated with these areas are mainly larval Diptera (mosquito and horse fly), Coleoptera, and Hemiptera. Emergent plants, such as *Spartina* sp., support certain grasshoppers and planthoppers. Intertidal regions are also harsh, and insect life is restricted to a few species living in or under rocks and crevices, mainly larval Diptera, Coleoptera, and Hemiptera.

TERRESTRIAL ENVIRONMENT

Water

Water exists in the terrestrial region of the earth in an equilibrium of solid, liquid, and gaseous states. Below 0°C, water is a solid; above 100°C, it exists as a gas. Evaporation can take place directly from either a solid or a liquid, the rate being proportional to the temperature, surface area, and concentration of water already in the air. The amount of water present when compared with the carrying capacity of air at a given temperature is termed *relative humidity.* High temperatures permit more water to remain as a gas than do low temperatures; therefore, air that has a relative humidity of 70 percent at 40°C has more water than air that has the same relative humidity at 20°C. When temperatures drop, relative humidity increases; if the saturation or dew point is reached, water condenses to form dew, rain, or snow, depending on the temperature and other existing environmental factors. Precipitation is not equally distributed in amount or time over the earth but varies with directional winds, their interaction, amount of moisture in the air, and presence of topographical features such as mountains. This variability, coupled with soil types, results in distinct habitats that influence the distribution of both plants and animals.

Because of the evaporation properties of air, water loss from insects is normal and takes place through four major routes: (1) transpiration from body surfaces, (2) loss through breathing, (3) excretory processes, and (4) defecation. Each of these routes has been discussed previously (Fig. 73), but a review here may be helpful. Transpiration from the external body surface is greatly reduced by the epicuticle. Spiracular valves reduce water loss through the respiratory system; nevertheless, transpiration accounts for approximately 60 percent of the total loss. Excretion of nitrogenous wastes requires water, but this fluid passes into the proctodeum where it is reabsorbed. Immediate restoration of lost water comes from the hemolymph where concentrations are maintained primarily by ingestion of free water, breakdown of food to release cellular fluids, and oxidation of food to produce metabolic water. A few African beetles in the coastal regions of the Namib Desert have an unusual means of trapping water.

These beetles either construct trenches perpendicular to fog winds or face the fog in a head-down position to trap and ingest water as it condenses into trenches or trickles down the head to the mouthparts (Seely, 1976; Seely and Hamilton, 1976).

Behavior is vital in water conservation. High temperatures and wind velocities must be avoided whenever possible by seeking refuge and restricting activity. If drought conditions are normal in the ecosystem, prolonged inactivity or *estivation* often preserves body water.

A nocturnal activity pattern aids in restricting water loss because temperatures are usually lower and the relative humidity higher than in daytime hours. Another less obvious effect of water is its mechanical impact upon insects. Because of their small size, insects can easily be trapped by rain droplets (Fig. 7) that have little impact upon larger animals. Many eggs also have *aeropyles,* which permit survival when submerged in water, either by flooding or by rain.

In addition to its impact on the individual, water has a vital impact on the reproduction potential of a given species. Evolution has been toward direct copulation where sperm may be transferred directly via liquids without contacting dry air. It is not unusual, therefore, that primitive insects are found in humid surroundings, whereas the more specialized species have been able to disassociate themselves from this requirement.

Temperature

From 80 percent to 90 percent of solar radiation, the primary source of earthly energy, passes through the atmosphere and is absorbed by the earth's surface. Much of this energy is radiated as heat, but transfer is slowed by the insulation of air, particularly by carbon dioxide and ozone. Daytime temperatures are high because more energy is received than is radiated, but the opposite is true at night.

Many animals, including most insects, depend on radiation to maintain body temperature for normal metabolism, but they often must avoid too high an exposure (Fig. 140). Some scientists refer to such animals as *poikilotherms,* meaning cold-blooded, but most scientists refer to them as *ectotherms,* meaning that they obtain heat from outside the body. Ectotherm is a better description here because the body temperatures of insects are often from 10° to 20° higher than ambient temperatures. Insects raise their internal temperatures by actively basking in direct sunlight (Fig. 141), by physiological activity, or by seeking warm sites. Conversely, they may stay cool or lower their body temperatures and also water loss by avoiding activity or molts during midday temperatures and/or by moving to and resting in shade and on cool substrates. Heat gain or loss is rapid because of insects' small size and can be lethal (Fig. 142). In the case of hairy caterpillars, the hair does not affect the rate of radiation heat received, as might be expected, but it does reduce the rate of heat loss (Casey and Hegel, 1981). In most species found in the tropics and subtropics where extreme cold is rare, sunning behavior and activity (Fig. 141) can produce the necessary warming, and the bases of wings often play an important role because of their pigmentation (Douglas, 1981).

At the other extreme of temperature, temperate and polar regions have great daily variation and freezing temperatures, often for extended periods, and many specialized adaptations become necessary for survival. *Homeostasis,* the equilibrating ability,

Figure 140. Short-horned grasshopper (*Melanoplus differentialis*) that has moved up the vegetation into a light wind to escape some of the summer heat.

Figure 141. Skipper (*Hesperiidae*) basking to raise body temperature for flight.

Figure 142. American cockroach (*Periplaneta americana*) unable to complete molt because of dry air and failure of ecdysial suture to fracture.

maintains internal temperatures higher than ambient temperatures in mammals and birds, but such activity would be detrimental in insects, for individuals the size of a mosquito would need to expend approximately 90 percent or more of their metabolic energy solely for heat. Nevertheless, there are some *endothermic* (raising body temperatures by basal metabolism) examples among five different orders of insects (all large insects), but only during limited periods of activity. Some scientists substitute the term *heterothermic* for these examples, indicating that endothermy is restricted to the periodic activity periods, while the insect becomes ectothermic during nonactive periods. Sphynx moths require body temperature near 36°C for flight and achieve this during cool weather by vibrating their thoracic muscles until sufficient heat is produced for flight. Bumblebees undergo similar behavior both before flight and during visits at flowers. Scales (in moths) and plumous setae (in bees) aid in slowing down heat loss through insulation. Nevertheless, because maintaining body temperature higher than the ambient temperature is a tremendous drain on energy (about 9 times that of ectotherms), bees must visit many flowers to replenish their energy supply.

What about insects living in temperate regions where freezing occurs regularly? Insects can move to protected environments by vertical or horizontal migrations to such insulated habitats as soil or wood. They also can restrict activity or enter a facultative diapause, providing the freezing is not intense or of long duration. However, for species that inhabit extremely cold regions over long periods of time, cold resistance during some stage is vital. This is achieved in one of two ways. First, and by far the most common, insects can avoid freezing by *supercooling* (the ability to stay in an unfrozen state below the melting point of water, often as low as from −20° to −60°C in some Alaskan species). Cryoprotective substances such as certain alcohols and sugars aid the process. The second way, uncommon among arthropods, is to permit freezing but under controlled biochemical methods that keeps ice growth slow and restricted to the extracellular spaces while cellular cytoplasm remains liquid.

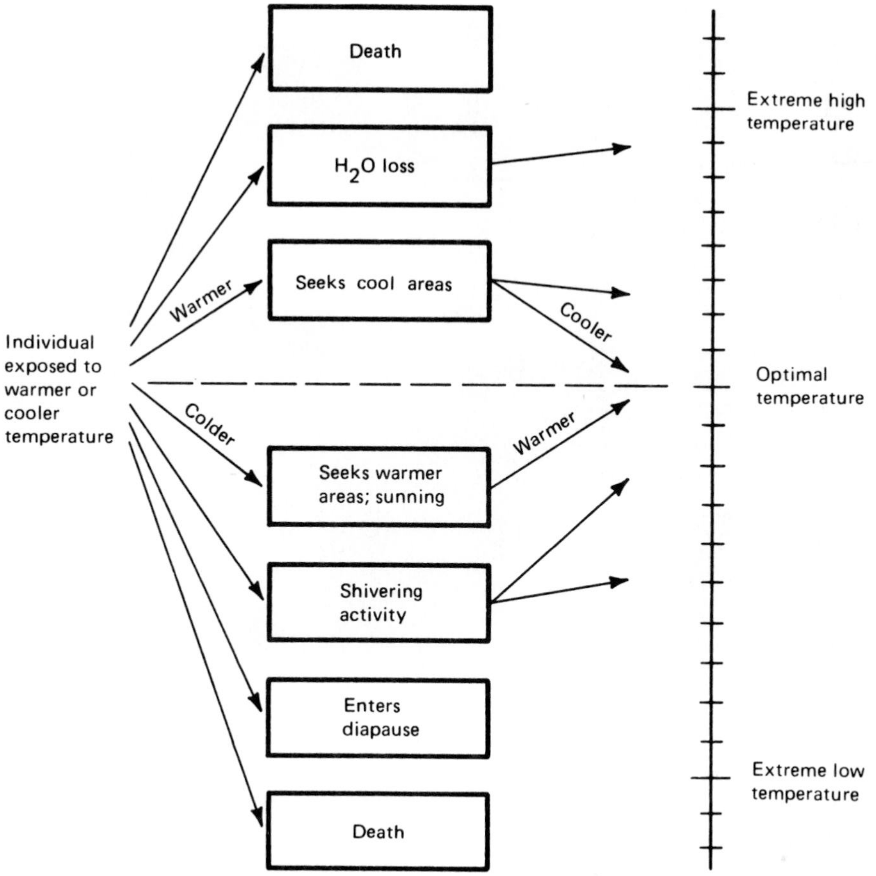

Figure 143. Some of the major homeostatic or feedback mechanisms made in response to environmental temperature. Behavior is the major means of maintaining the body near the optimal temperature for insects.

Species that inhabit hot regions face different problems. High temperatures produce increased metabolic rates, and food reserves are more rapidly depleted than at moderate temperatures. More serious are the tolerance limits that, when exceeded, result in death. Insects normally avoid upper limits by seeking shade, by orienting bodies so that exposure to direct sunlight is minimal, and by restricting activity to moderate temperatures (Fig. 143). In deserts, most activity occurs during the cool early morning or evening hours. Here, bees commonly collect pollen and nectar as the sky lightens before sunrise, and they cease activity several hours later because temperatures rise rapidly.

In summary, temperature is regulated both daily and yearly by these ectothermic invertebrates through a combination of (1) microhabitat selection, (2) structures such as wings, color, and body orientation, (3) physiological activity such as flight and shivering, (4) orientation of bodies to regulate the amount of solar energy received, (5) activity restricted to optimal temperature periods, and (6) periods of diapause.

Figure 144. Long-horned wood borer and burrow. This insect is one of the few insects capable of producing its own cellulase for digesting cellulose.

Gravity

Because air is approximately 700 to 1,000 times less dense than water, animals that inhabit land have problems of support. Because insects are small, their exoskeletons need not be extensive for muscles to operate effectively. Most of the adaptations to gravity have been covered in previous chapters.

Specialized Habitats

Some insects have been able to avoid the rigors of the terrestrial environment by invading plants, both living and dead. Long-horned wood borers (Fig. 144), for example, live from 2 to 3 years as larvae inside trees, literally eating their way through the wood. The environment within the resulting burrows is relatively stable, and interspecific competition is reduced. The physical state of the wood is important, as can be seen in the gradation of species' preferences ranging from live trees to nearly decayed logs. Also within a given log or tree, the depth of feeding and the specific tissues fed upon are important; for example, some borers prefer cambium and others prefer xylem or pith. When pupation is near, the last instar larva positions itself near the wood surface before undergoing metamorphosis. After emergence, the adult chews the remaining short distance to freedom, if the larva did not make an exit.

Figure 145. Leaf miner on *Phryma* sp. As the miner grows, the width of the tunnel enlarges.

The most common borers and miners (those that bore in leaves) are members of the orders Coleoptera, Diptera, Lepidoptera, Hymenoptera, and Isoptera. With the exception of termites and ants, borers and miners are larvae (Figs. 145, 146). Many of these larvae are legless, have reduced eyes and antennae, and possess well-sclerotized prognathic mouthparts for chewing. The borer literally chews its way forward. Wood passing through the digestive tract may be digested by enzymes produced by the insect itself, but in most instances protozoa or bacteria digest the cellulose, and the insect utilizes the metabolic products liberated by these symbiotic organisms (Fig. 147). In some insects much of the nourishment is not derived from the wood but from digesting fungi growing in it (Batra and Batra, 1967). Carnivorous insects also may be found in burrows feeding on borers (Fig. 148).

Feeding by most herbivorous insects causes few changes in the host plant other than destructive losses of parts. In contrast, some *monophagous* (feeding on one species of plant) and *oligophagous* (feeding on several plant species) insects evoke abnormal tissue development by their feeding or through stinging the plant. These tumorlike growths, or *galls,* form specific shapes and sizes, one of which is seen in Figure 149. More than 2,000 species of insects, including the orders Diptera, Lepidoptera, Hymenoptera, Thysanoptera, Coleoptera, and Homoptera (Fig. 150), in the United States induce galls. Galls may be found on the leaf, at the terminal end of a twig, or in stems or roots. Those species such as Homoptera that have piercing-sucking mouthparts cannot bore out of the typical or *closed galls* and hence are found in *open galls.* Advantages gained by insects in the system include a uniform environment with much food, little water loss, and some limited predator exclusion.

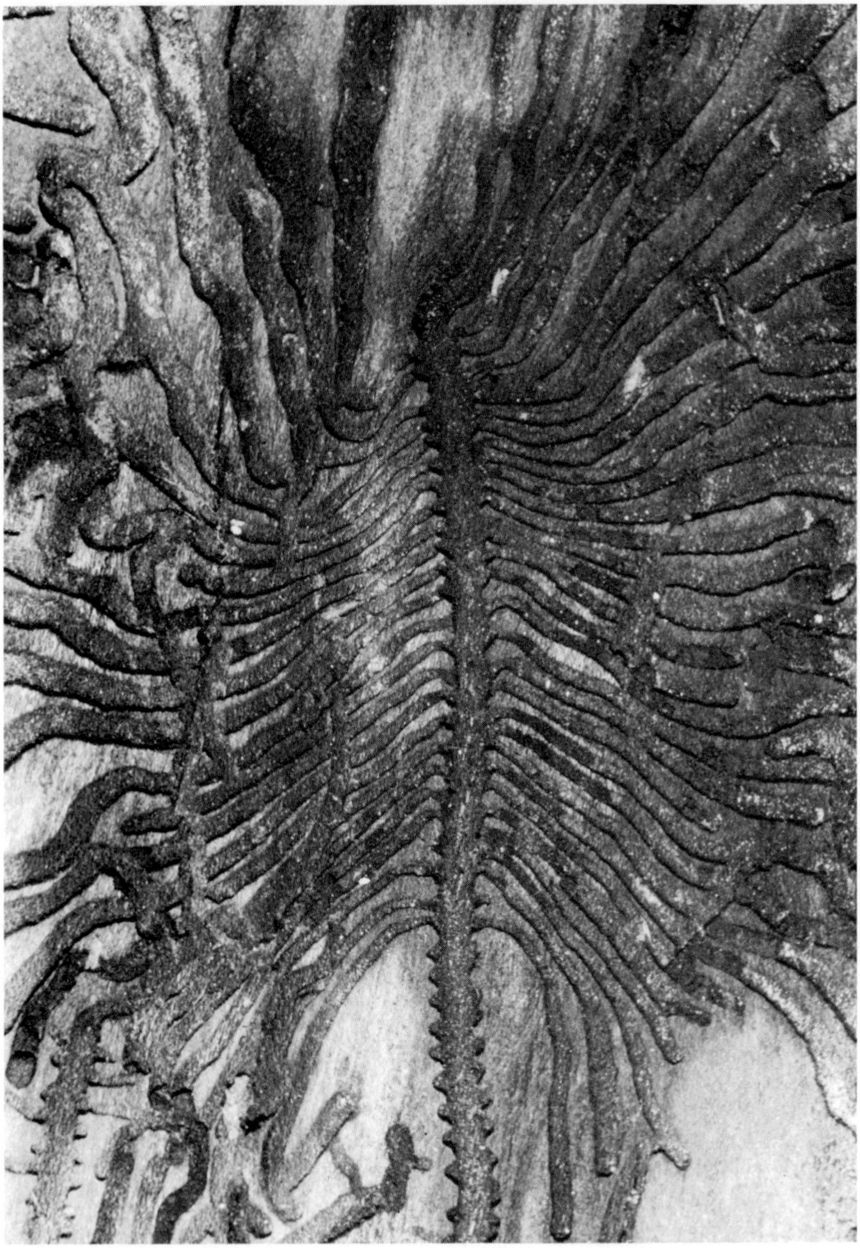

Figure 146. Galleries formed under the bark of elm by bark beetles (*Scolytus multistriatus*). The main central gallery was made by the female as she deposited her eggs. Galleries radiating out from the central gallery were produced by the developing larvae as they fed.

Figure 147. The presence of borers may sometimes be determined by fecal material or frass pushed from the burrow as seen on this sunflower.

Figure 148. Some of the insects found within tunnels in wood are predaceous, as is the case with this click beetle larva (*Alaus oculatus*).

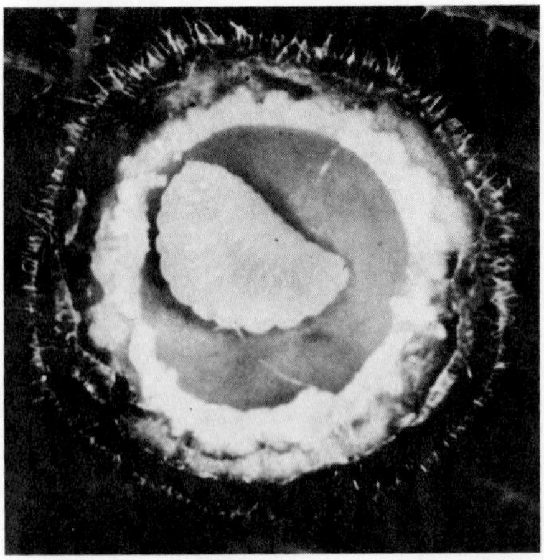

Figure 149. A gall opened up to expose the gall midge that induced its formation.

Figure 150. Spruce galls formed by aphids (*Adelges cooleyi*) feeding.

A most unusual habitat is the frothy mass of spittle in which spittlebugs reside (Fig. 151). This slimy material is secreted from Malpighian tubules via the anus, and air bubbles are forced into the liquid from the tracheal system. Apparently this material provides protection for the nymph as it feeds upon sap from the host plant.

Another interesting means of modifying the terrestrial environment may be seen in insects that roll or fold leaves (Fig. 152). Silk is spun across the leaf in parallel strands that shorten upon drying. The process is continued until the leaf is either folded or rolled to form a closed retreat. Some insects feed upon the leaf from within, whereas others venture out to feed. The most common species that roll leaves are Lepidoptera. Other Lepidoptera spin silk and enclose many leaves (Fig. 153).

A further modification of leaf utilization may be seen in bagworms (Fig. 154), in which the moth caterpillar incorporates leaves into a bag of silk. This enclosure can be moved about as the larva feeds. During periods of inactivity the bag is securely fastened by silk to a branch, and the opening is sealed for protection. Pupation and

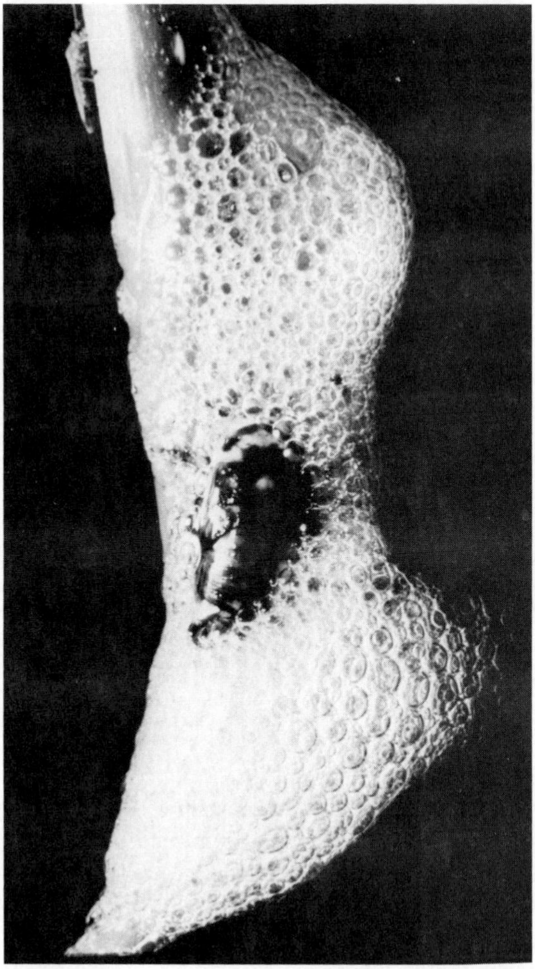

Figure 151. Spittlebug within the frothy secretions it produces for protection.

deposition of eggs by the wingless females also take place within the silken retreat, and only the winged males leave these protective retreats.

Burrows in the soil offer another variation in specialized habitats. Larval tiger beetles, for example, dig vertical burrows in sandy soil where they wait for prey to venture within range of their massive mandibles. Removing these larvae is difficult because the spines on their fifth abdominal segments [Fig. 357(A)] anchor them securely to the soil. Other insects such as the solitary bees and wasps and the dung beetles also utilize burrows in the soil, but it is the adult that digs. Female mining or sweat bees often dig deep burrows along the banks of streams, each of which has several lateral cells. The design varies with the species. Each cell is provisioned with sufficient honey and pollen to enable the single larva within to complete development. Pupation occurs also in this environment, after which the emerging adults seek their own burrowing sites.

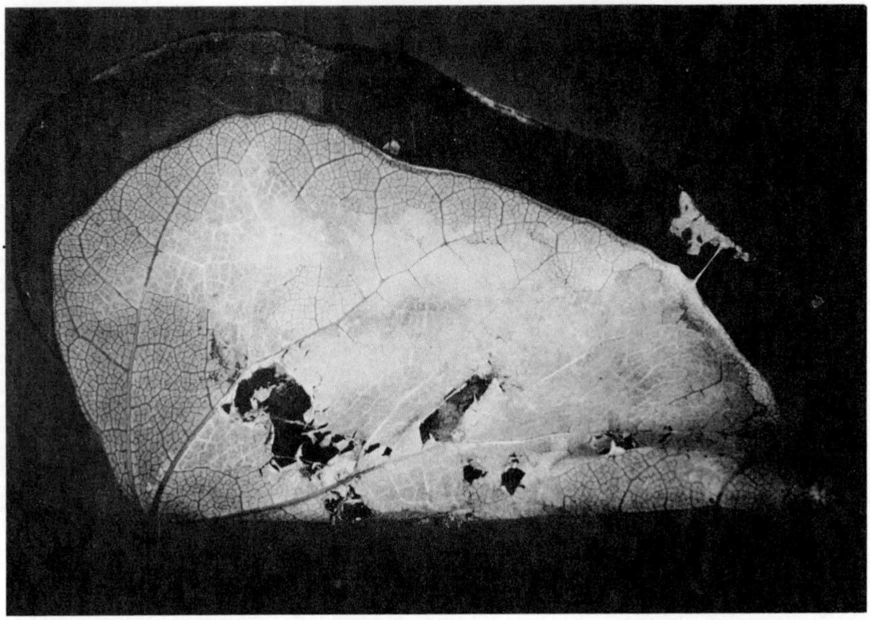

Figure 152. The redbud leaf roller (*Fascista cercerisella*) produces a moderated microhabitat by folding over a leaf, which also is used for food. It may also be a means of avoiding light, which induces toxic chemicals in the plant (Maugh, 1982).

Primitive nests are often manufactured by solitary wasps such as the *Trypoxylon* species (Fig. 202). Each individual cell is constructed from mud with subsequent cells added linearly. Prior to the departure of the female wasp, cells are mass-provisioned with spiders that serve as food for the developing larvae.

Certainly the acme of insects *directly* modifying the climate for habitation would have to be the social insects, particularly the fungus-eating termites and the social bees. In the termite genera *Macrotermes, Amitermes,* and *Nasutitermes,* we see the microclimate greatly modified within the highly specialized nests or *termitaria.* The *Macrotermes'* nests in Africa often reach 6 m high and 30 m in diameter and are constructed by gluing soil particles with saliva and excreta.

Lüscher (1961a) discussed the architecture of these fortresses, noting the air-circulation mechanisms and the temperature and humidity controls that maintain an internal temperature that varies less than 6°C, and a carbon dioxide concentration (a significant problem because of the respiration of the 2 million termites and the fermentation of grass by fungi used for food) that fails to reach critical levels due to the intricate air-circulation system. Air is warmed through released heat by the metabolic breakdown of grasses by fungi near the central core. This heated air rises to the "attic" and then spreads laterally into wall channels where it cools and descends. *Amitermes* species in Australia produce termitaria that are laterally flattened with the narrow profile facing the noon sun, thereby reducing solar radiation uptake during the heat of the day. Humidity often is maintained by burrowing deep into the soil, up to 40 m, to reach the water table. In contrast to these surface nests (Fig. 213), *Nasutitermes* species produce arborial nests

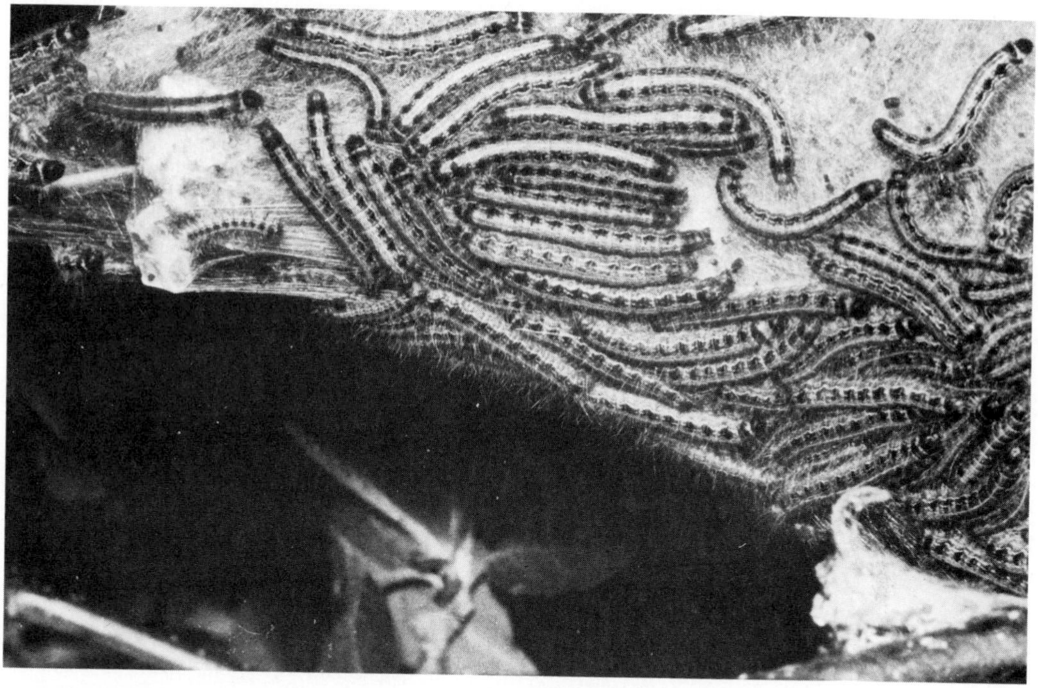

Figure 153. These eastern tent caterpillars (*Malacasoma americanum*) form an extensive webbing on plants, from which they gain protection as they feed.

Figure 154. A bagworm larva (*Thyridopteryx ephemeraeformis*) feeding on juniper. As the larva increases in size, the bag is also enlarged by attaching fresh leaves with silk.

(Fig. 212) from *carton,* a substance made by mixing masticated wood, soil, excreta, and saliva. In most instances, termitaria of any of the above types are enlarged just after a rain when the high humidity reduces transpiration, but repairs are normally completed soon after damage is incurred no matter what the environmental hazards are.

Honeybees nest naturally in hollow trees, but humans have, in the past 100 or so years, manufactured various types of artificial hives for ease in obtaining honey (Figs. 268, 269). Within these structures, climate can be modified somewhat but not as efficiently as in the termitaria. Humidity is maintained at about 35 to 45 percent. Temperatures vary with latitude and whether or not shade is available but can be cooled somewhat by workers fanning their wings (Fig. 216) and by evaporation of water that the bees spread over the cells during extremely hot weather. Further discussions on these and other social insects and their nest modifications may be found in later chapters.

POPULATION DYNAMICS

Gene Populations

Each species represents the summation of all genes present within its populations, or its *gene pool. Variation,* the different expression of these genes in individuals, results from mutations in gametes and their subsequent incorporation and expression through matings and genetic recombination. When the average amounts of variation (percent of gene loci that are heterozygous) are studied, insects have about 15 percent of the loci heterozygous; this is about three times the variation in mammals and birds.

Before proceeding further, it is best to understand what occurs to a population of genes if matings are at random (*panmictic population*) and if external pressures are absent. Let us view a hypothetical population of diploid individuals that has 30 percent of the gene pool with a dominant gene *A* and the remainder, or 70 percent, with its recessive allele *a* (recessive genes are often the most common). If these genes segregated independently into gametes and if matings occurred,

(1) $$p(A \text{ sperms}) + q(a \text{ sperms}) \times p(A \text{ ova}) + q(a \text{ ova})$$

(2) $$0.30(A \text{ sperms}) + 0.70(a \text{ sperms}) \times 0.30(A \text{ ova}) + 0.70(a \text{ ova})$$

then the following Mendelian genotypes and frequencies would be expected in the offspring:

(3)

		Sperm *A* (0.30)	Sperm *a* (0.70)
Eggs	*A* (0.30)	*AA* (0.09)	*Aa* (0.21)
Eggs	*a* (0.70)	*Aa* (0.21)	*aa* (0.49)

Genotypes	*Gene Frequencies*
$AA = 0.09$	$A = 0.09(AA) + 1/20.42(Aa) = 0.30$ or 30%
$Aa = 0.42$	
$aa = 0.49$	$a = 0.49(aa) + 1/20.42(Aa) = 0.70$ or 70%

Note that under these stable conditions, gene frequencies remained constant at 30 percent and 70 percent. Therefore, the genotypes and frequency, as summarized from all possible matings between genotypes, may be expressed as:

$$p^2AA{:}2pqAa{:}q^2aa = 1 \qquad (4)$$

where p = frequency of A
q = frequency of a

This formula, often referred to as the *Hardy-Weinberg formula,* defines a population under stable equilibrium conditions.

However, several factors act upon populations that tend to place pressures on this stability.

1. *Mutations.* Changes in DNA produce different expressions of a gene from the "wild type," each of which is termed an *allele.* Most alleles are recessive, such as *a,* and remain hidden in heterozygous individuals. They are expressed when homozygous and are usually deleterious because they make for less fitness.
2. *Migration.* Populations are not exclusive, and individuals that have more *A* genes may enter (immigration) or leave (emigration) more frequently than those that have *a.*
3. *Natural selection.* Not all individuals contribute genes equally to the next generation because some individuals are more fit to meet the environment and to reproduce more successfully than others.
4. *Genetic drift.* In small populations, allele frequency may change because of random occurrences and may fluctuate toward the homozygous condition. Even mutations with selective value are often lost because of random forces.

Natural Selection

Although various portions of the principle were known previously, the concept of natural selection is credited to Charles Darwin. Natural selection is based on the premise that individuals within a population have differential survival and reproductive success, and this is heritable. There are many ways of describing the process, but the following three propositions should be sufficient to show that the frequencies of diploid genotypes change, which in turn alters the gene frequency in the gene pool.

1. Individuals within a population vary genetically because of mutations and especially recombinations. Only in parthenogenetic insect species do we find some exception to this statement.

2. Organisms have the potential to reproduce at a logarithmic rate.
3. Fitness differences exist, and those individuals best adapted to the local existing environment, both biotic and abiotic, survive and pass their genes on to their offspring. If, for example, individuals with the homozygous dominant condition of *AA* should become immediately and totally unfit when the population of *A* genes is 30 percent, then 9 percent of the potential gene population would be lost in the first generation, and the gene frequency of *A* would drop to 23 percent (see formula 5).

(5)

		Sperm *A* (0.30)	Sperm *a* (0.70)
Eggs	*A* (0.30)	~~*AA* (0.09)~~	*Aa* (0.21)
Eggs	*a* (0.70)	*Aa* (0.21)	*aa* (0.49)

Surviving Genotypes	*Surviving Gene Frequencies*
AA = 0	$A = 1/20.42(Aa) = 0.21$ or now 23%
$Aa = 0.42$	
$aa = 0.49$	$a = 0.49(aa) + 1/20.42(Aa) = 0.70$ or now 77%

The previous example, although helpful to visualize, is a drastic oversimplification of what actually happens to gene frequencies in a natural population. Natural selection involves survival of entire genotypes consisting of many thousands of genes and their complicated interactions rather than simple gene effects. Phenotypic results are often modified by other genes. Deleterious mutations, therefore, are often "masked" in diploid animals, such as insects, by dominant genes and by modified complexes of genes in differing environments. Also, genes that are of little present selective value may be carried in high frequency because of linkage with selected genes.

Stable populations of insects have the best frequency of genes, and the best adapted individuals have the most harmonious interaction of these genes for that given environment and time. Therefore, under short-term circumstances, mutations and their subsequent recombinations are mostly detrimental. However, if the environment changes or if emigration has occurred, slightly different genotypes or developmental expression of the same gene might have increased selective advantage. For instance, many English moths evolved from the abundant light form, one that blended with the lichen-covered bark of trees upon which they rested, to an abundance of a melanic morph or variant in response to pressures imposed by the deposition of soot upon the landscape from nearby factories (Kettlewell, 1959; Bishop and Cook, 1975). This example will be discussed in Chapter 6 on protective coloration.

Population Density

The number of individuals of a species per unit area of space is referred to as *population density*. Population size has a great impact on those genotypes and variations that survive. The size of a population is constantly changing. *Growth of a population* is the number of individuals added minus the number lost during a unit of time and can result from interaction of these factors:

1. An increase in number of births (*natality*).
2. An increase due to immigrations into the area.
3. An increase in the number of generations during the season.
4. A decrease in time to reproductive age.
5. A decrease in deaths (*mortality*).
6. A decrease in emigrations.

The rate of growth of a population at a given instant is calculated as follows:

$$\text{Growth rate} = \frac{dN}{dt} = rN \qquad (6)$$

where r = instantaneous rate of increase
N = number of individuals
t = time

As an increasing number of individuals initiate reproduction, the rate of increase does not remain uniform. Most populations start slowly, and then undergo a logarithmic increase as new individuals are added to the population. If one assumes that each individual survives and reproduces, fantastic populations can be produced in a short period of time. Daly, Doyen, and Ehrlich (1978) calculated that if 100 grain beetles were reared under optimal environmental conditions, more than 300 million progeny would be present after 20 weeks. Obviously this does not occur. As populations increase exponentially, the number of individuals that can be supported by the environment and by competition reaches a plateau, the *carrying capacity* or K. This can be inserted into the above formula to produce the following simplified equation:

$$\frac{dN}{dt} = rN\left(\frac{K - N}{K}\right) \qquad (7)$$

The resulting rate of increase per unit time produces a typical sigmoid growth curve, as shown on p. 168.

When carrying capacity K is reached, population growth stops because death and emigration rates equal birth and immigration rates. Either the population crashes or a "steady state" is established; some fluctuations are possible because natural populations and environmental conditions rarely remain truly stable. As limiting factors change, the size of the populations again increase or decrease, thus producing periods of fluctuations in density. Also, variations from the curve that follows occur when the age structure of the population changes.

Animals are often categorized as *K*-strategists or *r*-strategists. The *K*-strategist devotes the least energy to reproduction and is believed to expend most of its energy into areas of competition and to escape parasites and predators. Populations predominantly with these specialists, therefore, closely approach the carrying capacity (*K*) of the environment. In contrast, *r*-strategist populations are adapted for higher reproductive rates and, hence, more rapid population growth during habitat disturbance when little competition exists. Therefore, *K*-selective forces favor efficient competitive use of the environment, whereas the *r*-selection favors a "gambling" strategy. Although *K*-selection and *r*-selection appear to be polar entities, they are not equivalent terms because *r* can be expressed in terms of life history traits whereas *K* cannot. Nevertheless, the terms can be useful when comparing organisms and populations if applied properly.

What are the major factors that influence density? Previous discussions have indicated that climate has a great effect on population and may be referred to as *density-independent;* that is, changes occur whether or not the population size is 10 or 100. Many stabilizing factors, however, are *density-dependent,* and the number of individuals has a direct influence on population size and growth (see polymorphism in previous chapter). For insects, the major density-dependent factors are interrelated; these may be categorized as *food, number of preferred habitats,* and *predators and parasites.* For example, if a pair of fruit flies is placed into a bottle with food and optimal microclimatic conditions, reproduction occurs at a standard logarithmic rate. Eventually, the population growth slows down (carrying capacity is neared) because increases equal losses. In this instance, food has become a limiting factor at high densities, whereas it was not during periods of lower densities. This trend also can be seen in the competition between species. For example, when the blowfly population in a rabbit carcass is reduced, the flesh fly density increases sixfold (Denno and Cothran, 1976).

As one tries to generalize, it is important to remember that population numbers for most endemic species are the result of an interaction between density-dependent and density-independent factors, although *K*-strategists may be more influenced by the former and *r*-strategists by the latter.

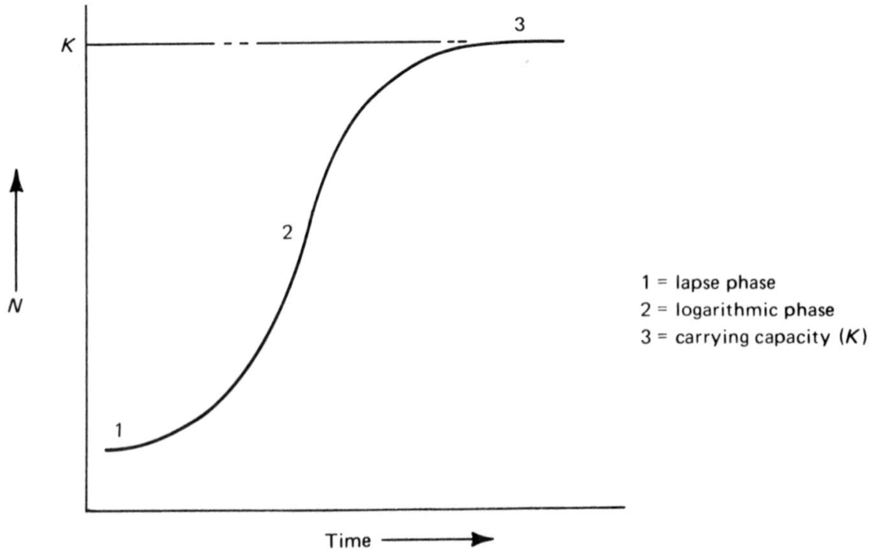

Communities and Energy Flow

Individuals in insect populations do not exist alone but are part of a complex of organismic interactions. All populations of plants and animals found in a given area comprise a *community.* Such assemblages are often named for the dominant plant or group of plants, such as Oak or Juniper. Boundaries between communities are often ill-defined and, hence, arbitrary.

Plants play a dominant role in communities, as in any ecosystem, because they convert solar energy into chemical energy, which enables the community to function. All other organisms are *heterotrophic:* they cannot make their own food and must feed on other organisms. Insects that ingest plants are termed *primary consumers.* These herbivores are eaten by carnivores or *secondary consumers,* which in turn serve as energy sources for other carnivores and parasites, the *tertiary consumers,* as illustrated in Figure 155. Such a sequence of trophic levels is termed a *grazing food chain.* What appears to be a relatively simple listing or classification can become quite complicated. For example, it is very common for holometabolous insects to have the immatures feeding at one trophic level, while the adults are at another level. In a few species, males and females also may exhibit trophic level differences.

The energy budget of each consumer population may be summarized as follows:

$$C = P + R + FU \tag{8}$$

where C = ingested energy
P = energy converted to tissue
R = energy used in respiration
FU = energy lost in feces, uric acid, etc.

Because most energy taken in as food by an individual is lost through cellular respiration, the amount of biomass produced at each trophic level diminishes rapidly. The efficiency of a trophic level can be calculated as follows:

$$\text{Ecological efficiency (EE)} = \frac{\text{calories ingested by carnivore}}{\text{calories ingested by prey}} \times 100 \tag{9}$$

Teal (1971), in his classic study of energy flow in a cold-water stream ecosystem (Fig. 156), found that 2,300 Kcal/m^2/year of plant tissue were eaten by herbivores with only 576 Kcal/m^2/year becoming new tissue. Carnivores ingested 208 Kcal of this 576 Kcal/m^2/year. When 208 Kcal are divided by 2,300 Kcal and multiplied by 100, an EE of 9 percent is obtained. A simplified diagram of the trophic relationships in cold-water streams may be seen in Figure 157.

In contrast to the relatively simple ecosystem studied by Teal, most terrestrial communities are much more complex, and significant energy goes into producing plant support to counter gravity. Where climax vegetation consists of large trees, approximately 90 percent of the plant energy goes into synthesizing trunk and stems, and little energy is available in leaves for herbivores. Eventually, however, this vast source of energy, as well as that tied up in consumers, becomes available to detritus

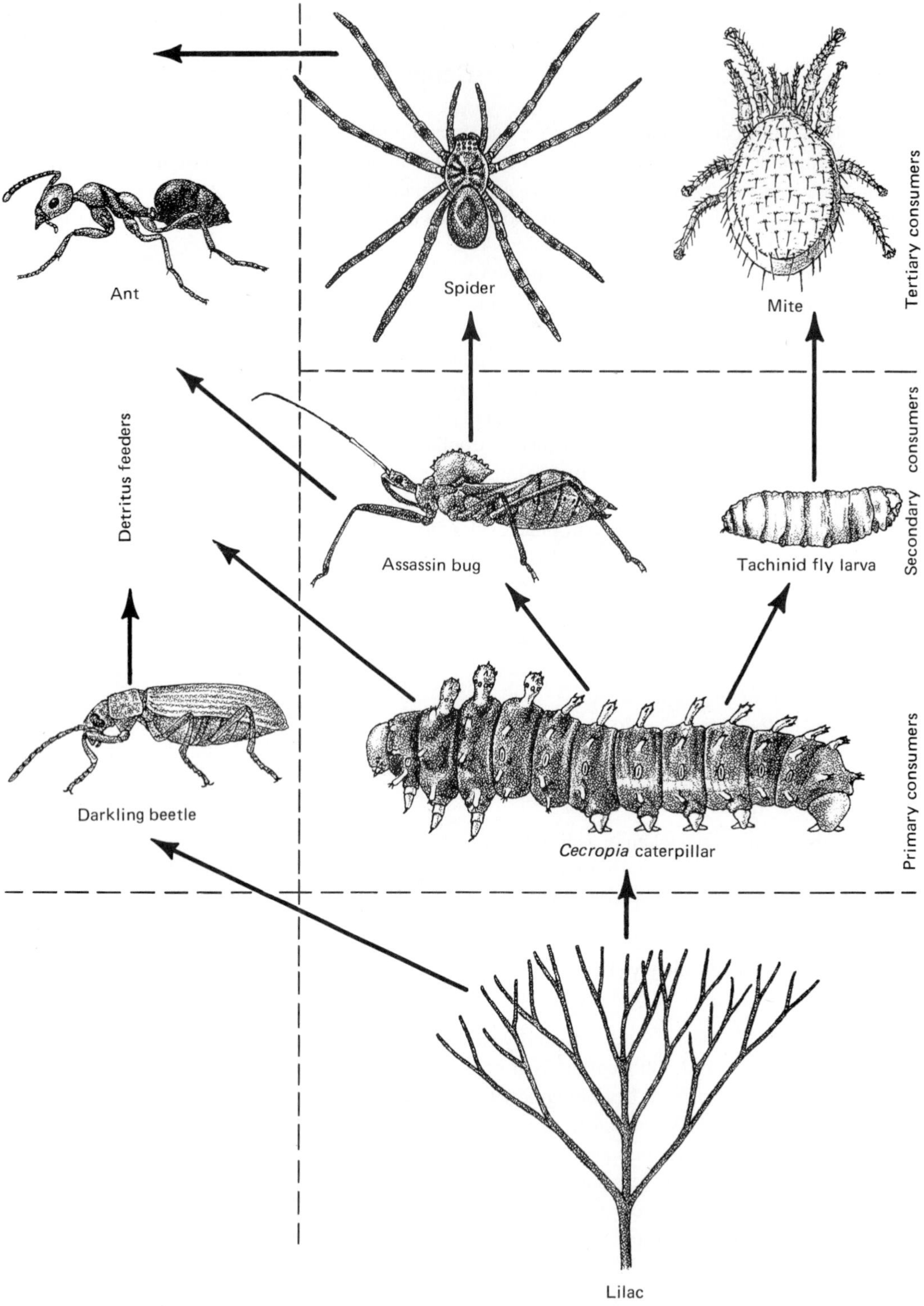

Figure 155. An example of steps in trophic levels through food chains.

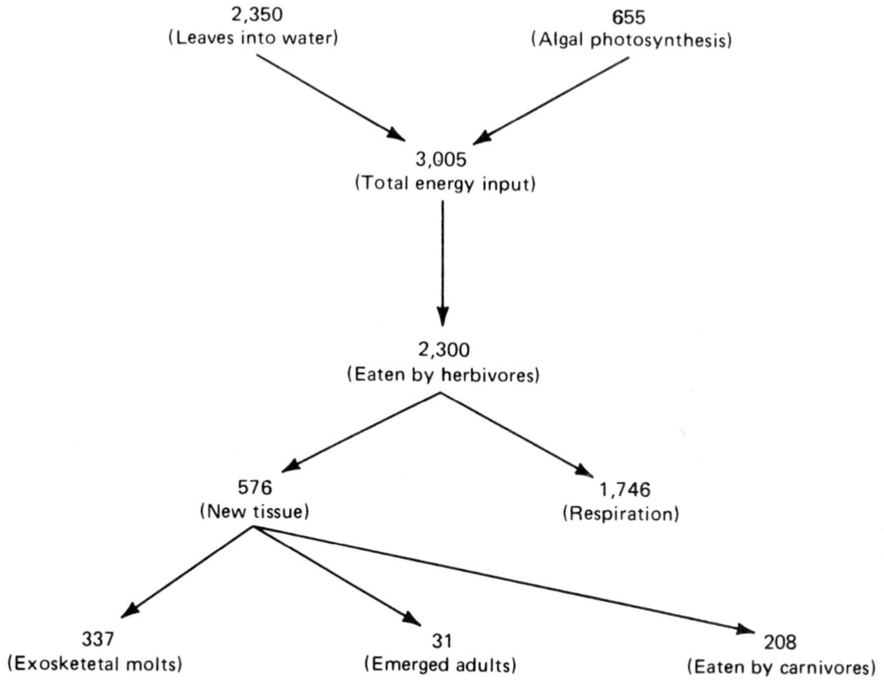

Figure 156. Productivity and energy flow through a cold-water stream ecosystem. Insects played a dominant role in the system energetics. All numbers represent Kcal/m^2/year. (Based on data from Teal, 1971).

feeders (Fig. 140) when death occurs, and a second direction of energy flow, the *detritus food chain* results.

The number of species of insects, as in all animals and plants, varies greatly with the type of community and the stage of ecological succession. Tropical rain forests have vast numbers of species, estimated at nearly 50 percent of those existing, and food chains become extremely complex, although with some localized instability. Recently disturbed areas or land cultivated by humans are at the other end of the terrestrial spectrum. Here, a few species are found, food chains are consequently simple, and great community instability exists.

QUESTIONS

1. What are the major differences between an aquatic environment and a terrestrial environment? How do these environments affect insects?
2. What is a *plastron?* How does it function?
3. What structures are characteristic of aquatic forms? Are these different in immatures and adults? What insect orders have high affinities for water?
4. How do aquatic-inhabiting insects become distributed?

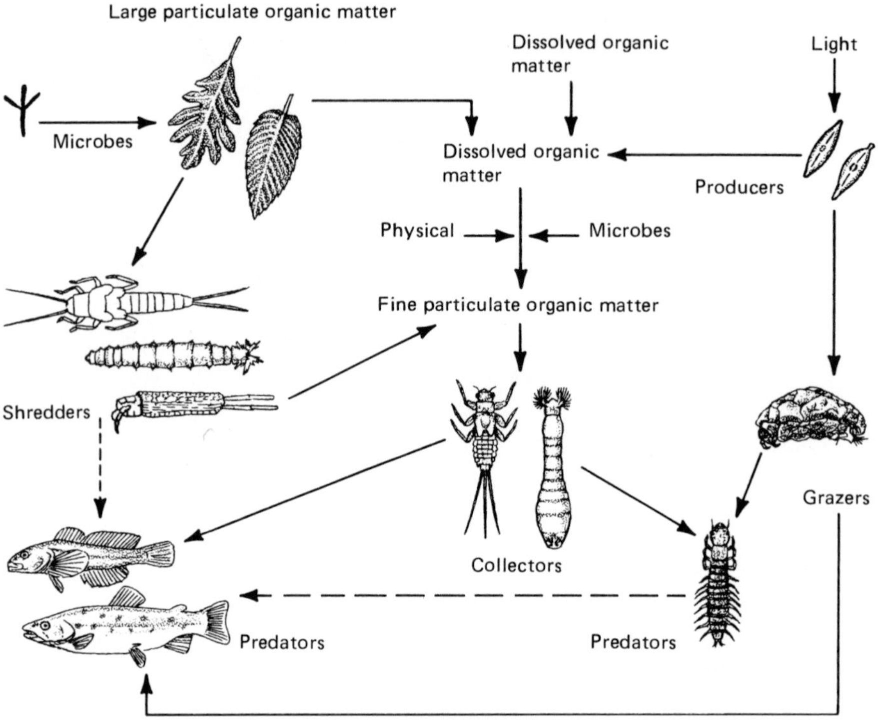

Figure 157. A simplified diagram of trophic relationships in a woodland stream. Solid lines indicate normal routes, whereas dashes are less frequent alternatives. (Redrawn from Cummings, 1973.)

5. What is an *ectotherm?* an *endotherm?* a *heterotherm?* Which is the most common in insects? Explain your reasoning.
6. What is *natural selection?* How are insects affected?
7. If the frequency of the recessive gene *t* is 20 percent, what would be the frequency of the genotype *TT* in the population?
8. What is *carrying capacity?* Express it mathematically. What is its significance?
9. A pair of fruit flies is placed into a bottle with an abundance of food and optimal temperature and humidity. The population increases to approximately 200 flies but fails to go higher. What has happened?
10. Suppose that a box of live insects being mailed from one country to another becomes damaged and the insects escape and establish a local population. Will the gene pool of this new population be identical with the original? What might happen to these two gene pools over the next 20 years? Explain your reasoning.
11. What is an *ecosystem?* How does a knowledge of food chains aid one in understanding its complexity?
12. What is *ecological efficiency?* How efficient are animals in converting food to tissue?

6 Behavior

In previous chapters we studied numerous adaptations that permit the organism to develop, survive, and reproduce. Directive action and reaction of an individual toward the environment using a repertoire of modifications is called *behavior.* Responses that benefit the insect permit survival and reproduction, but inappropriate behavior may result in death, inability to mate and deposit eggs properly, lowered food intake, and higher water loss. Behavioral patterns are coordinated by both hormonal and nervous system mechanisms. Further, when an animal is small and an invertebrate, the number of neurons available to coordinate potential responses is limited (an ant brain has about 1 million neurons, whereas a human brain has about 100 million), and the energy investment in functioning and maintenance is high; therefore, much of the insect behavior is programmed into the system genetically to handle the "predictable" aspects of the environment.

Fixed-action pattern behavior, or *instinct,* is a catch-all category for observed but poorly understood innate responses. A digger wasp illustrates many behavioral sequences of this type, and most responses require the previous one before being initiated, as illustrated in Figure 158. If the sequence is interrupted, disorientation occurs, and the female wasp either enters into abnormal behavior or initiates the sequence anew.

Two types of primitive innate behavioral responses are *kineses* and *taxes.* Taxes are nondirectional or random movements in which the insect does not specifically direct its body toward a stimulus but simply becomes active and moves. On the other hand kineses are directed orientation movements toward or away from a specific stimulus. In many instances the body is simply shifted until an equal intensity of stimulus is received on both sides of the body, resulting in either a forward movement toward or away from the stimulus. A prefix is often added to indicate the kind of stimulus being responded to, as in *photo*taxis and *geo*taxis, and the response also may be indicated as positive or negative.

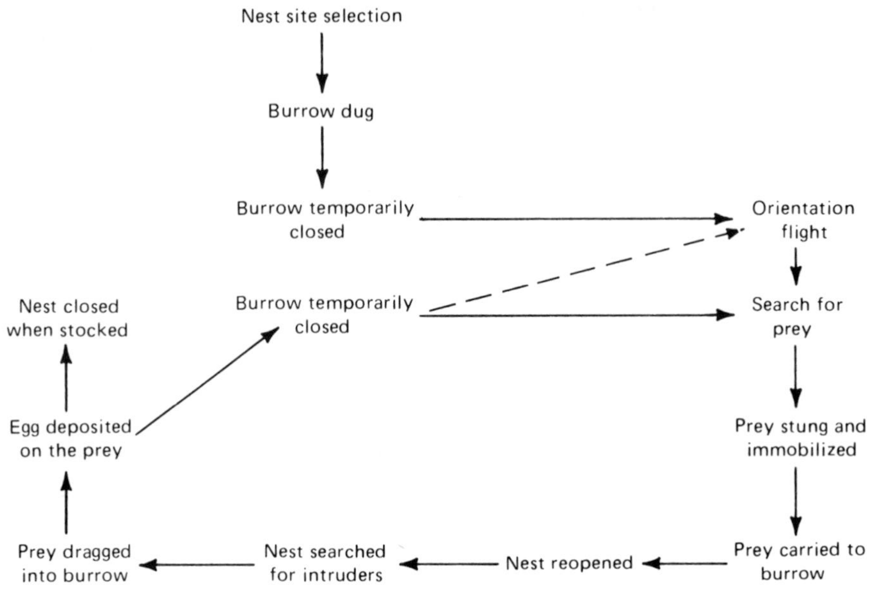

Figure 158. Digger wasp behavior in provisioning a nest.

Until recently all insect behavior was considered to consist of simple fixed-action, or instinctive responses, but limited amounts of learning and decision making have now been documented. Learning behavior involves unpredictable relationships between stimuli and responses; that is, an individual demonstrates a change in behavior that results from varying past experiences. Honeybees, for example, can be taught to take nectar only from certain aroma-scented or colored dishes or from sources placed on specific patterned backgrounds (von Frisch, 1971). Experienced bees can forage and even communicate food sources to other bees on cloudy days, using memory to keep track of time and sun movement during the day. Another example is the digger wasp (Fig. 158) that orients to landmarks before leaving the nest to search for prey. Therefore, learning involves sensing cues, transmitting these to the proper association center, encoding the information, and retrieving it at a later date. Insects have both short-term and long-term memory; at least this is true in a few studied cases. But the limited number of neurons, complex life cycles, and short life span are not conducive to select extensively for long-term memory.

Various traditional classifications of learning behavior are available. Starting from the least in complexity, *habituation* is a decrease in responses to repeated stimuli that are not the result of muscle fatigue. For example, mosquito larvae dive from the surface film in response to a sudden appearance of shadows (potential predators?); however, this response decreases if the shadows consistently move over the site (as would happen with tree shadow production). *Associative learning* involves associating a proper response between a stimulus and a reward. The previously mentioned studies by von Frisch (1971) of bee foraging involving color and nectar rewards are a good example of this learning. Many bees required only a single trial to learn, whereas "slow learners" required repeated exposures, and a few never learned. *Latent learning,* a still higher form of learning, involves behavior in which no immediate reward is obvious, such as the reconnaissance flights of the digger wasp. The most complex learning, *insight,* involves new adaptive responses or reasoning to solve problems, and may not exist in insects.

A number of types of behavior, involving mostly innate but also learned behavior, are classified, but only rhythms, locating food and feeding, locating mates and oviposition, orientation, and migration will be discussed here because of their obvious role in survival and reproduction.

RHYTHMS

Some examples of behavior rhythms are simple responses to periodic changes in the environment, known as *exogenous rhythms;* others are programmed internally and are followed with or without environmental changes, the *endogenous* or *circadian rhythms.* Both are very common in insects. The word *circadian* is derived from the Latin meaning "approximately" and "day." To indicate the precision of these responses, the term *biological clock* is used extensively in the literature. Once this "clock" has been synchronized by light (only a flash is necessary), an individual placed in constant light or dark will continue to carry out the approximate 24-hour physiological and behavioral changes characteristic of the species without further light cues. The "clock" can be altered somewhat, however, by temperature and latitude changes.

A number of behavioral patterns are closely linked to circadian rhythms. Advantages gained from such a system include such examples as initiation of molting in the early morning hours when the humidity is high, initiating activity when food sources are greatest, and synchronizing male and female development and mating periods. Rhythms also allow insects to share a common environment with other species with a minimal degree of competition. Insects that are active during some period of the day are called *diurnal,* those that are active at night are called *nocturnal,* and those that are active during periods of weak light such as at sunrise and sunset are called *crepuscular.* Activity is often restricted to very short intervals of several hours (Fig. 159), although some species may have several peaks of daily activity.

Many rhythms, in response to changes in the length of day and night, are seasonal. Some species develop during short-day periods, whereas other insects require long-day periods. Even species that have several-to-many generations per year have differences in some generations in response to day-length changes, often in the form of diapause (Fig. 116), reproduction, mating, or migrations.

LOCATING FOOD AND INITIATING FEEDING

Perhaps the most common activity that engages an animal is to find and consume food. Insects can ingest nearly all types of organic food. Their food habits may be classified by the type of food eaten: *phytophagous* (plant tissues), *carnivorous* (animal tissues),

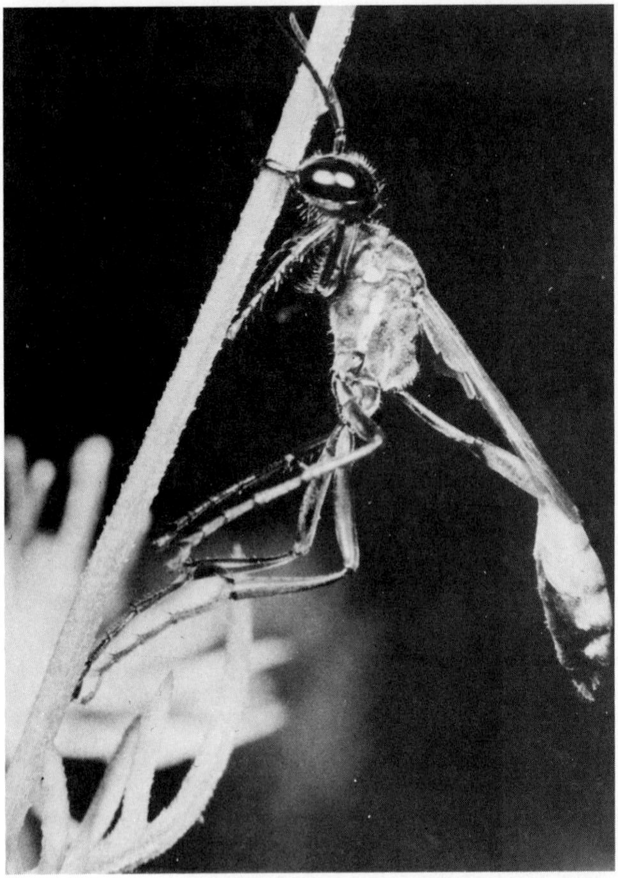

Figure 159. A solitary wasp (*Ammophila* sp.) sleeping. (Photograph by C. W. Rettenmeyer.)

saprophagous (dead organic and associated microorganisms), or *omnivorous* (more than one type of food). They also may specialize in specific areas of a food source, such as leaves, sap, roots, or pollen. The tactics of such exploitations are of obvious importance to understanding insects.

Ingestion of food normally requires at least two sets of behavioral patterns. First, the food must be located. Carnivorous insects, such as mantids, are alerted by prey movements; this usually involves vision, although reception of vibration may be important. Herbivores use vision or odors to locate potential food sources. Some parasites, such as horse flies, rely greatly on vision, but others, such as fleas and mosquitoes, use mainly temperature, carbon dioxide, or host odors. Second, after the potential food has been found, a final discrimination must occur, usually utilizing sensory receptors on the mouthparts (especially the palpi as shown in Fig. 160), antennae, or tarsi, depending on the species. Adult house flies, for example, taste with their tarsi, and a positive response results in lowering of the proboscis, a second discrimination by receptors on the labellum, and finally feeding. Herbivorous insects that have chewing mouthparts normally make a trial bite to taste the food (Fig. 161), whereas insects that having piercing-sucking mouthparts determine suitability after preliminary probes. Not all food is

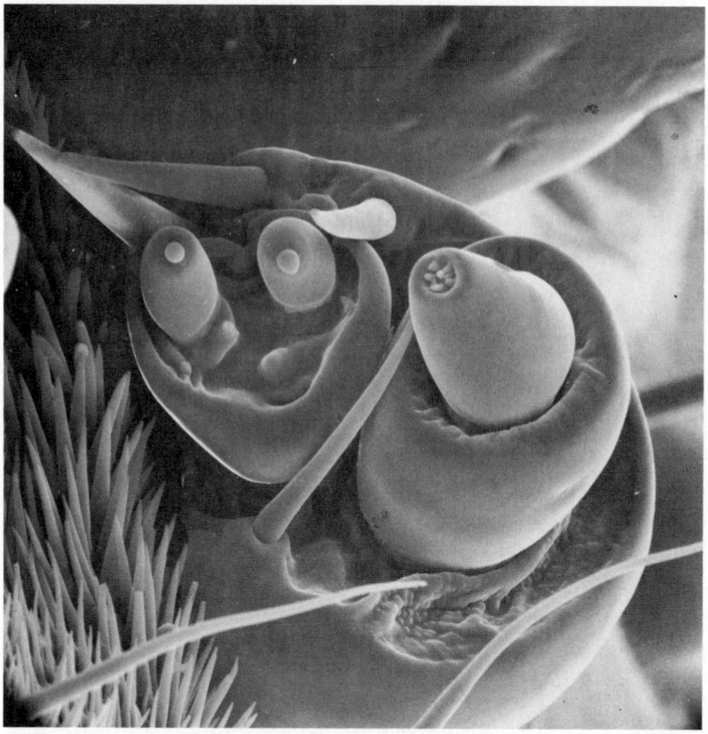

Figure 160. Palpal lobes on maxillae of a noctuid caterpillar containing chemoreceptors at their apex.

Figure 161. *Hyalophora cecropia* larva feeding on lilac. Because of its heavy body, the caterpillar feeds first on one side of the midrib to the tip, bending the leaf as it goes whereupon the other side of the leaf is eaten back to the leaf petiole. Such behavior prevents falling accidentally as the larva feeds.

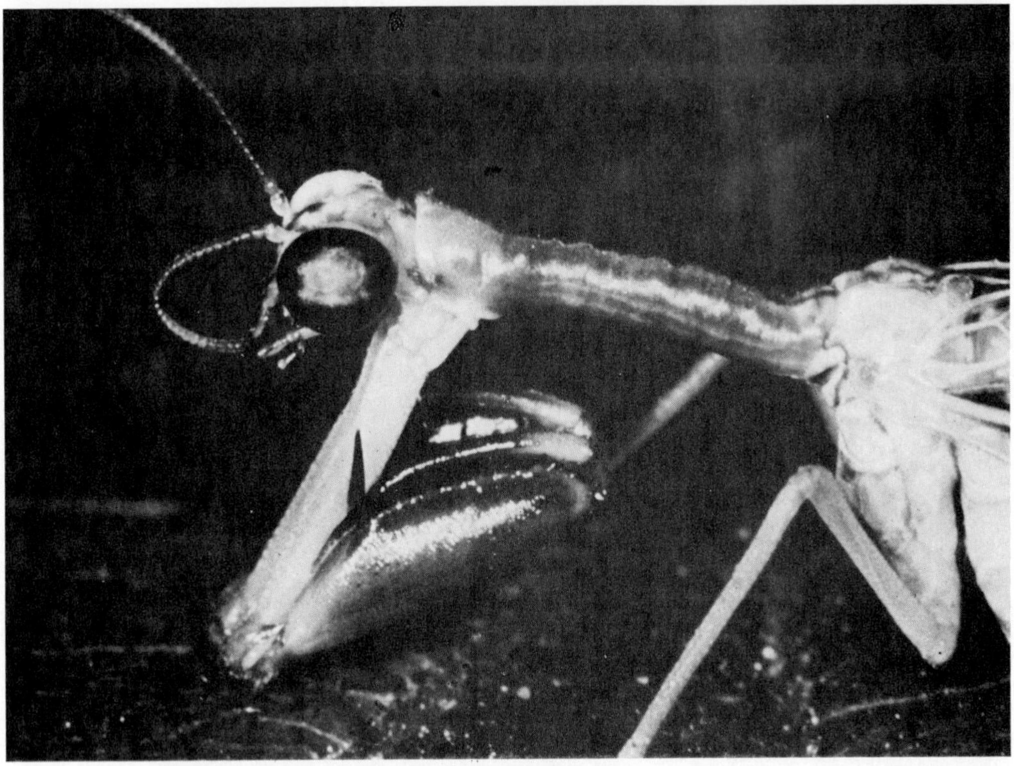

Figure 162. Preening behavior of mantispid, *Climaciella brunnea,* using its mouthparts to rub off dust and other contaminants from its antennae.

equally acceptable. If the potential food is low in preference or is repellent, the insect leaves and searches anew. The state of the insect (age, time of last feeding, habituation, etc.) has an influence on the acceptance or rejection phenomenon.

Some insects are specialists, whereas others are more general in their feeding. Food discrimination has an obvious role in such selection. Receptors needed in discrimination are kept clean by grooming behavior (Fig. 162). Insects with chewing mouthparts commonly use these structures and their legs to preen themselves, whereas insects with piercing-sucking mouthparts use only their legs to scrape detritus off. In other insects, specialized grooves and setae are adapted for grooming.

Further complicating the understanding of feeding, especially in immatures, is the oviposition behavior of the adult female. In butterflies, for example, the proper food for the larvae is normally preselected; that is, the eggs are deposited on the plant on which the larvae feed. Initiating feeding, however, still requires that the caterpillar taste and discriminate. If the food is nonpreferred, the individual wanders in search of the "right" food, but often starvation occurs before it is reached. Searching for food and actual feeding also are affected by the environmental factors of photoperiod, humidity, and temperature. Grasshoppers, for example, feed primarily in temperatures above 20°C and during daylight. Certain species of mosquitoes feed during the night

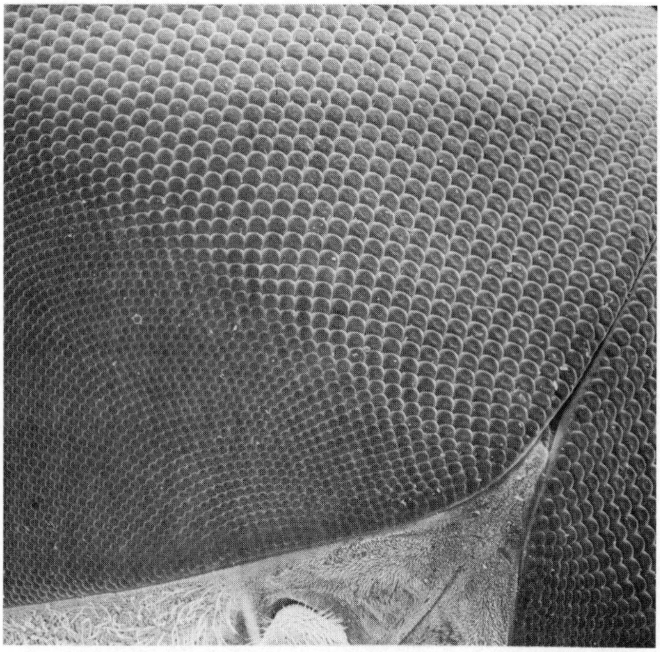

Figure 163. Compound eyes of a male horse fly (*Tabanus lineola*). The lower ommatidia are of normal size when compared to females, whereas the dorsal ommatidia are larger and more effective in locating females while flying rapidly.

hours, whereas other species prefer crepuscular periods; in all cases, however, mosquitoes are attracted to warm-blooded hosts (endotherms) that emit the proper blend of chemicals.

LOCATING MATES AND COPULATION

Males and females have different strategies toward copulation and reproduction. Generally, males invest energy in locating mates and producing sperm; therefore, they often search out and copulate with what appears to be any available and receptive female. Persistence increases the likelihood of a successful mating. Females, on the other hand, produce fewer gametes, provide most of the nutrients for progeny development, and are far more selective in mate selection because their investment in energy is considerable. Locating a mate that results in copulation and embryo formation involves a complex series of exacting conditions and behavior for both sexes. First, both sexes must be active at the same time and in the same locality. Second, at least one sex must be capable of locating the other, recognition must occur, and each must possess the correct genitalia and be physiologically ready to carry out the process. This section will deal primarily with the second set of conditions.

The opposite sex is located by responding to at least one of a variety of stimuli including vision, hearing, smell (including pheromones), or touch. These stimuli normally evoke mating behavior during certain restricted time periods of a 24-hour day. Although many diurnal species use visual clues in locating mates (Fig. 163), other stimuli often are necessary to initiate copulation. Swarming has been observed to play

(A)

(B)

Figure 164. Pierid butterflies, *Colius caesonia,* when viewed under different light. (A) male to the left and female to the right as seen by the human eye with full-spectrum light; (B) the same individuals when viewed under ultraviolet wavelengths as the butterflies would most likely see.

a major role in mayflies, caddisflies, and many primitive Diptera. These swarms, consisting primarily of males, form and orient toward conspicuous landmarks such as trees or roads. Females are attracted to and fly into these swarms where males immediately copulate with them. Various distinctive patterns of flight, in insects such as butterflies, initially attract mates, and then color patterns and odor often are used to complete the discrimination process. Although color patterns in butterflies can serve to attract mates, it is usually only specific localized patterns or reflectant ultraviolet patterns, many of which bear no correlation to "visible" features, that are the most effective. The females of *Colius caesonia* strongly absorb ultraviolet, whereas the males have localized regions that strongly reflect these light waves (Fig. 164). Behaviorly, the males are attracted to the ultraviolet-absorbing features of the female but are repelled from the males by the reflecting flash signals.

Vision also can be important in a few nocturnal insects. In fireflies, mates are located by flashes of light from abdominal light organs (Fig. 165). Two types of strategies are involved. In the first, the female broadcasts specific signals that attract the

Figure 165. Ventral view of firefly (*Photuris*) showing abdominal luminescent organ. Some females of this genus, in addition to attracting males of their own species, can mimic flashes of other firefly species and lure males, which are captured and eaten.

male. In the second, the male flies about flashing its specific pattern, and when a receptive female is near, she returns the proper flash. Each species has its own coded signal, color, shape of light organ, and height of flight to isolate species other than its own from potential matings. Some males of tropical species aggregate on trees and flash syncronously (in what may be termed a *lek*) to give a Christmas tree flashing effect.

With the exception of the previously mentioned fireflies, most nocturnal and some diurnal insects utilize nonvisual cues for discrimination. In certain moths, the virgin female assumes a stationary calling position (Fig. 122) and emits, from exocrine glands near the tip of the abdomen, *sex* or *attractant pheromones,* which diffuse into the environment. These pheromones are usually specific and are detected by antennal receptors located on the male antennae (Fig. 166) as the insect flies crosswind in a searching behavior. If a minimal threshold of molecules (10^{12} μg for the gypsy moth) comes in contact with the antennal receptors, the male becomes "excited" and flies upwind, the most likely direction of pheromone source. The environment in which these pheromones can be detected is often referred to as the *active space.* Olfactory attractants bring the male close to the female, at which point other sensory stimuli such as visual, tactile, or wing-beat clues become important. The distance of initial attraction varies. Male gypsy moths have been reported to fly nearly 4 km to a female, but effective attraction is probably less than 100 m.

(A)

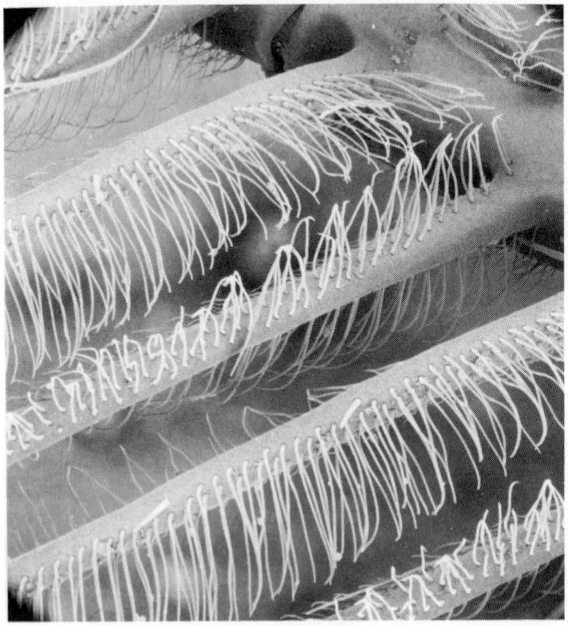

(B)

Figure 166. Pectinate antennae of giant silkworm moth (*Antheraea polyphemus*). (A) lateral view; (B) SEM micrograph of filaments showing sensory setae used in detecting female sex pheromones.

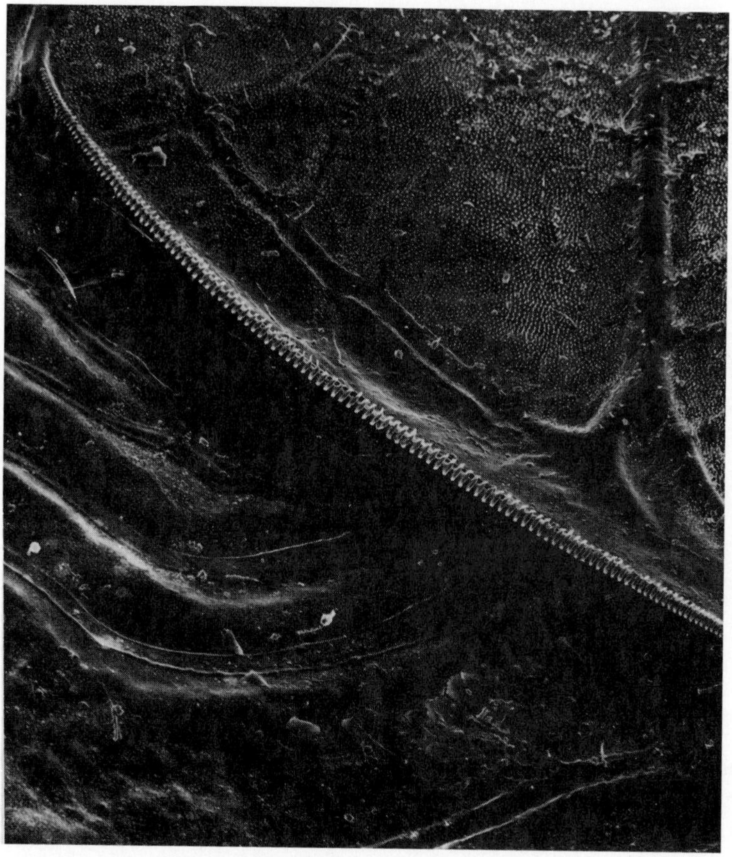

Figure 167. File on front wing of a male cricket (*Acheta assimilis*) that is rubbed across a scraper on the other forewing as the wings are moved. The result is vibration of the wings and a sound that attracts female crickets.

Species of Orthoptera, Homoptera, and Diptera may locate mates through sound. Calls are attractive by their characteristic pitch, pulse, or intensity (Ewing, 1990). In Diptera, such as mosquitoes, wing vibrations by the female produce the sound, and males perceive this sound through antennal receptors. In contrast, it is the male in Orthoptera and Homoptera that produces the attractant sound. Long-horned grasshoppers and crickets have a file on the large vein of one forewing (Fig. 167) that is rubbed across the scraper at the edge of the other wing. In addition, several species of mole crickets construct specialized burrows in the ground where they "sing"; the burrows greatly amplify the stridulated sound. Certain tree crickets in Africa chew holes in leaves in which they also stridulate to increase the volume of sound.

Another means of sound production is found in cicadas. Males possess an abdominal *tymbal* (Fig. 168) that vibrates in a resonant chamber formed from the metathoracic epimeron and first two abdominal segments. Large muscles pull on the tymbal causing it to buckle inward and produce a "click" sound, which is followed

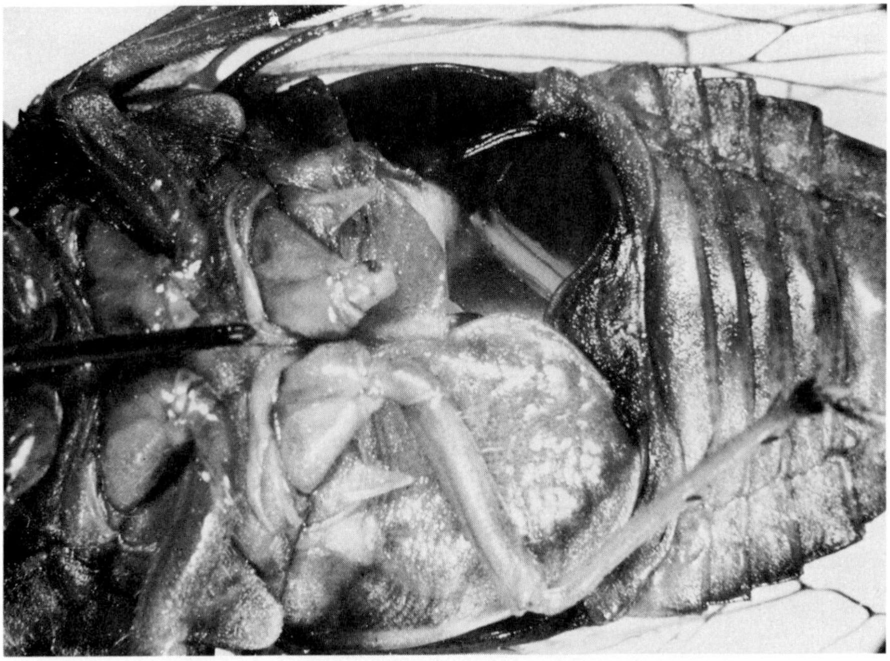

Figure 168. Tympanic organ in a male cicada (*Tibicen pruinosa*) that produces sounds that attract the female. The right metathoracic epimeron has been removed to expose the resonant chamber of the abdomen. Each species has its particular frequency and pattern of song, which maintain specific isolation.

by a second click as the tymbal returns to its original shape following muscle relaxation. The sound is amplified by a shiny membranelike mirror posterior to the tymbal, and it can be varied by regulating the opening into the resonance chamber under the metathoracic epimera. Each species produces songs different from others found in a given locality and time, and male cicadas of a single species may join together into local mating groups by synchronizing their calls.

Sounds are received by the "ears" or *tympana* of the female. These tympana are located on the fore tibia (Fig. 169) in crickets and long-horned grasshoppers, on the first abdominal terga of short-horned grasshoppers, and elsewhere in other insects. It is interesting to note that as sound production varies with temperature, receptor thresholds also respond correspondingly, thereby ensuring success with the system.

Attracting the opposite sex by sound is not without peril, however, because calling also attracts predators and parasites. Birds, for example, are known to locate male cicadas and long-horned grasshoppers by their calls. Likewise, those crickets that are consistent callers have a relatively high rate of parasitism by fly larvae, whereas noncalling males and females rarely are parasitized.

Once a potential mate has been attracted, other stimuli become important to ensure mating, including chemical and tactile stimuli. Certain male butterflies and moths have *androconia,* specialized secretory wing scale patches (Fig. 170), and

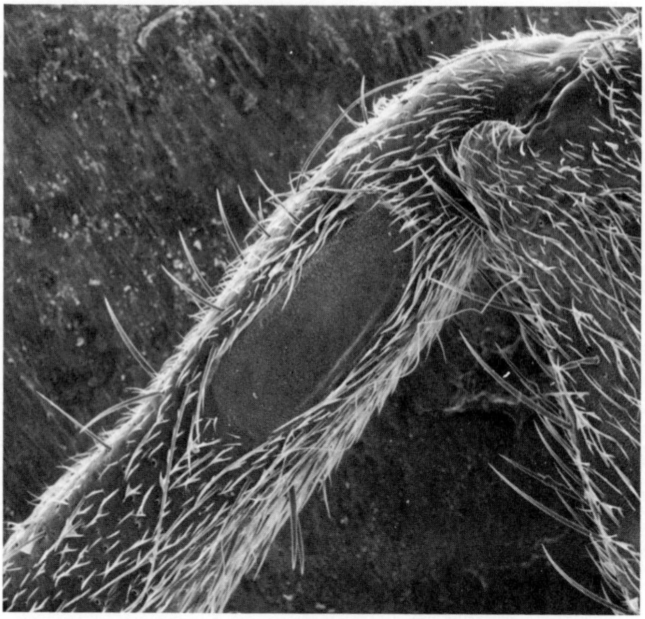

Figure 169. Tympanum on tibia of a female cricket (*Acheta assimilis*).

some butterflies and moths have *scent pencils* (Fig. 171), eversible tubular extensions from between abdominal segments 8 and 9, that secrete and disperse an *aphrodisiac* pheromone as part of courtship behavior (this pheromone also can act as an antimale and antispecies substance to terminate fruitless courtship pursuits, particularly where mimetic species are involved).

Danaidone, an arrestant and aphrodisiac pheromone produced by male monarchs and other danaid butterflies, is synthesized from pyrolyzidine alkaloids (allelochemicals) obtained when the adults feed on certain plant exudates. The male queen butterfly, a relative of the monarch, hovers over the female, after locating her by vision, and releases his secretions (from hairpencils) near the female antennae until copulation can occur. In many Orthoptera, substances produced from male metanotal glands are eaten by the female before or during mating. Gustatory aphrodisiacs also are found in beetles. Truly, the insects exhibit a multitude of variations in attracting and encouraging sexual contact.

Eventually, the pair attempts to mate. Many variations of copulatory positions exist (Fig. 172), especially in hemimetabolous insects. The male may mount the female, or the female may mount the male. Abdomens may be twisted to fit the copulatory organs together. Once coupled, the pair may remain together for less than a minute or up to several days, depending on the species and environmental interference. Silkworm moths have been observed to remain *in copula* for more than 3 days (Fig. 266).

After separating, males often search for additional mates. Females, however, have limited mating and quickly become "unattractive" to other males. Fluids from the male accessory glands, spermatophores, mating plugs, or anti-aphrodisiacs, all produced by males, often prevent reinsemination by other potential mates.

Figure 170. Hind wing of male monarch butterfly (*Danaus plexippus*) with typical dark patch of secretory scales, the androconia (*A*).

Multiple matings do occur, however, and seem to be influenced by, or at least are correlated with, sperm longevity and the amount of sperm transferred previously. Bedbugs, for example, often mate every 2 weeks because sperm is apparently short-lived. In contrast, sperm are able to survive for years within queen ants, and after initial matings the female becomes incapable of mating.

Some butterflies consistently mate several times, and some entomologists believe this is to ensure sufficient sperm for fertilizing all eggs. A growing amount of evidence, however, suggests that a disproportionate number of eggs are fertilized by the sperm received from the last mating (Smith, 1979). In many insects, sperm transferred by the last male apparently either are the first used for fertilization or displace sperm from previous couplings; therefore, it is of selective advantage to be the last mate prior to oviposition. This phenomenon is termed *sperm precedence.* Some male Odonata even have parts of their secondary genitalia, located on the anterior abdominal sterna, modified to "scoop out" sperm in the female spermatheca from previous matings before depositing their own. Some male insects ensure their paternity rights

Figure 171. Scent pencils everted from specialized internal pouches on male heliconian butterfly from Ecuador. (Photograph by C. W. Rettenmeyer.)

Figure 172. Brushfooted butterflies from Paraguay mating in typical lepidopteran end-to-end pattern, each facing the opposite direction.

by producing a mating plug or some chemical in the sperm that repels or deters other potential mates.

In addition to transferring sperm during mating, spermatophore fluids also contain certain compounds that either increase the fitness of the female or add nutritively to the eggs. A single spermatophore produced by certain male heleconian butterflies

Figure 173. This monarch butterfly (*Danaus plexippus*) egg has been glued to a milkweed leaf, the preferred host plant for the developing larva.

contains the equivalent nitrogen content to that found in 15 to 30 eggs produced by the female; the female needs this nitrogen content to produce the first egg batch. Similar findings also are known from certain orthopterans. In addition, defensive compounds may be transferred by the male to protect the female and her eggs in some butterflies and meloid beetles.

OVIPOSITION

Each species is usually very discriminating as to where its eggs are deposited. Many herbivorous insects oviposit directly on the food plant necessary for development of the young (Fig. 173). Similarly, parasitic species deposit eggs either on or in the host or in areas frequented by the host (Figs. 174, 175). Although some insects seem to be indiscriminate and place the ova only in a type of habitat, such as water or soil, undoubtedly there is more selectivity than that observed by humans.

In most instances, at least two responses are necessary. The first involves only general discrimination as to an area, shape of plant, or animal. The second requires specific sensory conditions to initiate the behavior needed for egg deposition. Chemoreceptors and tactile receptors on the tarsi and ovipositor, tasting the substrate by the mouthparts, and other senses are employed. The long-horned grasshopper illustrated in Figure 176, after locating the plant, made a taste bite, apparently found the plant to be suitable, and then split the leaf with its ovipositor. As the ovipositor

Figure 174. This great golden digger wasp (*Sphex ichneumoneus*) has immobilized a long-horned grasshopper nymph, which will serve as a food supply for a larva.

Figure 175. Ichneumonid wasp (*Megarhyssa macrurus*) with slender ovipositor to deposit egg on a burrowing insect larva. The hatched ichneumonid larva will feed on the borer host.

Figure 176. Female long-horned grasshopper (*Scudderia texensis*) feeds as it oviposits in euonymus leaf. The curved ovipositor splits the leaf where feeding has occurred.

was withdrawn, the egg was released. During the entire oviposition, the female continued to feed. Face flies locate manure by smell, alight, and sense the feces through receptors on the feet, labellum, and ovipositor (Fig. 65). If the feces have the proper moisture content, physical constitution, and aroma, eggs are laid.

Some of the more complete studies on oviposition have dealt with solitary wasps whose fixed action patterns are often inflexible and must be performed sequentially. The female usually digs a burrow in the ground, temporarily closes it, and then searches for some specific prey. Those wasps that seek caterpillars sting them in every body segment to prevent segmental reflexes and bite the head of the prey to prevent jaw movement. Others subdue crickets or spiders in very precise patterns. The prey is now carried or dragged to the burrow entrance (Fig. 174), the nest is opened and searched, the prey is pulled inside, and an egg is deposited on the prey. The sequence is repeated until the nest is sufficiently stocked for the future larva to complete development. Each of the previous sequences requires specific stimuli to initiate, and an interruption often results in either repeating the behavior or starting the entire sequence over again. For example, the artificial introduction of prey at the wrong time usually will evoke a behavior necessary to repel invaders instead of initiating a predatory pattern.

Once deposited, the eggs of most insects are abandoned and survival occurs only through such strategies as rapid embryology, wide separation of eggs to reduce predation, and effective hiding of eggs. However, a few insects do remain and protect their eggs. With the exception of certain Hemiptera (Fig. 137), such parental care by nonsocial insects is carried out by the female.

ORIENTATION

Little is known about orientation, but major reference points such as the sun and moon serve as important reference points for many insects. Most of the limited research has been done with Hymenoptera. Many wasps take short orientation flights around the nest before flying off in search of food. Objects such as trees, highways, rivers, and large rocks are used to fix their position. In addition to these landmarks, the sun is also used as a reference point by many bees and ants. The insect compensates for the movement of the sun across the sky and is able to return to the nest by using this reference guide. In honeybees, the direction of food sources may be passed on to other honeybees in the hive by means of a dance in which the sun is used as the reference point (Fig. 227).

Moths apparently use the moon to orient themselves. Normally, a certain angle between the eye and the distant moon is maintained during flight. If a moth approaches a light or fire, however, it attempts to maintain this angle to the new bright "moon" and starts a circling motion about the light (Wigglesworth, 1984).

MIGRATION

Unlike birds, relatively few insect species engage in dramatic geographical mass movements each year. Those that do are usually characterized by:

1. Undergoing this behavior soon after adulthood is reached.
2. Females with developing ovaries being the consistent migrant sex.
3. The migration often being a one-way movement, although some of the best-known examples, such as certain populations of the monarch butterfly (Fig. 177), do make at least partial return flights.

Migrations often start in an area where populations are increasing and where environmental carrying-capacity pressures are building. Under these conditions, nymphal *Schistocerca gregaria,* a Middle East grasshopper, becomes morphologically and behaviorally differentiated into a migratory phase. When these grasshoppers reach adulthood, there is an exodus flight that normally encounters and gains assistance from the wind (see polymorphism, Chapter 4). Big swarms may extend over 250 km, have more than 40 billion individuals, contain an estimated 70,000 tons of biomass, and eat its weight in food each day. Flight of these and other migratory insects is maintained until other stimuli, such as convergence of wind-flow patterns, matured ovaries, wing-muscle deterioration, energy exhaustion, length of day, and other settling responses, redirect behavior.

A major function of migration in most species is dispersal of individuals to new and suitable habitats. In many instances, because these new areas cannot support a species population during the entire year or the climatic conditions will not permit overwintering, reinvasion occurs each year. In temperate regions the peak direction taken by the spring and summer generations is toward lower temperatures, whereas

Figure 177. Monarch butterflies (*Danaus plexippus*) often migrate great distances. This specimen has been tagged as to its release point to determine distance, duration, and direction of flight.

this is reversed in the fall, when migrations are toward warmer temperatures. For example, North American populations of the monarch butterfly cannot withstand freezing and migrate southward to overwintering sites as winter approaches. The western populations fly to the mild regions south of San Francisco, whereas the central and eastern populations migrate thousands of miles across the United States and Gulf of Mexico to a rugged mountainous region west of Mexico City (Urquhart, 1976; Brower, 1977). An estimated 100 million to 200 million butterflies survive the journey. Enroute large stores of fat are built up by feeding on nectar so that they are essentially 50 percent fat on arrival. This fat is the only source of energy for the butterflies during the entire winter. In either site, massive aggregations build up; however, deforestation is reducing the number of potential roosting sites.

Overwintering monarch aggregations, because of their distastefulness (mechanics to be discussed later in section on Revealing Coloration) to many vertebrates, provide a massive group-warning display (Tuskes and Brower, 1978). This protection is not absolute, however, for certain birds can penetrate this defense; orioles eat only the soft tissues where the poisons are low in concentration, whereas grosbeaks can tolerate the toxins. Up to two thirds of the monarch mortality may be due to such bird predation, although winter storms and frost take up to 20 percent of the population of butterflies in any given year. Survival of the remainder is the result of a slowed metabolism in the cool, moist climate during this critical time of nectar scarcity. Following winter, the aggregations dissolve as females leave and migrate northward and deposit eggs on milkweeds, the specific larval food. Although some monarchs may make the complete return flight, most are believed to make only a partial return flight, and range expansion to the north is the result of the several succeeding generations. By the end of the summer most of the United States and southern Canada are repopulated. As fall arrives, a sexually depressed overwintering generation continues the cycle with its southern migration.

Johnson (1966), using the adult life span as his major criterion, lists three types of migrations. *Type I* are "short-lived adults that leave their breeding site for new areas and die within a season" (some locusts, butterflies, and aphids). *Type II* are "short-lived adults that emigrate to feed and return to reproduce" (some beetles, mosquitoes, and many dragonflies). *Type III* are long-lived adults that migrate either before or after hibernating or aestivating and may or may not return to the breeding site the following season (some beetles, monarch butterflies, and many noctuid moths).

PROTECTIVE BEHAVIOR

Insects, like all organisms, are eaten by many predators and parasites. For protection they possess many morphological structures that permit them to bite, sting, or flee from the assailants. Associated with the presence of these structures, however, must be the proper behavior to utilize these adaptations. Additionally, other less obvious behaviors also exist that increase survival.

Autotomy

A few insects such as walkingsticks have weakened areas at the trochanter that break under extreme stress, such as might occur when the appendage is grasped by a large predator. Reflexes within the removed appendage cause the leg to twitch, which diverts attention from the escaping insect. Specialized membranes prevent bleeding, and regeneration of the leg may be possible should future molts occur.

Reflex Immobility

Many beetles with hard exoskeletons feign death when disturbed and remain motionless for a few seconds or several hours. During this period of immobilization, normal overt activities cease, and stimuli fail to induce responses. The hardened exoskeleton and the lack of movement apparently protect these quiescent individuals because many predators require movement of prey as an attack stimulus. Eventually this reflex subsides, and the insect returns to normal activity.

Reflex Bleeding

When threatened, some insects are capable of squeezing drops of hemolymph from specialized skeletal weak points. Ladybird beetles and blister beetles are good examples. Usually this hemolymph contains toxic substances such as malachiines and cantharides or repellents.

Figure 178. Protective secretions are being emitted by this darkling beetle (*Eleodes* sp.).

Defensive Secretions

The use of chemicals in defense (*allomones*) is found commonly in such phylogenetically diverse groups as Thysanoptera, Hemiptera, Orthoptera, Lepidoptera, and Coleoptera. Generally, this strategy is found in insects that have extended longevity and, hence, a longer exposure to predators and parasites. These insects are generally aposematic and often are models for mimetic species (see next section).

Although some of these noxious substances are directed at specific predators, most are effective against a broad spectrum of enemies, both vertebrate and invertebrate. Allomones commonly are released from specific exogenous glands in insects; however, some are present in the hemolymph and are leaked out between sclerites with increased blood pressure. Anal glands, in many beetles such as Tenebrionidae (Fig. 178) and Carabidae, secrete volatile fluids, some of which are ejected with explosive force for distances up to 1 m in the latter insect. In many Hemiptera, grasshoppers, and some caterpillars (Fig. 195), noxious secretions are produced in tergal glands, and leaf beetles have glands at the edge of their elytra. Predators react to such fluids as either repellents or irritants, although few are killed.

Sonic Behavior

Many moths gain protection from birds by their nocturnal activity, but in turn they become exposed to predation by insectivorous bats. Some of these moths, mainly Noctuidae, Geometridae, and Arctiidae, have metathoracic tympana that detect the echo-location apparatus used by bats in locating and intercepting potential prey. Once the ultrasonic "chirps" have been detected, the moths immediately commence evasive flight behavior.

Another adaptation, a stridulatory mechanism consisting of a series of cuticular thoracic ridges overlying a cavity filled with air, is found in certain distasteful arctiids. When these ridges are rapidly bent and unbent by leg muscles during flight, a series of ultrasonic clicks are produced that can be heard by bats and apparently become a warning as to prey palatability.

CONCEALING COLORATION

Most insects are difficult to see, particularly when populations are low. They blend with their background, and unless movement (Fig. 179) betrays their position, they go unnoticed. This is called *concealing* or *cryptic coloration.* A number of grasshoppers, mantids, and walkingsticks are green and difficult to distinguish from leaves. Exceptional examples may be seen in the long-horned grasshoppers in which a few species even have brown areas on their wings that resemble disease patterns commonly found on host leaves (Fig. 180). Others, including walkingsticks and most mantids, resemble twigs (Fig. 181). Some treehoppers resemble thorns. A few butterflies in the tropics have transparent wings and continuously blend with the background while resting or during flight (Fig. 182). Certain walkingsticks change color from day to night, thereby blending more closely with their background. Countershading (one side darker than the other) or shadow minimization (with many hairs to disrupt shape (Fig. 183) may be present, and from the side, this results in a more flattened-appearing object. Concealing coloration is best suited for those insects that are sedentary during the daylight hours when movement would reveal their presence.

Inherent in the concept of concealing coloration is the axiom that cryptic coloration is protective only when the insect rests upon the proper background (Figs. 183, 184, 185, 186). Experimental data during the past 10 years suggest that insects do in fact search for suitable substrates. Moths that have stripes tend to orient themselves to correspond to lines on bark upon which they rest. Both tactile and visual stimuli are involved for this orientation.

Does cryptic coloration actually provide protection from predators, or is this anthropomorphic reasoning? Evidence exists that it does provide protection, at least

Figure 179. This South American grasshopper (*Proscopus* sp.), although nearly 10 cm in length, is inconspicuous to bird predators because of its sticklike shape and color and slow movement.

from vertebrate predators. Birds often pass by insects that are shaped and colored like twigs, leaves, and feces (Fig. 185) until one of the insects moves, whereupon the bird seems to become programmed and locates additional insects with relative ease. As a result, some birds may attack any object resembling this gestalt form, including twigs and leaves. If similar insects become scarce or if too many vegetative objects are encountered, the bird usually reverts to its original attitude and again passes by camouflaged insects. Adult giant silkworm moths, and probably most large insects with cryptic coloration, also tend to have short adult survival periods, apparently a selection that lowers exposure of the total population to predation (Blest, 1963; Ricklefs, 1973). One must be aware, however, that this protective color is not absolute; the phenomenon refers only to a differential survival rate, and many moths with the "proper" coloration are eaten, although not as many as with a more obvious coloration.

Figure 180. This long-horned grasshopper (*Pycnopalpa bicordata*) from Central America is somewhat conspicuous here, but when located on leaves that possess fungal infections, this insect is difficult to observe. (Photograph by C. W. Rettenmeyer.)

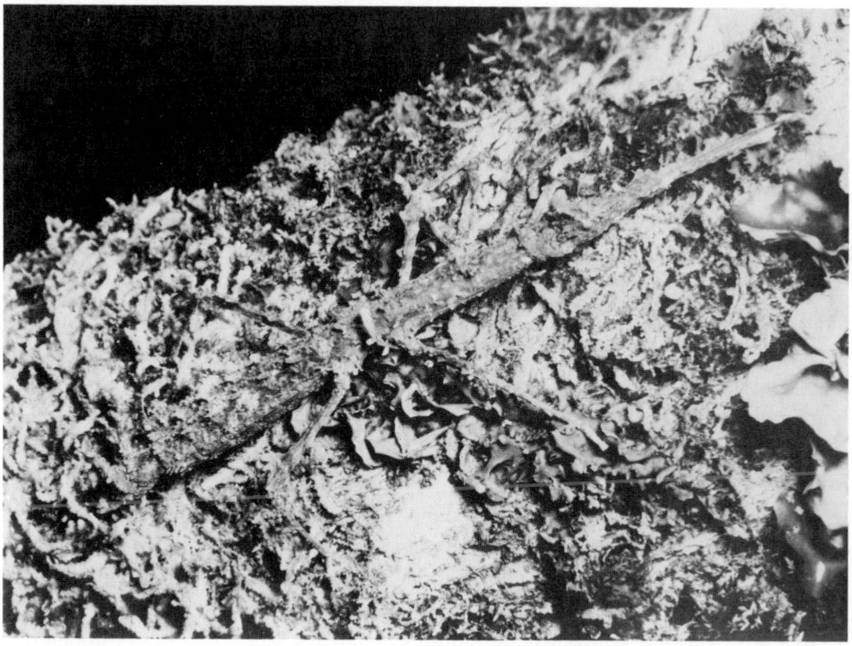

Figure 181. A Costa Rican walkingstick effectively hidden. (Photograph by C. W. Rettenmeyer.)

Figure 182. Butterflies (satyrid and ithomiid) from Costa Rica that are transparent and, with their slow flight next to the ground, very difficult to see.

As in all animals, differences exist in color and patterns within the same species (polymorphism). Very often variability includes distinct dark and light individuals, which brings up the question of what mechanisms are involved. What possible advantage can come from not blending with the background? Certainly the benefits would not seem to be to the individual easily located by a predator. Observations on the English peppered moth, *Biston betularia* (Figs. 186, 187), indicate that both light and melanic forms (differing by only a single dominant allele *C*) attempt to select suitable background color. Before industrialization, light forms (homozygous recessives) were the most successful because they were well concealed on the background of the numerous trees covered with lichens, whereas the dark phenotypes (*CC* or *Cc*) had fewer suitable substrates to rest on and hence were more easily seen and eaten. With the increase of industrialization and subsequent deposition of soot on the landscape, lichens became sparse or disappeared, and the light phase then became the conspicuous form. In the 50-year interval from 1848 to 1898, the melanic morph increased from about 1 percent to 99 percent of the population. *Industrial melanism,* as the process is termed, resulted in a reversal of selection advantage, and in addition to *B. betularia,* approximately 70 of 780 other English moth species evolved toward predominantly darker phases. A return to the original state is possible, however, and has occurred in many areas of England where laws have drastically reduced industrial pollution and where the landscape is returning to its original state (Bishop and Cook, 1975).

Figure 183. Protective coloration in the noctuid caterpillar (*Catocala* sp.), whereby the larva blends with the background by color and the form disruption produced by the long setae at the body sides.

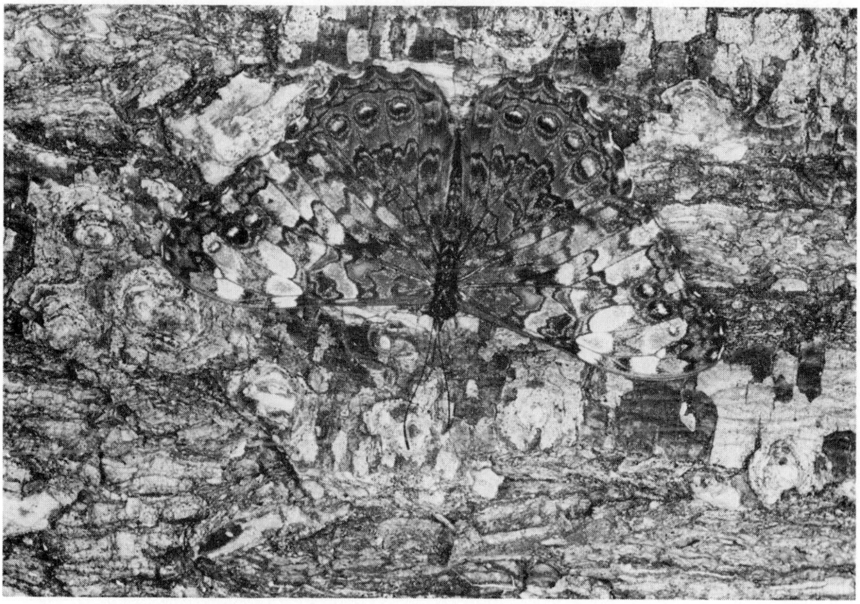

Figure 184. Butterflies are not as commonly cryptic on the dorsal surface as are moths. This Paraguayan brush-footed butterfly, however, blends in with trees with light bark, upon which it rests with its head in the downward position.

Figure 185. Some lepidopteran caterpillars, such as this swallowtail larva from Paraguay, resemble bird droppings, and predators apparently pass by them without recognizing the larvae as food.

Figure 186. Dark and light forms of the peppered moth (*Biston betularia*) were photographed on the trunk of an oak blackened by the smoke-polluted air of the English city Liverpool. The light form is clearly visible, whereas the dark form is almost invisible below and to the left. (Photograph by J. A. Bishop, University of Liverpool.)

Figure 187. Same two forms as in Figure 186 of the peppered moths were photographed where less soot and more lichens were present. Here it is the dark form that may be clearly seen. The light form is below the dark form. (Photograph by J. A. Bishop, University of Liverpool.)

In addition to morphological adaptations, some insects conceal themselves with available material from the environment. Certain leaf beetles (Fig. 188) cover their bodies with fecal material during their larval stages and become less conspicuous. Caddisflies, also as larvae, use pebbles and twigs from streams to form cases that blend with the background and also provide physical protection (Fig. 369). A caterpillar in Borneo attaches flower buds from its food plant to its body and replaces the buds as they become wilted.

REVEALING COLORATION

Some of the larger insects have forewings cryptically colored, but the hind wings are brightly colored or contain eyespots. When resting, the forewings cover the body and the hind wings and provide concealment. Such cryptic coloration is of no benefit once discovered, and birds, upon locating such prey, initially peck and often seize the insect by the head for easy ingestion (Fig. 189). At this point, the bright hind wings or eyespots are exposed by reflex action (Fig. 190) and startle the bird, often causing the insect to be released or to deflect the attack to the wings and away from the more vulnerable body. Examples of this type of wing coloration may be seen in many giant silkworm moths, hawk moths, some grasshoppers, and a few tropical mantids and planthoppers.

Figure 188. Larval leaf beetles (*Lema trilineata*) with fecal matter attached to aid in concealment.

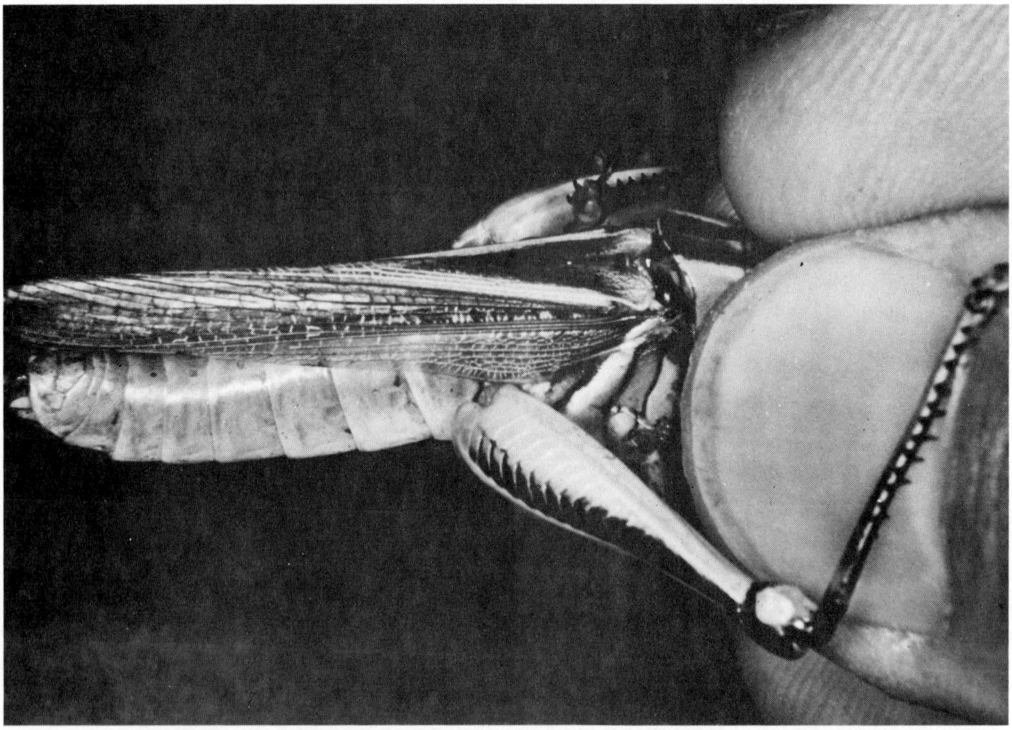

Figure 189. Short-horned grasshopper, *Melanoplus bivittatus,* grasped by its head, which initiated protective reflexes by the hind legs.

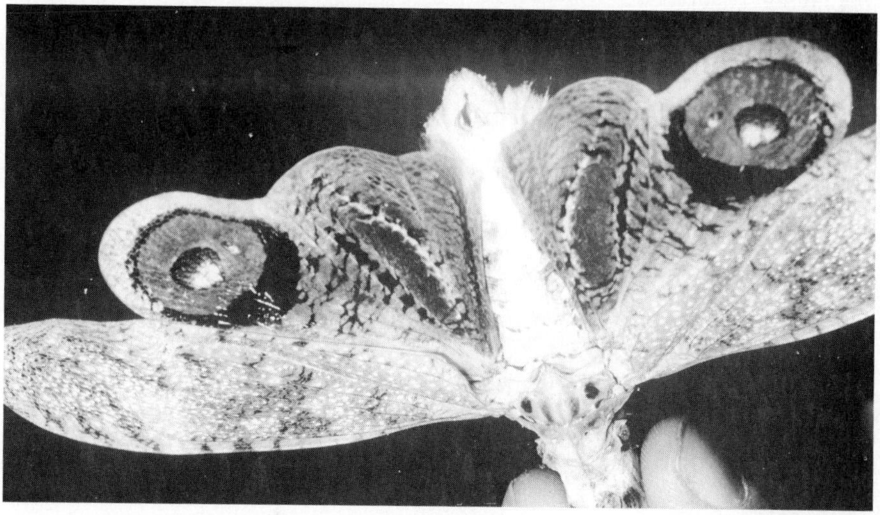

Figure 190. Many of the large tropical insects, such as this peanut bug (*Lanternaria pospho-rea*), have either bright colors or eyespots on their hind wings. When grabbed by the head, as commonly occurs when birds attack, these wings are exposed and often startle the predator sufficiently to permit escape.

In some butterflies, the dorsal surface of the wings is brightly colored and iridescent, whereas the ventral surface is drab (Fig. 191). Light may be reflected for great distances from such iridescent wings, and is often observable from low-flying airplanes. During the flight of these butterflies, the alternative bright-dull-bright sequential appearance of the wings is deceptive, and few predators are successful in their attempts to capture this prey. When not flying, the wings are closed dorsally in typical butterfly fashion, thereby concealing the reflective dorsal surfaces.

Of special interest are the great number of insects that are brightly colored over their entire body and appear to rest upon discordant colored backgrounds. How does such revealing coloration aid in survival when insectivorous birds can see red? Evidence shows that brightly colored insects are often rejected by vertebrate predators. If such species were unpalatable, a "linkage" of this trait with such a conspicuous color would be mutually beneficial to the bird and to the potential prey. The prey is avoided, and the color serves as a warning so that the predator does not waste time and energy on chasing nonpalatable forms. Experimental evidence indicates that many brightly colored insects are indeed distasteful, can sting or bite, or can emit disagreeable odors, and color becomes *aposematic,* warning vertebrate predators.

How do these distasteful or disagreeable chemicals originate? It appears that some are biosynthesized by the insect itself. Insects with this ability include certain grasshoppers, true bugs, and many beetles. A second source for these chemicals is the incorporation of toxins from plants, chemicals that have evolved to protect the plants from herbivores. These toxins are referred to as *allelochemicals,* and insects capable of using them for their own protection include certain lepidoptera larvae, lygaeid bugs, pyrgomorphic grasshoppers, a few aphid species, and several beetles, including the Colorado potato beetle.

Figure 191. The genus *Morpho,* a tropical group, includes species whose ventral surface is cryptic or drab (*top specimen*), whereas the dorsal surface is iridescent (*lower specimen*). The sudden appearance of the bright surface produces a flash effect and makes the butterfly difficult to capture.

Figure 192. Aposematic notodontid larvae (*Datana ministra*) often assume this characteristic pose when disturbed, thereby exposing ventral thoracic glands used for squirting acids.

Milkweed plants, for example, often contain cardiac glycosides (similar to digitalis), which cause vertebrates to vomit and eliminate toxins from their bodies so as to avoid serious damage. In contrast to the vertebrates, certain butterfly larvae, including the common monarch, use these plants exclusively for food, a very profitable adaptation because eggs and larvae are not lost to accidental ingestion by vertebrates when leaves are eaten. Experiments by Brower (1969) indicate that monarch caterpillars (Fig. 193) incorporate these allelochemical substances into their body and retain them into adulthood with the greatest concentrations in the wings, where birds discern them before killing or swallowing the butterfly (Brower and Glazier, 1975). The result is an insect that has conspicuous coloration and is unpalatable to vertebrate predation. Brower found a strain of monarch that could be fed cabbage and therefore lacked the toxin from milkweed. Blue jays were trained to feed upon these palatable butterflies. Then monarchs reared on *Asclepias curassavica,* a milkweed containing toxins, were introduced. After eating those reared on milkweed, the birds became violently ill and vomited many times over a half-hour period (Fig. 194), but all recovered. Subsequently, these birds refused monarchs on sight. Should the color pattern be forgotten, these birds are "reminded" by tasting the next monarch as they manipulate the butterfly prior to ingestion. Although not all North American milkweeds are toxic, apparently only a limited number of butterflies produced from toxic plants must be available (30 percent or more) for selective advantage to occur.

The higher the density of unpalatable individuals, the greater the protection. The aposematic coloration coupled with emetic qualities of the larvae permits this stage to bask in the midday sun with reduced predation, especially in the cooler areas of North America, thereby increasing body temperature and shortening developmental time. This shortened developmental time is useful in understanding why this essentially tropical insect can locate so far north and develop several generations prior to the annual southern migration to avoid freezing temperatures.

Figure 193. This monarch larva (*Danaus plexippus*) feeds on milkweed and incorporates toxic secondary plant substances, when present, into its body. Protection from predators is thereby gained using the host plant's biochemistry.

Figure 194. A jay vomiting after ingesting a nonpalatable monarch butterfly. (Photograph by Lincoln B. Brower.)

In the previous experiments by Brower, the individuals were consumed by the jays, thereby educating these birds about the monarch color pattern-emetic linkage. It is now known that an individual need not be "sacrificed" altruistically for a color pattern to become aposematic. Many aposematic butterflies have been collected with bird bill marks on their wings indicating an escape. Wiklund (1982) noted for the first time experimentally that more than 84 percent of the tested aposematic insects were released by four species of naive birds, indicating that insects commonly escape and most likely can pass their favorable genes on to their offspring.

Distastefulness is not the only factor for conditioning vertebrate predators. Wasps and bees are commonly avoided because of their stings, not because of their unpalatability; if the stings are removed, unconditioned birds readily eat these insects, but the birds will avoid these insects once they have been stung. The birds apparently associate the sting with the color pattern. Toads also demonstrate an ability to associate insect color and shape with stings. Cott (1940) fed honeybees to toads over a 28-day period. On day 1, 45 bees were ingested by the 34 experimental amphibians, but the number of bees eaten subsequently decreased until on day 7 only one slow-learning toad continued to eat and be stung. Insects with aposematic coloration are also avoided because of noxious odors, such as those produced by anal glands in certain ground beetles and the osmeteria of swallowtail butterfly larvae (Fig. 195), or because of stinging hair or setae sometimes found on certain lepidopteran caterpillars (Fig. 278).

Unpalatability of insects to vertebrates, however, is probably not directly correlated to unpalatability to other insects. In the African queen butterfly, those caterpillars that were distasteful to birds had high insect parasitization by certain wasps, whereas the palatable larvae were low in insect parasites.

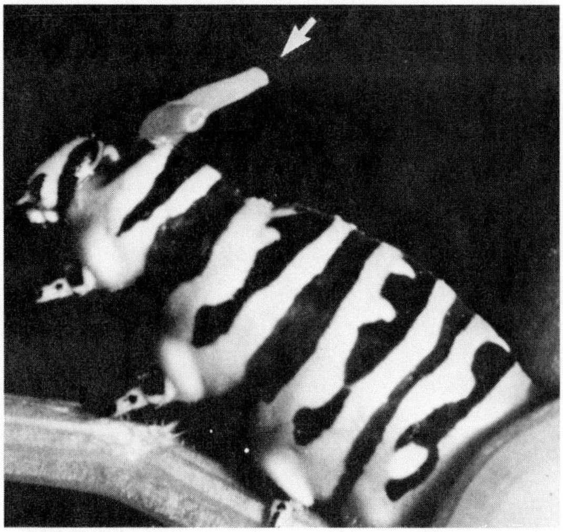

Figure 195. Swallowtail larvae have an osmeterium that is everted when disturbed and releases disagreeable odors. This larva is *Papilio polyxenes.*

MIMICRY

In the previous discussion we noted that bright and revealing coloration is often coupled with repellency of some type of predation. Other insects with similar coloration also gain limited protection. Over time, such a second species will be selected to resemble more closely this aposematic species, the rate depending on the degree of similarity and the intensity of predator pressure. The first or aposematic species is referred to as the *model,* whereas the species gaining from or reinforcing the similarity is the *mimic.*

Although other types that involve tactile and chemical cues are known, we will restrict discussion to visually selected mimicry (Figs. 196, 197). Behavior as well as morphology is involved; the mimic usually flies or runs similar to the model (Figs. 197, 198, 199). The mimic is usually less abundant and exists in the same habitat as the model.

Although mimicry has been categorized into many types, only the two most common will be discussed. The first is termed *Batesian mimicry* after the noted English naturalist, Henry W. Bates. Here, the mimic is restricted to those species that are palatable but gain benefit because predators are apparently deceived by the mimic's resemblance to the unpalatable model.

Hundreds of examples of apparent Batesian mimics may be found in the literature. The classic example of Batesian mimicry in the past were the monarch (model) and the viceroy (mimic) butterflies (Fig. 200). We now know that many viceroys are distasteful, so the mimicry type can no longer be considered to be Batesian.

The second type of mimicry is termed *Mullerian mimicry,* named after Fritz Muller, a German naturalist and proponent of Darwinism. Here both model and mimic are unpalatable, and the ingestion of either by vertebrate predators results in the avoidance of both species. Whereas Batesian mimicry is usually disadvantageous to the model if the palatable mimics are eaten and the predator actively commences searching for this

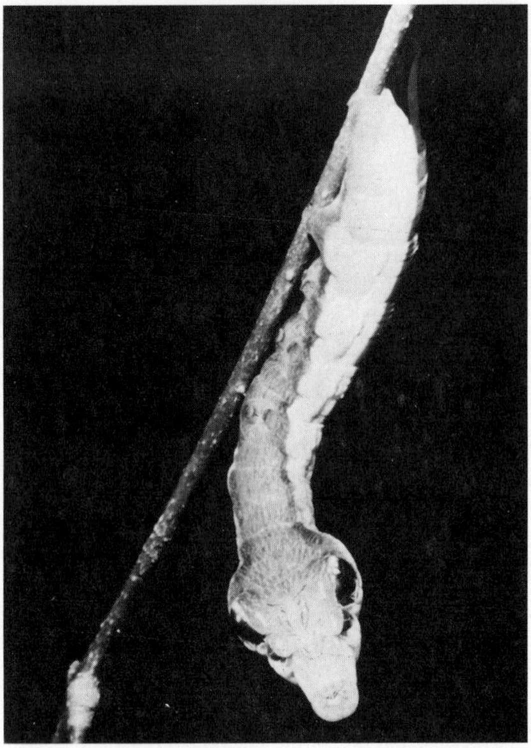

Figure 196. A tropical sphingid larva, *Leucorrhampha ornata,* which mimics a snake when disturbed, by swelling the anterior region of the body and swaying back and forth. (Photograph by C. W. Rettenmeyer.)

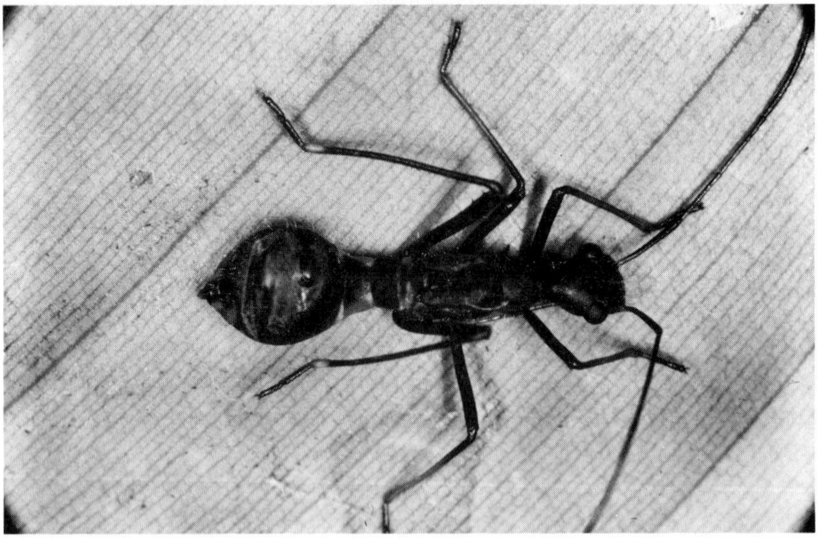

Figure 197. A hemipteran mimic of ants in Paraguay. Note the reduced wings, elbowing of the antennae, and the narrowing of the abdomen for the close resemblance.

Figure 198. A central American long-horned grasshopper (*Aganacris insectivora*) that is a mimic of a spider wasp. The antennal tips are black to give the appearance of shortened length. (Photograph by C. W. Rettenmeyer.)

Figure 199. A reduviid or assassin bug from Central America that is a mimic, as in Figure 198, of a spider wasp. (Photograph by C. W. Rettenmeyer.)

recognizable color pattern, Mullerian mimicry benefits both mimic and model because both are unpalatable and feeding on one species causes avoidance of both.

Brower, Brower, and Collins (1963) studied relative palatability of heliconian butterflies to silverbeak tanagers in Trinidad and noted that similar shape and size gave selective advantage to the Mullerian complex of butterflies. Holling (1965) indicated that the higher the density of models and mimics in a Mullerian complex (Fig. 201), the greater reinforcement on the predator search image and the greater the survival benefits until other density-dependent factors become dominant. Holling (1965) also noted that total rejection of the distasteful prey is not necessary to raise the predator attack behavior threshold and that, contrary to common belief, the mimic can outnumber the model in

Figure 200. The lower two butterflies are the models, the monarch, and the upper is the mimic, the viceroy (*Limenitis archippus*). Because the viceroy is often nonpalatable, the relationship is termed Mullerian mimicry.

this type of mimicry. In addition to butterflies, common examples of Mullerian mimicry can be seen among the lycid beetles, wasps, bees, and ants.

QUESTIONS

1. What is *instinctive behavior?* What benefits are gained by instinctive behavior over acquiring learning? Does the size of the organism have any effect upon its ability to learn? What are the types of learning in insects?
2. What advantages are gained by circadian rhythms?
3. Are there behaviorial differences between locating food and initiation of feeding?

Figure 201. Examples of similar color patterns and mimicry in six species of butterflies in Costa Rica. These orange and black butterflies represent species in up to four families, depending upon the classification system used.

4. Does behavior prevent interspecific matings? Explain your answer.
5. What are some of the advantages and disadvantages of utilizing insects as experimental animals in elucidating animal behavior?
6. What are *pheromones?* How do they affect insect behavior?
7. Do insects migrate? How might their behavior and migration differ from birds?
8. What types of defensive behavioral patterns have been identified in insects?
9. What are the advantages and disadvantages of concealing or cryptic coloration and revealing coloration? Observe some insects and attempt to determine why they are colored as they are.
10. Are there advantages for two species to have similar aposematic coloration? Explain your conclusions.
11. What is *mimicry,* and what advantages can the mimic receive from this interrelationship? Is this interaction equally beneficial to the model? Explain your reasoning.

7 From Solitary to Social

All animals are characterized by:

1. Irritability, or the ability to sense and respond to stimuli.
2. Locomotion.
3. Metabolism, the chemical reactions within protoplasm.
4. Growth, an increase in mass as a result of metabolism.
5. Adaptation, or the ability to survive in a changing environment.
6. Reproduction.

Most insects exist as *solitary* animals; that is, they live as separate, individual entities and are primarily concerned with satisfying the above listed biological requirements for self-survival. Interaction with others of the same species usually occurs by chance and is not normally "intentional," except for certain limited periods such as mating. Parents normally leave soon after egg deposition (Fig. 202); therefore, each individual is independent from the very beginning (Fig. 203), and its only parental legacy may be the specific environment into which it was deposited and a nearby food source.

In some species of insects, however, individuals associate with one another during certain stages, and this behavior cannot be explained by chance meetings as a result of high populations, eggs being deposited together, or ecological conditions causing localization of individuals. Certain caterpillars have an aggregative behavior (Fig. 204) and may have a cooperative foraging strategy that involves recruitment communication through silk and chemical signals. However, such togetherness is restricted to the larval stage.

Other forms of gregariousness may be seen in wasp sleeping aggregations, hibernating aggregations, and migratory aggregations of certain Lepidoptera. Some solitary bees may share a common entrance hole but build separate nests. Many other examples

Figure 202. Rearing cells constructed of mud by the solitary organ wasp (*Trypoxylon* sp.) showing exit holes by wasps and their parasites. Although many cells, each to feed a developing larva, are mass-provisioned by a female, there is no gregariousness exhibited by emerging wasps.

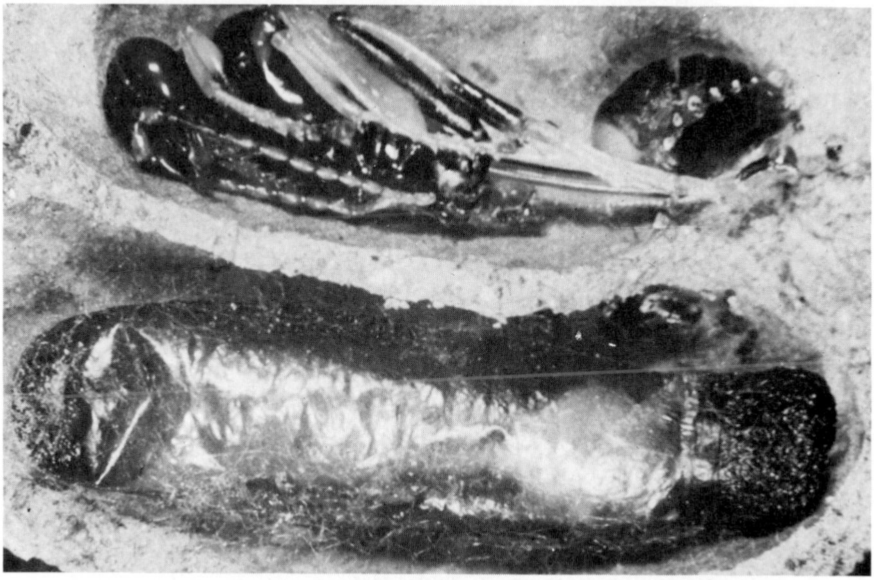

Figure 203. A cutaway view of a mud dauber nest (*Sceliphron caementarium*). Each cell contains a pupa. The pupa at the top has had the last larval exoskeleton removed.

Figure 204. Gregarious caterpillars in a Costa Rican rain forest.

could be addressed, but each is the result of limited microhabitat and macrohabitat suitability rather than true gregarious behavior.

Extended parental care is rare in insects; however, a few species stay and care for eggs or young for a short period of time (as in the case of earwigs). Passalid beetles have all stages living together in rotting logs, and certain burying beetles, after both male and females prepare a nest, remain with the immatures until the larval development is complete. In these examples we see circumstances that involve prolonged interactions between individuals of different generations, a prerequisite necessary for socialization.

What then are social insects? Insects are normally considered to be truly social (eusocial) only when the following conditions have been met.

1. Overlapping generations of parents and offspring live together as an integrated unit or colony.
2. Division of labor occurs within the colony, and reproduction is not carried out by all individuals. However, individuals may cooperate in caring for the offspring of the colony breeders.
3. Resultant population develops some form of containment structure or nest.

True societies are normally viewed as existing only in the orders Isoptera and Hymenoptera; however, some 20 species of aphids (Aoki, 1982) and 2 species of thrips (Crespi, 1992) are now considered to be social. We will concentrate only on the social Isoptera and Hymenoptera in this text.

Socialization in Isoptera probably evolved only a single time, because all termites are social in response to cellulose feeding. The most primitive termites are incapable of digesting cellulose except through symbionts, and because these are lost at each molt, a close proximity to other individuals that are infected and that can transmit these through proctodeal feeding is a necessity. The formation of colonies seems a logical direction for evolution.

However, selection and colony formation in Hymenoptera were probably different, and societies have arisen independently several times. What factors enhanced such evolution? Michener (1969) proposed that eusocial bees evolved from either *subsocial* groups, family groups in which the female rears the young but dies or leaves before maturity of the young, or from a *semisocial* group, a family of females of the same generation that cooperate but are not necessarily related. Hamilton (1964) suggested that societies were the result of kin-selection on the haplodiploidy form of reproduction (in which males are haploid and females diploid) and that there is a selective advantage for daughters to help their mother rear what would be their sisters. Daughter siblings are more genetically related to one another under haplodiploidy than they are to their mother (Hamilton, 1964b; Wilson, 1971; Eberhard, 1975). Hamilton's concept broke with tradition in that selection for societies was also on the group level by altruism as well as on the individual level. Alexander (1974) proposed that kin-selection benefited the parents rather than the siblings. Eberhard (1975) reviewed the varying theories of social behavior and suggested that each was not exclusive but that Michener's mutualism, Hamilton's kin-selection, and Alexander's parental manipulation were interrelated or functioned sequentially in colony and sterile worker evolution. Nevertheless, kin-selection remained a dominant force in explaining the evolution of sociality in Hymenoptera until the late 1970s when some serious challenges arose (Crozier, 1982).

Few scientists dispute the success of social insects, and some such as Wilson (1985) consider sociality to be one of the "landmark events in the evolution of insects." Wilson (1985) also states that as much as one third of the entire animal biomass in the Amazon area may be termites and ants. Wilson (1987) later found that a single tree in Peru harbored 26 genera and 43 species of ant; such figures approximate the entire ant fauna of the British Isles. The formation of colonies "frees" most individuals from expending energy for reproduction and permits a division of labor or specialization that greatly benefits the entire society. Attempts have been made to compare the individual from such a colony with the cell and the multicellular body of an organism; this "superorganism" concept is intriguing, but is highly controversial among most biologists (Wilson, 1989).

Much can be learned by contrasting primitive societies with highly evolved societies. When such comparisons are made, the following evolutionary trends appear.

1. Increased division of labor by castes of individuals that are behaviorally (polyethism) and usually morphologically (polymorphism) specialized for various functions within the colony, such as workers, soldiers, and reproductives.
2. Increased population size within the colony.

Figure 205. Thysanuran (*Trichatelura manni*) in an army ant column of *Eciton burchelli.* Because of the nomadic behavior of the ants, which move daily, the thysanuran must be and is capable of following the trail of the ant migration. (Photograph by C. W. Rettenmeyer.)

3. Daily provisioning or direct feeding of the young (Hymenoptera).
4. Production of a nest or physical structure to enclose the colony, which can moderate environmental conditions.
5. Complex variations of trophallaxis or mutual exchange of substances from the proctodeum and/or salivary glands.
6. Colony odor whereby individuals become restricted in their activity and normally cannot enter and join other colonies.

Along with highly evolved insect societies come opportunistic organisms; many thousands of arthropod species have become associated with social insects, especially termites and ants (Figs. 205, 206). Some of these *inquilines,* or guests, such as Coleoptera and mites, are tolerated and in some cases are defended by the societies and through time have come to resemble the hosts in both structure and behavior. This mimicry may be carried down to the molecular level where certain staphylinid beetles have the same cuticular hydrocarbons as their termite host. Other inquilines are scavengers or predators on the host population or are essentially social parasites of the host ants that feed and rear them. Certain insects, such as a few Homoptera, are mutualistic and provide some nutritive substances for the ants; these *trophobionts* are protected by their hosts and are analogous to domestic cows, for they yield food sugar solutions, "honey-dew," from the hindgut upon request.

We will now discuss sociality in the termites and bees, ants, and wasps. Emphasis will be on their evolution, what castes are present and what factors produce such spe-

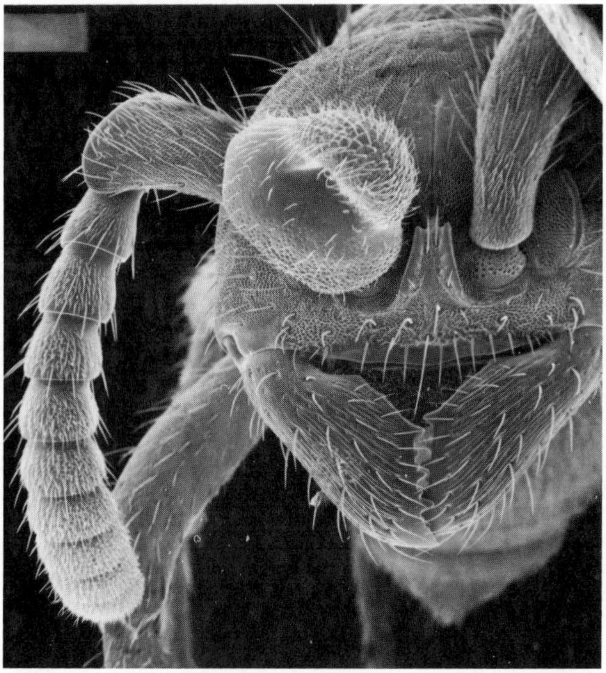

Figure 206. Mite (*Antennequesoma rettenmeyer*) attached to antenna of army ant (*Nomamyrmex esenbecki*). The "fit" is only for the antennal site, and the color and ornamentation are very similar to the ant host.

cializations, what type of nests are produced and what advantages are gained from such structures, and what impact these societies have upon the environment and humans.

ISOPTERA (TERMITES)

There are more than 2,000 (more than 50 in the United States) species of termites, all of which are social and form colonies from several hundred individuals to more than a million in some of the highly specialized species. Termites are unique among social insects in that (1) they undergo incomplete metamorphosis, with the nymphs after the second or third instars, at least in the primitive species, serving as workers in the colony, and (2) the colony is bisexual and consists of both males and females, although these may be expressed differently in the castes.

Termites may be found in all regions of the world between the 47° north and south latitudes, but they are most abundant in the tropics. These insects are believed to have evolved from cockroaches. Evidence for such ancestry includes observations that certain wood-boring cockroaches are monogamous and require intestinal protozoa to derive benefits from cellulose. Close relationships between these cockroach protozoa and those found in primitive termites strengthen this hypothesis (Krishna and Weesner, 1969). From this primitive type of association, termites have evolved into the complexity of social and morphological structure seen today, with most individuals being sterile, seldom or never exposed to the outside environment in temperate regions, and restricted to specific roles or castes that ensure the survival of the colony.

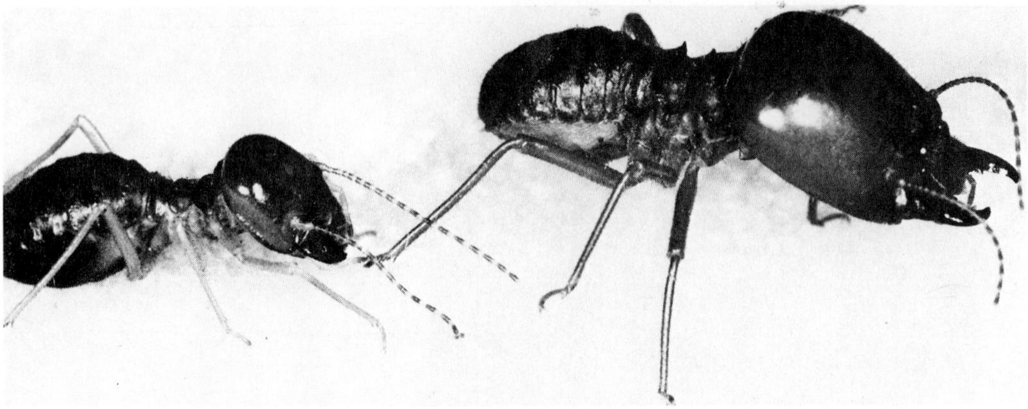

Figure 207. Two castes in a South American subterranean termite. The worker is to the left, the soldier to the right. This soldier's mandibles can easily cut through human skin and draw blood.

Each colony is based on *reproductives* (Fig. 324). Other castes develop along a neuter pathway and include *soldiers* and *workers* (Fig. 207); workers are often immatures that will differentiate into other castes, whereas soldiers represent a true caste of terminally specialized individuals.

New colonies usually result from the joint efforts of a reproductive male (king) and female (queen). At appropriate times, annual swarms of winged reproductives emigrate from the nest (Fig. 295). Following a dispersal flight, wings are lost, and a male and female court and then proceed to locate a suitable nest site. Next, a chamber or *copularium* is formed, the pair mate, and after several weeks to a month, a dozen or more eggs are deposited. Eggs are cleaned meticulously to prevent fungal infection, and after several weeks the eggs hatch. Early instar nymphs (referred to as larvae by some entomologists) either are fed by oral trophalaxis or feed upon fecal material, one form of trophallaxis, which infects them with symbiotic protozoa or bacteria.

In some primitive species, nymphs serve as the only workers within the colony. In the higher termites, however, a true worker caste of sterile adult workers develops, and nymphs carry out only minor activities. In *Macrotermes michaelsoni,* the caste development is along two lines. In one line, the fertile direction of development, some males and females become alates (winged) after undergoing six molts and leave the nest to form new colonies. The remaining males develop into large workers termed *major workers* after completing three molts, and female immatures develop either into small *minor workers* (after three molts) or *major soldiers* (after five molts). All soldiers are sterile and have hard heads and often large mandibles to protect the colony. In contrast, species of *Nasutitermes* have a complete opposite direction of development for the nonfertile individuals. In these termites, the workers are sterile females whereas the soldiers are males, and each possesses an elongate frontal protuberance (Fig. 208) through which sticky, toxic secretions may be squirted. The latter soldiers, the *nasutes,* usually lack mandibles, but they are very effective in repelling predators attacking the nest with their chemical warfare.

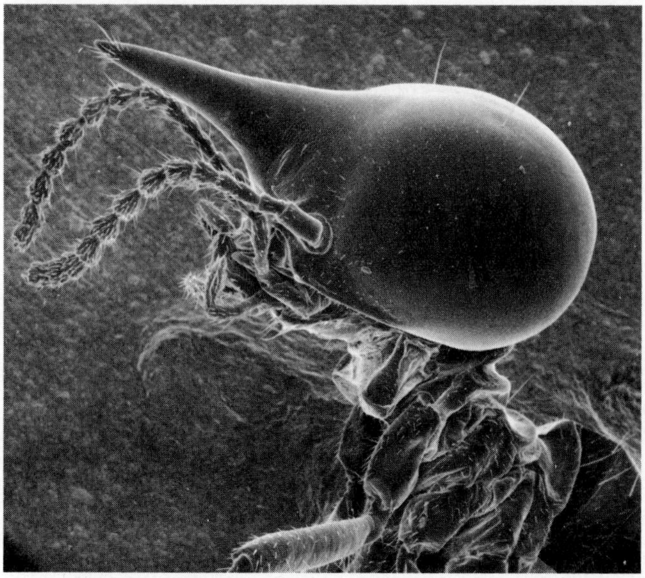

Figure 208. Scanning electron microscope micrograph of a *nasute.* This specialized soldier is able to eject sticky and noxious fluids from the frontal process, which are especially effective on ants, the major predator of this species.

Most swarming reproductives are not successful in establishing colonies because only 4 or 5 dozen offspring are produced the first year. After the first year, however, the success rate increases markedly because egg production by the queen increases from 3 to 6 eggs per day to about 30,000 eggs per day in physogastric (swollen with eggs) *Macrotermes* queens. With such an enormous increase in oviposition, colonies become immense, with more than 2 million individuals in one African species. Most termites have but a single king and queen, which carry out colony reproduction for 10 to 20 years or more (Fig. 209). If one or both of this royal pair die or if the colony size becomes large and dispersed, reproductive replacements may come from developing immatures or adults, particularly in tropical species. Such *supplementary* or *replacement reproductives* (Fig. 210) have been classified as *adultoids* (derived from adults remaining in the nest), *nymphoids* (derived from nymphs and having wing pads and nymphal anatomy), and *ergatoids* (being similar to workers). All of these begin reproducing within 6 to 8 weeks after a colony has been deprived of the queen or king, and they enable the survival of the unit. Most termite nests in the United States are reproducing by these supplementary forms.

The mechanism for caste determination is not completely understood, but current evidence indicates, at least in the lower termites, that it is probably under the influence of juvenile hormone and at least two primer pheromones. One of these pheromones is produced by the royal pair, transferred throughout the colony, perhaps by fecal trophallaxis, and apparently inhibits the production of primary and secondary reproductives from developing last instar workers, as seen in the dry-wood termite, *Kalotermes flavicollis* (Fabr.) (Lüscher, 1961b). When colonies get large so that the queen substance is not evenly distributed throughout (coupled with juvenile hormone [JH] in the eggs) or when the queen dies and the pheromone is no longer produced, pockets of immatures develop into either winged primary reproductives (Fig. 324) that will swarm from the colony or secondary reproductives (Fig. 210) that remain.

Figure 209. The greatly enlarged (*physogastric*) queen has become an egg-laying machine in her enclosed chamber. Note the many-worker retinue and the young nymphs above the queen.

Figure 210. Certain termites with extremely large colonies, such as this African species (*Amitermes hastatus*), may have supplementary reproductives. (A) adultoid queen; (B) adultoid king; (C) nymphoid queen; (D) ergatoid queen; (E) ergatoid king. (Redrawn from Skaife, 1954.)

Figure 211. A section of a dissected termite tunnel (*Reticulitermes flavipes*) formed next to the wood in a house in Kansas. This tube permits the termites to return to the soil, which is necessary in this species, and yet have access to the wood without encountering the desiccating action of the dry air.

The other pheromone is produced by soldiers, and when a certain concentration exists in a colony, it inhibits the production of JH needed for the development of most of the nymphs beyond the nymphal worker stage to soldiers. A sudden decline in soldiers because of death from an ant invasion, for example, would have an immediate effect on developing nymphs by reducing the inhibitory soldier pheromone concentration; therefore, early instar immatures or pseudergates soon molt into a presoldier instar, which becomes a soldier at the next molt because of high JH titers. This increase in soldiers would restore the proper balance of the colony castes and the soldier pheromone concentration.

Energy use within termite colonies starts with the worker caste. Food for workers normally consists of cellulose (such as in wood, grass, dung, leaf litter, humus, and paper products), although some species utilize living parts of plants and fungus growing on the plant matter brought into the nest. Most soil-inhabiting termites have burrows or tunnels to the food source. If the food is wood and it is not in contact with the soil, these tunnels may be extended to the source by means of tubes formed from cementing soil and wood particles with intestinal secretions and saliva (Fig. 211). If the food consists of grasses, such as in some African and Australian species, foraging may be carried out on the ground surface at night or through underground tunnels. Cellulose is digested by cellulase enzymes from protozoan symbionts in primitive species, but the capability to digest cellulose directly has been acquired by the specialized termites

Figure 212. A typical termite nest in South America. These *Nasutitermes* nests are usually found in trees and are manufactured from wood and soil particles mixed with termite secretions.

in the family Termitidae (O'Brien and Slaytor, 1982). Subsequent digestion of the released sugars, acetates, and butyrates by the termite produces the necessary metabolic energy, but proteins and vitamins probably come from either digesting the intestinal symbionts or by cannibalism (which is relatively common). Workers then feed the soldiers and early instar nymphs by trophallaxis; the more common type is probably salivary exchange, which often contains wood fragments, but proctodeal releases of wood fragments and symbionts are also needed. Reproductives receive salivary secretions only.

A variety of nests are constructed, varying from random shapes in soil forms, to oval nests on trees (Fig. 212) in tropical areas, to elaborate ground nests (Fig. 213) in semiarid regions, some of which in Africa may extend upward 20 ft (6 m) and measure 12 ft (3.66 m) in diameter. These latter *termitaria* are produced by cementing soil and fecal material with secretions either from specialized frontal glands in the termite's head or from the proctodeum. Termitaria are veritable fortresses and often survive as geographic landmarks for years after the colony has died. Because of their unique construction (Fig. 213), these nests also can control water loss (often the most critical restriction of termites) while still providing adequate ventilation and temperature regulation. Temperature has been found to remain a constant 29.4°C ± (85°F)

Figure 213. Mud termite nests in South America. These termites forage for grass and, when abundant, seriously compete with vertebrate herbivores for pastures.

in these *Macrotermes* nests. Such temperature stability is achieved by either a series of chimneys that can be opened or closed to regulate entering cool air flow, which counters metabolic heat produced within the mound, or by thin-walled ridges that serve to radiate the excess heat (Lüscher, 1961a).

HYMENOPTERA (SOCIAL WASPS, ANTS, SOCIAL BEES)

The evolution of societies has occurred in Hymenoptera on at least 11 independent occasions (Wilson, 1971). More than 11,000 species exhibit this way of life. All have evolved from ancestors having these hymenopteran characteristics: holometabolous development with legless larvae, males being haploid and females diploid, a modified ovipositor for stinging, and restricted matings with the female capable of storing sperm for extended periods of time. Hymenopteran societies have the following characteristics that termites lack:

1. Colonies are predominantly female with most tasks being carried out by the adults. Males are present only during a brief reproductive period.
2. Castes consist of only reproductives and workers (some protective ant workers are incorrectly termed soldiers), and these are determined by either reproductive dominance by the queens (which suppresses egg laying by other females) or by nutritional switches in food given to the developing larvae by workers.

Figure 214. A new *Polistes fuscatus* nest founded by this single fertile female. The nest is open and is being manufactured from masticated wood fibers mixed with saliva.

3. Colony origination can be by a single fertile female (*haplometrosis*), by multiple queens (*pleometrosis*), or by colony division (*swarming*). Males do not participate in colony formation. If a colony has or retains a single queen, the term *monogyny* is used. *Polygyny* refers to multiple queens within the nest.

Social Wasps

The least specialized social Hymenoptera are found among primitive wasps whose colonies are small, often numbering only several dozen individuals. Each colony may have several to many queens (*polygynous*), although one queen is normally aggressively dominant and the remainder tend to function as workers; high aggression is positively correlated with high production of juvenile hormone, and dominant wasps have the largest corpora allata. A sterile worker caste, when present, is difficult to distinguish from queens and consists of few individuals. Nests consist of an open layer of brood cells suspended by a single pedicel (Fig. 214). In the temperate region, each colony survives only for several months during the summer (Fig. 215) and disorganizes during the autumn. Colonies are initiated the following spring by single or multiple queens that have overwintered.

In the more advanced social wasps, such as yellow jackets, castes become distinct. In temperate regions and also in many tropical species, usually one queen is present per nest, and she is considerably larger than her daughters, the workers. Colonies may

Figure 215. The same *Polistes fuscatus* colony as in the previous figure 2 months later. Some of the cells contain eggs, others nearly mature larvae. Pupae are in the capped cells. Two central females are chewing up arthropod prey to be fed to the larvae.

be founded either by a single queen or by swarming. In the former, the queen selects a nesting site, begins nest construction, forages, and cares for the young until they reach adulthood as workers (Akre, 1995). In contrast, swarming, a process in which a queen and a group of workers desert the nest, is common in the tropics, and a new nest colony is formed rapidly by such a migrant group of wasps.

Colonies founded either by a single queen or by swarming soon number in the thousands and produce a large enclosed nest of papery material constructed from masticated wood mixed with salivary secretions. Nests start as a single layer of cells, or comb, similar to the nests of primitive social wasps, but usually have many layers of multicelled combs added during expansion, each suspended under the preceding one either by means of stems or fastened to the lateral envelope or covering. One or more entrances are constructed and guarded by workers (Fig. 216).

Like the nests of termites, wasp nests maintain a relatively uniform environment for the occupants, particularly the developing young. Covered nests shade the larvae and pupae from direct exposure to sunlight and are layered to give insulation. Nests constructed in the ground vary the least in internal microclimate, whereas the more common aerial nests become subject to greater climatic changes, particularly in the temperate regions. If temperatures rise too high, workers use their wings to create air flow to cool the nest. In addition to the climatic modifications, wasp nests also provide protection against their major predators, the ants. The petioles of primitive uncovered nests are smeared with ant-repellents (Fig. 214, see blackened stalk), and the covered nest somewhat restricts ant contact with the brood except for the specialized raids of army ants.

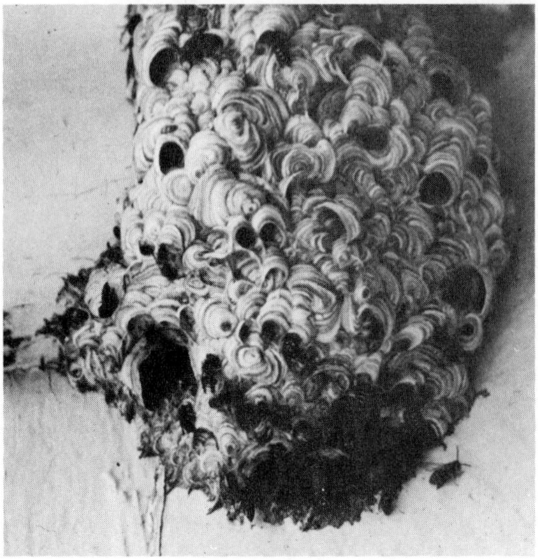

Figure 216. A paper wasp nest with many openings. Concentric rings are the result of different types of wood being added to the nest.

Social wasps feed their young daily. Food consists of pellets of masticated prey (usually insects) or sometimes dead material. In return for the food, workers and especially the queen receive a watery secretion from the larvae (trophallaxis), which is high in amino acids and aids in feeding the adults and in maintaining the social structure of the colony. After several weeks of growth, the larva spins a silken cap over the cell, pupates from 3 to 4 weeks, and then emerges as an adult.

Ants

More than 12,000 species of insects are ants. All are social. Primitive ants are wasp-like and exist in small colonies, often with only 2 or 3 dozen individuals, and little differences in polymorphism or polyethism exist between workers and queen. Eggs are not laid in cells, as wasp eggs are, but are deposited openly in the nest. After hatching, larvae are fed large pieces of unchewed arthropods, a feature unique to these ants.

Advanced ants have large colonies (up to 20 million) and highly varied diets. Reproductives are usually very different in size and shape from the worker caste (Fig. 217), and, although there is considerable overlap, the smaller worker forms (minor or minimus and media workers) tend to do more of the typical worker activities, whereas the large workers (often termed majors) generally defend the colony. Majors may have hard flattened heads to block entrances to the nest or long, piercing mandibles (Fig. 218) for active defense.

Reproductives have wings when they leave the nest to form swarms, in which mating occurs with reproductives from other nests. The males then die, and the mated females lose their wings and seek out suitable areas for nesting. In most higher species, the newly hatched larvae are fed saliva from the queen and not prey as in primitive ants. Nutrients in the saliva and energy for the queen during this early period are de-

Figure 217. The size disparity between the large queen leafcutter ant (*Atta*) and the small minor workers. The white background is fungus, used for food, which grows on leaves brought back to the nest by workers.

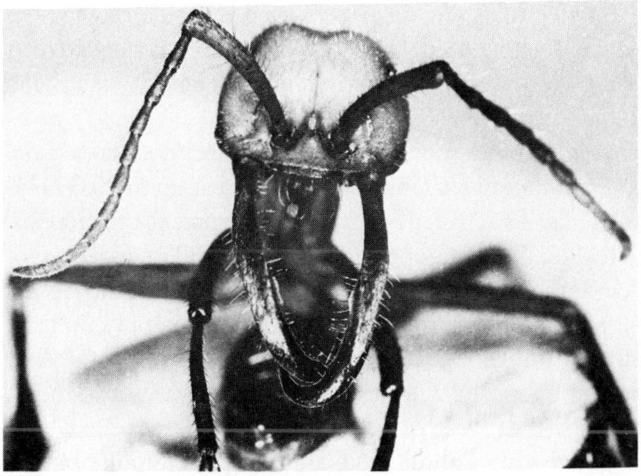

Figure 218. Army ant major (*Eciton burchelli*) with its ice tong type mandibles. Trophallaxis from workers is the only means of obtaining food for this defense specialist.

rived from the reabsorption of her flight muscles and fat body. Because the first brood receives a relatively poor diet, the resulting workers are small minor workers. Larger workers appear as the colony grows in population and more food becomes available. Fully grown larvae produce a cocoon (Fig. 219) in sharp contrast to the naked pupae in primitive species.

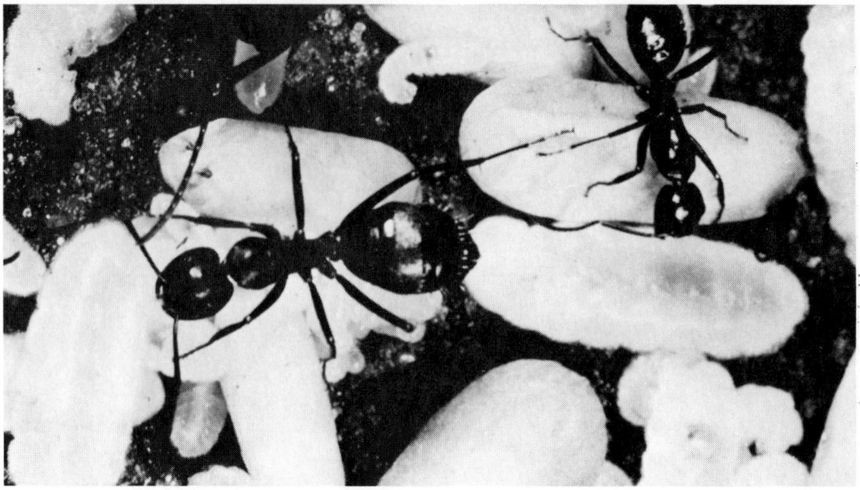

Figure 219. Eggs, larvae, enclosed pupae, and adult carpenter ant (*Campanotus pennsylvanicus*).

Ant colonies, unlike most termites, increase rapidly in population during the first year. A colony may be started by one or several queens depending on the species of ant. Queens may survive from 12 to 17 years, whereas the workers may only live 5 or 6 years. In some ants, when population size becomes excessive, the colony may divide, each with a separate queen.

A false impression that one receives from observing ants is the superficially busy atmosphere. In actuality, most individuals are inactive. If food is scarce, as many as 30 percent of the workers may be seeking food, but a "well-fed" colony may have as few as 5 percent searching at a given period. Some, however, may be participating in other activities, including colony protection (Fig. 220).

Ants have the widest range of diet of all social insects. Some are carnivorous, and a single colony of army ant (*Eciton burchelli*) may capture and eat up to 90,000 insects and other arthropods per day as it forages through neotropical rain forests. Others are scavengers, and a few tend various species of Homoptera for sugar secretions. Many species also have turned to plants and their products for food. Seeds, fruit, and nectar from special plant structures, the *nectaries,* are commonly utilized. A few plants produce specialized fruitlike structures that some ants use for food. Ants also may subsist on fungi raised in specialized nest chambers, or *fungus gardens.* Species of *Atta,* the leafcutter ants (Figs. 123, 217), can strip entire trees of leaves in several nights. These leaves, which are used for food, are transported to the fungus gardens within the nest. In the case of *Atta,* the fungus is specific, and departing queens take spores with them in head pouches for initiating new fungal gardens.

Trophallaxis plays a greater role in ants than it does in the social wasps. Ant larvae are fed by adults and in return produce fluids, from the body integument and not from saliva as in wasps, that the adults ingest. This interaction is believed to provide a firm base for maintaining the functioning of the social roles within the colony.

Figure 220. Worker carpenter ant (*Campanotus* sp.) in defensive pose. Formic acid, a potent irritant as well as an alarm pheromone, can be sprayed forward at an antagonist.

Each ant often deposits material during its movements that results in "trails." These trails can be followed by others and are important in finding food, direction to nests, and often in determining distance.

Nests of ants vary greatly, but in no instance are they comparable to the highly evolved finite structures of the specialized termites, bees, and wasps. In harvester ants, nests consist of a gravel mound that provides good water drainage. Optimal temperatures for the occupants are obtained by vertical migrations. In the morning or evening, eggs and larvae are carried upward in the cone for warmth from solar radiation. During the extremely warm afternoon heat, the brood is transported downward to cooler sites. In the tropics, leafcutter ants may have massive nests of more than 20 ft (6.1 m) in diameter that have multiple openings and that descend more than 10 ft (3.05 m) in depth. Some ants have become arboreal, and the entire colony lives in hollow twigs or thorns. The presence of these ants often provides protection for the plant by driving away potential herbivores. A few ant species such as the army ant lack nests; the permanent nest is replaced with a temporary structure or *bivouac* (Fig. 221) formed by interconnecting legs and bodies. In the New World, this bivouac contains the brood and remains formed for less than 24 hours except during periods when the queen lays eggs.

Social Bees

Although having evolved independently at least eight times, social bees parallel the social wasps, especially in the temperate regions. Differences between queens and worker bees in primitive species are often only that of dominance behavior. Unlike the food for wasps, food for bees usually consists of either nectar or pollen mixed with nectar. Also, certain brood cells are often used for storage of food (this is in contrast

Figure 221. An army ant bivouac (*Eciton burchelli*) formed by the intertwining of ant bodies and legs. (Photograph by C. W. Rettenmeyer.)

to the social wasps). The more conspicuous social bees are the bumblebees, honeybees, and the tropical stingless bees.

Bumblebees vary from most insects in being more numerous in the temperate than in the tropical region. They go through two phases in their adult history, a overwintering and inseminated female starts a new nest in deserted rodent burrows or bird nests. Initially, a honey pot is fashioned from wax and filled with nectar. Subsequently, brood cells are constructed, a few eggs are deposited, and the resulting young are fed pollen and/or nectar. Full-grown larvae (Fig. 363) spin cocoons and pupate for approximately 2 weeks. Adults that emerge are all females, usually smaller than their mother (Fig. 222), and are subordinated by her aggressive behavior. These workers stay in the colony and feed the young, convert the old brood cells to new storage tanks, or become foragers for nectar and pollen. Growth during the summer may result in several thousand individuals, but several hundred is more common. In large colonies, a few workers also may lay eggs, but these become males or drones (parthenogenesis).

The most advanced social bees are the honeybees, genus *Apis.* About nine species are known. Some species nest openly from tree branches, whereas others may have either exposed layers of wax cells in their nests (Fig. 223) or are located in caves, hollow trees, or logs. The opening into a tree hive of *A. mellifera,* the only honeybee in the United States, is small, and the bark outside the entrance is usually smoothed by the bees. The hollow cavity usually is elongate vertically (about 45 liters in volume) and is lined with *propolis* (plant resins). Combs are fastened to the side and top of the cavity walls, with waxy cells for honey storage located dorsally and the brood ventrally.

Apis mellifera has been associated with humans for hundreds of years, and most references within the text to honeybees will refer to it. Commercial hives (Fig. 224)

Figure 222. Queen bumblebee, *Bombus pennsylvanicus,* incubating a brood cluster. Smaller workers are in the process of filling honey pots.

Figure 223. An exposed nest of the honeybee, *Apis mellifera.* Such nests are common in the southern United States but do not survive winters in areas where long periods of freezing temperatures occur, as this photograph records.

consist of a series of open-ended boxes stacked upon one another, each containing vertical "super" frames with combs whose cells open laterally. A cover and bottom are added to the top and bottom box, respectively, to close the hive. As the population increases, additional boxes may be added. The hive entrance is located within the bottom box and normally faces either east or south.

Figure 224. Honeybees (*Apis mellifera*) kept under commercial conditions.

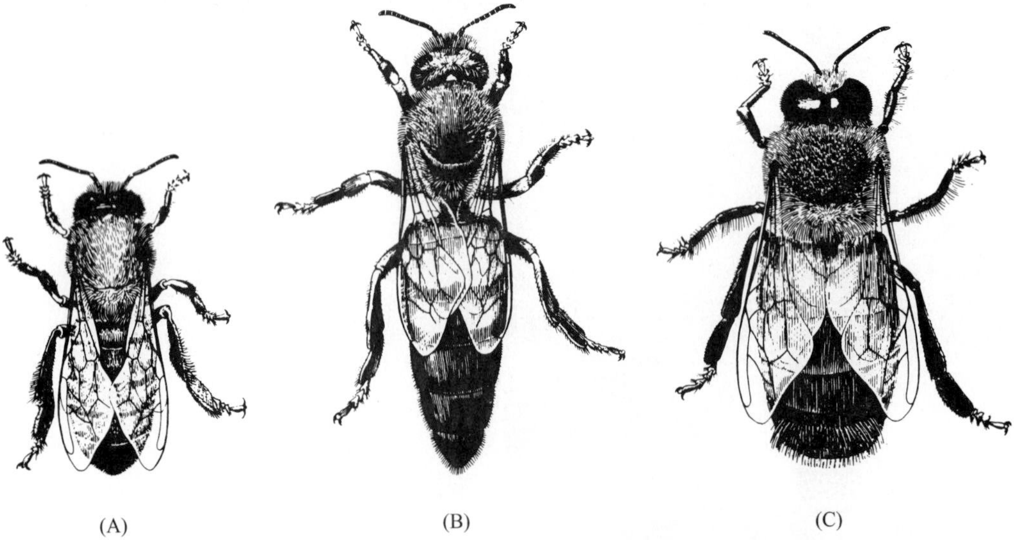

Figure 225. The honeybee castes (*Apis mellifera*). (A) worker or sterile female; (B) queen or fertile female; (C) drone or male. (Courtesy of Agricultural Research Service, U.S.D.A.)

The castes of honeybee (Fig. 225) consist of the fertile diploid female (queen), the somewhat stunted diploid females (workers), and haploid males (drones). A single queen and about 30,000 to 80,000 workers compose a colony during peak population periods, and about 200 drones are added just prior to and during colony division. Drones serve only one function: to mate with virgin queens from other colonies.

New colonies originate by division of the old colony. That takes place when the established queen leaves the nest with a swarm of up to 30,000 workers (queens can-

Figure 226. Capped brood cells in honeybee hive.

not survive alone). The new swarm often forms a cluster on vegetation as the bees await site selection by scout bees. When high numbers of scouts have visited a preferred site, they initiate dance movements into the swarm after returning to activate their sisters. Once airborne, the swarm flies directly to the site using both visual and pheromone clues and establishes a new nest.

After departure of the swarm, the old colony is queenless; however, help is on the way. Just prior to swarming, special queen cells that resemble peanuts were constructed and a newly hatched larva was placed into each. These larvae feed continuously on royal jelly that was mass-provisioned within each cell. Royal jelly is a substance that contains high amounts of worker mandibular and hypopharyngeal secretions and high amount of sugars; this rich diet induces high titers of juvenile hormone that in turn results in rapid development and fertility. After about 16 days, the first queen emerges and systematically destroys other queen pupae.

If two queens emerge at the same time, a battle ensues until only one survives. The virgin queen now exits the nest for several nuptual flights where she mates with about 7 to 17 drones (drones die after mating) before returning to the hive. Initially, the remaining workers are her sisters, but these eventually will be replaced with her own progeny.

Queens oviposit into wax cells of each comb. Larvae hatching within these cells (Fig. 226) are fed worker jelly, a secretion that has low amounts of hexose sugars mixed with mandibular gland secretions. After 3 days, beebread (honey mixed with pollen) is substituted. This diet switch alters the developmental program by lowering the juvenile hormone concentration and results in stunted, sterile females or workers. Development becomes longer than occurred in the queen and takes approximately 21 days. Newly emerged workers now enter into a series of age-dependent tasks during their short summer existence of approximately 6 weeks, progressing from the initial cell cleaning chores to brood duties, to food storage, before finally emerging from the hive to forage for nectar and pollen. Worker bees produced late in the summer are often called "winter" bees and possess increased longevity that enables them to survive the cold winter.

Honeybees, like ants, utilize the sun and polarized light for orientation during foraging activities. On cloudy days, navigation appears to be based on memory using any landmarks available (Dyer and Gould, 1981). Once a food source is located, workers return to the colony and communicate the direction (in relation to the sun) and distance to the food by a series of dances on the vertical surface of the combs. Each strain or species of honeybee has slightly different dances and interpretations. Figure 227 illustrates two such dances. A strong, vigorous dance with the accompanied sound of wing vibrations (Kirchner and Towne, 1994) creates excitement because the food source is near and abundant; a weak dance elicits little response, and hive energy is therefore conserved. This communication by dances is unusual and is not found in most bees, although some stingless bees may leave a scent trail to assist others in locating food.

Usually, a bee forages for only nectar or pollen, but a number of worker bees (17 percent) take both on the same trip. Pollen adheres to the body hair and is then combed off and packed into special baskets on the outer surface of the hind tibia. Nectar is sucked into the crop, or "honey sac." When the bee returns to the hive, the pollen ball is scraped off each hind leg and packed into a wax cell, and the nectar is either fed to others or regurgitated into cells for storage. Evaporation of water occurs, and the honey is capped [Figs. 228, 268(B)]. Normally from 15 to 100 lb of honey and from 1 to 15 lb of pollen are needed as a reserve for the commercial colony, depending on climate and colony size. Foraging flights for pollen or nectar require considerable energy. For every continuous 2 hours of activity, each bee consumes its weight in sugar.

Bees in the tropics inhabit a uniformly warm environment, with rain and predators acting as the major limiting factors. The honeybee in the temperate region, mainly *A. mellifera,* however, must also regulate temperatures in the nest to counteract both the heat of summer and the cold of winter. Warm temperatures (approximately 92°F or 33°C) are necessary for rapid brood development, and the brood dies or becomes abnormal as adults if the temperature is maintained below 30°C. If temperatures rise too high and become potentially detrimental for the colony, honeybees fan their wings and increase air flow and evaporation of water within the nest. Some bees orient themselves at the hive entrance to increase air exchange with the outside environment (Fig. 229). During cold periods (below 50°F or 10°C), honeybee populations drop significantly, and the bees uniquely form a massive aggregation called a cluster. Metabolism, insulation produced by the layered hairy bodies, and alternate movement into and out of the cluster coupled with shivering maintain heat far above the hive and ambient temperatures. Energy for this homeostatic system comes from the fat body and stored honey. Species other than *A. mellifera* are not as effective in regulating

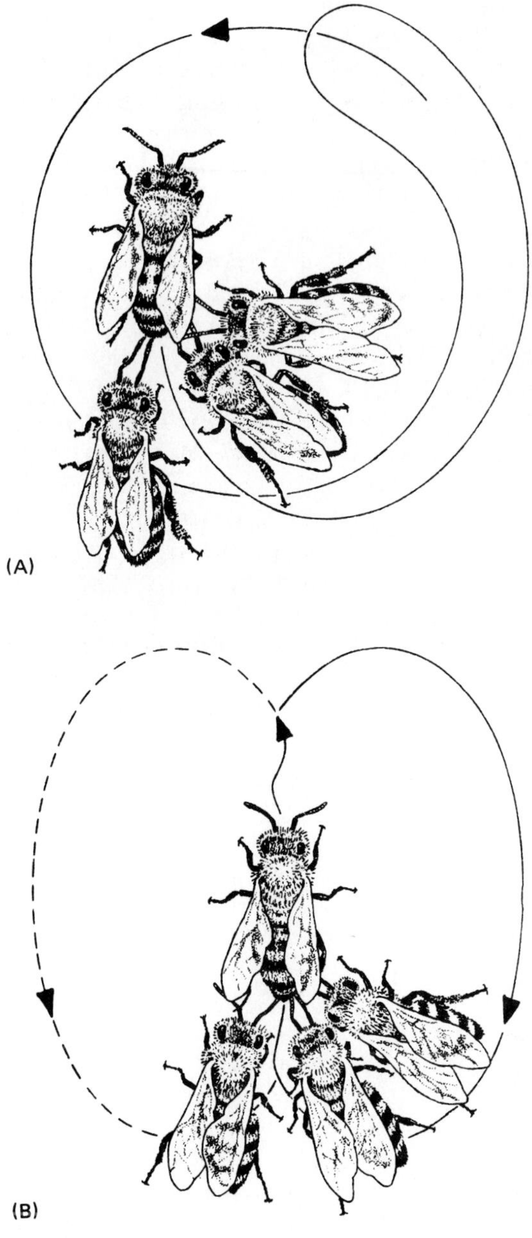

Figure 227. Types of dances by honeybees used in communicating nectar sources. (A) round dance indicating a source near the hive; (B) wagging dance indicating direction (angle of the straight run to gravity is same as angle of nectar to sun) and distance farther away than in the round dance. (From "Dialects in the Language of the Bees" by Karl von Frisch. Copyright 1962 by Scientific American, Inc. All rights reserved.)

temperatures and are usually unable to survive temperate climates. In summary, surviving temperate climates requires that (1) the bees cluster in a protected nest site, (2) the temperature of the outer surface of the bee cluster be above 9°C, (3) at least 20 kg of honey be stored as an energy reservoir for winter, (4) populations become reduced, and (5) brood production initiates in midwinter to produce maximal numbers of bees for exploiting the spring nectar flow prior to spring colony division.

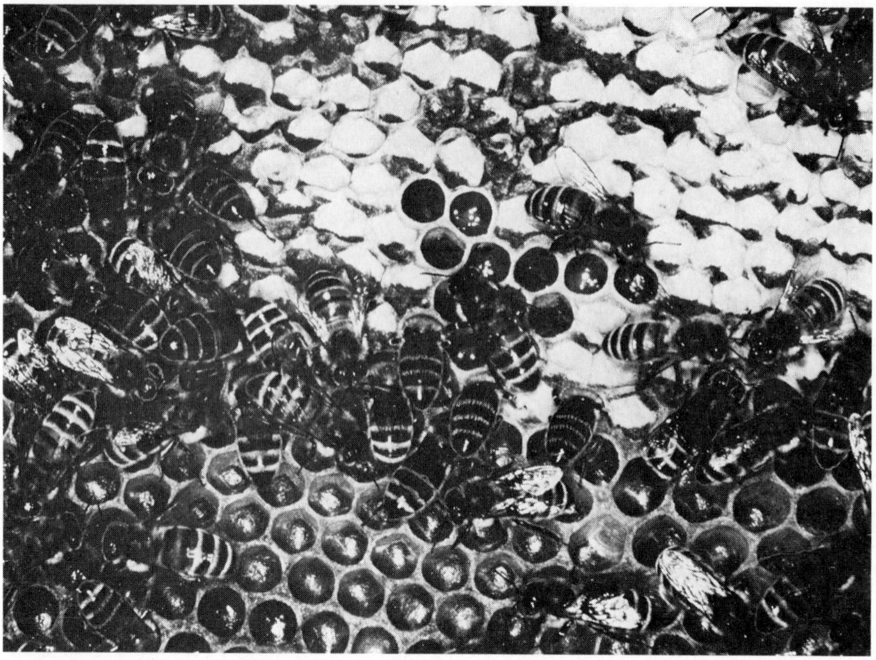

Figure 228. Honeybees (*Apis mellifera*) and honey production. Some cells are capped, and others are in the process of being filled.

Figure 229. Worker honeybees (*Apis mellifera*) fanning wings at hive entrance to cool the hive (ambient temperature more than 95°F).

Beginning in the early 1960s, concern has been expressed about an aggressive strain of honeybee, *A. mellifera scutellata*, which has been dubbed the "killer bee" by the New World press. It was introduced into Brazil from Africa in 1956 to hybridize with resident European strains that were poorly adapted for the existing tropical conditions. The strain, however, has several traits that bring it into conflict with humans. This African strain frequently absconds, a behavior in which the entire colony abandons the nest, and engages in high stinging activity when disturbed, about 10 times that of European strains.

Unfortunately, several dozen colonies from the original introduction either escaped or were released in 1957. Those bees that spread northward have remained relatively pure (little hybridization with other honeybees), and they reached Venezuela in 1978, entered Guatemala in 1985, and reached the United States in 1990. Various predictions have been made concerning their eventual distribution and impact on the environment and humans, but there is much disagreement (Spivak et al., 1991). There is no question that some nonprofessional beekeepers in the south will cease functioning. For those beekeepers who remain, some modifications from their past routines must take place: (1) better protective and more conscientious use of protective clothing, (2) commercial colonies located in more isolated areas away from livestock and human dwellings, (3) colonies distanced from one another to reduce vibrations while working the bees, (4) more efficient and powerful smoke blowers, (5) queens obtained from reliable sources where lineages can be certified, and (6) good insurance coverage to cover the inevitable lawsuits resulting from accidental stinging incidences of the public.

QUESTIONS

1. What criteria are used to define a social insect? Are social insects common? Are their sex numbers equal in a colony?
2. What are the advantages and disadvantages of each individual's being a member of a colony? Do these differ from the advantages and disadvantages gained by the colony as a whole?
3. What determines castes in termites and honeybees?
4. What selective pressures are exerted upon a colony as a result of individuals having to undergo complete metamorphosis?
5. What means of defense do insect colonies possess?
6. What are *inquilines*? How might this relationship have arisen?
7. Are social insects better able to modify the environment than solitary species?
8. How are new colonies formed by termites, ants, honeybees, and social wasps?
9. Are termites beneficial or detrimental in nature?
10. What is the African honeybee? Of what importance are these bees to the people in North America?

8 Parasitism by Insects

Not only do individuals aggregate within their own species to form colonies, as discussed in the previous chapter, but they also may become associated with other species to form symbiotic relationships. *Symbiosis* means "together" (*sym*) + "life" (*bios*). Although some authors use the word interchangeably with *mutualism,* most biologists use symbiosis in a broad sense to designate a prolonged and intimate relationship between two different species in which at least one species benefits. This association or interaction can be divided into mutualism, commensalism, and parasitism. In *mutualism,* both species benefit from the presence of the other. In *commensalism,* the second species neither derives benefit nor is harmed. In the third type of symbiosis, *parasitism,* the relationship is one-sided, and the second species (*host*) is harmed but not immediately killed as in the case of predation. However, there is a form of parasite, the *parasitoid,* that does cause death to its host but only after completing its development. Discussion in this chapter will be restricted to parasitism, a phenomenon found in 10 insect orders (Table 3) and in about 15 percent of all insect species (Askew, 1971).

After viewing the multiplicity of parasitic niches, one sees a continuum of interactions, from parasites that differ little from carnivores or scavengers to intimate relationships between host and parasite. Adaptations for different feeding sites on or in the host also have to be selected for over time. Categorization is useful to illustrate the interactions and adaptations of parasitism. One commonly used classification system is based on the degree of interaction between host and parasite. Those species that cannot survive and reproduce without a specific host are *obligate parasites.* These individuals usually spend most of their life cycle, at least during the parasitic phase, upon the host. Other species that visit a host periodically for food may be described as *intermittent parasites,* and different individual hosts can be utilized. Finally, *facultative parasites* are organisms that normally exist as free-living individuals but become parasitic when the opportunity presents itself.

TABLE 3. Orders of Insects with Parasitic Species (on animals)

Order	Percent Species	Stage	Host
Dermaptera	1	Nymphs, adults	Mammals
Mallophaga	100	Nymphs, adults	Birds, some mammals
Anoplura	100	Nymphs, adults	Mammals
Hemiptera	<1	Nymphs, adults	Mammals, birds
Neuroptera	4	Larvae	Arthropods
Coleoptera	2	Larvae and/or adults	Invertebrates, mammals
Lepidoptera	<1	Larvae or adults	Insects, mammals
Diptera	12	Larvae or adults	Invertebrates, vertebrates
Siphonaptera	100	Adults	Mammals, some birds
Hymenoptera	50	Larvae, some adults	Arthropods

A second classification system is that of *ectoparasites* and *endoparasites,* the former feeding on the host externally and the latter internally. Again we note a cline of variation as well as the difficulty of designating the host habitat as external or internal. For example, are the nasal sinuses of vertebrates internal or external? A third system (Askew, 1971) separates those that feed parasitically as adults (and sometimes their immatures) from insects where only the immatures are parasitic (*protolean parasites*).

Delicate balances often exist between the host and parasite, particularly when the parasite becomes *host-specific,* or requires a specific species or genus of host to complete its life cycle. First, the host must be available and locatable. Various visual, chemical, tactile, and temperature cues have to be sorted out by the parasite itself, as in the case of intermittent parasites such as mosquitoes, or by the adults of protolean parasites, as in Hymenoptera and Diptera. Second, the parasite or parasitoid must be capable of utilizing some host resource for food to survive and reproduce. Those parasites that ingest blood are referred to as *hematophagous,* those that feed on tissues are *histophagous,* and others that utilize mucus are *mucophagous.* Third, feeding without or with minimal irritation and damage to the host is advantageous. The ideal parasite will flourish without affecting the host's ability to develop and reproduce.

In contrast to parasites that are host-specific, most species are somewhat more liberal in their preferences for hosts. For example, specificity of many fleas is low, and the majority feed on a wide variety of mammalian and avian hosts. Whenever mammals and/or birds share a domicile, they commonly share fleas. However, hosts (birds) with very similar types of nests and behavior are quite often varied in the number of parasites feeding on them; for example, 17 species of fleas are specific to swallows or martins, yet swifts do not have a single flea peculiar to themselves. Many ectoparasites, such as lice, are quite specific as to their host species as well as to regions upon the host itself, and the host appears to be partitioned geographically to reduce interspecific competition. Also, life cycles of highly modified parasites are often synchronized to changes in the physiology of their host, a fact illustrated by certain rabbit fleas that are induced to reproduce at the same time as the host by ingesting hormones from their rabbit host.

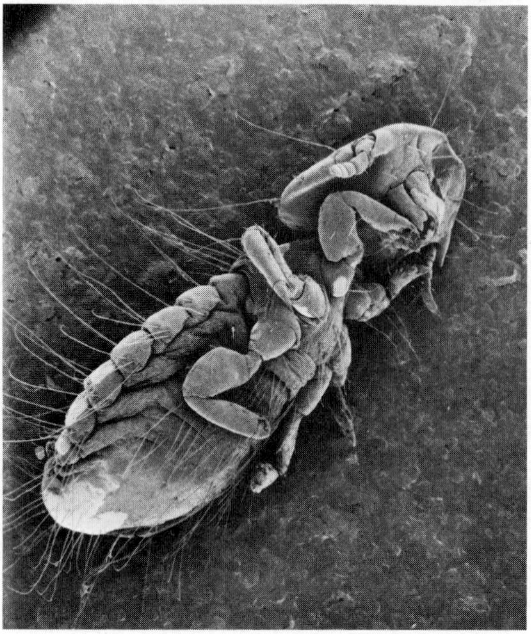

Figure 230. A chewing louse in the large family Philopteridae as seen under the scanning electron microscope.

To explore the great variation of insect parasites as logically as possible, parasitism will be categorized first into ectoparasites and endoparasites and then illustrated through life cycles and other important biological aspects of selected examples. The discussion will proceed in a taxonomic sequence starting with hemimetabolous orders and ending with the holometabola. Also, the important nonparasitic stages found in many of the intermittent parasites' life cycles will be included.

ECTOPARASITES

The large majority of parasitic insects feed through the outer surface of the host. Ectoparasites with chewing mouthparts usually ingest cells or blood seepage, whereas others with piercing-sucking mouthparts either penetrate and take blood directly from blood vessels (*solenophages*) or lacerate blood vessels and feed from the resulting blood pool (*telmophages*). Some feed only as adults (fleas, mosquitoes), others only as larvae (some Hymenoptera), and certain species, normally those with incomplete metamorphosis, are parasites both as adults and as immatures. Structural adaptations often include the absence of wings in most of the obligatory forms and in some intermittent parasites, and the presence of spines to deter removal by the host (principally vertebrates) and of mouthparts for piercing tissues.

Mallophaga

Chewing lice are continuous inhabitants of birds (Fig. 230) and a few mammals. They are usually flattened dorsoventrally, lack wings, have reduced eyes, and feed upon

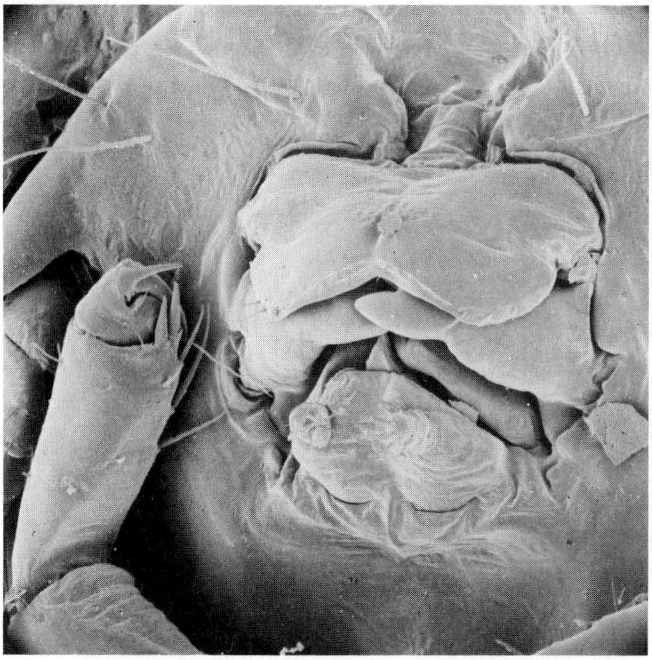

Figure 231. Enlarged view of Figure 230 showing the dorsal labrum, lateral mandibles that are pointed and ventrally positioned, and reduced maxillae and labium.

feathers or hair, skin debris, serum, secretions of sebaceous glands, or sometimes blood from developing feathers. Several genera ingest only blood and serum and have pointed mandibles for piercing the skin (Fig. 231). Mallophaga that infest birds have two pairs of claws, whereas mammal inhabitants possess only a single pair but have a modified thumb-claw complex to anchor themselves to the hair. Eggs are glued either to the feathers on the inside of the quill or to the hair in mammal-infesting species. Three molts are required, usually within 30 days, to reach adulthood. These lice rarely leave the host, and individuals have been collected from preserved museum skins long after the death of the host. Correlations between louse shape and preferred sites on the host are high. Chewing lice found on the necks and heads of birds are normally round-bodied with large heads and are the most host-specific, whereas flattened and elongated species are located on the back and wings where they can move sideways rapidly between feathers to avoid preening by the host. Species located elsewhere on the bird have low specificity and their shape varies extensively. Transfer from one host to another occurs during mating, brooding, and roosting periods. The presence of these parasites often induces dusting and preening by the birds, and selection has resulted in certain birds possessing specialized combs on their claws for removing these lice.

Anoplura

Sucking lice are restricted to mammals. Structural adaptations include highly modified piercing-sucking mouthparts that are retracted into a slender head, poorly sclerotized bodies, absence of wings, and well-developed grasping legs (Figs. 232, 233)

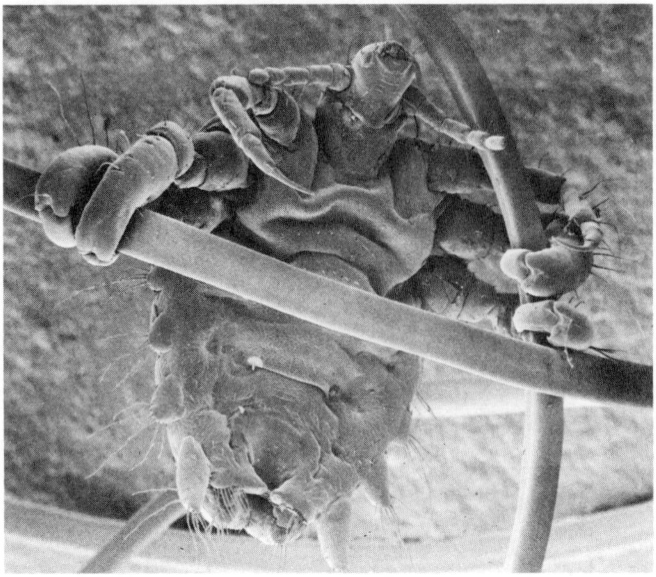

Figure 232. A sucking louse, *Pthirus pubis,* attached to a human pubic hair.

Figure 233. An enlarged view of Figure 232, indicating the adaptation of the legs for grasping hair.

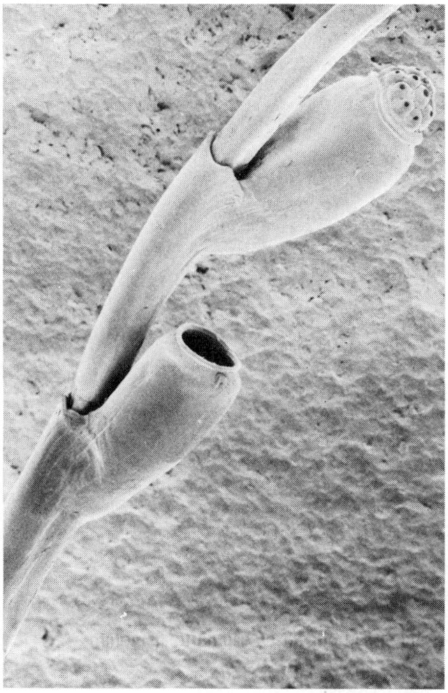

Figure 234. Two eggs of a sucking louse, *Pthirus pubis,* attached to a human pubic hair. The lower egg has hatched, whereas the upper one contains an unhatched nymph beneath the detachable operculum.

for anchoring to the host hair. With the exception of the human body louse, lice eggs are glued to the host's hair (Fig. 234), and, as in Mallophaga, the host's body heat produces rapid embryology. Nymphs emerge from the eggs by compressing air behind their body, which forces them against the egg cap or *operculum* until the cap snaps open (Fig. 234). Feeding begins shortly thereafter and occurs intermittently until the normal third or fourth nymphal instars are completed. Females are more common than males, with the latter virtually unknown in a few species. Host and site specificity are high as illustrated in the family Echinophthiriidae, which is restricted to seals (the lice utilize air trapped in the dense underfur when the host is in the water). Most lice can endure only short periods off the host, so transfer is usually restricted to mating and periods of suckling of young by the host. *Mycetomes,* peculiar internal structures containing symbiotic microorganisms, are seen in these parasites; these structures remain in the male but are passed on to the young through the eggs by the females and probably provide needed vitamins for the lice.

Hemiptera

In contrast to the obligate parasites Anoplura and Mallophaga, intermittent parasites are found in the order Hemiptera. Bedbugs are a complex of 74 species classified as Cimicidae. These wingless bugs are reddish brown, have broad and dorsoventrally flattened bodies, possess a three-segmented beak, and are avoided by many potential predators because of their disagreeable odor. Eggs are laid in cracks, under wallpaper

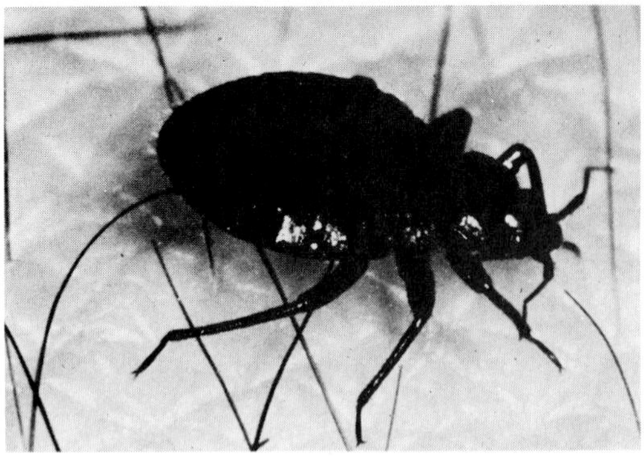

Figure 235. An adult bedbug, *Cimex hemipterous,* on a human hand.

Figure 236. *Triatoma sanguisuga,* a reduviid, is a bloodsucking bug that feeds on mammals including, occasionally, humans.

or mattresses, in crevices, or in bird nests. At night both nymphs and adults leave these hiding places and feed upon warm-blooded hosts. Most species prefer birds; however, *Cimex lectularius* in temperate regions and *C. hemipterus* (Fig. 235) in the tropics seek humans. Both nymphs and adults require blood, the former to molt and the latter to reproduce. A blood meal can be obtained within 3 to 15 minutes, whereupon the bug returns to hiding. If food is abundant, development can be completed within 10 weeks; however, if a host becomes unavailable, bedbugs can survive for more than a year without food, although development is delayed.

A second group of Hemiptera, the reduviid bugs of the subfamily Triatominae, includes about 100 parasitic species, most of which exist in the New World. These bugs are often brightly colored, have a narrow head, and are good fliers, but they differ from other assassin bugs in having a straight beak (Fig. 236). Most species live in the burrows or nests of their mammal hosts, but a few associate with humans. Eggs are deposited either in the nest of the host or, for those found with humans, in cracks in walls and roofs.

Figure 237. *Platypsyllus castoris,* a beaver parasite. Note the flattened body and enlarged spines that are adaptations for parasitism on mammals.

Coleoptera

A few species of this holometabolous order are ectoparasitic. The adult *Platypsyllus castoris,* a species of beetle inhabiting beavers, is flattened dorsoventrally (Fig. 237), possesses many long hairs and spines to reduce removal by the host, has an enlarged pronotum, has elytra fused medially, and lacks eyes. This obligate ectoparasite deposits eggs in the fur of the host, where both larvae and adults (rare for holometabolous parasites) probably feed on skin debris.

Siphonaptera

Fleas are parasitic only as adults. They are flattened laterally and lack wings (Fig. 238) for ease in moving between hair, have piercing-sucking mouthparts (Fig. 239), possess strong bristles and often enlarged spines or *ctenidia* for snagging hair during removal

Figure 238. Lateral view of a flea, *Cediopsylla simplex,* normally parasitic on rabbits. Note the enlarged spines or ctenidia on the gena and pronotum that deter removal by the host.

attempts by a host, have reduced eyes, and have enlarged legs for jumping and contacting their mammal or bird hosts. Unlike lice, most species leave the host frequently between blood meals and often oviposit in the nest. The legless larvae (Fig. 379) feed mainly on organic debris accumulating in the host nest, although a few species, such as the rat flea, *Xenopsylla cheopis,* also require blood squirted to them by the adults as the latter feed. The larval stage is completed in from 1 week to several months depending on the species and climatic conditions. Larvae then spin silken cocoons and either emerge within a month or remain dormant as adults within the pupal case until a host appears. Adults survive from several weeks to a year or more.

Fleas that inhabit restricted areas, such as rodent nests, find hosts more readily than do nonnest fleas, but adults of both types locate and jump toward vibrations, heat, and high carbon dioxide concentrations, such as endothermic hosts normally emit. Leaping ability varies from poor in nest fleas to good in others such as the human flea, *Pulex irritans,* which has been observed to vault up to 30 cm horizontally and 20 cm vertically. Such extended jumps are possible only because of a protein, *resilin,* which stores compressed energy in the pleural exoskeleton for release through the hind legs. After leaping, the flea tumbles forward with its legs extended for grappling hooks.

Most fleas can feed on a wide variety of hosts (*polyhaemophagy*), but some require specific blood to produce viable eggs. Transfer to other hosts occurs when a new host enters the nest or when direct contact between hosts occurs. The apex of parasitism in fleas is seen in "stick-tight" fleas, a group that attach themselves to the host for extended periods and are capable of swelling to the size of a pea as blood is ingested. *Tunga penetrans* seeks humans and other mammals, whereas *Echidnophaga gallinacea* attaches to poultry and various animals. Both species often cause extreme irritations and infections.

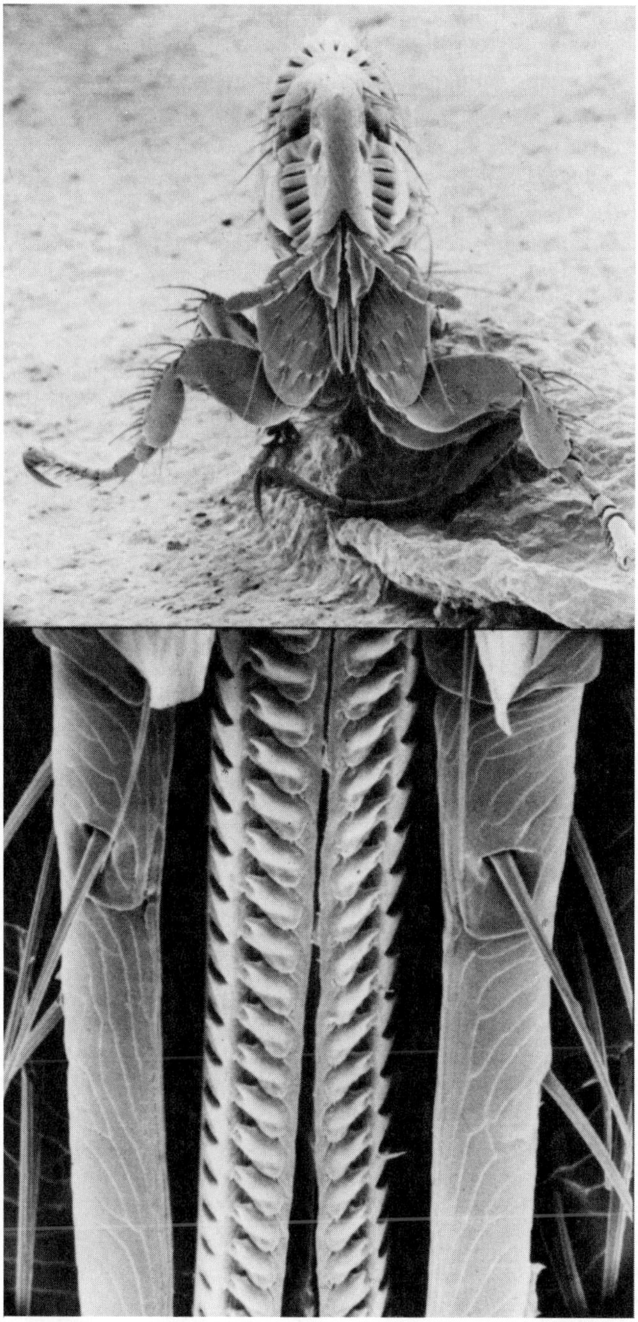

(A)

(B)

Figure 239. Frontal view (A) of the flea, *Cediopsylla simplex,* and enlargement of the lacinial stylets (B).

Diptera

Intermittent ectoparasites are common in the order Diptera, although usually it is the adult females that ingest blood. For convenience, these are categorized as biting flies and mosquitoes. The former includes horse flies, black flies, biting midges, stable flies, and tsetse flies. In contrast to fleas and bedbugs, most flies have wings and can search over extensive areas, up to several miles, to locate suitable hosts. Once located, tsetse flies (Fig. 38) and stable flies use well-developed teeth at the end of a labial proboscis to rasp and penetrate the skin until blood is released. Others, such as horse flies, deer flies, and black flies, possess cutting-sponging mouthparts [Fig. 36(A)] that lacerate the skin and blood vessels to release blood. During ingestion, a hungry fly may swell up to twice its original size. Larvae of biting flies usually feed either as scavengers (black flies in water, stable flies in manure) or as predators (horse flies in water and soil). A notable exception is the tsetse fly; larvae are retained and nourished viviparously within the female until ready to pupate, at which time they are deposited into moist soil.

The second group, mosquitoes, are slender-bodied with piercing-sucking mouthparts [Fig. 36(D)]. Females feed intermittently on blood, but they can survive long periods without blood, and some species produce viable eggs by ingesting only nectar and fruit juices (males never take blood). Hosts are located by odor, heat, carbon dioxide, and movement. The mosquito stylets are inserted through the skin into capillaries (Fig. 41), from which as much as 3 mg of blood may be withdrawn within several minutes (Fig. 240). Blood is routed directly into the ventriculus (only nectar goes into the crop). Eggs are usually oval and are deposited either singly or in rafts on water, under floating leaves, or singly in moist soil. They hatch from within 24 hours for those laid directly in water, to several years later in some *Aedes* eggs oviposited in depressions in arid habitats. The larvae or "wigglers" are free-living and usually feed on plankton or other suspended food particles. Air is obtained through a terminal siphon in both the larvae [Fig. 385(A)] and the nonfeeding but motile pupae (Fig. 132).

ENDOPARASITES OF INVERTEBRATES

Many thousands of insect species, usually Diptera or Hymenoptera, are protolean parasites or parasitoids. Most parasitize free-living insect hosts and are therefore *primary parasites,* but a few utilize other parasites or parasitoids a condition termed *hyperparasitism. Multiple-parasitism,* a condition in which more than one species infest a host, may also occur.

Diptera

Hosts for dipteran parasites are located either by the free-living adults or by the recently hatched larvae. Some bee flies and flesh flies drop eggs as they hover over their host, whereas others attach their eggs directly to the body of the host. A few, such as *Stylogaster* (Conopidae) and a few tachinids, pierce the host exoskeleton with their tubular

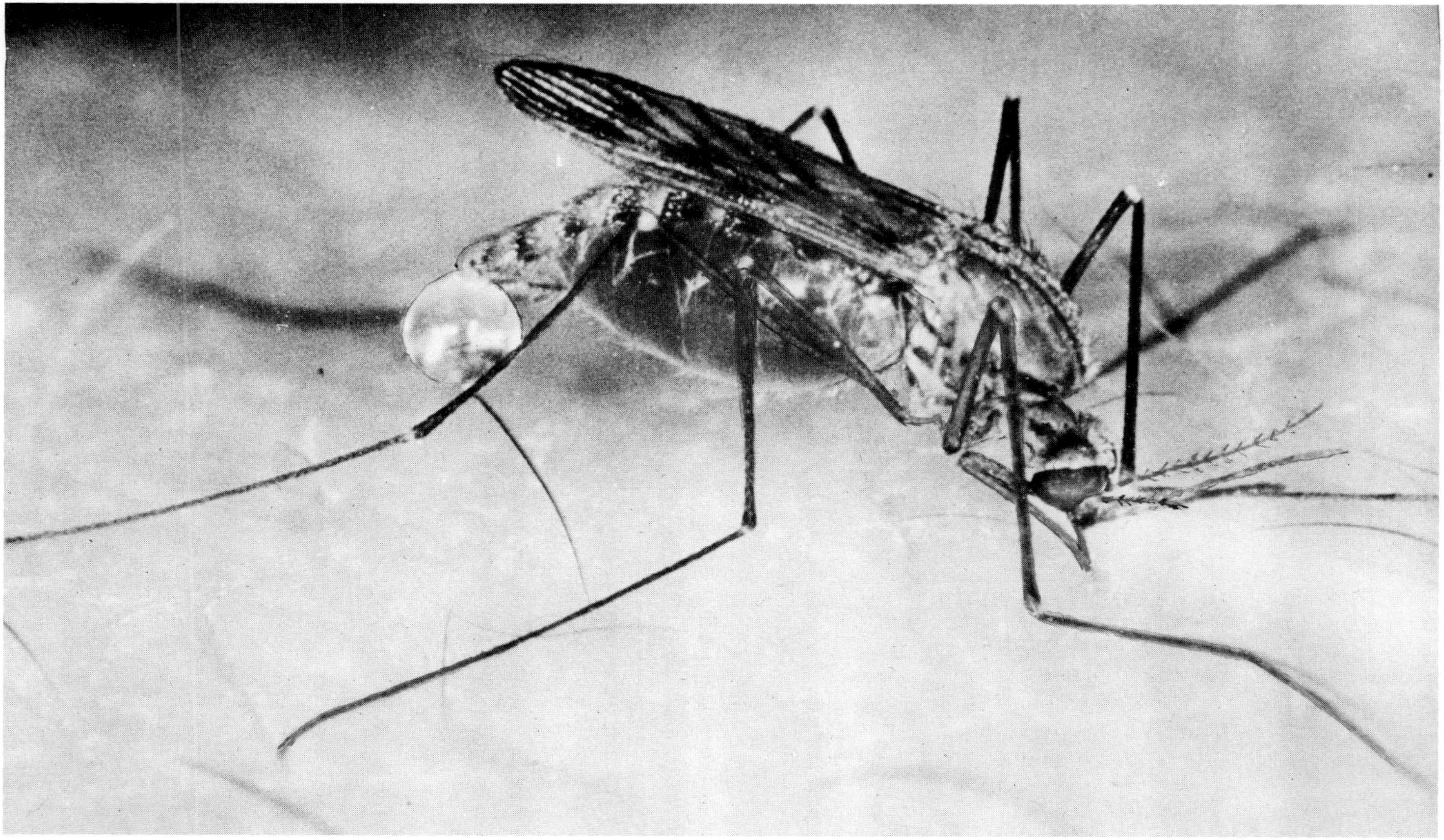

Figure 240. A female mosquito, *Anopheles quadrimaculatus,* engorging upon blood. Note the large excretory droplet forming during the feeding process and the abnormal feeding position (most *Anopheles* assume a 45° angle when feeding, but this individual failed to do so).

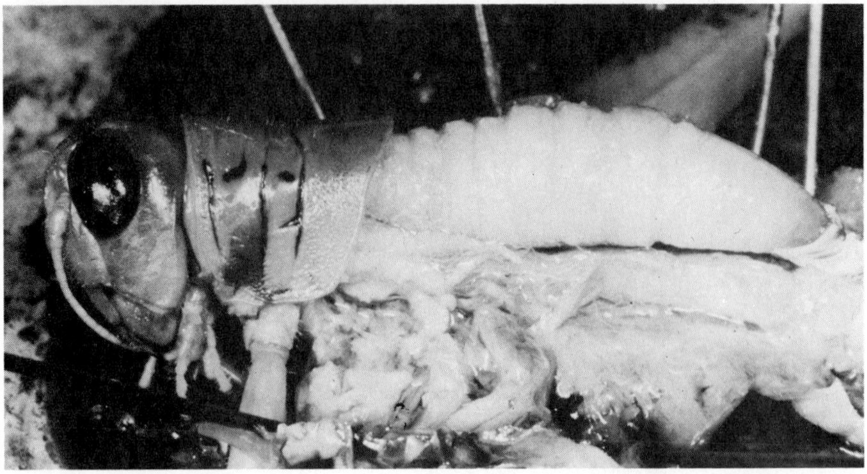

Figure 241. A short-horned grasshopper (*Melanoplus differentialis*) dissected, revealing dorsally a larval nemestrinid fly. Air to the parasite is obtained by attaching to the host's body wall through a long respiratory tube.

ovipositor. Many Diptera, however, locate their host as larvae. The Nemestrinidae, a family resembling bee flies, utilize a specialized first instar larva, the *planidium,* to contact the host. Tachinids may also have a planidium, but many oviposit on leaves where the eggs are ingested by the host. Once the host has been penetrated, the larva can withstand the antibody system and the encapsulation attempts of the host, although a few protolean parasites actually use the encapsulation mechanism to their own advantage and live within a chitinous sac produced by the host. Host tissue and suspended material in the hemolymph are utilized as nutrient sources. Oxygen is often absorbed from the host, and some large dipteran larvae attach to the host tracheal system or body wall (Fig. 241), where they may be encapsulated. Pupation then occurs outside the host, often in the soil.

Hymenoptera

Hymenopteran protolean parasites and parasitoids include Braconidae, Ichneumonidae, and Chalcidae, as well as many other species (Askew, 1971). Hosts for these parasites and parasitoids include eggs and larvae of Lepidoptera, Coleoptera, Diptera, sawflies, and spiders. Adult female Hymenoptera possess an appendicular ovipositor, and some, such as Ichneumonid wasps (Fig. 175), use this adaptation in piercing extensive substrates to reach a host. In such instances the chorion is normally elastic, thereby permitting the egg to be constricted and lengthened as it traverses the narrow oviposition canal to the host. Other Hymenoptera use the ovipositor as a sting to temporarily immobilize the victim, after which eggs are either deposited on the host or are inserted into the hemocoel of the host. The braconid wasp, *Lysiphlebius testaceipes,* deposits ova directly into aphids. After feeding on and developing within a host, the larva cuts a small hole in the aphid exoskeleton through which the adult will emerge after completing pupation

Figure 242. *Lysiphlebius testaceipes,* a braconid, emerging from the exoskeletal remains of an aphid.

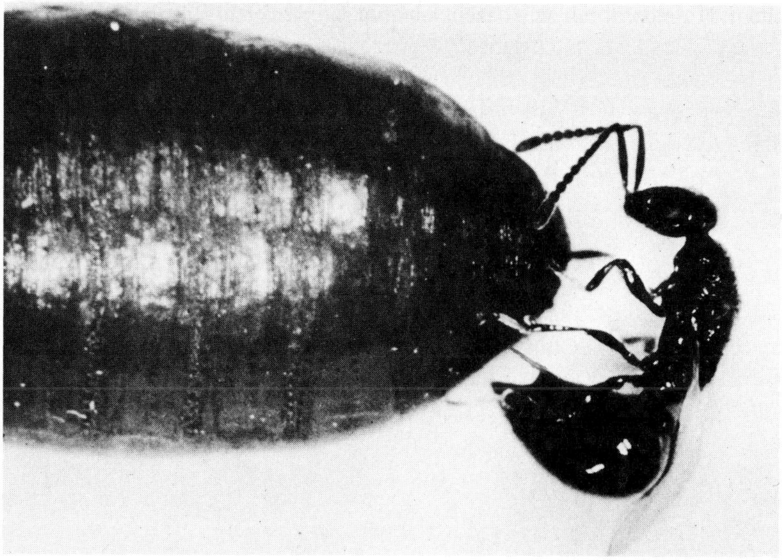

Figure 243. *Splangia endius,* a pteromalid, ovipositing on a house fly puparium. About 10 to 12 minutes were required to drill the hole through the exoskeleton of the puparium.

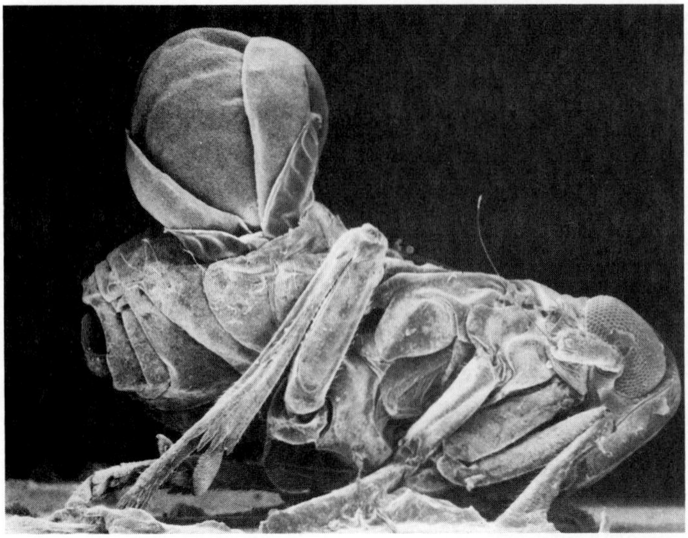

Figure 244. A fifth instar dryinid larva parasitizing a planthopper. Although the larva exists as a typical parasite until this stage, it will now ingest extensive internal host structures (prior to the departure), and the host will normally die.

(Fig. 242). Another example is the pteromalid wasp, *Splangia endius,* which parasitizes house fly and stable fly puparia (Fig. 243). Feeding by the wasp larva slows the host developmental period so that food will be available for its own 30- to 40-day development. A final example, one in which the parasite begins as an endoparasite but completes development as a "near" ectoparasite, is the wasp family Dryinidae. Eggs are inserted into the body of a homopteran host, and the larva increasingly bulges from the host with each instar until the fifth when it is mainly outside the host's body (Fig. 244).

Coleoptera and Strepsiptera

In the coleopteran Meloidae and Rhipiphoridae and in the order Strepsiptera (sometimes classified as beetles), a few species are also protolean endoparasites. Although adults oviposit in appropriate host habitats, it is the *triungulin larva* that locates the host. These first instar larvae have legs for rapid movement. After penetrating the host, a molt occurs, and subsequent parasitic instars are legless and grublike.

ENDOPARASITES OF VERTEBRATES

In contrast to invertebrate hosts, vertebrates possess much biomass and normally can survive moderate to high protolean parasite invasions. The most common endoparasites of vertebrates are larval Diptera of the suborder Cyclorrhapha. Parasitization by fly larvae or "maggots" is referred to as *myiasis.* Some larvae feed on blood, others live in the host's alimentary canal, and still others live in wounds or boils in the skin or nasal cavities. Zumpt (1965) postulated two evolutionary beginnings for myiasis: the first developed from saprophagous larvae that extended their feeding from dead tissues of wounds to the living, and a second arose from predaceous larvae living in

Figure 245. Horse bot larvae, *Gasterophilus intestinalis,* attached to the linings of the stomach of a horse. Note the mouthhook attachment that resists an attempt at removal by forceps.

rodent burrows or birds' nests that started feeding on these larger animals by puncturing tissues and sucking blood.

Obligatory species include some Muscidae, Calliphoridae, Sarcophagidae, and all Oestridae, Cuterebridae,and Gasterophilidae. In Gasterophilidae, eggs of the horse bots (genus *Gasterophilus*) are deposited on grass or in the hair of horses from which they are ingested with food or by licking. Larvae hatch within the mouth, penetrate oral mucous membranes, and migrate through the host's body to preferred sites—the stomach (Fig. 245), intestine, or rectum. Eventually they pass out with feces and pupate in the soil. Warble flies (Oestridae) oviposit on cattle. Larvae bore into the back of the animal and migrate for about 4 months, at which time they reappear as lumps or warbles on the back. Holes in the host's skin are torn to breathe. Growth continues for an additional 3 months until the larvae emerge or drop to the ground to pupate. Other endoparasites (may be also ectoparasites) are screwworms, which infect cattle, goats, sheep, hogs, and even humans; they are capable of extensive damage, even death, as they feed in wounds. Eggs are deposited initially at the edge of skin breaks; then the hatched larvae ingest blood and scabs, gradually extending their feeding out into living tissues. Full-grown larvae occur after 4 to 8 days. Pupation by these calliphorid larvae also occurs in the soil.

A more complicated life cycle involving *phoresy* (utilizing another animal for transportation without feeding) is found in *Dermatobia hominus,* the human bot fly or *torsalo* of tropical America. Females glue eggs to mosquitoes or other bloodsucking flies, which carry them to humans and other mammals. In response to host heat as the mosquito feeds, the larvae hatch and leave the insect. Using their sharp mouthhooks, the larvae bore into the subcutaneous tissues and commence feeding. A small orifice extends to the outside for breathing. After completing a developmental period of 5 to 10 weeks, the last larval instar leaves the host to pupate in the soil.

Some saprophagous larvae can withstand the digestive enzymes of vertebrates. If these free-living larvae are ingested accidentally, they either pass through unharmed (*pseudomyiasis*) or remain within the gut as facultative parasites. Many species of Sarcophagidae, Calliphoridae, and Muscidae exhibit this capability.

A few insect larvae other than Diptera parasitize vertebrates. If they are beetle larvae, the condition is called *canthariasis.* Lepidopteran infections are referred to as *scholechiasis.* Both, however, are rare occurrences.

SOCIAL PARASITES

Some social insects live in a parasitic state with ants, bees, or termites. Using the terminology of Wilson (1971) and Mathews and Mathews (1978), these relationships are as follows. One interaction involves a small species robbing a larger species of caches or robbing returning workers of food (often two ant species). This is *kleptoparasitism* (or *trophic parasitism*). The small opportunistic species is known as a *nest parasite,* and it survives because the host is larger and unable to pursue the parasite through its smaller nest burrows. Another interaction is *temporary parasitism,* a situation in which an ant or wasp queen enters an established colony of another species and eventually replaces the invaded queen and workers with its own progeny. The host species is needed only as the foundation for starting the colony. A more advanced relationship is *slavery* or *dulcosis.* Here one species of ant becomes dependent on another throughout the existence of the colony, specifically to provide it with a worker caste. The queen invades another ant colony, as in temporary parasitism, but both her offspring and the original species offspring are required for continued colony survival because the invading queen and her offspring are unable to forage for food and perform other worker type tasks. As needed, the slavemaking ants venture forth and raid colonies of the host ant species to obtain developing larvae or pupae. These "kidnapped" workers carry out most menial tasks of the mixed-species colony, and all ants collaborate as if they were the product of a single queen.

The ultimate of these interactions is *complete inquilinism* (Wilson, 1971), also known as *social inquilinism* (Mathews and Mathews, 1978). In this interaction slavery is not necessary because the parasitic ant lacks a worker population and survives within the host ant colony as a reproductive parasite. For example, *Teleutomyrmex schneideri,* a small "degenerate appearing" ant that lacks a worker caste, is usually found riding on the queen of another ant, *Tetramorium* sp. These inquilines are fed and groomed by workers of the host species as if they were a component of the colony. After mating, adult male and female *T. schneideri* either remain or seek a new colony. The host colony is affected adversely because the larvae are unable, for some unknown reason, to develop into reproductives and only develop into workers.

QUESTIONS

1. What are *parasites* and *parasitoids?* How do they differ from carnivores or scavengers?
2. How do obligate and intermittent parasites differ in their structure, behavior, and life cycles?
3. What are *protolean parasites?* How does their niche differ from that of fleas or lice?
4. Contrast the adaptive structures between ectoparasites and endoparasites.
5. What advantages are there to utilizing large animals, such as vertebrates, as hosts? What disadvantages might there be?
6. What is *host specificity?* What advantages and disadvantages can you see in being highly host-specific?
7. How does metamorphosis influence parasitism?
8. What are *social parasites?* How do they differ from other types of parasites and parasitoids?

9 Predation by Insects

Predators are free-living organisms that kill more than one prey organism for food in order to complete their development. Often they are larger than the organism being eaten. This form of foraging is common and found in all insect orders except for the Protura, Embioptera, Zoraptera, Isoptera, Mallophaga, Anoplura, Homoptera, and Siphonaptera. Although all stages in the life cycle may be predaceous (Odonata, Mantodea, most Neuroptera, some Coleoptera), in many instances this form of feeding may be restricted to a single stage. The following discussion will be restricted to examples from the orders Odonata, Mantodea, Hemiptera, Megaloptera, Neuroptera, Coleoptera, and Diptera. Each of these examples illustrates a notable variation in morphology, type of prey sought, development, habitat restrictions, and impact on other insects and organisms. Several will be of special importance to humans and will be discussed additionally in the biological control section of Chapter 11.

ODONATA

All species within this order of about 5,000 species are predaceous both as nymphs and as adults. Nymphs exist in the aquatic habitat and obtain oxygen either through their rectal gills (dragonflies) or through three caudal gills (damselflies). Prey consists of insects, crustaceans, and sometimes tadpoles and fish. A unique elongated and extensible labium with terminal-toothed lobes serves as a grasping organ; although normally folded under the head (Fig. 315), this structure can be rapidly extended forward, up to an inch in some species, to grasp prey (Fig. 138); then it is retracted to permit the mandibles to tear it apart. In dragonfly nymphs, water is taken into the rectum to encounter the rectal gills where oxygen uptake occurs, but this water can be ejected rapidly, thereby forcing the entire body forward in a similar manner to jet

TABLE 4. Common Predator Styles

Order-Family	Predation Abundance (species)	Nymphal Habitat Zone	Larval Habitat Zone	Adult Habitat Zone
ODONATA	All	Aquatic	—	Aerial
MANTODEA	All	On vegetation	—	On vegetation
HEMIPTERA				
Belastomatidae	All	Aquatic	—	Aquatic
Gelastocoridae	All	On shorelines	—	On shorelines
Gerridae	All	Water surface	—	Water surface
Naucoridae	All	Aquatic	—	Aquatic
Nepidae	All	Aquatic	—	Aquatic
Notonectidae	All	Aquatic	—	Aquatic
Pentatomidae	Some	On vegetation	—	On vegetation
Phymatidae	All	On vegetation	—	On vegetation
Reduviidae	Most	On vegetation	—	On vegetation
Velliidae	All	Water surface	—	Water surface
PLECOPTERA	Some	Aquatic	—	Nonfeeding
NEUROPTERA				
Chrysopidae	All	—	On vegetation	On vegetation
Myrmeleontidae	All	—	Subterranean	Nonfeeding
MEGALOPTERA	All	—	Aquatic	Nonfeeding
COLEOPTERA				
Cantharidae	All	—	Subterranean	Nonfeeding
Carabidae	Many	—	Subterranean	Soil surface
Dytiscidae	All	—	Aquatic	Aquatic
Cicindellidae	All	—	Subterranean	Soil surface
Coccinellidae	Most	—	On vegetation	On vegetation
Gyrinidae	All	—	Aquatic	Water surface
Hydrophilidae	Most	—	Aquatic	Nonpredaceous, aquatic
Lampryridae	Most	—	Subterranean	On vegetation
HYMENOPTERA				
Formicidae	Some	—	Prey provided by adults	Social predators
Pompilidae	All	—	Prey provided by adults	Provide prey for young
Sphecidae	Most	—	Subterranean	Provide prey for young
Vespidae	Most	—	Prey provided by adults	Provide prey for young, some also eaten
DIPTERA				
Asilidae	Most	—	Subterranean, rotting wood	Aerial
Empididae	Most	—	Subterranean, aquatic	Aerial
Rhagionidae	Some	—	Subterranean, aquatic	Aerial
Syrphidae	Some	—	On vegetation	Nonpredator

Figure 246. The male damselfly (*Calopteryx maculata*) clasping the female by the "neck" to hold her while the female extends her abdomen and genitalia forward to receive sperm from the male secondary copulatory organ.

propulsion, to escape other predators. Nymphal development requires up to 3 years to complete, at which time the immature exits the water by scaling emergent vegetation, molts to the adult stage, and becomes an aerial predator. These are strong fliers (although paleopterous), and aerial insects such as mosquitoes, bees, and butterflies are grasped and eaten. Territories are often well defended against other odonates, and interlopers are seldom tolerated except for mating. Mating is unique among insects because the male copulatory organ is located at the anterior portion of the abdomen and requires the female to extend her abdomen forward to receive sperm (Fig. 246).

Eggs are usually deposited either directly into water or inserted into aquatic vegetation or mud. In a few instances the female may actually submerge herself to oviposit, but usually she only inserts her abdomen or releases the eggs as she "brushes" the water surface during flight. The male may remain attached to the female during this time, even if she submerges herself, to prevent other males from mating with the female prior to oviposition. This is vital because the male secondary copulatory structure is modified to scoop out sperm from any previous matings.

The area covered searching for prey varies from where the predator emerged from the water to regions many miles distant. A few species also migrate, and they may do so in vast swarms.

Figure 247. A mantid (*Stagmomantis carolina*), having grasped a snowy tree cricket with its raptorial forelegs, commences to feed.

MANTODEA

All of the approximately 2,000 species of mantids (Fig. 247) are terrestrial predators, both as nymphs and adults. Their color is normally brown or green, which permits them to remain concealed from both other predators and their prey. Mantids usually ambush prey as it moves into striking distance and grasp it with the raptorial legs located at the front of an elongate prothorax that increases the effective capture distance. Feeding commences immediately, often while the victim continues to struggle (Fig. 247). All body parts are consumed by the sharp and efficient chewing mouthparts.

Mantids have large compound eyes that provide extremely good visual acuity and depth perception. The head can be rotated nearly 300° so little of the nearby environment goes unnoticed. As in most predators, mating is a hazardous undertaking, particularly for the smaller male, and he may fall victim to the hungry female, both before and during the process.

More than 80 eggs are deposited together in a frothy textured, yet hardened, oötheca that becomes the protective overwintering stage. Nearly all eggs hatch at the same time in the following spring. Nymphs undergo up to 12 molts to complete development. A single generation is normal for temperate climate species.

Figure 248. A predatory stink bug (*Stiretrus* sp.) nymph feeding on a chrysomelid larva.

HEMIPTERA

Twenty-two families within this order have species that are predatory, seven of which are fully aquatic (reside beneath the surface film), two that inhabit the surface film, and the remaining which are terrestrial (Table 4). Most temperate species have a single generation per year, and they overwinter either as eggs or as adults. All Hemiptera undergo hemimetabolous development; hence, both nymphs and adults are predators. All have piercing-sucking mouthparts in an elongate beak and ingest liquids, either blood and/or lysed tissues of the prey. Many have raptorial forelegs and ambush their unwary victims.

One example is the predatory stink bug. As the name indicates, these true bugs possess large stink glands located in the abdomen of nymphs and on the thorax in adults, and these glands effectively deter many predators. Although most stink bugs ingest sap, a surprising number are carnivorous. The preferred prey of these bugs are often soft-bodied insects such as lepidopterous caterpillars and beetle larvae. Prey is located usually by searching the vegetative habitat. Once located, a toxin is injected though the beak stylets, which effectively immobilizes the prey and permits the hemolymph and lysed cells to be sucked out (Fig. 248). In a few species the first instar nymphs feed on plant juices before reverting to the predatory instars and the adult diet.

A second example is the assassin bug. These bugs are effective predators, both as adults and immatures. When insect prey is encountered, hypodermic-like stylets are inserted and a neurotoxin is injected that rapidly subdues the attacked insect. The digestive enzymes that are also injected aid in predigesting the tissues to be taken into the body (Fig. 249).

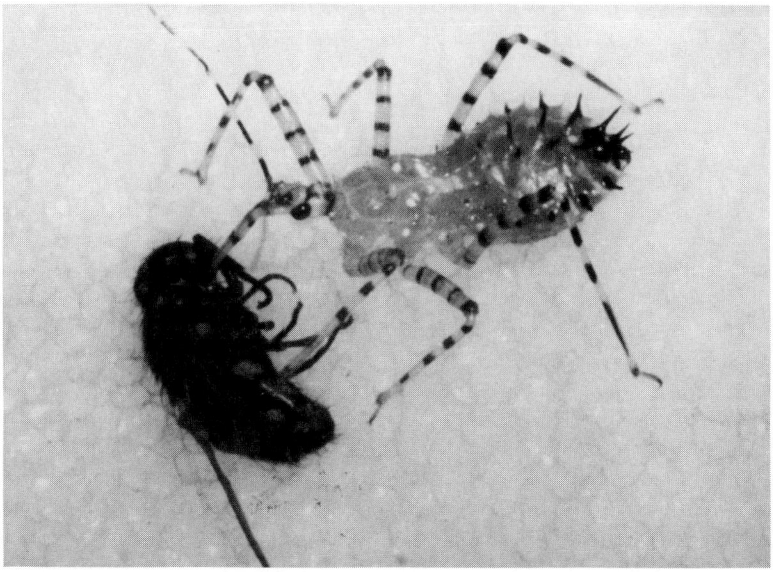

Figure 249. An assassin bug (*Pselliopus* sp.) nymph subduing a house fly.

Eggs of both of the above examples are either deposited singly or in clusters on the vegetation. Some species guard the eggs, but most leave them to hatch. Overwintering is usually in the adult stage under stones, bark, or litter.

MEGALOPTERA

This order is characterized by aquatic predaceous larvae (Fig. 345) and terrestrial nonfeeding adults [Fig. 344(A)]. About 200 species are known in the world. The larvae, often called hellgramites, possess lateral appendages with associated filamentous gills [Fig. 305(A)] and powerful chewing mouthparts. Most larvae cling to rocks and often rely on discovering and utilizing stonefly and mayfly nymphs for food. Few observations have been made as to how prey is captured and eaten.

Development takes 2 to 3 years, passing through 10 to 12 instars, at which time the larvae leave the water and pupate under rocks or debris (Fig. 139). Adults emerge within 3 to 4 weeks, mate, and deposit large aggregates of eggs on a streamside vegetation.

NEUROPTERA

Both the larvae and adult are normally carnivorous in this order, although the prey is often quite different. The specialized larval mouthparts consist of two hollow,

Figure 250. A lateral view of an antlion (*Myrmeleon immaculatus*).

soda-straw-like structures formed by uniting the mandible and maxilla on each side. The adult mouthparts are sharp chewing structures, although they may be non-functional in some species. With the exception of the Mantispidae, raptorial legs are absent.

An obvious example of neuropteran predators is the antlion (Fig. 250). Many of this group of more than 600 species construct conelike depressions in dry sandy soil (Fig. 347), occupying the narrowed bottom area. Here they wait for small insects, often ants, to wander accidentally into the unstable depression. As the prey struggles to exit—and this struggle often increases as the antlion catapults sand particles at it with its mouthparts—the slope sides give way and the victim slides downward. Once at the bottom, the prey is grasped with the sickle-shaped mouthparts and pulled into the sand where it is subdued and sucked dry. The adult [Fig. 344(C)] is a fragile organism that resembles a damselfly; it is a poor flier and is not predaceous. Eggs are deposited singly in appropriate sandy soils. The larva passes through three instars, each of which often produces a new pit after molting. Development requires 1 to 3 years but can be extended farther by unfavorable environmental conditions. Pupation occurs in the last pit within a cocoon formed from silk secreted from the Malpighian tubules and passed out through the anus.

A second example is the green lacewing. About 1,600 species exist. Here both the adult (Fig. 346) and larva (Fig. 251) are predators. Eggs are deposited on slender elongate stalks, often in groups. The larva, after hatching and climbing down onto the leaf, feeds upon aphids and other soft-bodied and small insects. Prey is impaled on the hollow mouthparts and fluids ingested. Development requires the entire summer in our common species. As in antlions, pupation takes place within silken cocoons,

Figure 251. Anterior portion of a green lacewing larva (*Chrysopa oculata*).

not in the soil but on leaves or under loose bark. The adult is nocturnal and also feeds on aphids. Interestingly, the adult turns brown as winter approaches, and it overwinters with this color.

COLEOPTERA

More than 30 families within this order have species that are predaceous. Most are terrestrial, but they inhabit many types of habitats, from the soil (Fig. 252), to within trees and logs (Fig. 148), to leafy and other aerial surfaces. In most cases, both the larva and adult are predators, but usually on different groups of insects or habitats.

Our first example is the tiger beetle, with more than 2,000 world species. The larva (Fig. 252) forms a narrow shaft into moist soil where it waits near the surface to ambush prey as it passes over the entrance. The unsuspecting organism is no match for the bulldoglike head, and it is pulled into the burrow to be consumed. A group of dorsal hooks on a lobe of the fifth abdominal segment [Fig. 357(A)] prevents the larva from being extracted by the struggling prey or predators. Pupation is within the burrow. The adult is a rapidly moving and diurnal beetle (Fig. 253) with long legs and powerful sharp mandibles. Prey is normally captured by rapid pursuit. The adult is often brown with either cream-colored markings on its elyra or with shiny metallic coloration. These beetles are active runners and fliers, particularly when approached. The large compound eyes are indicative of its excellent vision.

Figure 252. The larval stage of a tiger beetle (*Cicindela* sp.) in burrow waiting for prey.

A second example exists in the water both in the larva and adult stage. It is the predaceous diving beetle [Fig. 349(G)]. About 3,000 species are known. The larva, often called *water tiger,* is streamlined and can move rapidly through the water when prey is perceived. Its mandibles are long and curved, with either a groove along the inner edge or with an internal duct that opens at the mandible tip. It is through the duct or groove that a toxin and digestive enzymes are injected into the prey, consisting of insects (Fig. 254), fish, or tadpoles. The predigested "soup" is then ingested. These larvae are voracious and may consume up to 30 individuals. Air enters the body through two terminal spiracles at the end of its abdomen, and larva often hangs from the surface while waiting for prospective food. Development takes place in 5 to 6 weeks, at which time the larva crawls onto land to pupate. After metamorphosis, the adult either reenters the water or seeks a new area. The adult is also an aquatic predator and feeds on a wide range of organisms similar to the larva. Oxygen is obtained from the atmosphere through abdominal spiracles that open under the elytra; air enters this subelytral cavity when the elytra part the surface film. Males have suckers on their front pair of legs [Fig. 61(E)] that aid in grasping the female during courtship. The other legs have swimming hairs to aid in swimming after prey or avoiding predators. Adults may survive several years.

Figure 253. The adult tiger beetle (*Cicindela* sp.).

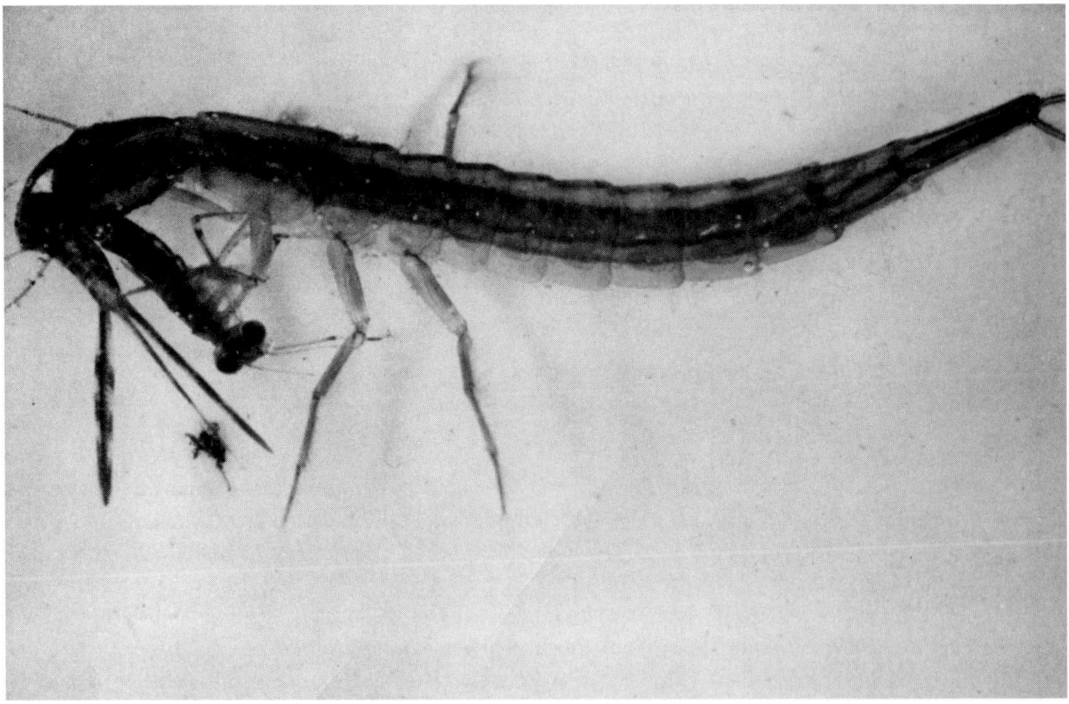

Figure 254. The larva of the predaceous diving beetle with its damselfly nymph prey.

Figure 255. A mating pair of ladybird beetles (*Hippodamia convergens*).

A third example of coleopteran predators is the ladybird beetle (Fig. 255). These beetles are brightly colored with red, orange, or yellow elytra and often also with black spots. Larvae are blackish in color and often have projecting spines. Both stages feed arborially on aphids, scales, and eggs of insects, and both stages do so diurnally. These insects oviposit batches of eggs in aphid or scale infested areas. Both stages are voracious feeders and are considered to be one of our most beneficial insects. Development is completed in about a month, and pupation occurs on vegetation. Adults are active fliers and use this ability to locate new prey sites and overwintering habitats. More than 5,000 species are known.

HYMENOPTERA

The previous examples have highlighted solitary insects. Although many Hymenoptera are also solitary predators, our example here will be the neotropical army ant. About 147 species are known. These ants form colonies of up to an estimated 700,000 adults. Prey capture is by group effort. Daily forays from the bivouac (temporary nest formed by the aggregate bodies of the ants) explore the nearby environment, either above ground in some species or below the surface in others. If on the surface, these blind ants move out either with a broad front called a swarm-raid (Fig. 256), killing whatever arthropod is unable to escape, or they move in a column to raid wasp and other social insect nests. In either case, prey is overcome by sheer numbers coupled with stings and bites. Large subdued prey is torn apart and only fragments are returned to the nest, whereas small animals are carried in their entirety. Returning ants follow a scented trail

Figure 256. A swarm raid of the neotropical army ant *Eciton burchelli.*

deposited by the outgoing ants. *Booty,* the term used to indicate captured prey and dissected pieces, is fed to any developing larvae within the bivouac and to those adults, including the queen, that remain behind. A vast amount of prey must be obtained each day to maintain each large colony. Except when the queen lays a vast number of eggs (numbers may reach more than 50,000 over several days of oviposition), the colony emigrates daily to new fresh sites where the food supply is abundant. During both the daily foraging raids and migrations, the ants are protected by specialized workers, the majors, with their long mandibles (Fig. 218). New colonies arise by a division of the old colony after rearing an additional queen.

DIPTERA

About 30 families within this order have predatory representatives, with many in the legless larval stage. The example chosen to represent this order is the robber fly (Fig. 257). This group of about 5,000 species are predators both as adults and as larvae (only a few of the immatures have been studied, however). The few larvae studied burrow into the soil or rotting logs and feed upon insect larvae such as scarabaeid larvae. Their sharp mouthparts are inserted into the prey and salivary secretions injected to predigest protoplasm prior to sucking out the body fluids. In contrast, the adults are swift flying insects that either ambush flying insects or pounce upon resting species. The prey is grasped and the proboscis inserted (Fig. 257) to inject an immobilizing venom, and then the prey is drained of its body fluids. Some adults appear to prefer bees and flies, but many appear to be general feeders.

Figure 257. Robber fly (*Efferia* sp.) feeding upon a damselfly.

QUESTIONS

1. What is a *predator?* How does it differ from a parasite or parasitoid?
2. What orders of insects include predators?
3. How does the type of metamorphosis impact the larval predator as it becomes an adult? Can there be predaceous adults when the larva is not of this lifestyle and vice versa?
4. What structures and behavior aid the predaceous insect to be successful?
5. How does digestion in many predators differ from most herbivorous insects?
6. How might insect predators be beneficial to humans? How would you determine this?
7. What advantages and disadvantages would a social predator have over solitary predators?

10 Insects, Plants, and Humans

Humans are becoming aware that the increasing human population and the associated food production have both direct and indirect impacts upon ecosystems and that changes must be carefully evaluated as to their short-term and long-term effects. We, as humans, tend to view the components of nature and changes within the ecosystem as beneficial or detrimental to ourselves. Although this often creates problems, such as the erroneous belief that insects must be categorized as either good or bad, there are distinct advantages to be gained from such an approach, and some of the major aspects will be discussed in this chapter.

BENEFICIAL ASPECTS

Pollination

This phenomenon is the transport of pollen from the anthers (male part of the flower) of one flower to the stigma (female part of flower) of another plant of the same species. More than 65 percent of the flowering plants (angiosperms) are pollinated by animals such as insects, birds, snails, and bats, but insects play a dominant role, especially in the tropics where individual plants of each species are widely separated. Coleoptera, Hymenoptera, Lepidoptera, and Diptera are the major insect orders involved.

The interrelationships between insects and flowering plants probably existed back in the Cretaceous period (more than 125 million years ago), and there is little doubt that each of these two groups has had a profound effect on the species evolution of the other. Baker (1968) and Crane et al. (1995) indicate that most primitive angiosperms were pollinated by insects (*entomophily*), especially beetles. These primitive flowers were essentially saucers with many pollen-receiving stigmas in the center. Pollen was distributed as the beetles crawled about the flower eating petals and stamens. From these flowers have evolved the many complex forms we see today. *Anemophily* (wind pollination) is a secondary development but is still vital to pollinators because entomophilous flowers may not be able to provide sufficient food during portions of the year to maintain abundant populations of those insects.

Some plants dependent on insects utilize promiscuous visitors and have few specializations to attract pollinators; others have specific visitors, and often reproductive isolation between related plant species is maintained by these insect-flower relationships. Four major categories of plant adaptations that ensure or increase the efficiency of insect pollination are: (1) those that attract the insect to the flower so that pollen will be encountered, (2) those that have sticky pollen or pollen packets (*pollinia*) to adhere to the insect body, (3) those that "reward" the insect with either pollen or nectar, and (4) plant-density factors.

In the first category, insects are attracted to flowers by different stimuli, and slight modifications result in specificity of this relationship. Petal colors of yellow, blue, blue-green, purple, or those that reflect or absorb high amounts of ultraviolet light are especially attractive. Flowers that appear one color to humans may actually appear differently to insects because they can see different wavelengths, especially ultraviolet (Fig. 258). *Nectar guides* [Fig. 258(B)], pigmentation arrangement near the center of the flower, are often vital in bee-pollinated flowers. One can experimentally remove petals and reverse them so that the normal attachment points are toward the periphery. This rotation causes bee pollinators to move to the outside of the flower where the guides are located, rather than to the center where they should be. Flowers also may be located by smells, some of which are specific. An example of this is an odor that resembles decaying meat produced by certain flowers (Fig. 259), which attracts carrion-feeding insects. Nocturnal flowers attract insects primarily by releasing strong perfumes, and they normally are closed during day hours. A few rare, but interesting, flowers also resemble female wasps; pollen is picked up by the male wasp as it attempts to copulate with this mimic.

The second plant adaptation is in the pollen itself. By adhering to an insect, pollen can be transported from one flower to another with maximal efficiency and minimal loss, at least when compared with wind pollination. Orchids have reached a high plateau of specialization here. Many of these flowers are "architecturally designed" so as to direct the pollinator into exacting positions whereby the pollinia (sticky masses of pollen) can be ejected and attached to specific sites on the insect (Fig. 260). As the orchid bee visits another flower, the pollinia are dislodged onto the female pistil.

The third plant adaptation to ensure insect pollination is that of "rewarding" insects that visit the flower. Historically, pollen itself probably served as the initial benefit received by the insect (beetle-pollinated flowers), but nectar-producing structures, the *nectaries* (Fig. 261), have evolved and have greatly supplemented pollen as a reward in bee-, moth-, and butterfly-pollinated flowers. Plants gain through secreting an attractive nectar, but the nectar is sufficiently low in volume to prevent satiation of the insect. Early satiation reduces the number of flowers visited, and hence pollination frequency drops. Because nectar is often located deep within the flower, only insects that have long probosci, such as in Lepidoptera and bees, can feed (Figs. 262, 263). In Madagascar, one species of orchid has a flower that is about 30 cm deep and

(A)

(B)

Figure 258. What an insect sees is not necessarily what humans visualize. (A) photograph of daisy as a human would see it; (B) the same flower to a bee. The inner parts of the petals, the invisible (to us) nectar guides, absorb ultraviolet light, whereas the lateral portions reflect these wavelengths.

is pollinated when the hawk moth *Xanthopan morgani* inserts its equally long proboscis down into the tubelike flower. In some instances openings are cut by some insects through the sides of elongate flowers to bypass this adaptation, and no pollination occurs. Some orchids have progressed beyond normal nectar and reward their visitors with narcotic fluids; the "addict" bees become behaviorly modified as they increasingly seek to satisfy their addiction. Finally, some plants are pollinated by insects that receive no nectar or pollen rewards (Fig. 259). These putrid-smelling flowers mimic carrion oviposition sites sufficiently that eggs may actually be deposited in the flowers; however, most of these carrion-attracted insects leave and search out additional plants, carrying with them pollen from previously visited flowers.

Plant density is yet another plant adaptation. Pollinators develop "search images" whereby particularly attractive rewards for any given day are sought. Because this strategy must be cost-effective (more energy obtained than expended in the search), the more abundant the species of flower in localized areas, the more likely the pollinator will visit and specialize upon this plant. Although specialists depend upon one or two

Figure 259. Carrion flower *(Stapelia)* has a brownish-purple color and a rotting aroma, both of which attract blowflies and flesh flies that are unable to distinguish between such flowers and animal carcasses.

Figure 260. Orchid bee (*Euglossa* sp.) with a pollinia attached. This pollen-bearing structure will rub off on the next flower. (Photograph by C. W. Rettenmeyer.)

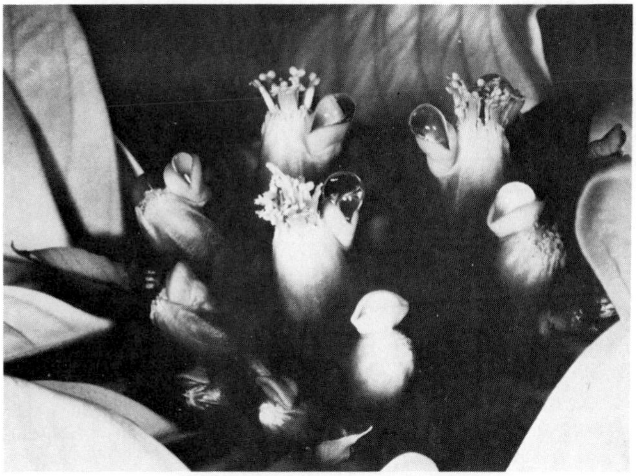

Figure 261. Nectaries produce nectar that attracts many insects. Most plants produce minimal amounts of the sugar solution, but the above poinsettia exudes copious amounts.

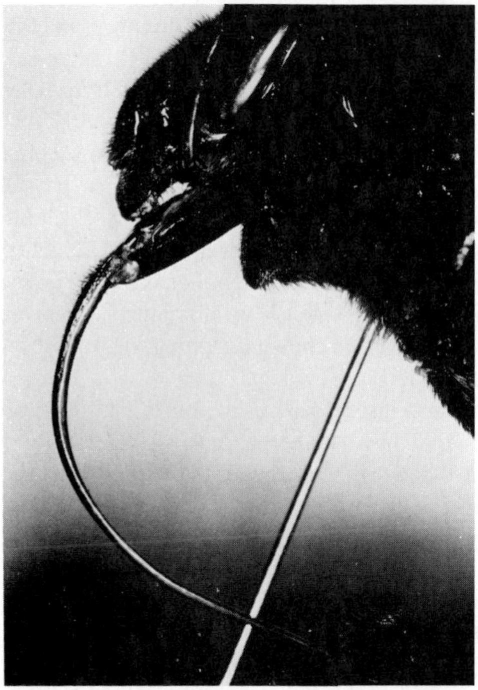

Figure 262. Orchid bee with elongated mouthparts for entering deep-chambered orchids to obtain nectar.

plant species, most switch to other plants or leave the area when these resources become unavailable or when the environment lowers the average reward of the nectar or pollen.

Insect efficiency in utilizing flowers may be both structural and behavioral. First, the seeking behavior must occur during periods when flowering and nectar flow occur. For generalized feeders, such as bumblebees and many beetles, this may involve activity during several hours of most days, but for the more specialized insect, this

Figure 263. Anthophorid bee taking nectar from snow-on-the-mountain (*Euphorbia* sp.) flower.

also involves having life cycles synchronized with the plant. The orchid bees (Fig. 262) and yucca moths are interesting examples of specific pollination and the mutualistic relationship between insects and plants. The yucca moth gathers a pollen ball and carries it to another flower where it is forced onto the stigma. An egg is then laid on the flower, and the resultant larva feeds upon some of the developing seeds. Pollination, therefore, ensures food for the caterpillar, and surplus seeds allow plant propagation.

Second, pollen and/or nectar must be collected and ingested efficiently. Such insects as beetles often chew pollen from the anther. Bees, however, illustrate a higher degree of specialization. As honeybees move about the flower, pollen is picked up by plumose body hair; then the pollen is brushed off by their legs into pollen baskets on the hind legs. Insects that feed on nectar usually have chewing-lapping, siphoning, or sponging mouthparts, although many exceptions are evident.

The co-evolution of flowering plants and insects has resulted in a multitude of species for each group. Insect orders that have numerous species (Coleoptera, Lepidoptera, Hymenoptera, Diptera) have been associated with flowering plants for thousands of years. Beetles are attracted mainly by odors to flowers that are dull in color and often lacking in nectar. Short-tongued (sponging mouthparts) flies are attracted to flowers in which nectar volume is low or absent, and food consists mainly of pollen and extrafloral nutrients. These fly-pollinated flowers often have traps to ensure pollen attachment to the body of the insect. In contrast, bees and long-tongued flies are attracted to brightly colored flowers where both nectar and pollen are abundant. Hawk moth–pollinated flowers are elongated, pale to white in color, and open during early evening to night hours. Butterflies pollinate similar types of flowers as those pollinated by hawk moths, but the flowers are more brightly colored and open during daylight hours.

Of the 300 species of agronomic plants, only 12 provide 90 percent of the world's food (grains, coconuts, bananas), and none of these are insect pollinated. However, many insect-pollinated plants are important supplements and luxury foods to both hu-

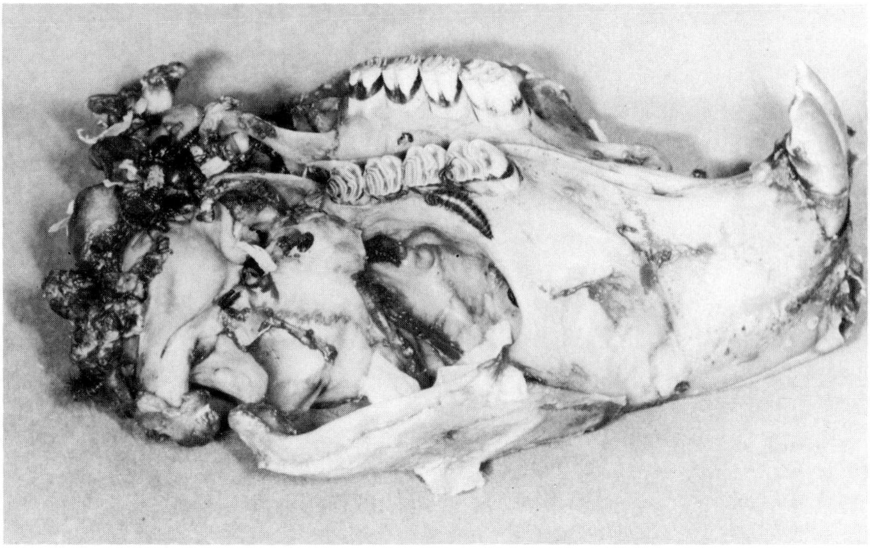

Figure 264. Dermestid beetle larvae feeding on dried flesh of beaver skull. Museum mammalogists often use dermestid beetle larvae to clean skulls.

mans and livestock, including beans, peas, tomatoes, many fruits, cotton, alfalfa, tea, asparagus, peppers, lettuce, melons, many nuts, cabbage, onions, garlic, carrots, and cocoa. In a few instances the interrelationships have become obligatory, for neither the plant nor the insect can reproduce without the other (e.g., figs and fig wasps), although such relationships are potentially hazardous to both if populations of either partner become reduced in number.

Trashburners and Soil Builders

The word *trashburner* has been used by many biologists to indicate organisms involved in degradation and recycling dead plants and animals, the detritus feeders discussed in Chapter 5. This process is particularly important in preventing a permanent tie-up of necessary nutrients. The most common examples of such decomposers are the bacteria and fungi; however, many insects feed directly on and burrow into both plant and animal cadavers (Fig. 264). The insects not only directly assist in the recycling but also break the outer barriers of the cadaver and enhance invasion by bacteria and fungi. This is important because in dry areas a dead animal untouched by multicellular burrowers or scavengers will mummify long before much decay can occur. Certain carrion beetles inter dead mice many times their size (Fig. 265) by excavating soil from beneath the bodies and then covering them. Once the cadaver is covered with soil, hair is removed and digestive juices are regurgitated to soften the carrion until a large reservoir of predigested food is available to feed the emerging larvae. Eventually the mouse is eaten completely.

Figure 265. This carrion beetle (*Nicrophorus* sp.) buried the mouse, which serves as a food supply for developing larvae.

Larvae of other insects such as Diptera and Coleoptera recycle feces through burrowing and feeding. In some instances involving dung beetles, the feces are formed into balls and are rolled to suitable sites; in other instances the feces are buried directly at the place of deposition. Dung beetles have been observed to completely cover human stools in less than 3 hours. Extensive areas of Australia now control cow dung with the introduction of certain dung beetles from Africa. Further evidence of the importance of insects in the recycling process may be seen in Australia, where an estimated 25 percent of all wood is decomposed by the feeding of termites.

Soil building is closely related to decomposition. The breakdown of plant and animal remains and the mixing of this material with soil aid in developing layers of humus. The role of earthworms in soil building is widely recognized. Although no complete studies comparing earthworms with insect activity in soil formation and renovation are known by the author. Undoubtedly the role of insects is significant and, according to a few scientists, must at least approach the effectiveness of earthworms in such areas of the world as the tropics. Earthworms are absent in many African savannas, and termites assume the role of recycling woody material and enriching the arid clay soil. Preliminary information from studies of termites and soil formation in Australia (Lee and Wood, 1971) indicate that these insects are unique in several ways. First, instead of turning over the A horizon as earthworms do, they selectively bring up wet soil particles from the B horizon to incorporate them into their nests. Second,

Figure 266. Silkworm adults *(Bombyx mori)* mating. The male was attracted to the female by sex pheromones released from everted lobes near her genitalia. They immediately coupled and remained *in copula* for more than 3 days.

nutrients such as calcium, magnesium, and potassium are often concentrated in and around the colonies because of their feeding activities, although these may become tied up for long periods because the nests do not decompose readily and thus become disadvantageous.

Products

Silk. Once referred to as the "cloth of kings," silk is a commercial product of the silkworm caterpillar, *Bombyx mori.* Silk cultures are believed to have originated in China about 1700 B.C., and this product remained a Chinese secret for more than 2,000 years, mainly because persons attempting to steal or export these insects were put to death. Attempts to use other insects, mainly saturniid and other lepidopteran caterpillars (Peigler, 1993), on a commercial basis have proven unsuccessful, and, in the case of the gypsy moth, resulted in the introduction of an insect pest that is currently ravaging forests in the United States.

The silkworm is one of the few "domesticated" insects, for it has become essentially dependent on humans for survival. This moth (Fig. 266) is mainly reared in the Orient, the Mediterranean, and South America, but the industry is located in more than 35 countries. In some areas, technology now permits mass rearing in silkworm factories. Eggs are processed under sanitary conditions, and workers entering rearing rooms must comply with strict regulations of clothing and personal cleanliness. After hatching, larvae are transferred to rearing trays, and mulberry leaves or artificial diets

(A)

Figure 267. Commercial silk is now mass produced. Killing the pupae is vital to prevent adult metamorphosis and emergence that would damage the cocoon. (A) silkworm cocoons moved by a conveyer from drying oven to packaging area; (B) cocoons of silkworms opened and compared. The one on the left is before entering the oven, whereas that on the right demonstrates the results of oven exposure.

(B)

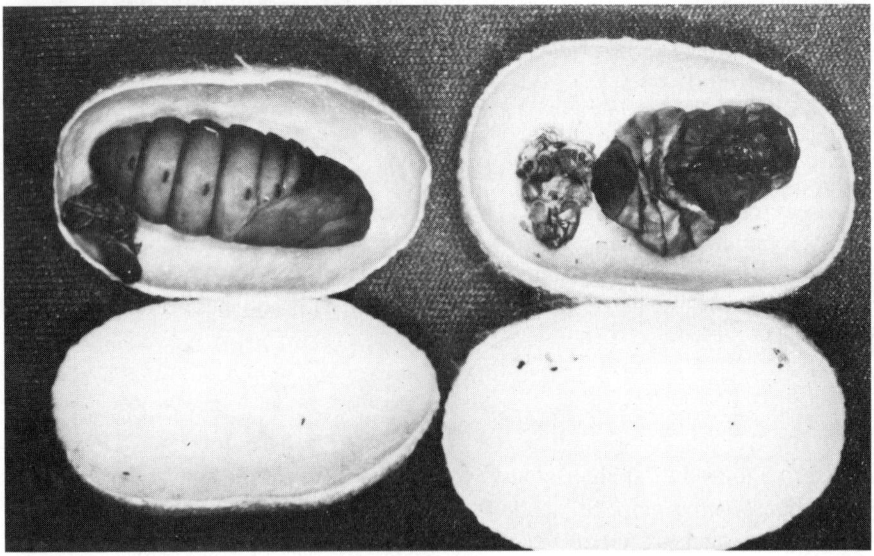

consisting of mulberry leaves, soybeans, and corn starch are added until caterpillars reach the maximal size of 3 in. This takes about 25 to 30 days, and their body weight increases some 10,000 times during this period. Within a week, the cocoon is spun and placed in a high temperature drying oven to kill the pupa (Fig. 267). The silken cocoons are then bagged and delivered to a processing plant, often thousands of miles distant, where they are unwound and processed into commercial thread and cloth. Approximately 1 ton of mulberry leaves is thus transformed into 12 lb of silk. That translates to 110 cocoons for a tie or 630 for a blouse.

Honey. Honey, a product of honeybees, represents concentrated and partly hydrolyzed nectar. The principal sugar in nectar is sucrose, which is broken into glucose

Figure 268. Honey production using manufactured nests of stacked supers. (A) supers are periodically removed and the combs checked for the stage of honey production; (B) comb in the process of being filled with cells uncapped.

and fructose by salivary enzymes as the bee returns to the hive. This fluid is regurgitated from the crop into comb cells where the solution is concentrated by bees fanning their wings and evaporating water from the mixture. Once the mixture is concentrated to approximately 80 percent sugar, the cells are sealed with wax for future use by the bees, although humans usually intervene at this stage and reap the rewards.

Most areas in the United States have at least 100 different species of plants that are visited by honeybees for nectar, but only a small fraction are used extensively. Specific-tasting honey can be produced by placing hives near appropriate nectar sources because each source has its own aroma and influence on the taste of the product. Alfalfa, citrus, cotton, and clover are the most common sources of honey.

To harvest honey, special hives are maintained. A series of separate movable sections, each containing vertical wax combs, or *supers,* are stacked one on top of another. These supers are inspected periodically to determine the time of harvesting (Fig. 268). People usually wear protective clothing to prevent being stung; one never knows when hypersensitivity to bee stings may develop. This reaction can result in anaphylactic shock and death unless promptly treated. Once the combs are full and the honey is sealed by the bees, the supers are removed and the honey is processed.

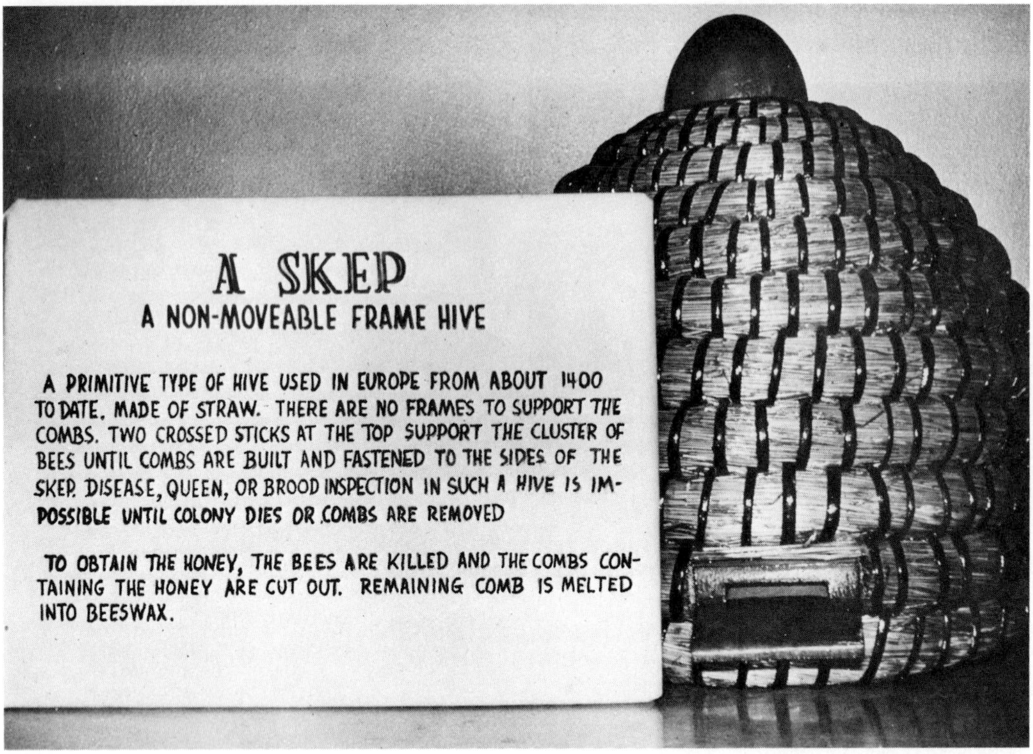

Figure 269. Skeps were formerly used in maintaining honeybee colonies. This one was manufactured from coiling strawlike material. The major disadvantage to their use was the need to kill the colony in order to enter and remove the honey.

Grade A honey must contain less than 18.6 percent water. Fermentation in commercial honey is low if the honey is kept in sealed containers because it is also pasteurized by heating to 160°F (71.1°C) for 1 minute.

Early American farmers either had primitive beehives or *skeps* (Fig. 269) or, more often, relied on wild bees for their honey. In the latter instance, searches were conducted until a wild hive was located, usually in hollow trees or caves. The combs were removed, cut into small pieces, and either squeezed or enclosed in some porous container over a fire or in the hot sunlight. After the majority of honey had been thus extracted, the remaining residues were soaked in water and then allowed to stand and ferment with other additives into mead for drinking.

Beeswax. Honeybees produce beeswax to build their combs. About the twelfth day as an adult worker, specialized hypodermal glands on the fourth to seventh abdominal segments start to secrete thin flakes, which are scraped off by the hind legs and molded by the mandibles. Energy comes from metabolizing honey, and it is estimated that from 3 to 20 lb are required to produce 1 lb of wax. Beeswax is utilized in innumerable products from smokeless candles and carbon paper to cosmetics.

Other substances. Additional materials synthesized by insects and used in our society include royal jelly (from honeybees), cantharidin (from certain blister beetles), lac used in shellac (from certain scale insects), and cochineal dye (from certain mealy bugs). The substances are used less and less each year as many new synthetic materials reach the market. Of importance to the future, however, are the many other compounds synthesized by insects that await discovery and application.

Food

It has long been known that insects are an important component in the diet of birds, comprising from 50 to 60 percent of their food. The actual number of insects eaten varies with the bird species; Reed (1943) estimated that 100 ingested per day is average, but flickers have been found to have more than 5,000 ants in their stomachs. Although most species utilize this source of food only during their juvenile development, a large number of birds, including warblers, woodpeckers, swallows, and vireos, continue to feed upon insects throughout their life. If insect populations decline, birds migrate or alter their diet. Meadowlarks, for example, feed on these arthropods primarily during the spring and summer but eat seeds and other food in cold weather. Many persons, perhaps unintentionally, have attempted to persuade the uninformed that birds should be protected because of their destruction of pest species; however, the role of avian predators has been greatly exaggerated except in isolated situations. Normally, birds do not specialize solely on pest species, and their impact upon such populations is only minimal.

Mammals also include insects in their diet. Some animals are occasional consumers, such as bears, whereas others, including moles, bats, and shrews, have become nearly specific insect feeders. Control recommendations for moles, for example, often include soil treatment for beetle grubs, the reasoning being that once the primary source of food is eliminated, these insectivores either starve or migrate to other areas. Most bats feed exclusively on flying insects and locate them by echoes. Bat cries bounce off the prey, and position is determined by the strength and time of the returning sound vibrations. Certain moths have specialized sense structures, tympana on the thorax, that hear the high-pitched bat cries, and the moths immediately initiate a series of evasive flying maneuvers. Typical "herbivores" such as deer mice also feed extensively on insects.

Predation of insects by cold-blooded vertebrates is considerable, although qualitative and quantitative data are incomplete when compared to that on birds and mammals. Insects make up more than 75 percent of the diet of the common toad, *Bufo* sp. Lizards such as *Sceloporus* sp., chameleons, horned toads, and green glass snakes are effective insectivores. An estimated 40 to 90 percent of the diet of freshwater fish is of insect origin (Fig. 270). Trout, for example, ingest great quantities of insects, including mayflies, stoneflies, and caddisflies. One fish species, *Gambusia affinis,* has been used extensively in controlling mosquito larvae in city parks and in extensive marshes where other control measures are impractical.

Although largely ignored by the Western world, insects are also a good source of food for humans. Many areas of the world have and still derive part of their protein from insect sources. This is not merely to avoid starvation; insects are prized additions

Figure 270. Insects are important components in the diet of game fish. This bullhead has captured a large cockroach, found it to be positioned incorrectly to swallow, and ejected it. Moments later the roach was taken in correctly, with the insect's head first, and swallowed.

to their diet. Insects are eaten raw or roasted or are dried and added to other foods. Ants, beetle larvae, and termites are a favorite among many Amazonian tribes. Some insects are eaten raw during gathering, but others are added as supplements to other food, such as soups and flour. Palms are often cut down to serve as a rearing source for desired cerambycid larval borers. About 3 to 4 lb of these larvae are obtained per decaying tree. Silkworm pupae are fried and eaten in many areas where silk is produced. The cocoons are immersed in boiling water to kill the pupae (to prevent damage to the silk), and the pupae are then eaten by workers as the silk is unwound. Indonesians harvest large rhinoceros beetle larvae as a delicacy. Corixid eggs are collected on special sheets in canals and are sold at markets in Mexico City. Large crickets are sought in Uganda and Thailand. History tells us that even the Romans ate certain wood-boring caterpillars as hors d'oeuvres. The practice of picking and eating body and head lice is well documented among Eskimo and other cultures. The American Indian often had recipes for insect preparation, and immatures of the common brine fly, *Ephydra hians,* and adults of rhagionid *Atherix* flies have been harvested extensively by Indians in California. Grasshoppers are also eaten by many cultures worldwide, as are termites and large caterpillars. Even toward the end of the last century, one could order cockchafer (beetle) bouillon in the finest French restaurants. Insects contain few human parasites, are of high food value (Table 5), and are readily obtained. Only a change in our modern thinking is needed in order to use insects to feed a portion of the growing population, but this change is not anticipated in the near future in so-called civilized societies.

In addition to the above deliberate ingestion of insects, vast numbers are eaten accidently as plants are eaten by herbivores. The impact of such food supplements upon these herbivores is unknown, but it must have at least a minimal impact.

TABLE 5. Comparative Value of Selected Insects and Other Foods

Food	Percent					*Calories of 100 g*
	Water	*Protein*	*Fat*	*CHO*	*Ash*	
Living termites*	44.5	23.2	28.3	—	—	347
Silkworm pupae*	60.7	23.1	14.2	0.5	—	207
Grasshoppers*	70.6	18.7	4.1	—	—	—
Turkey	64.2	20.1	14.7	0	1.0	218
Beef (standard)	60.0	18.0	21.0	0	0.9	260
Codfish	81.2	17.6	0.3	0	1.2	78
Crab	78.5	17.3	1.6	0.5	1.8	93
Lobster	78.5	16.9	1.9	0.5	2.2	91
Hen's eggs	73.7	12.8	11.5	0.9	1.0	163
Clams	81.7	12.6	1.6	0.5	1.8	76
Cow's milk	87.4	3.5	3.5	4.9	0.7	65

*Some of the data are not available for comparative purposes.

Control of Other Organisms

Fifteen of the 26 orders of insects have species that naturally parasitize or prey on other organisms, organisms that humans classify as detrimental. Hymenoptera and Diptera contain the greatest number of such species. Our deliberate use of these insects, as well as our use of other organisms such as bacteria in management programs, is termed *biological control.* We have much to learn in this area.

One of the best examples of successful biological control by insects occurred in Australia. By 1920 more than 20 species of the cactus *Opuntia* had invaded more than 60 million acres and had become sufficiently established to make many rangelands nearly useless. Mechanical and chemical control proved unsucessful in preventing their spread. Of the 50 species of cactus-feeding insects tested, 12 indicated sufficient promise to be introduced, but only the lepidopteran caterpillar, *Cactoblastis cactorum,* and several cochineal mealybug species proved effective. Their success was astounding, and several of the invading cacti species disappeared, with the remainder now found only in sparse, isolated populations.

A second example is the classic control of the cottony-cushion scale in California. This homopteran pest threatened the citrus industry in the late 1800s until two natural enemies of the scale, a ladybird beetle and a parasitic fly, were imported, released, and became established. The scale has been under control ever since.

The use of insects to control insect pests will be covered further in Chapter 11, Pest Management. We will see that there have been noteworthy successes as well as many failures in our attempts to duplicate nature.

Scientific Experiments

Until recently the use of insects for scientific experiments has not been widespread because of their small size and attendant requirement for delicate and precise instrumentation. Genetics has been a discipline in which some species have been used extensively.

Several fruit fly species *(Drosophila)* have been valuable test animals. Much of our current knowledge of mutations, sex-linked inheritance, and other classical genetic studies was obtained from experiments with these flies. *Drosophila* also is one of the leading experimental animals in learning how genes produce embryo development and how this compares to humans and other animals.

Environmental quality often can be monitored, to a degree, through observing insects, particularly the aquatic species. Stonefly nymphs, for example, require high concentrations of dissolved oxygen and relatively pure water; their presence along with a high diversity of other aquatic species, therefore, provides a measure of low pollution. Conversely, the presence of only chironomid larvae indicates severe pollution.

Population dynamics is another area that utilizes insects. Research on the migratory locust and on many of the stored product pests has added much to our understanding of the effect of density-dependent factors upon population growth.

Insects are used to reveal information in forensic studies through determining the waves of insect infestations of the corpse. The insects and their developmental stages aid investigators in determining time of death and other vital inferences. They also can serve as surrogates for toxins or drugs in the person prior to decomposition. Perhaps the future will see analyses of DNA found in blood-sucking insects, such as lice, that can be compared to suspects and victims.

With more refined instrumentation, the scientific community will continue to turn to insects as experimental animals because of their rapid life cycle, relative structural and behavioral simplicity, low cost of rearing, and diversity.

Aesthetics

Although some humans view insects with abhorrence, many find these arthropods to be of considerable aesthetic, cultural, and religious value. For example, the scarab beetles, wasps, bees, cicadas, grasshoppers, mantids, and flies had considerable importance to ancient Egyptians, and much jewelry and religious artifacts contained sculptured replicas. The scarab beetle, for example, in Egyptian mythology dates back more than 4,000 years. It portrayed the god Khepri and was responsible for pushing the sun across the sky (Cherry, 1985); therefore, it represented life itself. In 3100 B.C. King Menes made a hornet a symbol of the lower kingdom. Rings designed to resemble the scarab were used to officially seal buildings, and ornaments with it were placed in tombs as symbols of resurrection. In Egyptian beliefs, another insect, the mantid, carried a person home to the gods. The ant was considered as the example of wisdom and activity in the Bible and, to medieval Christianity, the scarab was considered to be a symbol of a sinner. Insects are referred to in the written word of the Muslim's Qur'an.

Insects have had songs and literature written about them: "Flight of the Bumblebee," "La Cucaracha," "To a Louse," and *Charlotte's Web*. They have starring roles in such movies as *Mothra* and *The Fly*. The role of insects in songs and literature is usually positive, whereas the cinema usually portrays them as villians or at least negatively.

An assemblage of modern artwork also exists either using insects as models or incorporating actual structural parts into the work. Scenes in trays and jewelry constructed from the iridescent wings of butterflies are incomparable. Even live insects

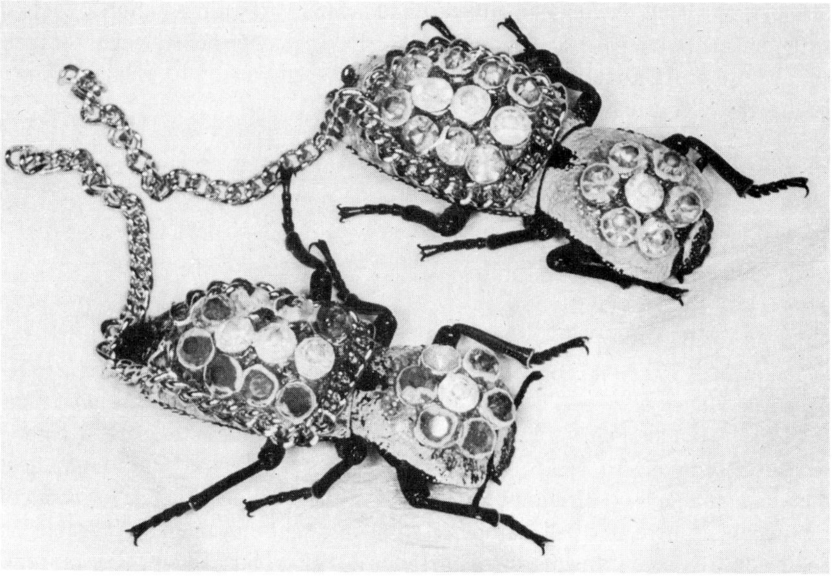

Figure 271. Jewels have been glued to these live tenebrionid beetles, which are sold in regions of Mexico as jewelry. Tourists often purchase these beetles and attempt to bring them into the United States without special permission.

Figure 272. The Mexican jumping bean is actually an arrow plant seed pod and not a bean. Movement is generated by the internal lepidopteran caterpillar *(Laspeyresia saltitans)* as it violently throws itself from side to side within the hollowed seed pod. Such curiosities are often imported into the United States and sold in novelty stores.

are used as jewelry in some parts of the world (Fig. 271). Butterfly ranching exists in Papua, New Guinea, where more than 500 villagers from 10 provinces are saving pupae, either reared in their gardens or collected in the wild; and about 2,000 to 3,000 specimens are sold each month on the world market for from 20¢ to more than $250 per specimen. A sufficient number of adults are returned to nature to prevent depletion of the species.

Stridulation sounds have special interest and meaning to various cultures, such as in the Orient and South America. For example, crickets may be kept in cages for their songs. The use of fighting crickets in gambling and the curiosity of Mexican jumping beans (Fig. 272) are other examples of insect use.

DETRIMENTAL ASPECTS

Human Disease Transmission

Although an estimated 1 of every 6 humans has suffered from some insect-carried disease (Southwood, 1977), it is only in the last 70 to 80 years that we have come to realize the importance of insects in disease transmission. Insects serve as *vectors* (carriers) of the major human diseases listed in Table 6 as well as of certain enteric diseases of lesser concern. Most of these major diseases have had a profound effect upon humans and their civilizations. Some outstanding examples include the death of approximately 25 percent of all Europeans in 1348 because of the plague (Table 7); the ending of Napoleon's dreams of conquest because of typhus (and the Russian winters); the delaying of the completion of the Panama Canal because of malaria and yellow fever (they cost the French alone some $200 million and 50,000 lives); and the exclusion of humans from thousands of acres of land in Africa because of African sleeping sickness during much of recorded history. Insect transmitted disease (malaria, dengue, yellow fever, etc.) caused enormous casualties in the U.S. Civil War, especially in the South where they where endemic. A short synopsis of the major insect transmitted diseases will now be presented; the numbers of cases cited probably represent low figures because many countries have poor reporting procedures.

TABLE 6. Major Human Diseases Transmitted by Insects

Disease	Causative Organism	Carried by Certain	Order
Human malaria	Protozoa	Mosquitoes	Diptera
Yellow fever	Virus	Mosquitoes	Diptera
Dengue	Virus	Mosquitoes	Diptera
Encephalitis, Arborvirus groups A and B	Virus	Mosquitoes	Diptera
Filariasis	Nematode	Mosquitoes	Diptera
Onchocerciasis	Nematode	Black flies	Diptera
Tularemia	Bacteria	Deer flies*	Diptera
African sleeping sickness	Protozoa	Tsetse flies	Diptera
Leishmaniasis	Protozoa	Sand flies	Diptera
Bubonic plague	Bacteria	Fleas	Siphonaptera
Murine typhus	Rickettsia	Fleas*	Siphonaptera
Epidemic typhus	Rickettsia	Human lice	Anoplura
Chagas disease	Protozoa	Assassin bugs	Hemiptera

Note: Specific names of pathogens and vectors are omitted.
*Causative organisms may be carried by other vectors.

TABLE 7. Major Epidemics of Bubonic Plague

Country or City	Year	Deaths
Western Europe	1348	25,000,000*
London	1603	33,347
London	1625	41,313
Messina	1656	48,000*
London	1665	68,596
Marseilles	1720	88,000
Moscow	1771	60,000*
Egypt	1834	32,000
Canton	1894	40,000–100,000*
Punjab	1906	675,307

Note: Because complete records of most epidemics are unreliable, only a few of the more reliable records are listed. Also, much of the widespread transmission of this disease was through secondary pneumonic action rather than through fleas.

*Estimated.

Human malaria is caused by four species of protozoa in the genus *Plasmodium,* and it is transmitted by the female mosquitoes of about 70 species in the genus *Anopheles.* The sexual stage of *Plasmodium* occurs in the mosquito, whereas the asexual stage is in humans (Fig. 273). The discovery of the cause of this disease was made by A. Laveran, and the proof of its transmission was made by R. Ross; these discoveries occurred in the late 1800s and netted both scientists Nobel prizes. Malaria remains one of the major world health problems, particularly in the tropics. In this broad band around the earth, over 50 percent of the human population in more than 100 countries are potential victims. Until the World Health Organization (WHO) initiated an attempt to eradicate this disease in 1947, about 5 to 15 percent of the human population contracted this disease annually. The program produced marked reductions in disease incidence, and many countries, including the United States, were declared disease free; however, the concern for environmental pollution by pesticides and the development of resistance in the mosquitoes to them resulted in formally abandoning the program in 1969 except for some local activity. Since 1969, the number of cases has remained low in some areas but has essentially returned to pre-eradication levels in many Third World countries. In the United States, approximately 961 cases were reported in 1983 and more than 1,100 in 1992. Most of these were believed to be imported, but such cases potentially could serve to infect mosquito vectors already native to this country. American citizens traveling abroad should be aware of malaria incidence where they are traveling and take the recommended precautions and prophylactic drugs to avoid becoming infected; however, there is no medication that guarantees suppression of the disease.

Another insect-transmitted disease that has great impact on humans is the plague (Table 7), often referred to as the "black death." Plague is an endemic disease of rodents, including rats, in many areas of the world, and it is transmitted to humans by fleas, primarily *Xenopsylla cheopis* or the Oriental rat flea, but also *X. brasiliensis* and *Nosophyllus fasciatus.* These insect vectors ingest infected blood, whereupon a blockage of the flea alimentary canal occurs. Subsequent feeding attempts result in

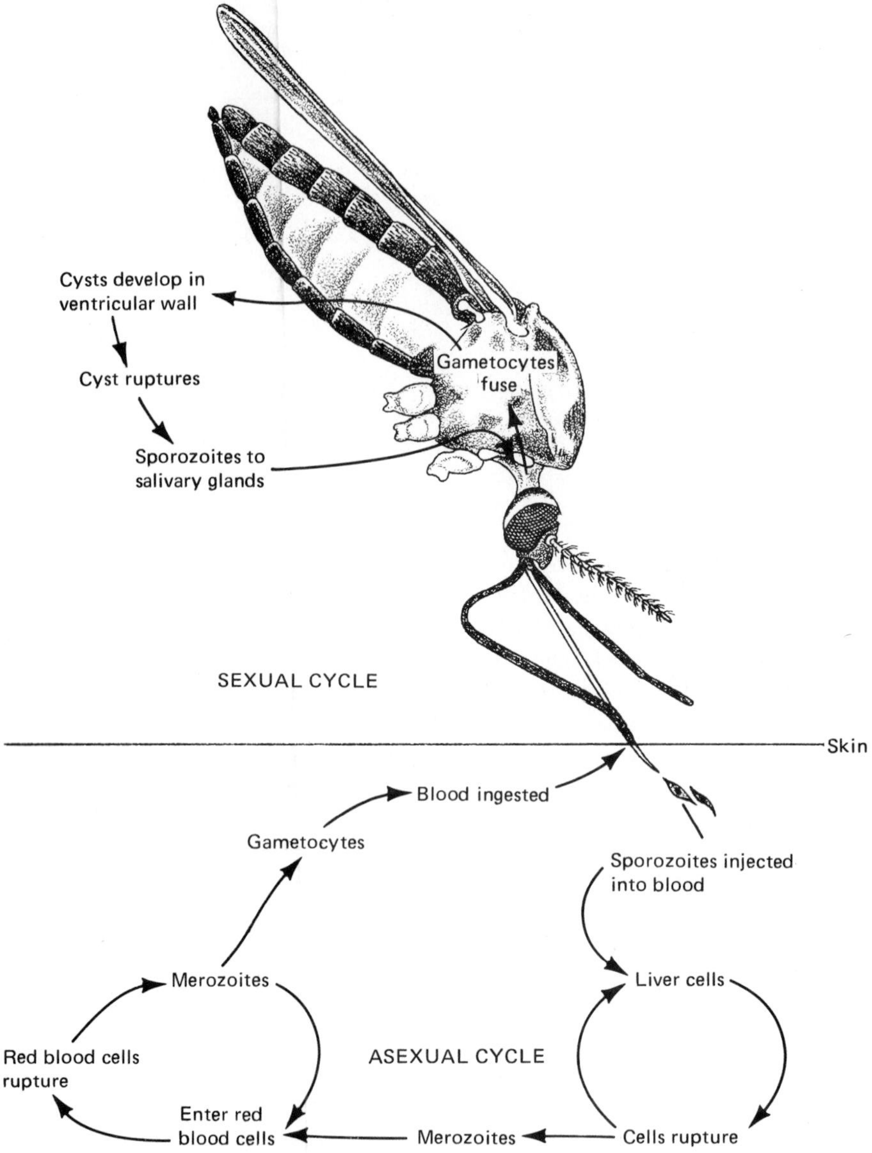

Figure 273. Life cycle of *Plasmodium* sp., the causative protozoan of human malaria. Sexual phase is in the mosquito; the asexual in humans.

regurgitation of the infected blood into the new host. Humans enter the cycle accidentally by being bitten by infected fleas, and, when the pathogen localizes in the human lymph gland, bubonic plague results. On occasion the blood and lungs become secondarily infected and septicemic and pneumonic plague develop, the latter situation permitting a new means of transmission by sneezing and coughing. The "ring o' roses" (black spot surrounded by a bright red ring in the skin) from the nursery rhyme "Ring o'roses, a pocket full of posies, ashes to ashes, we all fall down" had special meaning in 17th-century London, as it was a sure sign that the victim had the plague. Epidemics often have been traced to large-scale rat die-offs and the resulting release of infected fleas. Deaths result in about 80 percent of the infected humans, usually within 5 days. Three major pandemics are known: the plague of Justinian (A.D. 542), the black death of the Middle Ages (1346–1355), and the last pandemic from 1870 to 1906. More than 800 cases of plague were reported in 1979 and more than 1,000 in 1986 although these figures are very suspect. In 1994, India experienced an outbreak in the city of Surat, which had world health officials concerned to the point of suspending air travel. Hundreds of thousands of people fled the city. In the United States, about 13 cases were reported from 1983 to 1994, mostly among ecology-minded people in Arizona and New Mexico who had homes built with minimal disruption to the environment and kept cats as pets; hence, they shared their property and cat fleas (good vectors) with an inordinate number of native rodents, some of which were reservoirs of plague.

Yellow fever, after its introduction with slave importations from Africa, spread throughout the tropical New World and as far north as New England. In 1879 an estimated 15,000 died in the southern United States, but no epidemics have been reported since 1905. Human infections either result in death or life-long immunity. The decline in infections, as in malaria, probably was linked to intensive mosquito-control programs. The major mosquito vector has been the *Aedes aegypti.* Today the urban type is increasing worldwide but the sylvatic form, a disease found primarily in monkeys, exists as the most common yellow fever throughout the tropical New World and is transmitted by the *Hemogogus* species of mosquitoes. The cutting of tropical forests and migrations into those habitats is expected to produce an increase in this disease.

African trypanosomiasis, or sleeping sickness, occurs widely south of the Sahara Desert. More than 20,000 cases were reported in 1983 and 1990, but about 50 million people live exposed to this illness. The disease is caused by several trypanosome protozoans and is often fatal when the brain and central nervous system become infected; however, the central nervous system may not be invaded for several years after the initial infection. Transmission of the protozoans is by several species of tsetse flies in the genus *Glossina.*

Filariasis, another disease of the tropics, infects 300 million to 400 million people, but 900 million are at risk. Two species of filarial worms are the responsible parasites for this mosquito-borne disease (Table 6). Adult worms lodge in lymphatic vessels and, in extreme cases (about 5 percent of the infections), pathological changes occur in the host, resulting in the much talked about elephantiasis and hydrocoel.

Onchocerciasis, sometimes referred to as river blindness, is a disease caused by a nematode transmitted by black flies. More than 20 million people are infected, most in Africa (99 percent) but some cases are reported from parts of Latin America and Yemen. The disease localizes in the skin and lymph glands; however, worms may invade the eyes and cause blindness. An estimated 2 million humans are blind today from this disease, including up to 50 percent of some villages in Africa.

Dengue and dengue haemorrhagic fever, viral diseases located mainly in southeast Asia, the Caribbean, and Africa are transmitted by *Aedes* mosquitoes. History records numerous outbreaks, including an epidemic in 1920 of about 600,000 in the coastal Texas region and which may have involved more than 2 million in the adjacent Caribbean area. More than 350,000 cases were reported in 1981 in Cuba, more than 170,000 in Thailand in 1977, about 420,000 in Ecuador in 1988, and more than 20,000 in Puerto Rico in 1994.

Several types of encephalitids are transmitted by mosquitoes. For example, western equine encephalitis (WEE), a viral disease of wild birds in the United States, occasionally infects both horses and humans. WEE virus is clinically isolated from humans each year; more than 3,000 cases were diagnosed in 1941 in the western states. The fatality rate in humans runs about 3 to 4 percent and is highest in young children. The largest outbreak in horses was in 1937 to 1938 with more than 300,000 cases and 20 percent mortality reported.

Chagas disease, a protozoan-caused illness transmitted by certain bugs of the family Reduviidae, infects more than 10 million in Central and South America; and the list goes on and on. Interested students are encouraged to locate further readings in medical entomology and parasitology texts.

In each of the previously discussed vector-parasite associations, advantages gained by the pathogens include:

1. Selectivity and locating the host by the vector.
2. Protection of the pathogens during the transmission phase.
3. Relatively rapid dispersal.
4. Utilization of the insect for multiplication purposes.
5. Inoculation through the protective skin barrier of the host.

After a period of decline in some of these diseases, we seem to be entering a time of concern as the number of cases increases. No doubt better diagnostic methods and reporting procedures reflect part of this increase; however, the major reasons probably are demographic (population growth, urbanization, etc.) and societal changes (greatly increased travel, lifestyles, treatment rather than preventative medicine, etc.). If global climatic warming occurs as predicted, the range of many insect vectors will widen, and a further increase in the number of cases can be expected.

Stings, Bites, and Allergies

People generally perceive insects as annoying creatures and invaders of human privacy and property. Studies indicate that children often learn to avoid insects and other arthropods by a fear-provoking episode or through conditioning by others. Not surprisingly, we cope best with animals in general and insects specifically when we are not personally involved—when the insect is not alighting on our bodies or food. One only has to remember episodes of enjoying the sight of the setting sun and suddenly feeling the onslaught of newly activated mosquitoes or being stung while on a picnic to formulate adverse opinions. Derogatory views of insects appear to be most prevalent in the so-called advanced cultures where decreasing positive interactions between humans and insects through urbanization have occurred over the past 2 to 3 decades. Formal education, however, can have dramatic influence on these attitudes if directed correctly, and generally college graduates are more tolerant of these invertebrates than those who only have a high school education.

Insects, however, can be more than just annoying as they protectively use their stings and mouthparts. About 30 percent of the persons stung by hymenopterous insects develop at least a transitory sensitivity to the inserted venom, and about 3 percent suffer serious reactions. In 1983, 100 persons apparently died from insect stings, and the number appears to be increasing. Social wasps, fire ants, and honeybees pose the greatest threat, especially in the southern and southeastern United States. Those who died from these stings usually did so within the first hour, and they probably were allergic from previous exposures, although nearly all of them had never demonstrated sting reactions previously (most people fail to seek medical help for stings, however). Early allergic symptoms include swelling of the windpipe and larynx causing breathing difficulty and a lowered blood pressure, the combination of which sends the afflicted individual into *anaphylactic shock.* Immediate administration of adrenalin and transportation of the patient to a hospital are vital, because most violent reactions take place in less than 2 minutes after stinging. Although these figures should not produce panic, they nevertheless represent factual occurrences and should invoke precautions when dealing with insects, especially if a person has a record of allergic responses.

In addition to stings by Hymenoptera, the following insects commonly bite humans: Diptera (mosquitoes, black flies, stable flies, horse flies, deer flies, tsetse flies, sand flies); Hemiptera (bedbugs, assassin bugs, backswimmers, giant water bugs, nabid bugs); and Siphonaptera (fleas). Because biting or stinging insects are normally encountered outdoors on picnics or hikes, simple precautions such as insect repellents and proper clothing markedly reduce the incidence. Bites also occur indoors, and infestations of sucking lice (Anoplura) continue to be reported each year; it is not uncommon for an entire school to close temporarily for a treatment period.

In addition to stings and bites, numerous allergic responses also have been reported. These include some respiratory reactions commonly diagnosed as hay fever and asthma. Some scientists have gone so far as to state that insects are a major cause of many of these respiratory allergies. For example, about 40 percent to 60 percent of asthma cases in some areas of the United States are responses to cockroach exoskeletal fractions. Apparently disintegrated portions of the bodies of these insects become airborne and inhaled and then trigger allergic reactions. Other allergic responses are confined to the skin. As an example, lepidopteran caterpillars with urticating hairs (Fig. 274) affected 200,000 people in Japan in 1955, and an estimated 500,000 suffered from dermatitis in 1972 near Shanghai, China. Only a small percentage of "hairy" caterpillars are poisonous, however.

Entomophobia

In some individuals the perceived threat of insects goes beyond annoyance and can cause psychotic behavior termed *entomophobia* or *delusory parasitosis.* Victims of entomophobia imagine that insects are jumping at them, feeding on them, making loud noises while they move around at night, ruining their food, crawling on their

Figure 274. This tropical *Automeris* larva has urticating hairs and should be handled with care.

skin, and so forth. Distressed individuals are in earnest about their "observations" and require immediate psychiatric aid, although it is normally difficult to convince these people that they need help. Pomerantz (1959) and Waldron (1962) review a few typical case histories.

Livestock and Domestic Animals

Insects are frequent parasites of domestic animals. For example, lice are parasitic in all postovarian stages and remain on the host during their entire life cycle. Other species spend only a developmental stage upon the host, usually as larvae, such as bot flies. Some species feed on domestic animals only as adults, such as biting flies. Especially in high populations, parasitic insects cause irritability and produce decreased eating by the host, more susceptibility to other diseases, and sometimes death. The major insect parasites of livestock in the United States follow.

The screwworm has had a history, as mentioned previously, of parasitism leading to death. Horn fly populations sometimes reach levels of several thousand flies per cow, and more than $200 million in livestock losses through reduced weight gain are estimated annually from their bites. Stable flies, another severe biter, also have been known to reduce milk production in dairy cattle by 40 to 50 percent. Face flies, a nonbiting species, caused nearly $70 million damage in 1965 through lower weight gains resulting from their irritating feeding around the eyes of cattle. Grubs produce several hundred million dollars in losses to cattle through weight loss, milk production

TABLE 8. Examples of Parasites Where Insects Serve as Intermediate Hosts

Parasite	Causative Organism	Host	Insect Intermediate
Haematoloechus medioplevus	Fluke	Frogs	Dragonfly nymphs
Crepidostomium cooperi	Fluke	Fish	Mayfly nymphs
Prosthogonimus macrorchis	Fluke	Birds	Dragonfly nymphs
Dipylidium caninum	Tapeworm	Dogs, cats	Fleas, lice, beetles
Hymenolopis carioca	Tapeworm	Chickens	Beetles, house flies
Hymenolopis fraterna	Tapeworm	Rats, mice	Beetles
Moniliformis dubius	Acanthocephalid	Pigs	Cockroaches
Diplotriacnoides translucida	Nematode	Birds	Grasshoppers
Skrjabinoptera phrynosoma	Nematode	Lizards	Ants
Dirofilaria immitis	Nematode	Dogs	Mosquitoes, biting flies
Habronema megastoma	Nematode	Horses	House flies

Note: A wide divergence of insects are involved in these diseases.

decreases, and damage to hides. Lice also can become serious pests of livestock, and fleas cause much discomfort with domestic pets and sometimes with dairy cattle.

In addition to livestock losses caused by insect feeding, insects also inflict damage by pathogen transmission. More than 50 livestock diseases are carried by insect vectors. The tsetse fly species transmits *nagana,* a disease similar to African sleeping sickness in humans, to cattle and many native animals. This disease has kept more than 4 million square miles of otherwise ideal grazing land out of cattle production (except for some breeds of shorthorn which have low susceptibility), resulting in billions of dollars of losses and, more importantly, causing the loss of protein for needy people. However, some scientists have viewed this exclusion as beneficial because it prevents the mass introduction of domestic livestock and the resultant overgrazing and desertification of the fragile semiarid savannas.

An example of disease transmission in the United States would be the epizootic of Venezuelan equine encephalitis (VEE) in 1971. More than 1,500 horses died of this mosquito-borne disease in Texas during a massive campaign to isolate and control this disease introduced from Central and South America. Only after hundreds of thousands of acres were aerially sprayed, after more than 3 million horses in 19 states were vaccinated with a relatively untested serum, and after quarantine restrictions were placed on movement of horses was this serious viral disease controlled.

In addition to the transmission of virus and unicellular organisms, insects also serve as intermediate hosts of parasitic worms (Table 8). For example, immature dog heartworms, *Dirofilaria immitis,* are transmitted from dog to dog by more than 70 species of mosquitoes. The microfilaria circulate in the blood in infected dogs, are picked up by feeding mosquitoes, reside in the insect for 8 to 16 days, and then escape from the mosquito as it feeds again. This disease is currently rampant throughout the United States.

The ingestion of certain toxic insects also can injure livestock and other domestic animals. For example, horses and sheep that accidentally eat blister beetles either directly or in alfalfa or hay bales may die or suffer serious injury from the beetles' potent body toxin, cantharidin.

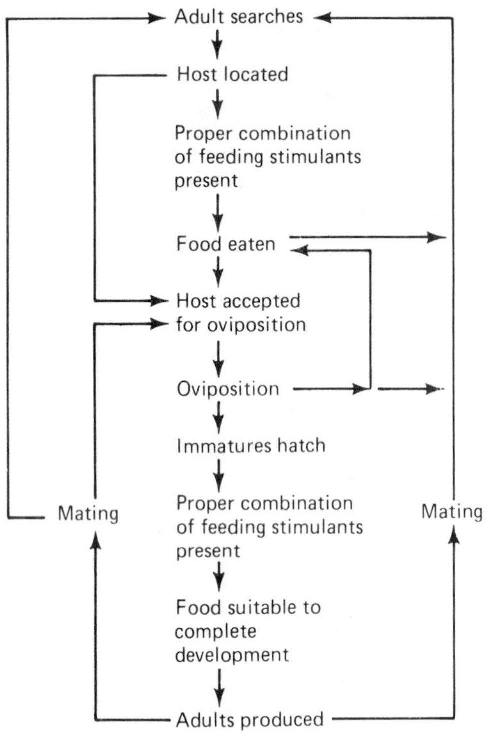

Figure 275. Summary of some of the behavior during the life cycle of a herbivore. Defensive activity and responses to climate are omitted for simplification.

Plants

Previous discussions in this chapter have indicated the long associations of flowering plants and insects. Many of these relationships are beneficial, as in pollination, but many are harmful, at least on a short-term basis. For example, migratory locusts have plagued the sub-Saharan region of Africa for centuries, and, although eaten as food by many organisms, they essentially destroy many plant resources eaten by other organisms. A single 160 sq.K migrating swarm of these short-horned grasshoppers can eat an estimated four times as much food as consumed in either London or New York in the same time frame.

Excluding pollen and nectar, an estimated 50 percent of all insect species feed on plants. Twelve orders have phytophagous species: Collembola, Coleoptera, Diptera, Hemiptera, Homoptera, Hymenoptera, Isoptera, Lepidoptera, Orthoptera, Phasmida, Psocoptera, and Thysanoptera. Insects vary as to which plants they eat. Selectivity may be so specific and may require such exacting stimuli to initiate feeding that only a given species (*monophagy*) or species in a given family of plant (*oligophagy*) are fed upon. Most herbivorous insects tend to be finicky feeders. *Polyphagus* insects, in contrast, feed upon a wide variety of plants and are less restricted; for example, Japanese beetles feed on more than 250 kinds of plants, and gypsy moth larvae are known to eat leaves of more than 500 trees and shrubs. A summary of herbivore feeding responses is shown in Figure 275.

Figure 276. Some larvae feed only on the outer layers of leaves, a process termed *skeletonizing*. The photograph illustrates extreme skeletonizing of elm leaves by the elm leaf beetle (*Pyrrhalta luteola*).

As insects have become more efficient herbivores, resistant plants (those capable of withstanding the insect pressures) have increased in numbers. Breeding plants for crops, however, has often produced the opposite selection result, and no plant used for human food is completely immune to insect attack. For example, corn is fed on by 200 species of insects, and apple trees by 400 species. In India and Bangladesh, 80 percent of the human food consists of rice, but more than 50 percent of this crop is lost to pests, including insects.

Nearly all plant structures are eaten by some insect (Figs. 145, 146, 147, 276, 277, 278, 279, and 280). If feeding is extensive, the plant may die, but most healthy plants can withstand insect infestations under normal conditions, at least as far as to propagate. What defenses do plants have? (1) Plants may regenerate lost structures at a fast rate. (2) Some plants have structures including spines or trichomes, pubescens, and thick cuticles as well as leaf shapes that physically interfere with insect feeding or, in certain instances, entangle the herbivore sufficiently so as to cause death. (3) There is relocation of certain vital nutrients to reduce insect survival. Nitrogen and starch are often in low percentages in wood and old leaves, especially in climax vegetation. Because insects use both proteins and carbohydrates for body support, whereas plant support is from carbohydrates, it is not surprising to find that insects require at least 1.1 percent to 1.5 percent nitrogen to survive (except species with symbionts), and wood has only 0.3 percent to 1.0 percent. (4) The timing of growth may be adjusted so that plants develop early and withstand the effects of nature to escape the full attack of insects whose appearance is

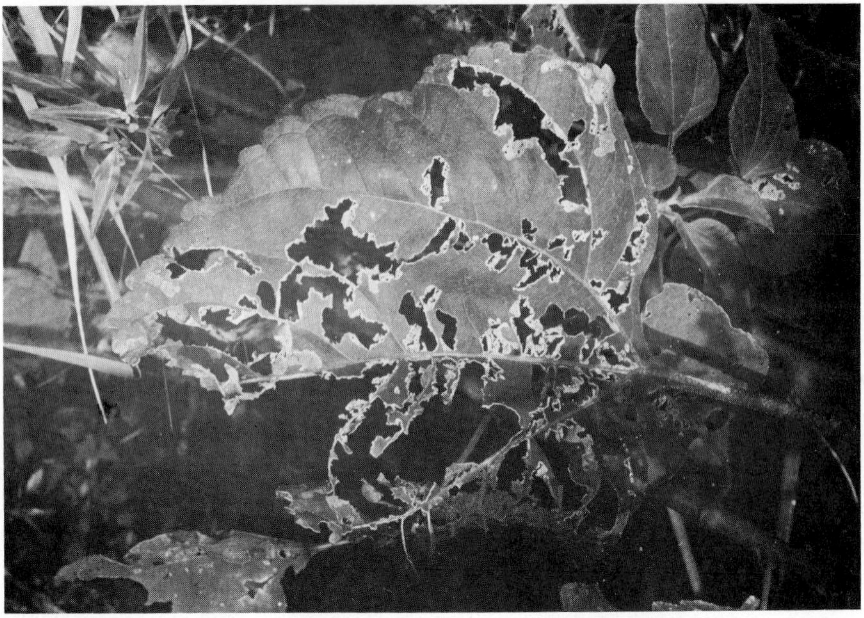

Figure 277. Damage on sunflowers occurs when the insect eats the entire cross section of the leaf.

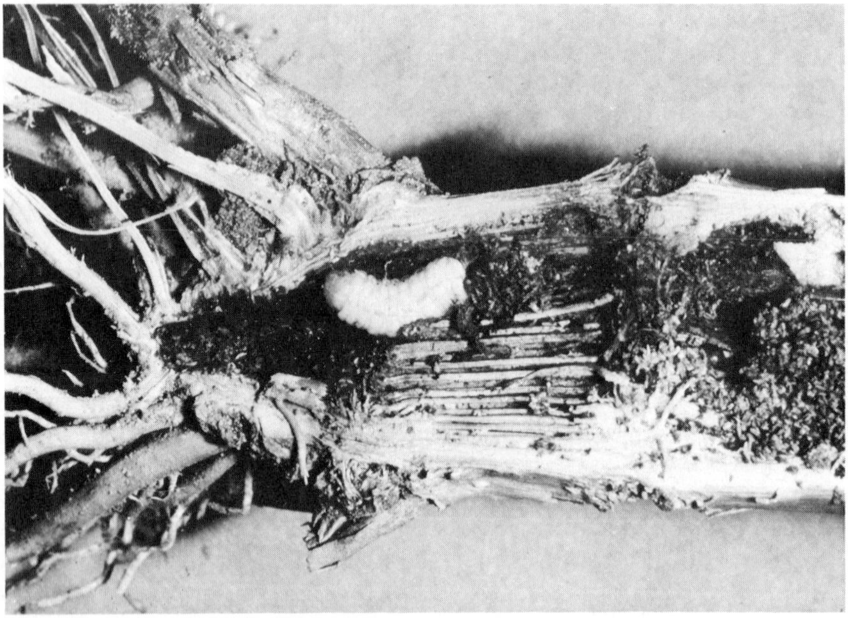

Figure 278. This southwestern corn borer (*Diatraea grandiosella*) has severely damaged and caused the corn plant to lodge under high winds. The larva overwinters in the base of the corn stalk.

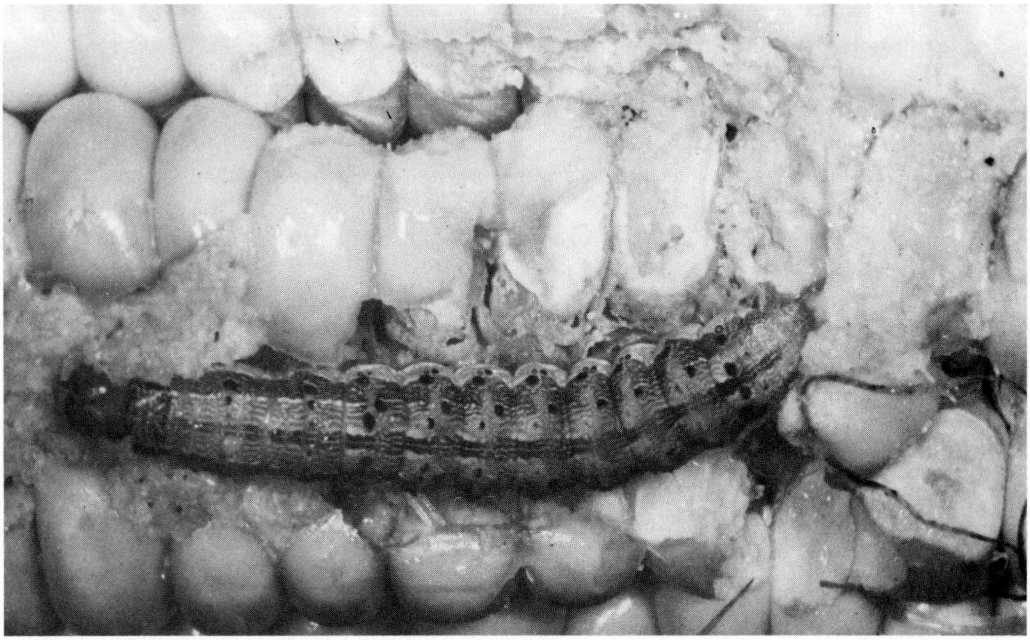

Figure 279. The larva feeding on this corn is the corn earworm (*Helicoverpa zea*). A voracious feeder, this pest is very repulsive to consumers.

Figure 280. Coleopteran grubs do considerable damage to subsurface structures of plants. This scarab larva has hollowed out a large cavity in a potato, resulting in a nonmarketable product for the farmer.

synchronized with the main plant population. (5) Wide dispersal decreases the likelihood of insects locating individual plants. Some scientists believe this to be a major factor in plant distribution in the tropics. (6) Allelochemicals often discourage some insects from feeding or ovipositing. These chemical defenses vary from indigestible compounds (tannins, resins, and silicones) that interfere with normal feeding to poisonous compounds (steroids, pyrethrins, mustards, atropines, nicotine, and saponins) that are often lethal to insect herbivores that have not specialized upon these plants. Also, some plants produce insectlike growth substances that affect development in some insects. Most of these allelochemicals are constantly present, but feeding by herbivores may induce their production in others, such as potatoes, tomatoes, clover, and lupines.

There appears to be a taxonomic correlation between plants with certain allelochemicals and the insects feeding upon them (Ehrlich and Raven, 1963). Because plants cannot afford to channel extraordinary energy into producing these substances, different plant families have become specialized for only several chemical adaptations. In turn, insects have evolved countermeasures, with generalist feeders having to have a broader range and higher levels of detoxifying enzymes than specialist feeders. The continual evolution of these plant substances and the responses by insect herbivores might be summarized as follows:

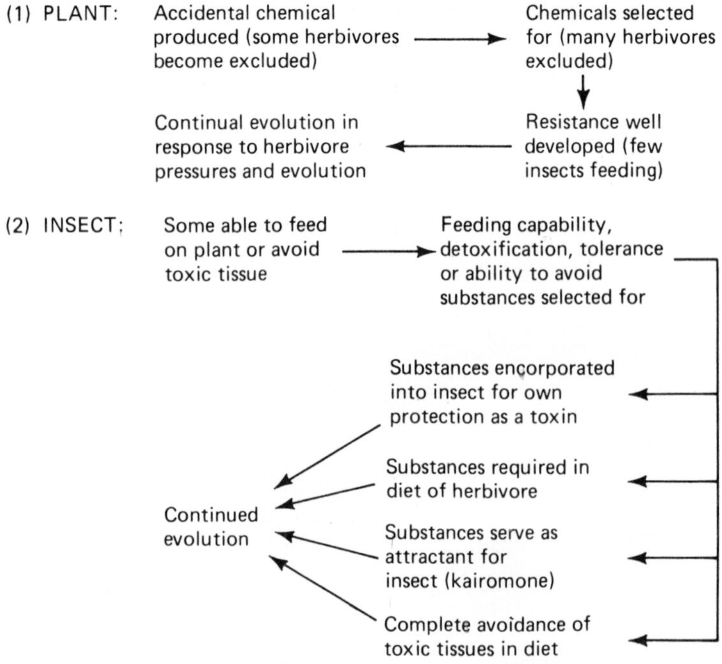

Protection of one species of plant from attack by herbivores is also gained by associating with other species of plants. These relationships may be divided into three types: (1) *insectary,* where nectar production in insectivorous plants is synchronized with oviposition of pests of neighboring plants, thereby lowering the latter's pest problems; (2) *repellent,* where preferred plants are not located by an insect herbivore

TABLE 9. Examples of Plant Crop Diseases Transmitted by Insects

Disease	Plant	Causative by Organism	Carried by Certain	Order
Brown rot	Apples, peaches, etc.	Fungus	Beetles, flies	Coleoptera, Diptera
Dutch elm disease	Elms	Fungus	Beetles	Coleoptera
Cucumber wilt	Cucumbers, melons	Bacteria	Beetles	Coleoptera
Tobacco mosaic	Tobacco, tomato	Virus	Aphids	Homoptera
Leaf roll	Potato	Virus	Aphids	Homoptera
Necrotic yellow	Lettuce	Virus	Aphids	Homoptera
Cucumber mosaic	Cucumber, weeds	Virus	Aphids	Homoptera
Sugar cane mosaic	Sugarcane, corn, sorghum	Virus	Aphids	Homoptera
Curly top	Sugar beets, tomato	Virus	Leafhoppers	Homoptera
Rice stunt	Rice	Virus	Leafhoppers	Homoptera
Aster yellows	Aster, celery	Virus	Leafhoppers	Homoptera
Corn stunt	Corn	Mycoplasma	Leafhoppers	Homoptera
Spotted wilt	Tomato	Virus	Thrips	Thysanoptera

because of sensory interference by chemicals released from adjacent repellent flora; and (3) *decoy,* where several plants have an attraction, thereby diluting the impact of the herbivores. Variability of plant species within a given habitat certainly affects plant-insect interactions.

Damage to the plants may come from causes other than insect feeding. Insects may carry plant disease, primarily viral and microplasmal diseases. The difficulty of a virus disseminating itself from a sessile plant is obvious. Insects, therefore, have become a ready means of distribution because of their intimacy and host species preferences, especially those insects that have piercing-sucking mouthparts and ingest sap containing the disease agent. More than 260 plant viruses are known to be transmitted by insects, with the majority having species of Homoptera as vectors; aphids alone transmit more than 160 of these viruses (see examples in Table 9).

Stored Products and Household Pests

Any harvested product, whether it is of animal or plant origin, is subject to attack by insects. Common materials fed upon include grains (Fig. 281) and their products, seeds, nuts, fruits and vegetables, meats, dairy products, furs, leather goods, wood and its products, wool, and tobacco.

Figure 281. Granary weevils (*Sitophilus granarius*) can cause considerable damage to stored grain left undisturbed for some length of time. Eggs are laid in cavities of corn kernels, where the larvae often complete development within 4 to 7 weeks.

Damage to stored foodstuffs by direct feeding of insects or by contamination may be great, with an estimated 5 percent to 10 percent of the world's produce damaged or lost annually. Such losses are equal to the amount of food or grain required for the survival of approximately 130 million people. Losses are greatest in the tropics where food, especially grains, cannot be adequately dried, and hence massive infestations of both insect (Fig. 281) and fungal species may be seen. To prevent this, some grains such as corn are left in the field until used. When stored, grain must be maintained at or below the 12 percent moisture level to deter infestations, but fumigation and/or spraying with short-lived insecticides are usually necessary. Of major concern in the United States, both in large facilities and in home storage, are approximately 50 insect species belonging primarily to the orders Coleoptera and Lepidoptera. The beetle species produce damage during both the adult and larval stages, but moths do so only as larvae. A knowledge of the development and ecology of each insect species pest is needed before specific infestation problems can be solved.

Books may be damaged by a variety of insect pests, including termites (Fig. 282), book lice, and Thysanura, the termites feeding mainly on the paper and book lice and Thysanura preferring the glue and bindings. Damage is seldom observed until the bindings give way, except in cases of termite feeding. In addition to books, other

Figure 282. This book damage is the result of termite feeding.

household items such as wallpaper and pastes can be eaten. Termites feed on newspaper and other wood or wood derivatives.

QUESTIONS

1. What structural and behavioral modifications do insects have to benefit from plant associations? What advantages and disadvantages have plants gained from this relationship?
2. Do you think that humans will utilize insects for food in the near future? Why or why not?
3. How are insects beneficial to humans? How are they detrimental?
4. What are some of the major human diseases transmitted by insects? How much of an impact have they had upon the human race? What can we expect in the future?
5. What insect orders are of major economic and medical concern to humans?

11 Insect Pest Management

Pests, by definition, are organisms that come into conflict with humans. For some species, just one insect is all it takes to cause physical trauma in the form of biting, stinging, or disease transmission or to produce psychological fear. Other species cause conflict in great numbers, competing with humans for food or damaging dwellings and products that we value economically.

An estimated 3,000 plant species are used for food by humans; 300 are used abundantly and 12 supply more than 90 percent of our food. Yet, millions are starving or undernourished. Why? Part of the answer lies in the competition for food with insects. Nearly 50 percent of the acreage of various parts of the world's countries, including India and Bangladesh, is in rice, but often more of this food has been lost during production and storage to insects and other pests than was available for human consumption. In the United States, about 8,000 species of insects are injurious to crops, and about 200 are serious pests. During our early history, farms were small and farmers very mobile and often moved from one area to another when conditions became intolerable, but this strategy is no longer possible with the swing to commercial agriculture and with most land already privately owned. Under the present system of *monocultures* (a crop grown on the same acreage for at least two crop cycles), the presence and activity of pest species is inevitable. Also, importation of high yield but genetically uniform cultivars into new areas of high pest susceptibility, high use of fertilizers and other chemicals, and irrigation practices have produced many pest problems that were not present on the typical early small farm that used only plants and their seeds that had survived past infestations. Transportation has become increasingly refined and crops marketed worldwide, resulting in pests becoming translocated into new areas. For example, many of the severe pests in the United States are not native species (Table 10).

In order to raise more food as well as to protect ourselves from biting and disease-carrying insects, we have devised methods to alter normal population growth of many

TABLE 10. Some Major Documented Introductions of Insect Pests into the United States

Date	Pests	Date	Pests
1779	Hessian fly	1900	Alfalfa weevil
Early 1800s	Cabbage maggot	1903	Brownbanded cockroach
1832	Pear psylla	1913	Southwestern corn borer
1834	Elm leaf beetle	Before 1915	Oriental fruit moth
1857	Imported currantworm	1916	Pink bollworm
Mid 1800s	Diamondback moth	1916	Japanese beetle
1860	Asparagus beetle	1917	European corn borer
1860	Grape phylloxera	1922	Asiatic garden beetle
After 1860	Imported cabbageworm	1929	Black imported fire ant
1869	Gypsy moth	1929 (+many times)	Medfly
After Civil War	Harlequin bug	1930s	Red imported fire ant
1870	Grape root borer	1936	Whitefringed beetle
Before 1880	Spinach leaf miner, spotted asparagus beetle	1940–60	European hornet
1882	Greenbug	1951	Face fly
1885	Carrot rust fly	1954	Spotted alfalfa aphid
1887	Horn fly	1962	Cereal leaf beetle
1889	Mediterranean flour moth	1966	Formosan subterranean termite
1892	Boll weevil	1986	Russian wheat aphid
1893	Confused flour beetle	1987	Asian tiger mosquito
1897	Browntail moth		

insect pests by reducing their chance for survival. Early practices were aimed at minimizing damage, but with the advent of synthesized insecticides in the 1940s, attempts became directed toward eradication. Despite these efforts, however, successful eradication of any pest species is unknown except in the case of localized areas such as islands. In the late 1960s and early 1970s a new era began when, as the result of concern for the environment and observed resistance of pests to insecticides, *pest management* reemerged as a dominant concept. Pest management is the practice of regulating insect pest populations to prevent outbreaks instead of attempting to eliminate them. Effective regulation measures involve a knowledge of the pest life cycle, number of generations per year, population densities and growth potential, density-dependent and independent factors, and all phases of crop production, including economics.

Pest species may be classified along a mathematic continuum from r to K of population stability. At the r extreme are those species that have high reproductive rates and short generation times, reproduce early and often only a single time, and are therefore highly adapted for fluctuating environments with little competition. Such insects generally put their energy into reproduction to guarantee finding the limited resources depended upon. These pests, such as the migratory locusts, are often best controlled by insecticides. At the opposite end of the spectrum are the K-strategists, those insects that are less likely to migrate, who reproduce many times but generally have low reproductive rates and long generation times, and hence are good competitors and adapted to high habitat stability. An example of the *K*-strategist would be the tsetse flies. Cultural and biological controls may be sufficient to control these pests if the environment is not disturbed during the desired period. Most insect pests, however, exist somewhere intermediate between the *r* and *K* extremes and remain at tolerable densities except for occasional population explosions when natural controls become inadequate. Techniques under these circumstances vary, depending upon how close to each extreme the pest is, but several methods of regulation are often needed. Reducing damage only slightly may often result in a profitable operation.

The current philosophy of contending with insect pests is to minimize damage using as many different techniques as possible without injuring the environment.

What methods are available to manage pests? How can these be used to supplement one another in an integrated program? Answers to the preceding and following questions are necessary for intelligent decision making but may not be available during early stages of developing management policies: (1) What is the pest? Is it a severe pest, one that is in every field every year, or only an occasional pest? (2) What is the economic damage? This is often very difficult to determine, especially where multiple pests and great weather fluctuations occur. (3) What population densities are necessary to produce economic damage and at what stage of the host's development does this occur? During the past several decades the concepts of *economic threshold* (ET) and *economic injury level* (EIL) have been used extensively. The former (ET) is the point in population density of the pest where control measures need to be applied to prevent EIL from being reached. This is a critical but flexible point and the most widely used index in making management decisions, but it requires constant monitoring of the pest and the dynamic environment (Fig. 283). When surveys indicate that populations have reached the EIL, economic damage to the crop is already occurring and the response must be immediate. (4) What is the value of the crop or product? Many techniques are not cost-effective in low-cash crops. (5) Must the crop or product be cosmetically "perfect"? Little damage to leaves of early lettuce drops the market value significantly, but leaf damage has little or no affect on alfalfa sales. (6) What manipulations achieve the best results in management and under what conditions? (7) What will be the effect of control measures on other potential pests, predators, parasites, and the ecosystem? (8) How do the pest management techniques interact with other aspects of crop production, such as weed control, crop varieties, nitrogen use, water stress, and temperature variations?

To formulate appropriate experiments for obtaining adequate information and then to apply the data to specific circumstances in a total crop production package often requires mathematical models and computers because of the factor interaction complexity. A model, when developed, will mimic closely the natural environment and could eventually permit a pest management program to function with adequate sampling inputs and predictive outputs and with the widest possible assortment of control choices.

The use of any single management technique nearly always results in selecting for those individual pests that can overcome that particular control strategy. This can be termed *resistance,* but it is really natural selection at work. As a result, the more efficient the technique in lowering the pest population, the higher the selection pressure for resistance. Therefore, a judicious use of a combination of techniques into a man-

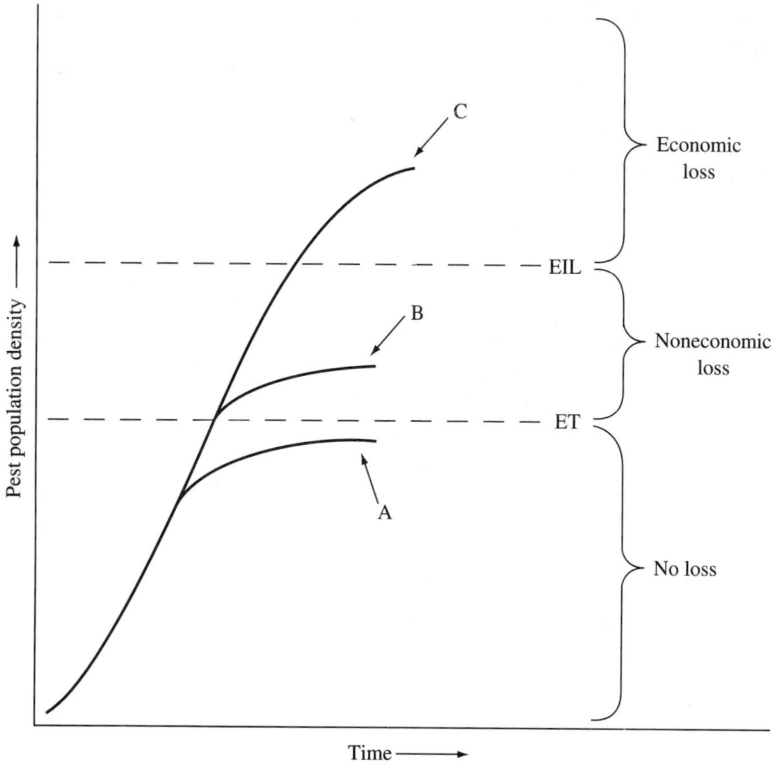

A = Noneconomic pest
B = Pest population controlled prior to causing economic damage
C = Pest population causing economic damage

Figure 283. Pest population growth and effect when correlated with economic threshold (ET) and economic injury level (EIL).

agement program appears to be the best solution to this dilemma. For example the discovery in 1980 of fruit infected with the exotic Medfly (Mediterranean fruit fly) in California was a potential multimillion-dollar threat until the combined techniques of fruit quarantine, sterile-male release, pheromone trapping, handpicking and destruction of infected fruit, spraying with the insecticide malathion, and distribution of poison baits resulted in control and finally a perceived eradication of what became a political "football" (Marshall, 1981; Angier, 1981; and Rogers, 1981). Again, however, management rather than eradication of a pest remains the realistic goal. An overview of major control or management techniques is now in order.

LEGAL AND CULTURAL ASPECTS

Obviously, insect pests cannot be expected to comply with governmental legislation or rulings. However, regulations dealing with the manufacture of goods, raising of crops, and marketing practices can, and often do, play an important role in the introduction and distribution of insect pests and the diseases they carry.

Regulations governing the importation of goods and food stuffs are the first line of defense against exotic pests. An estimated 280 pest species have been restricted from the United States by border inspections (Sailer, 1977). However, even the best system experiences periodic breakdowns, and about 40 percent of all insect pests in the United States requiring control practices are not native. Most entered prior to the current inspection system, however. A few of the major insect pests that are not endemic are listed in Table 10. The problem of exotic introductions will continue as tourism and travel among the public and as trade between countries increase.

In light of the potential for the accidental entry of serious pests, what defenses are in place to isolate and eradicate introduced species? Upon detection, efforts are mobilized to isolate the pest. Quarantines often involve different levels of government, depending upon the perceived threat. Some efforts are highly successful. For example, Venezuelan encephalitis, a serious viral disease of horses in Central and South America, appeared in Texas in 1971. By restricting interstate shipment of horses in the Southwest, this mosquito-borne disease was limited, and an intense program of mosquito control and horse immunization was employed with high effectiveness. Even with these stringent measures, more than 1,500 horses died in a month. Other attempts to eradicate exotic introductions have failed. Failure often comes when the eradication components, particularly those that use chemicals, come into direct conflict with values of the public.

Legislation is having an increasing effect upon control practices, particularly in the use of insecticides. The Federal Insect Pest Act of 1905 permitted regulation of articles that might transport insect pests. Then in 1910 the Federal Insecticide Act was passed to protect consumers against the fraudulent selling of misbranded or adulterated products. The Plant Quarantine Act of 1912 provided regulations to control possible introductions of disease and insects from foreign countries. In 1947 the Federal Insecticide, Fungicide, and Rodenticide Act was passed, which transferred to the manufacturer the responsibility of proving insecticide effectiveness; this effectiveness had to be demonstrated by adequate research before the approval and release of the product onto the market. The Miller Amendment (1954) to the Federal Food, Drug, and Cosmetic Act introduced the safety factor of residues. Subsequent concern over carcinogenic effects have entered the picture in the Delany Clause of Williams' Bill (1958), and chemicals that induce cancer in animals must be taken off the market. Because of these legislative changes, even more intensive screening by companies and university experiment stations for insecticidal effectiveness, dosage requirements, and safety factors must precede registration. In December 1970 the Environmental Protection Agency (EPA) was created, and in 1972 through the Federal Environmental Pesticide Control Act, the agency was charged with the regulation of all pesticide usage, including registration of the product and of applicators using certain insecticides. The Federal Pesticide Act of 1978, in response to a burgeoning of registration problems, simplified the regulatory interpretations imposed by the 1972 act, and the individual states received the enforcement responsibilities. In 1988 Congress required the EPA to reevaluate and reregister chemicals used as pesticides. In 1992, EPA issued the Worker Protection Standard (WPS) in an attempt to reduce pesticide-related illness and injury. Other regulations also have been developed but are not listed because they, as well as many of the above

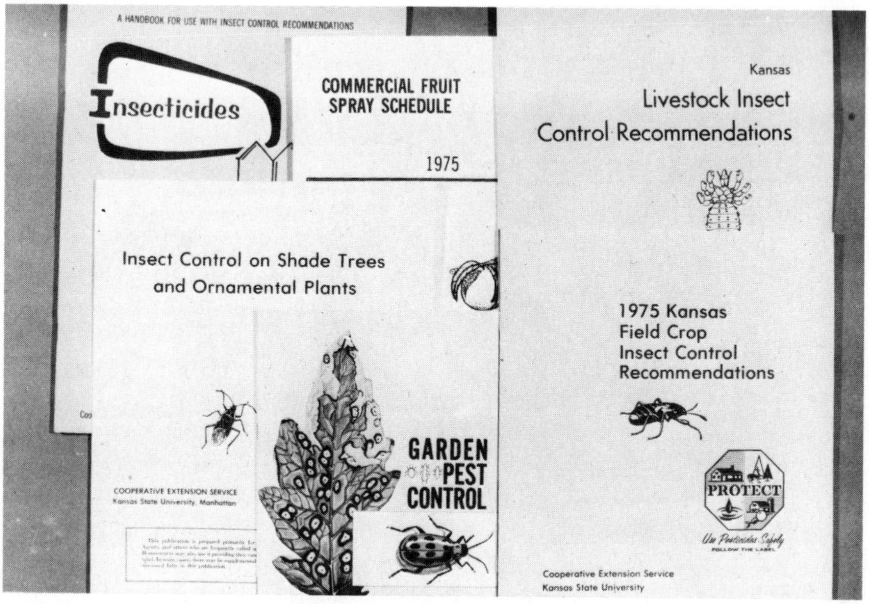

Figure 284. Many states regularly publish recommendations for insect control. This photograph includes a few examples of those previously distributed in Kansas.

long-standing enactments, are currently being reviewed. The current mood of much of the public, businesses, and legislators is for deregulation. Hopefully, this will not come at the expense of standards that have produced an increasingly better environment and better health for the individual.

In addition to federal influences, state and local agencies have generally been concerned with the local distribution, storage, control recommendations, and enforcement practices, and they sometimes have been more stringent than the federal mandates. Land-grant universities have been in the foreground here in providing vital data, and nearly all conduct research and publish yearly recommendations for the use of insecticides (Fig. 284) and other control methods for the crops in their areas. The development of IPM (integrated pest management) systems is also being undertaken in many states through agricultural experiment stations and extension services.

Cultural techniques are also vital in managing insect pests. Strip-cropping, destruction of infested crop residues, rotation of crops, soil tillage at proper times, and the timing of planting and harvest methods may mean the difference between an average or good crop. Such methods are designed to alter the environment sufficiently so that population buildups of pests are retarded. Most of these practices are cheap and are vital agricultural practices.

HOST-PLANT RESISTANCE

As discussed in Chapter 10, plants must be capable of resisting the effects of insect herbivores to survive; however, crop breeding has had a great influence on the ability

Figure 285. Plant breeding experiments on wheat indicating those plants resistant (*left*), tolerant (*middle*), and susceptible (*right*) to Hessian fly infestations.

of domesticated plants to survive such onslaughts. Some varieties are more resistant than others (Figs. 285, 286). Painter (1968) categorized plant resistance into three major types: tolerance, antibiosis, and nonpreference. *Tolerance* means the ability of a plant to survive pest infestations that would normally injure or kill similar plants. Environmental factors have a great influence upon this type of resistance to pests because growth rates are so temperature-dependent. *Antibiosis* is the ability to induce detrimental effects on the pest and thereby reduce damage by the insect. Effects on the insect pest include causing death, lowering fecundity, lowering survival rates, affecting diapause, causing behavior or physical aberrations, decreasing size, and lowering food reserves available to the pest. *Nonpreference* (now called *antixenosis*) resistance is somewhat intermediate between the first two types; plants in this category seem to be "ignored" or much less preferred by the pest. This may be caused by modifications in the substances that attract the pests, lack of or modified substances used by the pest as a feeding stimulant, by the use of repellents, or by changes in the interaction between the previous factors. A cline exists between strong resistance and susceptibility.

Breeding for resistance to pests, especially when coupled with other control measures, has achieved good results with certain crops. The Hessian fly, *Mayetiola destructor,* which nearly wiped out wheat in some Midwest states several decades ago, can now be controlled by combining resistant varieties of wheat with late planting in the "fly-free" period. Hessian fly larvae can cause considerable damage if the wheat variety is not resistant. More than 13 varieties of resistant wheat (Fig. 285) have been released in the past 12 years, but many are not in current use. The number of susceptible cultivars is increasing, partly due to the past success of this technique and partly because of selective breeding for other factors such as yield. Destruction of volunteer (growing from self-sown seed) wheat in the summer also reduces a potential fall

Figure 286. Photograph of various alfalfa plant strains developed to resist pea aphid attacks. Those on the left are much more susceptible than those on the right as determined by height and lateral growth. (Photograph by E. L. Sorensen.)

brood. Combining this practice with plantings in the late fall, during the "fly-free" period, greatly reduces the fly populations and potential damage.

In addition to wheat resistance to the Hessian fly, resistant germ plasm is available in the following crops: wheat (wheat stem sawfly, cereal leaf beetle), alfalfa (potato leafhopper, pea aphid [Fig. 286], spotted alfalfa aphid, alfalfa weevil), soybeans (Mexican bean beetle), potatoes (potato leafhopper, potato flea beetle), sorghums (greenbug), onions (onion thrips), barley (corn leaf aphid, greenbug), and tomatoes (potato aphid, flea beetles). Others are on the way.

Early efforts to manage insect pests used these techniques extensively, but during the period from 1960 to 1970, crop plant breeding focused on other factors such as yield, drought resistance, storage ability, higher protein, and better taste, and let insecticides take care of the pest problem. Because of insecticide resistance, the higher cost of insecticide development and manufacture, and other concerns over use of chemicals, resistance programs in pest management are again receiving much attention, particularly with low market value crops. Such an approach, however, is not without difficulties. Problems in developing and implementing a host-plant resistance program include:

1. The difficult and time-consuming task of screening and breeding (often 10 to 20 years in the past) to obtain the high resistance coupled with high yield and other desirable plant qualities.

2. The fact that the pest species is continuously evolving, resulting in the need for new plant breeding programs to overcome the pest change (resistant plant varieties often become obsolete in 3 to 10 years).
3. Requirements that select for resistance neither reduce the palatability of the food to the human population nor prove harmful when eaten. For instance, potatoes naturally contain high amounts of alkaloids as protection against insects and at least one variety has been withdrawn from production after optimistic breeding trials produced potatoes with dangerous levels of these compounds.
4. The fact that the plant uniformity that often results makes future breeding difficult without introducing "wild type" genes. These latter genes can come from various sources including closely related plant species; for instance, goat grass and rye have provided genes for resistance in wheat.

With the possibility of genetic engineering, the speed of transferring genes has increased significantly. Genetic engineering is the incorporation of specific genes for a given trait from one organism, often distantly related, into another organism. At present, genes from the bacterium *Bacillus thuringiensis*, which produce an insecticidal toxin have been incorporated into various plants including cotton and corn and can protect these plants from certain major pests. A variety of potato has been altered to produce a naturally occurring toxin to kill the Colorado potato beetle, an insect highly resistant to most synthetic insecticides. Similar possibilities may exist for other commodities.

BIOLOGICAL CONTROL

Insects, as previously indicated, are food for a wide spectrum of organisms, including other insects (Figs. 287, 288, 289), viruses, bacteria and fungi (Figs. 290, 291), vertebrates, and plants (Fig. 292). It seems logical, therefore, to utilize these naturally existing predators or parasites in reducing pest populations. One way is *introduction* where an exotic beneficial is introduced (after careful evaluation as to its impact upon other organisms) into a new area to control a specific pest. A second way is to *augment,* or release laboratory reared beneficials at times when natural beneficial populations are low. A third method is termed *inundation*, which is the release of large numbers of predators and/or parasites at critical times for short-term results.

The use of biological control is most likely to be effective when coupled with other techniques of regulating pest populations. Also, some serious obstacles face the development and implementation of this technique. Biological control programs require considerable scientific information on population potentials, fluctuations of both pest and control agents, possible ecological ramifications, expected variations from locality to locality, the capability to rear and transfer the agents, and practical economics. The following requirements must be satisfied if effective long-term pest management is desired:

1. There must be close synchronization of host and parasite cycles.
2. Predators and parasites must be effective during the growth phase of the host population.

Figure 287. This aphid infestation on milo is being brought under control by wasp parasites. Note the white mummies, which are the exoskeletal remains of parasitized aphids.

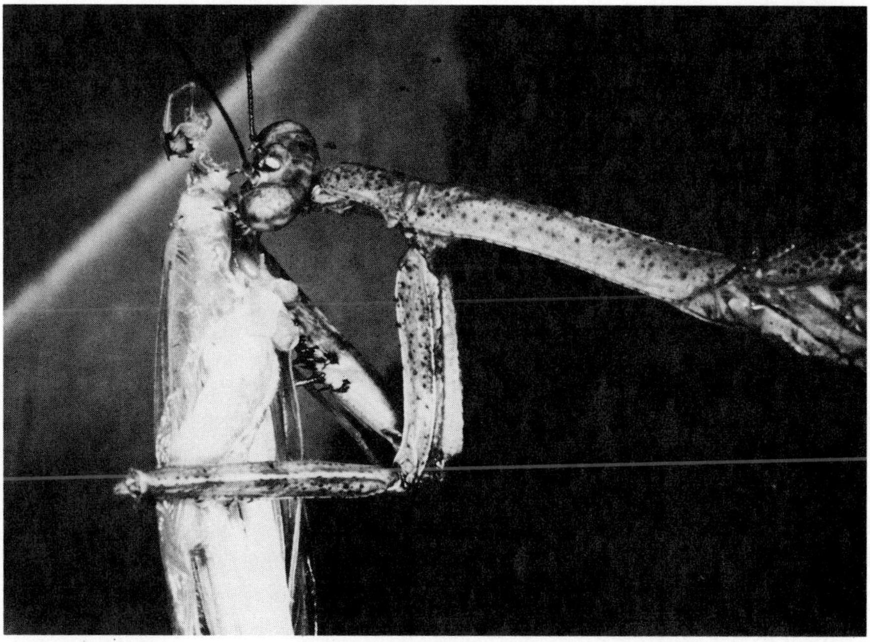

Figure 288. Carolina mantid (*Stagmomantis carolina*) feeding on a tree cricket (*Oecanthus niveus*).

Figure 289. This sphingid larva will not complete development because of braconid wasp parasites. These wasps, after feeding within the host, bore out and pupate in silken cocoons, as shown.

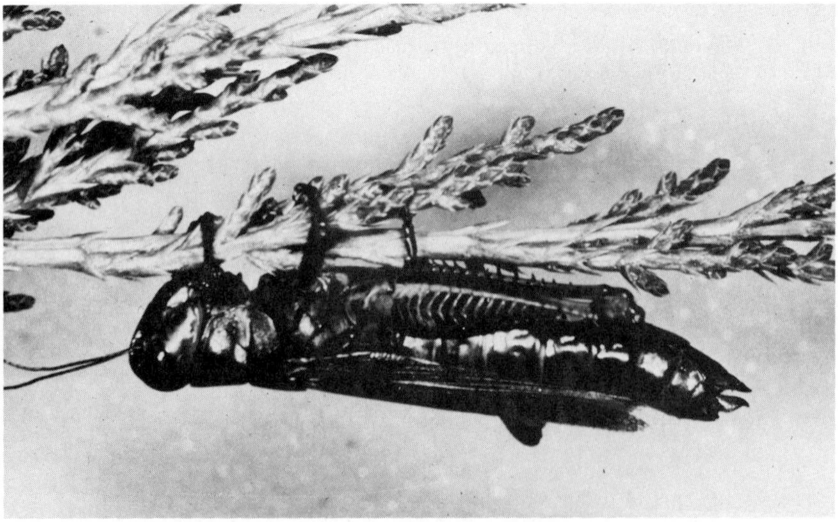

Figure 290. A fungus, *Entomophaga calopteni,* has killed this grasshopper, *Melanoplus differentialis.*

3. Host populations must be controlled before economic damage occurs.
4. The climate must be conducive for the survival of predator or parasite.
5. Alternative hosts or a residual of pest individuals must be left to maintain predators or parasites unless continual mass releases of predators or parasites are contemplated.
6. The use of insecticides should be minimal or the agent must be resistant to these chemicals, and cultural practices should not be disruptive.

Figure 291. Biological control of insect pests has proved to be an alternative for use of chemicals in certain instances. In some circumstances the environment is "seeded" with a specific pathogen of the pest. The above *H. cecropia* larva, although not normally considered a pest, is dying from a virus that has caused extreme dehydration through diarrhea, a situation of lethal consequence in caterpillars, which must maintain a turgid condition to locomote.

Figure 292. A few insects, such as this house fly (*Musca domestica*), fall victims to plant predation. This Venus's-flytrap has just reopened, exposing the fly remains.

The advantages of introducing parasites and predators instead of insecticides are: (1) these agents often are very selective, (2) insect resistance is less evident than in the case of pesticides, (3) the ecosystem is less affected, and (4) parasites and predators are less dangerous than pesticides to humans. Disadvantages include: (1) difficulty in achieving the requirements for effectiveness (as listed above), (2) difficulty in meeting government regulations for viruses, fungi, and bacteria because these organisms at times have had to meet some of the standards set under the pesticide laws for labeling and use, (3) viral and bacterial pathogens must normally be ingested by the target insect, (4) timing of spraying pathogens is often critical, and (5) introduction of polyphagus insect parasites could affect both native and pest species indiscriminately and result in undesired reduction or elimination of native species if done without great care, (6) predators and parasites are more susceptable to insecticides than are pest species although genetic engineering may be able to assist in this dilemma, and (7) rearing beneficials and maintaining them to have them available at "just the right time" in augmentation and/or inundative releases is difficult.

Of the 600-plus species of parasites and predators that have been released as potential biological control agents, only 30 percent to 40 percent have been judged to be successful. The first major success, and it still is used as the best example, is the control of the cottony-cushion scale in California by the importation and release of the vedalia ladybird beetle. The scale was unwittingly introduced in about 1868, and, after searching its native habitats in Australia, the beetle was found and introduced into California. Within several years, the scale was brought under control and remains so to this date. Another example, the Japanese beetle, was accidentally introduced into the United States in 1916 on nursery stock in New Jersey. This beetle now inhabits most areas east of the Mississippi River, feeding as adults on fruit and leaves while the larvae are destructive on the roots of grasses. Neither traps nor insecticides have been successful by themselves, but a bacterium coupled with several parasitic wasps is now used in a management program. Other notable successes include control of the olive scale in California by a parasitic wasp, the European spruce sawfly in Canada by a virus, and the Oriental fruit fly in Hawaii by three species of parasitic wasps. These examples illustrate that several different biological agents may be required. As research permits us to gain better understanding of the varied ecosystems and biotic interactions, most likely the success rate of implementing this technique into management programs will improve.

GENETIC CONTROL

The goal of this technique is to convert a pest insect to the nonpest status. Three avenues of accomplishing this are by sterilizing native pest populations through chemosterilants, by mass sterilizing males and then releasing them, or by conferring beneficial advantages such as insecticidal resistance upon natural predators or parasites that utilize certain pests. In the first avenue, experimentation has shown that success comes only when sterilized insects can compete effectively with their unaffected and fertile conspecifics and drastic reductions in populations are not frequent. In contrast, mass sterilization and releases have their greatest potential against pests when populations are low and isolated. These mass-reared insects also must be competitive and must be released at the proper time and place. In addition, the females should mate only a single time.

The screwworm fly control program in Florida, the southwestern United States, and Central America serves as the best example of genetic control of a pest. Screwworm fly larvae cost the cattle industry an estimated $100 million each year until the mid-1960s. Observations that the number of matings of female flies is limited, often to a single event, suggested success through male sterilization if coupled with diligent treating of infested wounds with insecticidal smears and cultural practices such as ensuring calves are born during cold months of the year, particularly since this species of insect is restricted geographically to a warm winter habitat. Initially, these flies were successfully eliminated from Florida, and concentration was shifted to Texas where a facility capable of rearing 200 million sterile flies per week (Bushland, 1975) was constructed. This also proved successful, and in 1982 the last native fly was captured in the Southwest. Efforts were next shifted to Mexico and Central America with the erection of a new facility in Mexico. Mexico was declared fly-free in 1991, and efforts are currently being placed in Central America and creating a barrier between this area and South America. The facility in Texas is now closed. Programs such as the screwworm fly are costly; therefore, the pest must be of high economic or medical significance to warrant such high expenditures.

PHEROMONES AND GROWTH REGULATORS

If there is a means of selectively controlling insects, it would lie logically in the manipulation of some factor that the pest requires for survival or reproduction. Pheromones appear to have such control implications, especially when combined with insecticides and traps. The strategy is either to confuse the mating behavior or to use different concentrations of pheromones to attract specific pests to a centralized area, thereby reducing the need to "blanket" spray areas and kill other insects, including predators and parasites. The synthetic compound Disparlure, which is a sex attractant for male gypsy moths, a severe pest of forests in the eastern United States, has had success for isolated areas or regions of new infestation. Cotton fields in the southeastern United States have used pheromone traps to detect and suppress boll weevils with success. Traps baited with attractants were also useful in eradicating the Mediterranean fruit fly from Florida in 1957. In addition, *parapheromones,* substances that mimic pheromones, or *antipheromones,* chemicals that interrupt responses, may be used in management programs.

A recent twist of strategy is the potential to lure beneficial parasites and predators to crops needing protection. Although more than 700 sex attractant pheromones are known, only a few are for beneficial insects. Once the pheromone of a needed predator or parasitic insect already existing in an area becomes known and can be manufactured synthetically, it should be possible to spray these attractants on the crop to be protected and attract and receive benefits of their activity. Much research lies ahead before such a technique can become an alternative in a management program. Most likely it will be restricted to low acreage crops unless tied to mass releases of a biocontrol program.

Because insects require juvenile hormones during their development, what are the possibilities of manipulating concentrations of these hormones during critical periods? Can pest species and not other insects be affected in a specific locality? One experiment proved that a species of European Hemiptera could not be reared in certain United States laboratories because the paper toweling contained a juvenile hormone mimic. Other examples of juvenile hormones and their synthetic analogs are known to have been effective in limited testing trials. Because retarding or maintaining a larval pest in its destructive stage would be counterproductive, the use of these growth-regulating compounds would be most appropriate if the insect is a pest during its adult stage and the larvae develops in a restricted habitat, such as livestock droppings or pools of water. Methoprene used in flood-water mosquito, horn fly, and flea control is one example of the successful use of these growth-regulating compounds. The discovery of insect antijuvenile hormones (precocenes) in plants (Bowers et al., 1976), substances that interfere with the titer of juvenile hormones by destroying cells of the corpora allata, could prove monumental if these substances were universal; however, screening has located them in only a few plants, and only Hemiptera and a few species of beetle adults seem susceptible.

Chitin-formation inhibitors are another recent method with some prospects. Diflubenzuron is registered for gypsy moth control and is also being tested in conjunction with pesticides in the boll weevil eradication program. Research data indicate that this compound can be absorbed by female weevils and passed into their eggs, thereby inhibiting the formation of chitin in the embryo.

In summary, the possibility of utilizing growth regulators and pheromones in controlling insect pests does exist, but further imagination and much testing must be conducted to refine these techniques for integration into pest management programs.

MECHANICAL AND PHYSICAL CONTROL

Except for screens on windows, light traps (Fig. 293), and certain instances of handpicking and use of fly swatters, mechanical or physical control probably has the most limited potential of any method discussed, especially if these are not used in conjunction with other techniques. Microwaves and other radio-frequency energy devices have been used experimentally to kill some pests, for example, those on stored grain, but their cost for large-scale operations is currently prohibitive. Freezing also has been used to control certain pests; for example, a few libraries have large walk-through freezers into which they rotate their old classic books to periodically expose the guild of book pests to lethal temperatures.

INSECTICIDES

Insecticides are chemicals used to kill insects with no apparent or at least only minimal effect on animals at the recommended dosage levels and conditions of use. The word *pesticide* is often used synonomously with insectide (but incorrectly as pesti-

Figure 293. Traps, such as this ultraviolet lamp associated with a high electrical discharge mechanism, are often useful in reducing insect annoyance in areas with high human use. In some instances, however, they attract more pests from neighboring areas than are removed from the immediate local population.

cides also include herbicides, nematocides, and other chemicals). These materials have proven to be extremely effective in controlling many pests. The use of chemicals to kill insects dates back some 3,000 years to the use of sulfur and arsenic compounds. In general, however, the use of chemicals in controlling insect pests is a function of the 19th and more specifically the 20th century.

Each insecticide must be registered for specific use, and a label must be developed that states detailed instructions for use, including the pest, habitat or substrate, and any restrictions. Obtaining this data involves much scientific research on the insecticide's reliability, ability to "selectively" kill insects, safety to humans, and rates of use. The process of registering takes approximately 7 years, and it is expensive for manufacturers who must spend millions of dollars just to test and register a new insecticide. It should be reemphasized that although these insecticides appear to selectively kill insects, they also have a varying effect upon other animals, especially when: (1) the concentrations are increased by improper formulation of sprays, (2) carelessness causes continual exposure, (3) the pesticides are used in prophylactic regimens, (4) buildups occur in the soil because of poor drainage and the use of long-lived residuals, (5) there are runoffs from sprayed areas into streams, and (6) insecticides concentrate in some food chains as higher trophic levels are reached.

Insecticides have different modes of entry into the insect. Some act after being ingested with food. These poisons are often used on insects that have chewing mouthparts and ingest toxicants sprayed on host plant surfaces. A few insects have piercing-sucking mouthparts and take in toxins carried within the plant sap. A second type is the contact insecticides that enter the body through the integument and often affect the nervous system. This latter type is often the choice for insect pests with piercing-sucking mouthparts. A third type enters the body through the tracheal system, but this often requires the environment to be enclosed as in fumigation. Once inside the body, these compounds attack in numerous ways, the mode of action. In addition, each insecticide has a different toxicity level for each pest; this is measured by the point at which 50 percent of the population is killed (the LD50).

What insects are the target of insecticides? Much of the use is in agriculture with lepidopteran species the most common target (35 percent), followed by Homoptera (20 percent), Coleoptera (15 percent), Diptera (10 percent); and the remainder among scattered groups. Insecticides also are used against medically important insects (mainly Diptera, Anoplura, and Siphonaptera) to prevent disease transmission, on insects of veterinary and domestic animal importance, and the remainder on household, garden, and horticultural insects.

An insecticide can best be used for only a year and achieve its maximal result, but over 95 percent of a population of insect pests must be killed to have a decrease in next year's numbers. Because normally this is not possible, yearly spray schedules were often instituted in the past. However, insect pests that survived such a regimen were genetically better adapted for withstanding subsequent applications, and thus *resistance* developed within populations, particularly when high levels of kill were achieved. In many instances, resistance has come not from new mutations but because of the close chemical similarities between the pesticide and naturally existing insecticides produced by plants, the allelochemicals, which the herbivorous insect has had to cope with during the evolutionary past. More than 500 insect species have been tested and show resistance, including 20 of the worst pests that are resistant to nearly all types of insecticides. Also, other insect species that have been inconsequential in the past may survive this treatment and increase abnormally in numbers because of less predation and competition and sometimes become more detrimental than the original pest, at least during the agroecosystem recovery period. For example, the cotton bollworm and western corn rootworm are now very damaging pests, whereas they were only minor pests 50 years ago.

Insecticides are often grouped into *botanicals, inorganics,* and *synthetic organics.* Of the botanicals, tobacco was used in France as early as 1793, the major active agent being nicotine, which comprised approximately from 2 percent to 5 percent of the leaves. Pyrethrins, a group of naturally occurring botanical insecticides, were used in the early 1800s in Asia. Both nicotine and pyrethrins are contact poisons and are available on the market today. Another contact poison from plants, rotenone, was first used about 1850.

The use of botanical insecticides was complemented by the addition of inorganic compounds in the late 1800s and early 1900s; inorganic compounds were far more toxic, but they had to be ingested to kill. First Paris green and then the arsenates, fluorides, sulfur, and so forth were introduced, but no controls were placed on them until 1926 when their lethal potential to humans, animals, and plants alike was finally acknowledged. Until the discovery of DDT, Paris green continued to be the major larvacide in mosquito control, and undoubtedly it aided in the decline of malaria in the United States. These inorganic poisons have been banned from use.

The last group of insecticides are the synthetic organics. DDT, the first of the new synthetic organic insecticides, was introduced extensively at the end of World War II.

Although synthesized in 1874 and proven to have insect-killing qualities in 1936, this chlorinated hydrocarbon required the seriousness of malaria and the epidemic typhus in World War II to bring about its experimentation and introduction. Its effective action, long life, and apparent safety were immediately hailed as the possible end to our insect problems. Soon it was used extensively. The impossible seemed probable, and such things as eradication of malaria and safe spraying of crops appeared just over the horizon; the *Science News Letter* (April 1948) carried the headline "Flyless Age Now In Sight." More than 65,000 tons were used worldwide in 1962, the peak year of its use. Then came reality, and opinions changed as this insecticide was reevaluated, spurred on by Rachael Carson's book, *Silent Spring.* Resistance of many insect pests to its action, its impact on wildlife, particularly the thinning of the egg shells of certain bird raptors, and the detection of increasing residue concentrations all over the earth reversed public opinion of this once viewed "miracle" chemical.

Misuse, accidental poisonings and contaminations, the number of inexperienced users of these chemicals (more than 15 percent are homeowners), lack of scientific information, overuse through set spray schedules, and many other reasons have created a reevaluation of the philosophy and use of all insecticides. Their use, as in the use of any chemical, is based on the concept of *risk.* Although humans eat much more of the naturally occurring insecticides each day than the residues found on food of the synthesized ones used in pest management, caution remains a good strategy. The viewed risk of using insecticides varies depending upon the person and his or her bias. In general, farmers and and many homeowners view insecticides more favorably than do environmentalists (Whittmore, 1977); college students voice more concern as to the safe use of these toxicants than professional and business club members and the League of Women Voters; and all the previous groups are significantly more concerned than the recorded listing of actual causes of deaths (Upton, 1982).

Advantages in using these chemicals include: (1) they act quickly and result in rapid reductions in pest populations; therefore, they often have become a substitute for inadequate in knowledge in regulating pest buildups; (2) a variety is available to select from; (3) their use can be effective regardless of what control measures are practiced by adjacent users; (4) suitable application equipment is readily available even under relatively primitive situations (Fig. 294); and (5) their use permits development of crops with "high cosmetic" value, which the public unfortunately requires.

The following difficulties arise when we rely too heavily on poisons to regulate pest numbers: (1) the proximity to water and human dwellings can make safe application of insecticides impossible; (2) "insurance" applications are often made to do a better job; (3) weather conditions often prevent satisfactory application of the toxins; (4) delays often occur in management attempts until the situation becomes "uncontrollable" and insecticides become the last resort; (5) resistance to the chemicals results faster when the greatest number of pests are killed; (6) resistance to one insecticide often induces cross-resistance to others (Roush and Tabashnik, 1990); often natural predators and parasites are eliminated with their use; (8) secondary pests often become as bad or worse than the original pest of concern; and (9) little is known about their buildup in soils. Such difficulties, coupled with perceived risks to humans and wildlife, have resulted in the removal of many of these toxins from the market or restrictions on their usage. Each year sees the number diminishing with but a few replacements.

Figure 294. Spraying can be done using minimal equipment when more sophisticated machinery is not available. This makeshift sprayer has been put together by enterprising farmers in South America. (Photograph by T. A. Granovsky.)

SUMMARY

It is now recognized that most endemic pests must be managed rather that attempting to eradicate them. This concept is referred to as integrated pest management (IPM), and it involves four major concepts. First, natural and cultural control techniques must be maximized to prevent or slow down the buildup of pest populations. Second, the pest populations and the effectiveness of the natural and cultural control efforts must be continually monitored and evaluated. Third, a combination of techniques must be applied depending upon the weather and other circumstances. Last, the use of insecticides should be limited and occur primarily to prevent economic damage and crop loss or for medical reasons. All too often we have resorted to the use of insecticides because of their ease of use and cost-effectiveness without considering all of the above four concepts and how chemicals can and often do affect the environment. Nevertheless, it would be folly to eliminate the availability of insectides as an option when the pressure to feed the human race continues to increase as well as the need exists to protect ourselves from insect-transmitted diseases.

QUESTIONS

1. What is insect pest management? What are economic threshold (ET) and economic injury level (EIL)? Of what use are they in the process of raising crops?
2. What questions should one ask when relying upon pest management?

3. How does pest management differ from eradication?
4. Compare the advantages and disadvantages of each of the major techniques of pest management. Do you feel that a combination of several of these is better than utilizing only one type? Explain your position.
5. How might a knowledge of the type of mouthparts, life cycles, predators and parasites, diapause, migration, host preferences, and time of activity be important in pest management?
6. Can pests of humans be beneficial to the ecosystem in general? Clarify your answer.
7. What is an insecticide? How do they kill insects? What are the advantages and disadvantages of these chemicals? How do you feel about them?

12 Household Insects

As dwellings are erected, they are soon invaded by many native animals from the surrounding environment, many of which are insects. The definition of true household inhabitants is "any insect pest that consistently feeds and often breeds in houses." The definition will separate this guild of creatures from *facultative household pests,* those that are "just passing through" when doors are left ajar or windows open.

Obviously specific household pests vary from country to country, from culture to culture, and from house to house, depending on climate, house architecture, economy, and the physical and mental impact they have upon the house and inhabitants who may perceive them either as pests or beneficials. What can be ignored in nature or even in the backyard suddenly can become alarming when the organism enters the sanctity of the house. The author has made several unofficial surveys, and the results correlate well with published results. Of major concern in the United States (at least sufficiently to require consulting extension workers and pest control personnel) are termites, cockroaches, fleas, silverfish, ants, stored grain insects, flies, and wasps. With the exception of flies and wasps, which are facultative household insects, some of the major pests within each of the other groups will now be discussed.

TERMITES

About 50 species exist in the United States, 30 of which have been mentioned as pests, whereas 6 are considered to be serious threats to dwellings. The most common pest species are *Reticulitermes flavipes* (the eastern subterranean termite), *R. hesperus* (the western subterranean termite), *Coptotermes formosanus* (the Formosan subterranean termite), *Incisitermes minor* (the western drywood termite), *I. snyderi* (the eastern drywood termite), and *Cryptotermes brevis* (the powderpost termite). The

Figure 295. Winged alates of the eastern subterranean termite, *Reticulitermes flavipes,* emerging from a house to swarm.

subterranean species are classified as Rhinotermitidae, whereas the drywood and powderpost species are in the Kalotermitidae. Locally, a few other species may be pests. Readers should refer to Chapter 7 for a general discussion on colony structure, castes, and the many other characteristics of these social insects.

In many areas of the United States, termites are synonymous with the subterranean genus *Reticulitermes.* At least one species is found in every state except Alaska, although they may be only sporadic in the northern states. As many as 5,000 supplementary reproductives may exist in each colony, so colony size can be 200,000 or more. *R. flavipes* is found from Mexico north and east. The alates are brown-black to black with sooty colored wings and usually swarm (Fig. 295) in April or May. This species is probably the most economically important species in this country. The second pest species, *R. hesperus,* is restricted to the Pacific coast, but it is found as far north as northern Washington. The alates of this species are dark brown to nearly black with brownish-grey wings. Swarming usually takes place in the spring, although some fall flights have been recorded. This species is less dependent on soil water than is *R. flavipes.*

Coptotermes formosanus is a species introduced into both Hawaii and the Gulf states. Alates are pale with a yellowish band at the wing tip. Swarming takes place in early summer during evening hours. This subterranean species is unique in that it also can nest away from the soil, for example, under leaky roofs if the building structure has a high moisture content. Colony populations are high, often of more than 1 million individuals, and their foraging can cover extremely wide areas. Of all the termite pests in this country, this species is perhaps the most destructive locally; a house can be destroyed within several years.

Figure 296. The foraging tube characteristic of most subterranean termites within houses.

Cryptotermes brevis is narrowly distributed along the Gulf states from Florida west to Texas. It is sometimes called the "house termite" because it has on occasion infested furniture, boxes, and woodwork. Alates have colorless wings, and the soldiers have black heads.

Incisitermes minor is responsible for most of the drywood termite damage in the western United States, and its range extends farther north than any drywood species. Colonies are long-lived but small (about 2,000). Alates are dark to black with grayish wings and swarm usually in the fall during day hours. A second species of *Incisitermes, I. snyderi,* also has small colonies and causes similar damage to *I. minor,* but it is located in the eastern states. The alates are yellowish to yellow-brown and usually swarm at night in the spring.

Subterranean termites usually locate their nests in the soil and reach their feeding sites through a series of tunnels (Figs. 211, 296). These species are responsible for most of the economic damage in the United States. In addition to the structure of the house, feeding damage can occur to anything located within the dwelling that contains cellulose, such as newspapers and books. However, not all materials are equally preferred, and it is common to observe boards or joists adjacent to infested ones apparently ignored. Over time damage can be extensive, and many municipalities require certified inspections of buildings prior to their sale.

Drywood termites locate their nests directly within their food source and, therefore, the long tubes from nest to food seen in subterranean forms (Fig. 296) are absent and cannot be used in diagnosing their presence. After swarming, alates enter the house structure through cracks and develop nests there. Nests are normally small, and supplementary reproductives are rare. The colony requires little moisture, so only occasional contact with the soil takes place.

A tenant often becomes aware of these social insects when hordes of winged reproductives suddenly appear (Fig. 295). Such swarms can emerge from just about any imaginable location, such as holes in the wall, cracks in the floor or woodwork, or light fixtures. This appearance often occurs on a warm spring day following a storm. At this point in time the colony will have been present for 4 or 5 years, actively enlarging itself and causing damage inobtrusively within the hidden world of the wood structure. Termites can be mistaken for ants, but termites lack the elbowed antennae and narrow waist of the ant.

What can be done with an infestation? Eradication efforts vary with the termite species, but normally they should be carried out by professionals with the proper training and equipment. For *Reticulitermes* infestations the foundation must be exposed and treated. Holes may need to be drilled and insecticides injected into the underlying soil. Baits and insect killing fungi may be useful. The bottom line is to destroy the queens and to place a barrier between the nest and the house to prevent reinfestation. Such barriers are best laid down when new houses are constructed and are called *pretreatments.* Other steps that a homeowner can take in dealing with subterranean species that may prevent infestations include: (1) having clean gutters, grading soil away from the house and ventilating crawl space to prevent moisture buildup, (2) eliminating any wood to foundation contacts of the house with tree roots, untreated wooden steps and posts, (3) removal of scrap lumber, sawdust piles, and firewood from around the house, and (4) sealing gaps and cracks in the foundation and around entering water pipes. The previous methods are designed mostly for subterranean termites. Drywood termites, however, do not require contact with the soil, and they must be handled differently. Unfortunately, eradication may involve treating the infested wood and walls with insecticides, or fumigation of the entire house may be needed. The Formosan termite requires a combination of the subterranean and drywood termite methods.

COCKROACHES

These insects are closely related to the previously discussed termites, but they are solitary insects. Of the approximately 4,000 known species, about 50 are regarded as household pests. The most common pest species in the United States (Fig. 297) are the German cockroach (*Blatella germanica*), the American cockroach (*Periplaneta americana*), the brownbanded cockroach (*Supella longipalpa*), and the oriental cockroach (*Blatta orientalis*). They are listed in the order of nuisance and reproductive potential. A few other cockroach species are pests in limited localities, including the Australian cockroach (*Periplaneta australasiae*) from Florida and southern California; the smokeybrown cockroach (*Periplaneta fuliginosa*) from various localities, the brown cockroach (*Periplaneta brunnea*) from the Gulf states, and the Pennsylvania wood roach (*Parcoblatta pennsylvanica*), an outdoor species that enters homes only occasionally.

The German cockroach [Fig. 297(D)] is light brown and about 10 to 15 mm in length. It can produce the greatest populations in the shortest time at room temperatures. Each female deposits about 37 eggs in each oötheca, and development to adulthood varies from 60 to 250 days; therefore, the population potential for the year can be more than 35,000 individuals. This species is seen most commonly in the kitchen, pantry, and bathroom, but, when food becomes scarce, it extends its range. Because they are small and abundant, they are easily introduced. Other important characteristics are that both

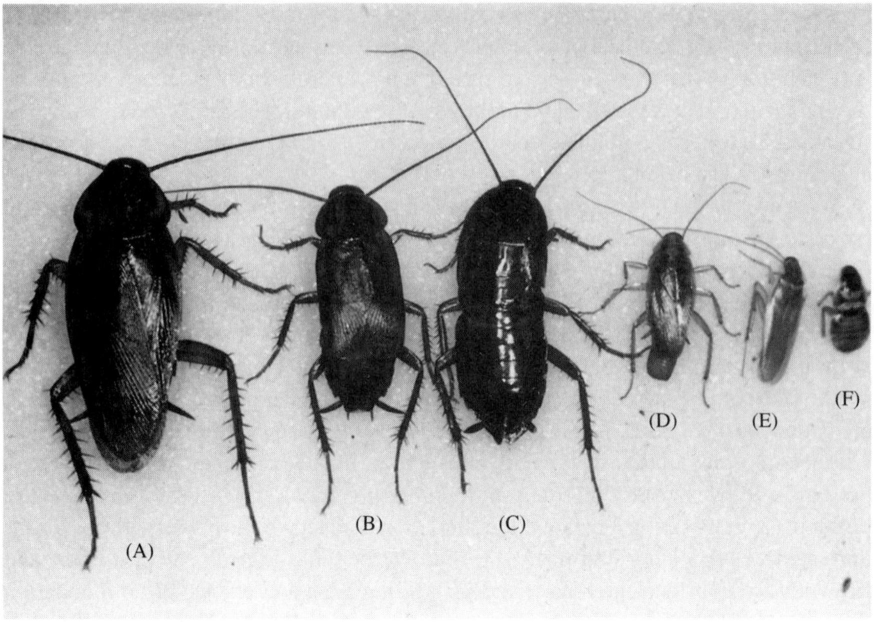

Figure 297. Common pest cockroaches in the United States. (A) American; (B) Oriental male; (C) Oriental female; (D) German; (E) Brownbanded male; (F) Brownbanded female.

sexes are fully winged, females carry their oötheca until just prior to egg hatch, they are gregarious, and the species is resistant to many insecticides.

The American cockroach [Fig. 297(A)] is the largest of the household pests, about 35 to 50 mm; it is light to medium brown except for yellow pronotal bands, and requires a long 320 to 1,000 days at room temperature to complete development. About 15 eggs form each oötheca, which, when coupled to the long developmental time, results in only 800 individuals from each original female each year. Both male and female are capable of flight and often live outdoors in southern states. Preferred sites within dwellings include kitchens, bathrooms, and steam tunnels in apartment buildings. Oötheca are not carried by the female.

The brownbanded cockroach [Fig. 297(E), (F)] is small (about 10 to 15 mm) and therefore easily introduced into houses. About 16 eggs are in each oötheca, and development requires from 140 to 380 days. The potential from each female is about 700 individuals per year. This cockroach occupies the widest range within the house (anywhere secluded), and it is the least gregarious of these pests. The species is dimorphic, with the female having short wings, which permits the yellow bands to be seen easily, whereas the male is macropterous, making the bands poorly visible. The oötheca is not carried by the female.

The oriental cockroach [Fig. 297(B), (C)], often called a "water bug" by landlords attempting to minimize the presence of cockroaches, prefers damp areas such as those found in basements. They are dark brown to black and from 15 to 20 mm in length. These pests are also dimorphic, with the males brachypterous and the females micropterous (neither sex can fly). The oötheca is not carried by the female. Development is long, from 210 to 990 days; therefore, populations build up slowly, about 200 per year per female. This species is also common outdoors during warm weather and often infests trash and garbage.

The saying that "one cockroach seen indicates 1,000 unseen" has not been scientifically tested but could be approximately true. Large populations can exist where feeble efforts at management exist. Of some medical concern is the discovery that many airborne antigens originate from these insects and cause high amounts of asthma in slum areas.

To exclude or minimize the number of cockroaches is a goal of most tenants. This requires a combination of the following: (1) immediate or early inspections of parcels and commodities introduced into the dwelling, (2) cleanliness within the house or apartment to remove or reduce available food scraps, (3) periodic inspection of out-of-the-way hiding sites, (4) use of roach traps, (5) chemical treatment (preferably by trained operators), and (6) concurrent treatment of neighboring homes and apartments if these are attached to one another to prevent back-and-forth migrations when treatments are administered.

FLEAS

These insects have been discussed previously in Chapter 8, so the material here is limited. The adult is wingless and parasitic upon mammals and birds, whereas the larva scavenges upon organic materials, often in the nest of the host.

Normally their presence in a dwelling is correlated with the presence of pets, but upon occasion rodents may serve as hosts. The major pest species in houses in the United States is the cat flea, *Ctenocephalides felis* (Fig. 298), although local populations of other fleas may be a problem, such as the oriental rat flea in some southern areas.

Fleas require attention to both the adult and larval stages for effective management. Managing adults requires regulating pet activity, bathing often and constant pet hygiene, cleaning pet bedding, using the new generation of flea traps with intermittent flashes of ultraviolet light, sterilizing through monthly flea pills containing growth regulators, and treating with insecticides when needed. Larvae are often present in pet bedding and in carpeting, especially wall-to-wall, and techniques of control of this stage involve laundering, vacuuming, and shampooing, plus insecticide use when needed. Larval populations obviously become reduced through the indirect action of the growth regulators that sterilize female fleas. Again, concurrent attention to both stages of the flea are necessary to reduce and/or eliminate the problem.

ANTS

These hymenopterous insects are social (see Chapter 7), have narrowed waists and elbowed antennae, have castes, and often start nests by single fertile females yet may number in the millions when colonies mature. Diets vary as widely as in any group

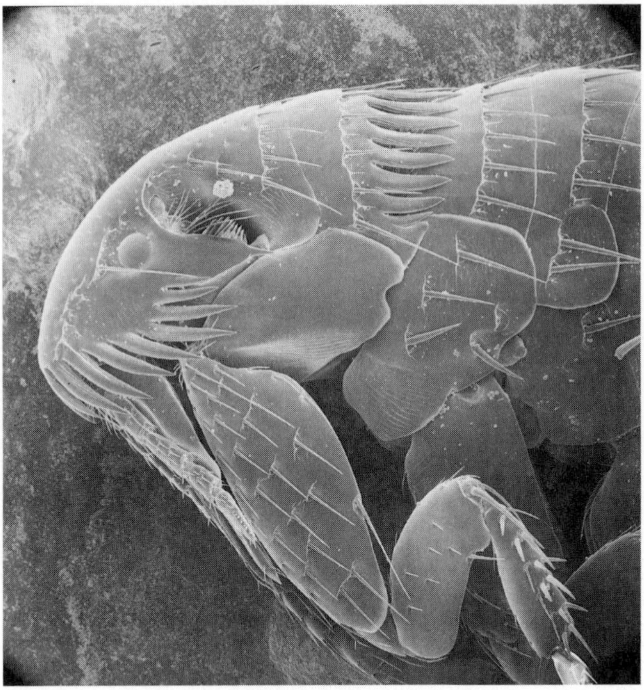

Figure 298. The cat flea, *Ctenocephalides felis.*

of insect. About 9,000 species are known with local diversities being substantial (Holldobler and Wilson, 1990).

Some of the troublesome species include the Argentine ant, *Iridomyrmex humilis;* various carpenter ants, *Camponotus* spp. (Fig. 219); the odorous house ant, *Tapinoma sessile;* the pharaoh or pavement ant, *Tetramorium caespitum;* and the thief ant, *Solenopsis molesta.* Other ants may be pests locally. The above-mentioned ants vary in size from about 2 to 12 mm; the smallest is the yellowish thief ant, and the largest is the black carpenter ant.

The pharaoh or pavement ant is probably the most widely distributed of the ant pests. It is yellowish-brown and, unlike other ant pests, colonies consist of multiple nests within house structures. It is especially attracted to fatty and sweet foods (Table 11). There is also some evidence that this species can transmit pathogenic organisms during their wanderings. This pest is difficult to control because of their multiple nests.

The large black carpenter ant enters homes throughout the country. Their presence, particularly if found nesting, indicates structural decay of the wood elements of the dwelling. Normally, however, this ant enters houses primarily during rainy weather when normal foraging is affected. Nests are usually located in nearby trees, logs, or posts. More than one queen may be present in the colony. Other carpenter ants may be pests in localized areas.

The odorous house ant, a small black ant, has extensive range throughout the United States and may, on occasion, also nest in house walls. It is especially visible during rainy weather when colonies invade the house. A nauseous odor results when individuals are crushed.

TABLE 11. The Most Common Household Insects

Order	Common Name	Preferred Habitat	Food
Thysanura	Firebrat Silverfish	Hot areas Cool moist areas	Glue, book bindings, wallpaper paste, starchy materials
Blattaria	German cockroach American cockroach Brownbanded cockroach Oriental cockroach Australian cockroach Smokeybrown cockroach	Various locations	Organic debris
Isoptera	Eastern subterranean termite Western subterranean termite Formosan subterranean termite Eastern drywood termite Western drywood termite Powderpost termite	House structure, soil around house	House structure and furniture, boxes, paper, glue
Psocoptera	Booklice	Various locations	Book bindings, wallpaper paste
Coleoptera	Cigarette beetle	Tobacco products, upholstered furniture	Tobacco, pepper
	Carpet beetle	Carpeting, clothing	Wool, leather, grain
	Flour beetle	Stored foods, kitchen, pantry	Stored foods
	Mealworms	Stored foods, kitchen, pantry	Cereals
	Saw-toothed grain beetle	Stored foods, kitchen, pantry	Stored foods
	Weevils	Stored foods, kitchen, pantry	Stored foods
	Clothes moths	Closets, stored fabrics	Wool, feathers, animal products
Hymenoptera	Argentine ant	Crawling through room, kitchen counter	Any organic food
	Carpenter ant	Crawling through room, decaying siding	None or any organic food
	Odorous house ant	Crawling through room, walls, counters	Sweets
	Pharaoh ant	Crawling through room, counters, walls	Sweet foods, some fatty foods
	Thief ant	Crawling through rooms, food containers, counters, walls	Greasy and some sweet food
Siphonaptera	Cat flea	Pets, pet bedding, carpeting	Blood
Lepidoptera	Angoumois grain moth	Stored food	Stored food
	Clothes moth species	Clothes, fabrics, rugs	Wool, hair, fur
	Indian-meal moth	Stored food	Stored food

The thief ant is the smallest of the house ants. It is yellowish and may nest in walls. This ant is commonly attracted to greasy or sweet foods left unattended around cupboards and counters.

The Argentine ant is a small brown ant that is distributed mainly in the southern United States and California. Colonies located out of doors actively invade houses in vast numbers searching for sweet foods, but they also can form nests within the dwelling itself.

Ants can be separated from termites by the presence of elbowed antennae and narrow waists (termites have straight moniliform antennae and broad waists). When certain that the pests are ants, the following measures can be applied: (1) locating the nest, (2) destroying the nest when possible, (3) removing unattended attractive foods such as crumbs and scraps, (4) employing other kitchen sanitation and proper food storage, and (5) intercepting ants at entry points with chemical barriers such as dusts and sprays. For nests located within the house, eradication is difficult and often requires professional assistance.

STORED PRODUCT INSECTS

Many insects adapted to feed and breed on naturally existing seeds readily enter the dwelling to feed on these foods and their manufactured products or, more often, are introduced with food items brought into the house. Such pests are usually certain beetles and moth larvae. The most common species include the sawtoothed grain beetle, *Oryzaephilus surinamensis* (Fig. 299); various weevils, *Sitophilus* spp. (Fig. 300); flower beetles, *Tribolium* spp.; mealworms, *Tenebrio* spp.; and the Angoumois grain moth, *Sitotroga cerealella.* Many other pests of stored products are known but are too numerous to discuss.

Stored grain pests sometimes appear abruptly when newly purchased or stored foods are opened, but more often they appear more gradually as populations build up; for instance, a candy bar left in a drawer or on a shelf may be an abundant source of food for these pests until either they eat it or they are discovered. Needless to say, proper food and dwelling sanitation, including periodic inspections and elimination of infested materials, can reduce this problem to one of the minimal concern. The use of insecticides around food to control these pests is *not* recommended.

OTHER INSECT PESTS

Many other insects become unwanted guests. Some set up "permanent residence" if they can reproduce within the house and, although reclusive, may induce reactions of hostility from whomever discovers them. Examples of such species include the silverfish, *Lepisma sacharina* (Fig. 310), the firebrat, *Thermobia domestica,* and several species of booklice, *Licoscelis.*

Silverfish have a uniformly silvery color because of their body scales and prefer damp areas of the house such as in basements, near sinks, or near water pipes with

Figure 299. Sawtoothed grain beetle, *Oryzaephilus surinamensis.*

Figure 300. The rice weevil, *Sitophilus oryzae.*

condensing water. In contrast, the firebrat has a mottled color, because of patches of white and dark scales, and they prefer furnace rooms or the upper parts of the house where heat readily builds up. Both species of Thysanura undergo ametabolous development, but they may require up to 24 months to reach the adult stage if the house is kept cool. Food includes book bindings, wallpaper adhesives, flour, and other starchy foods (Table 11). The elimination of these insects is difficult because they often exist within walls, ceilings, and insulation with glued backings. Insecticides have been the main method of managing their numbers. How can these creatures be managed? A combination of the following procedures will help: (1) remove food wastes and spills as soon as possible to reduce food sources, (2) caulk cracks and other openings into the living space, (3) store trash outside in regularly cleaned containers, (4) regularly "disturb" books and stored items to disrupt the normal reclusive habits of these insects, and (5) carefully apply insecticides (follow label directions exactly) to cracks and crevices and other preferred regions within the house, but not around food, pets, or children.

Booklice are minute and are rarely seen when present except for their damage. Damage is usually to book bindings and wallpaper adhesives. Management methods are similar to those used with silverfish and firebrats.

Other insects enter houses only accidentally when doors or windows are opened or when cracks exist in the structure. Flies, wasps, wood roaches, and beetles represent such facultative intruders, but they usually do not survive for extended periods of time or reproduce within such dwellings; they represent only temporary nuisances and are not true household pests. Such facultative pests can be managed by screens, caulking holes and cracks, weatherstripping doors and windows, taking care not to leave doors open for extended periods, and observing general house sanitation.

QUESTIONS

1. What is a household insect pest? What are examples of permanent and facultative pests? Why will they vary from country to country?
2. How can termites and ants be identified? What is the food of each? How do management techniques differ for each?
3. What cockroach species are the most likely to be in your dwelling? How is each identified? In what part of your dwelling would you expect each to localize? Populations increase more rapidly in which cockroach species? Why?
4. What flea is most likely to inhabit houses? How can it be managed?
5. Why are stored product insects present in houses? How can you regulate their numbers and presence?
6. What type of damage is caused by silverfish, firebrats, and booklice? How does this type of damage compare to that produced by termites?

13 Classification

One has to take only a cursory look at nature to conclude that there are many types and degrees of resemblances and differences in organisms. Many systems have been devised to put tags or names on these groups of plants and animals. The use of common names has been extensive and has the advantage of easy communication because common names often illustrate obvious biological peculiarities, but these names may also indicate different organisms depending on the geographical locality. For instance, potato bugs can refer to either a crustacean or several species of insects.

Out of the many diverse naming systems came the one used today, which has as its basis the 1758 work of Linnaeus, *Systema Naturae.* This early enumeration included more than 4,000 species of animals and classified insects into seven orders. One hundred years later, Agassiz and Bronn counted 129,370 species of animals. The estimate today varies, but it usually enumerates more than 1 million kinds with approximately 75 percent of these classified as insects (Fig. 8). Estimates of as high as 30 million kinds of insects have been printed (Erwin, 1982) but are probably unrealistic. The number of insect orders varies also, depending on the classification, ranging from 26 to 33, and insect families number more than 600.

The naming of this great assemblage pioneered by Linnaeus has the scientific name as a binomial (consisting of two parts). The first part indicates the *genus* and the second the *specific.* The adoption of this procedure has resulted in a standardization of names throughout the world; for example, *Musca domestica* has the same meaning in Russia as it does in Japan or Brazil. As a further refinement, species have been assembled into a hierarchy of similarities, a process referred to as *classification.* Classification is a means of reducing the bewildering diversity of species to a limited and manageable number of categories or taxa. This hierarchy can be illustrated as follows for the house fly, *Musca domestica:*

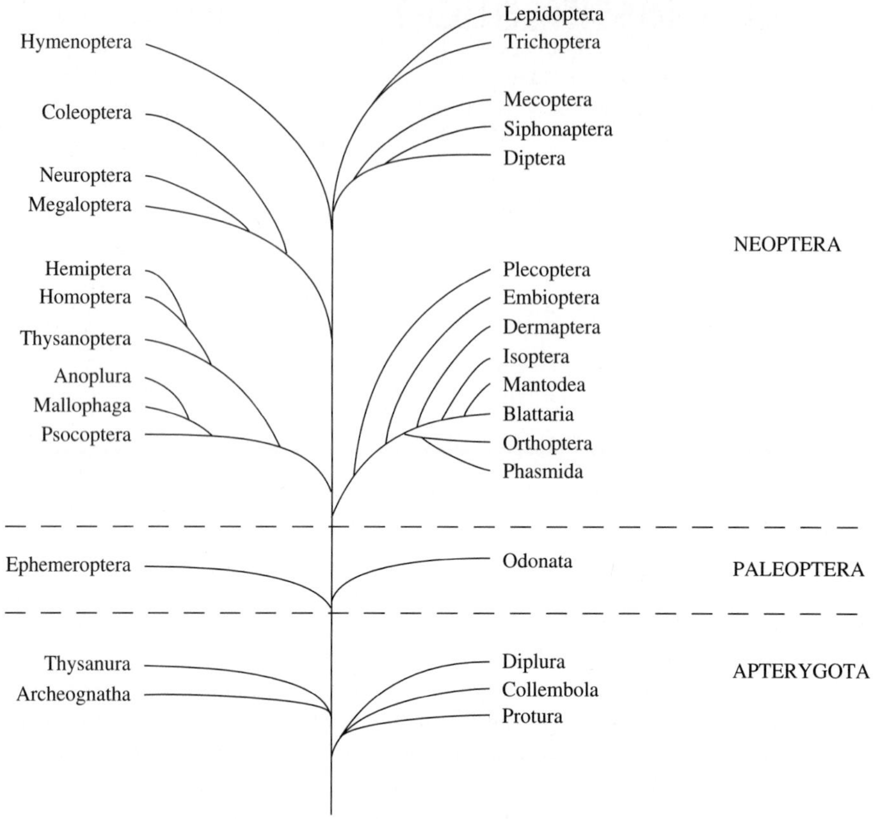

Figure 301. A suggested phylogenetic tree of insect orders.

Kingdom	Animalia
Phylum	Arthropoda
Class	Insecta
Order	Diptera
Family	Muscidae
Genus	*Musca*
Specific	*domestica*

Closely related or similar species are grouped into the same *genus* (plural, *genera*). Similar genera are placed within the same family. This procedure is carried on up the hierarchy until the classification is complete. The common families, orders, and so forth have well-established common names, and these also have been used throughout the text. A knowledge of the classification system is important, for it is vital to data interpretation, storage, and retrieval.

By studying closely related groups or taxa, it is possible to develop phylogenetic trees that visually display relationships. These trees can reflect evolution at any level from species within a genus to classes within a phylum. Figure 301 illustrates a possible evolutionary sequence in the most commonly seen insect orders based on three major evolutionary events. The first was the appearance of wings; this separated the Pterygota from the more primitive Apterygota. The ability to fold or flex the wings over the insect's back was the second major adaptation and gave a tremendous selective advantage to the Neoptera. A third major event was the ability to undergo holometabolous development; this permitted the extensive adaptive radiation seen in the "higher" Neoptera. Additional structural and ecological adaptations produced genealogical groupings, which we now refer to as orders and families. Each of these taxa have unique characteristics and biologies.

Because classification is based on grouping species into taxa with close relationships but with many unique attributes, identification of an unknown insect should be possible if done systematically. *Keys* serve this purpose. Usually they consist of a series of numbered couplets that have two choices. After deciding which choice is correct for the given specimen, proceed to the proper couplet as directed by the right-hand number. By observing carefully, answering each choice correctly, and proceeding through the key, an unknown insect can be identified and classified according to the various levels within the hierarchy. For introductory students, most keys are restricted to the higher levels of the classification such as the phyla, classes, and major orders and families.

The following two keys should prove useful as an introduction to the major insect orders commonly found in the United States (see also Fig. 301). The beginning student in classification must remember that other systems exist due to differences of opinion among systematists. The classification and keys presented here are designed for the most common orders and families seen by an introductory biology student and are to facilitate learning with a minimum of names. More detailed keys, which include more diverse and less common species, are available in advanced taxonomic texts; if these atypical or uncommon species were run through the generalized keys of this introductory text, incorrect identifications would result. The uncommon orders Grylloblattodea, Rhaphidiodea, Strepsiptera, and Zoraptera are conspicuous omissions, but they were not included because few professional entomologists have ever collected them.

KEY TO THE ORDERS OF ADULT INSECTS COMMONLY FOUND IN INSECT COLLECTIONS

1. Wings present 24
   Wings absent or reduced to small pads 2

2. Antennae absent; body slender and whitish color . . . . PROTURA
   Antennae present 3

3. Usually possess abdominal forked spring (furcula) [Figs. 303(B), 308]; minute size . . . . COLLEMBOLA
   Furcula absent 4

4. With long multisegmented cerci (if cerci short, then abdominal ventral styliform appendages present) 5
   Cerci short (Fig. 324) *or* lacking *or* forcepslike (Fig. 325) in shape; styliform appendages absent 7

5. Three terminal “tails” present (Fig. 310) 6
   Two terminal “tails” present, filiform or forcepslike (Fig. 307). . . . . DIPLURA

6. Compound eyes large (Fig. 309) and touching dorsally. . . . . ARCHAEOGNATHA
   Compound eyes reduced, not touching . . . . THYSANURA

7. Large insects, usually 1 in. (2.54 cm) or more in length 8
   Small insects, usually 1/2 in. (1.27 cm) or less in length. 10

8. Hind legs enlarged, modified for jumping [Fig. 302(D)]. . . . . ORTHOPTERA
   Hind legs similar in size to other legs 9

9. Body shortened and flattened dorsoventrally; coxae nearly touching . . . . . BLATTARIA
   Body elongate and tubular [Fig. 323(F)]; coxae usually well separated. . . . PHASMIDA

10. Large unsegmented forcepslike cerci [Fig. 303(D)] . . . . DERMAPTERA
    Cerci (when present) not forcepslike or unsegmented. 11

11. Cornicles present [Fig. 303(C)] or body covered with waxy filaments or scales. . . . . HOMOPTERA
    Cornicles absent; no scale or waxy filaments covering body 12

12. Abdomen constricted into narrow waist at juncture with thorax [Fig. 303(A)] . . . . HYMENOPTERA
    Abdomen not constricted into narrow waist 13

13. Enlarged basitarsi on forelegs (Fig. 326) . . . . EMBIOPTERA
    Basitarsi on forelegs not enlarged 14

14. Piercing-sucking mouthparts present in short conelike beak [Fig. 303(E)]; tarsi with eversible apical bladders; abdomen often terminally pointed. . . . . THYSANOPTERA
    Mouthparts not as above; tarsal bladders lacking 15

15. Legs modified for jumping; body flattened laterally [Fig. 239(A)] . . . . SIPHONAPTERA
    Legs not modified for jumping. 16

16. Mouthparts elongated into long piercing-sucking beak (Fig. 334) 17
    Mouthparts not as above although head may be prolonged. 18

17. Antennae hidden in grooves in head. . . . . DIPTERA
    Antennae long and easily seen . . . . HEMIPTERA

18. Body with dense hair and scales . . . . LEPIDOPTERA
    Body lacking dense hair and scales 19

19. Antennae moniliform (Fig. 324); short cerci present . . . . ISOPTERA
    Antennae not moniliform; cerci absent 20

20. Antennae long and slender [Fig. 302(B)]. 21
    Antennae short 22

21. Head prolonged and beaklike (Fig. 367) . . . . MECOPTERA
    Head not prolonged. . . . . PSOCOPTERA

22. Each tarsus four- to five-segmented . . . . DIPTERA
    Each tarsus one- to three-segmented 23

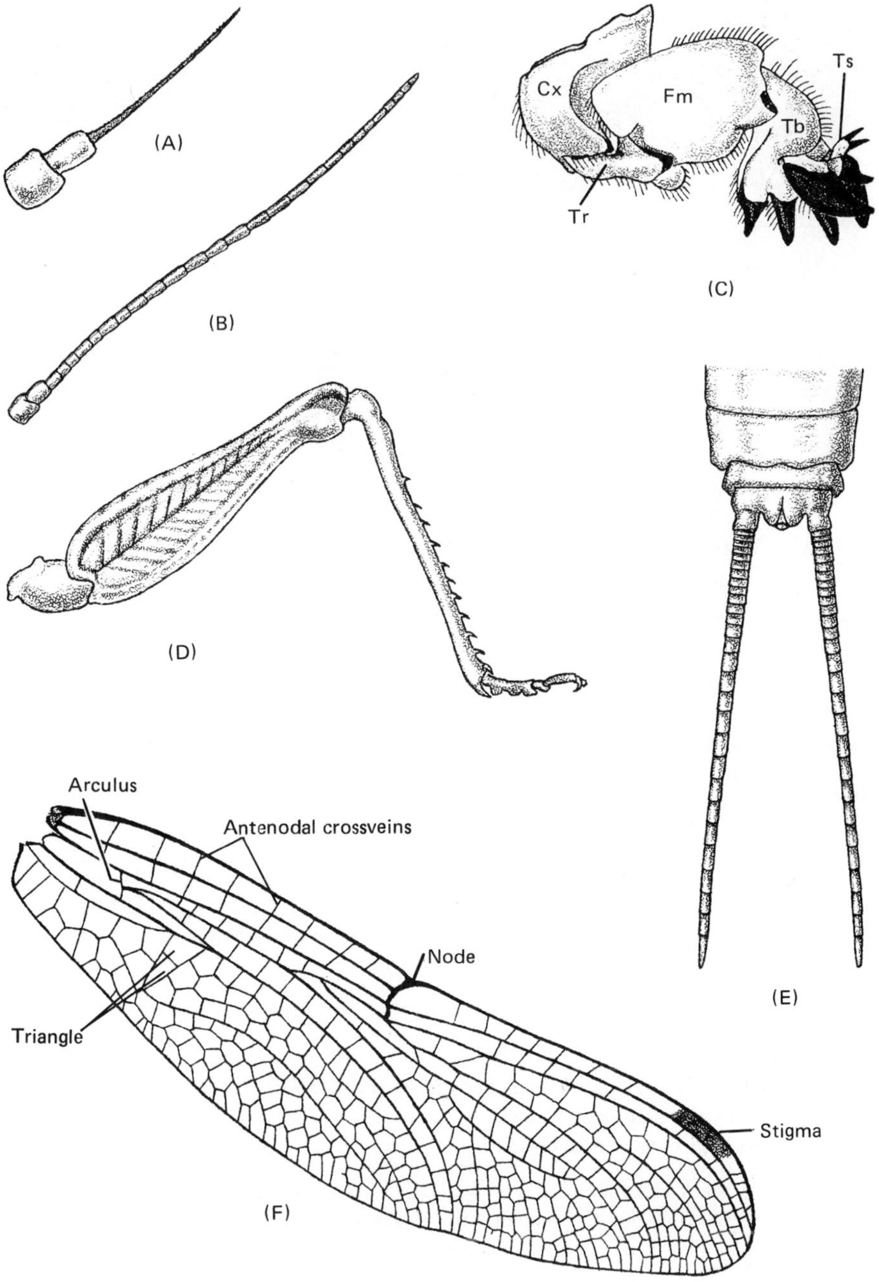

Figure 302. Diagnostic characters used in identifying adult insects. (A) setaceous antenna (Ephemeroptera); (B) filiform antenna (Orthoptera); (C) digging leg (Orthoptera); (D) jumping leg (Orthoptera); (E) long cerci (Plecoptera); (F) node, triangle, antenodal cross veins, stigma, and arculus of anterior wing (Odonata).

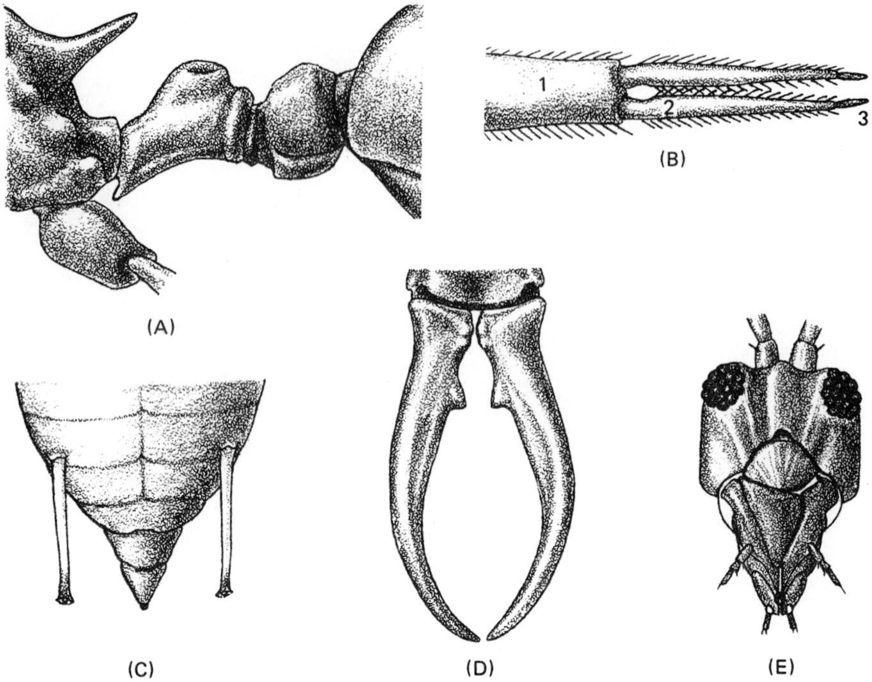

Figure 303. Diagnostic characters used in identifying adult insects. (A) slender waist (Hymenoptera); (B) ventral view of furcula (Collembola); (C) cornicles present on abdomen (Homoptera); (D) cerci forcepslike (Dermaptera); (E) conelike beak (Thysanoptera).

23. Head usually broader than long (Fig. 329); if same size or longer than broad, then with chewing mouthparts .... MALLOPHAGA
    Head usually longer than broad (Fig. 330); piercing-sucking mouthparts retracted into head; legs greatly enlarged for grasping .... ANOPLURA

24. With an anterior notch or node on each wing [Fig. 302(F)] .... ODONATA
    Wings lacking such notch.. 25

25. Multisegmented cerci often long as body (Fig. 312) and with setaceous antennae [Fig. 302(A)] .... EPHEMEROPTERA
    Not with above combination of traits..................................... 26

26. Hind legs enlarged for jumping [Fig. 302(D)] or front legs adapted for digging [Fig. 302(C)]; antennae filiform [Fig. 302(B)]......... ORTHOPTERA
    Hind legs similar in size to middle pair of legs; antennae various................. 27

27. With large unsegmented forcepslike cerci [Fig. 303(D)] .... DERMAPTERA
    Cerci filamentous [Fig. 302(E)] or absent 28

28. One pair of wings present.. 29
    Two pairs of wings present... 30

29. Halteres present behind wings [Fig. 387(A)] .... DIPTERA
    Halteres if present anterior to wings .... HOMOPTERA

30. Piercing-sucking mouthparts, usually in elongate beak (Fig. 334); palpi absent...... 31
Mouthparts other than above; palpi present.................................... 33

31. Hind legs without tarsal claws, adapted for swimming................ HEMIPTERA
Hind leg with tarsal claws 32

32. Beak arises at anterior part of head [Fig. 336(B)]; hemelytra [Fig. 335(A)] usually present .... HEMIPTERA
Beak appears to arise between front pair of legs; forewings with uniform texture (Fig. 338) .... HOMOPTERA

33. Piercing-sucking mouthparts in short conelike beak [Fig. 303(E)]; wings fringed with long hair (Fig. 331) .... THYSANOPTERA
Not as above.. 34

34. Front pair of wings horny or of different textures from hindwings................ 35
All wings membranous... 38

35. Front pair of wings thickened and usually hard, meeting at meson [Fig. 356(C)], normally without cross veins; cerci absent.......... COLEOPTERA
Front pair of wings with obvious cross veins, overlapping one another at least partially; cerci present 36

36. Front legs modified for grasping [Fig. 323(B)]...................... MANTODEA
Front legs similar to other legs.. 37

37. Body flattened dorsoventrally; wings usually covering entire body ...... BLATTARIA
Body sticklike; wings short stubs [Fig. 323(F)] .... PHASMIDA

38. Cerci at least length of last 3 abdominal segments [Fig. 302(E)]....... PLECOPTERA
Cerci short or absent.. 39

39. Siphoning mouthparts coiled under head (Fig. 372); wings and body usually covered with scales .... LEPIDOPTERA
Mouthparts not coiled under head; scales absent or few in number, primarily on wing veins.. 40

40. Fore basitarsi enlarged (Fig. 326)............................... EMBIOPTERA
Fore basitarsi not enlarged 41

41. Tarsi four-segmented; wings of equal length and venetion............... ISOPTERA
Tarsi not four-segmented [Fig. 323(C)]; fore and hind wings usually different 42

42. Head prolonged and beaklike (Fig. 367) .... MECOPTERA
Head not prolonged and beaklike.. 43

43. Many cross veins in wings, particularly at anterior edge; if few cross veins present, then wings and body with waxy coating.................. 46
Few cross veins in wings; body and wings lacking waxy coating................. 44

44. Mouthparts reduced, only palpi obvious (Fig. 368); hair often present on wings .... TRICHOPTERA
Not as above.. 45

45. Tarsi five-segmented; wings usually folded flat over body......... HYMENOPTERA
Tarsi two- or three-segmented; wings folded rooflike over body....... PSOCOPTERA

46. Hind wing with broader base than front wing and with wing area that folds fanwise when at rest; few or no marginal branching of longitudinal veins. . . . . MEGALOPTERA
    Hind and front wings similar in size and shape and lacking folded anal area on hind wing; longitudinal veins with many branchings near margin . . . . NEUROPTERA

KEY TO THE ORDERS OF THE MOST COMMON LARVAL AND NYMPHAL INSECTS FOUND IN INSECT COLLECTIONS

1. Antennae absent; 3 pairs legs always present; body slender and whitish color. . . . . PROTURA
   Antennae present or legs absent. 2

2. Six abdominal segments; minute; usually with furcula [Figs. 303(B), 308] . . . . COLLEMBOLA
   Not with above combination of characters. 3

3. Abdomen with paired ventral styliform appendages; antennae long and filiform. 4
   Abdomen lacking styliform appendages 6

4. Three terminal filiform "tails" (Fig. 310). 5
   Two terminal filiform or forcepslike "tails" (Fig. 307) . . . . DIPLURA

5. Compound eyes large, touching above head . . . . ARCHEOGNATHA
   Compound eyes small, far separated. . . . . THYSANURA

6. External wing pads present [Fig. 304(A)] and/or compound eyes well developed (nymphs). 32
   External wing pads absent or compound eyes reduced or absent 7

7. Piercing-sucking [Figs. 303(E), 336(B)] mouthparts present; if retracted into head, then all thoracic legs grasping type (Fig. 330) (nymphs) 33
   Piercing-sucking mouthparts lacking. 8

8. Cerci* usually present; 2 tarsal claws present per leg *or* all legs modified for grasping (nymphs) 45
   Cerci absent; 1 (rarely 2) normal-sized tarsal claw(s) per leg or legs reduced or absent (larvae) 9

9. Body lacking thoracic legs [Fig. 304(B)] or legs reduced to small lobes 10
   Body with well-developed thoracic legs 19

10. Head present, may lack pigmentation or be retracted into prothorax [Fig. 384(A)] . . . 12
    Without distinct head [Fig. 304(B)] 11

11. One pair of large posterior spiracles [Fig. 304(B)] or breathing tube [Fig. 304(D)] present . . . . DIPTERA
    Lacking terminal breathing tube or prolegs. . . . . HYMENOPTERA

*Do not confuse urogomphi (eighth or ninth segment appendages) or gills with cerci (eleventh segment structures).

12. Terminal breathing tube [Fig. 304(D)] or prolegs [Fig. 389(B)] present . . . . . DIPTERA
    Lacking terminal breathing tube or prolegs 13

13. Adfrontal area on head [Fig. 304(F)] . . . . . LEPIDOPTERA
    Adfrontal area lacking 14

14. Parallel-acting mouthparts [Fig. 389(D)], suckers, or mouthbrushes [Fig. 304(E)] present . . . . . DIPTERA
    Above structures absent 15

15. Several long setae at caudal end of abdomen (Fig. 379); antennae single-segmented; all spiracles inconspicuous . . . . . SIPHONAPTERA
    Not with above combination of characteristics 16

16. Uniformly elongate [Fig. 304(C)] or with fleshy filaments present [Fig. 384(A)] 17
    Body not uniformly elongate; never with fleshy filaments 18

17. Ninth abdominal segment with solid terminal spine or paired urogomphi [Fig. 67(D)] 18
    Ninth abdominal segment lacking above characteristics; fleshy filaments may be present [Fig. 384(A)] . . . . . DIPTERA

18. One pair of thoracic spiracles [Fig. 353(A)] . . . . . COLEOPTERA
    Two pairs of thoracic spiracles . . . . . HYMENOPTERA

19. Abdominal prolegs present [Fig. 304(G)] 20
    Abdominal prolegs absent 22

20. Five or fewer pairs of abdominal prolegs [Fig. 304(G)] . . . . . LEPIDOPTERA
    More than 5 pairs of abdominal prolegs (Fig. 362) 21

21. One pair of stemmata present . . . . . HYMENOPTERA
    Three or more pairs of stemmata . . . . . MECOPTERA

22. Gills or lateral appendages present on abdomen [Fig. 305(A)] 25
    Gills absent on abdomen 23

23. Body C-shaped [Fig. 353(C)] 24
    Body not C-shaped (some slight curving may occur due to alcohol preservation) 29

24. Distance between mesothoracic legs farther than between legs on prothorax . . . . . MECOPTERA
    Distance between mesothoracic legs same as between those on prothorax [Fig. 353(C)] . . . . . COLEOPTERA

25. Distinct lateral appendages on abdomen [Fig. 305(A)] 26
    Distinct lateral abdominal appendages absent, but threadlike gills may be present 28

26. Labrum absent; chewing mouthparts, although each mandible may be hollow . . . . . COLEOPTERA
    Labrum present; if absent, then mandibles and maxillae on each side form sucking mechanism [Fig. 305(B)] 27

27. Labrum and abdominal gills present . . . . . MEGALOPTERA
    Labrum absent; mandible and maxillae form sucking mechanism [Fig. 305(B)] . . . . . NEUROPTERA

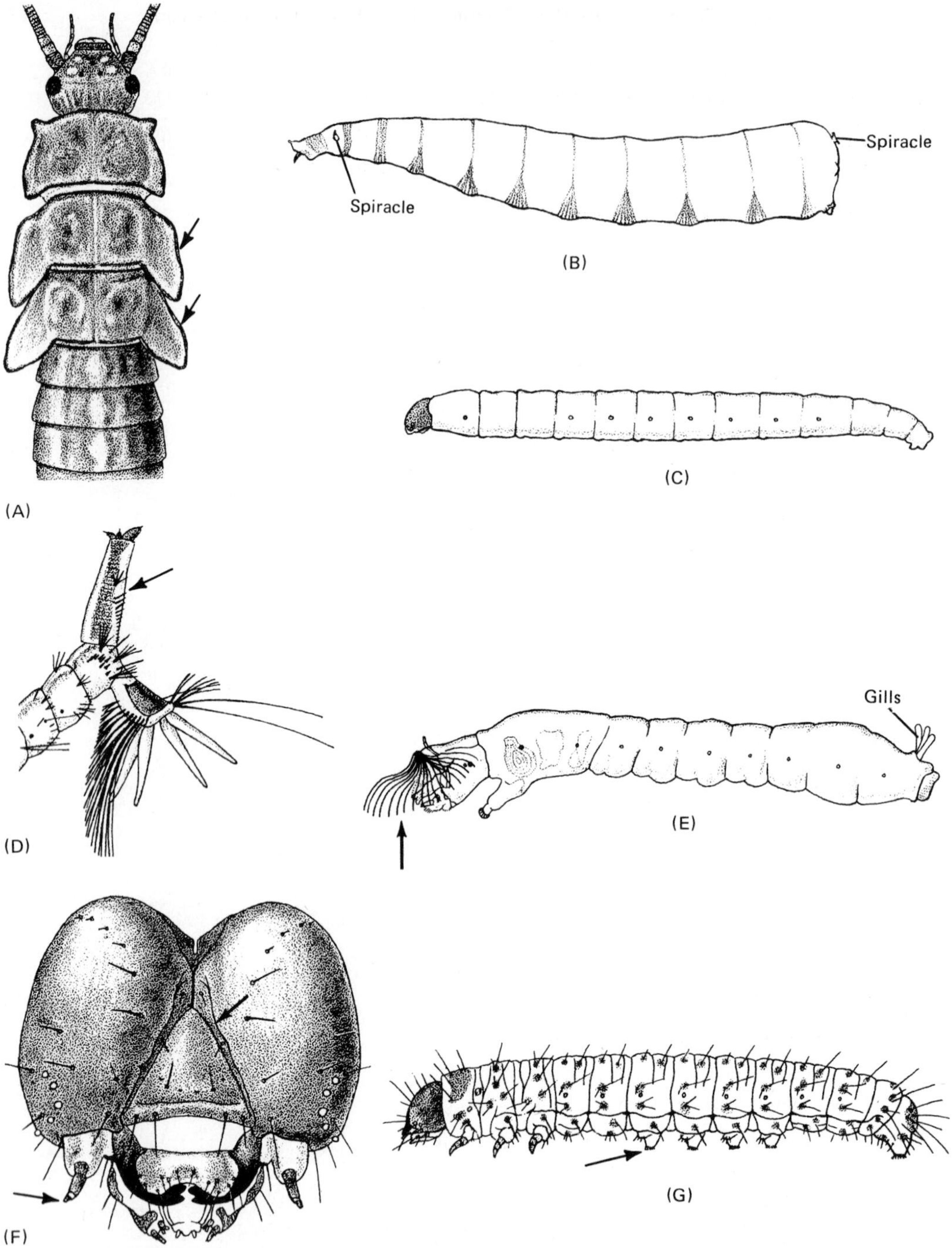

Figure 304. Diagnostic characters used in identifying immature insects. (A) external wing pads (Plecoptera); (B) distinct head absent (Diptera); (C) body uniformly elongate (Diptera); (D) terminal breathing tube (Diptera); (E) mouth fans (Diptera); (F) antennae short, adfrontal area (Lepidoptera); (G) five pairs of abdominal prolegs (Lepidoptera).

28. Posterior abdominal claws present [Fig. 305(A)];
    larvae often in cases (Fig. 369) .... TRICHOPTERA
    Posterior abdominal hooks absent 29

29. Abdomen without extensive sclerites (Fig. 347); 2 pairs thoracic spiracles 30
    Abdomen with extensive hard sclerites [Fig. 353(B)]
    or 1 pair thoracic spiracles.................................. COLEOPTERA

30. Mandibles and maxillae on each side form sucking mechanism
    [Fig. 305(B)]; labrum absent.................................. NEUROPTERA
    Mouthparts chewing type; labrum present.. 31

31. Mouthparts hypognathous .... HYMENOPTERA
    Mouthparts prognathous.. RAPHIDIODEA

32. Piercing-sucking mouthparts [Fig. 336(B)]; if retracted into head,
    then all legs grasping (Fig. 330) 33
    Chewing mouthparts 36

33. Mouthparts retracted into head (Fig. 330) .... ANOPLURA
    Mouthparts modified into beak [Fig. 336(B)] 34

34. Mouthparts in short conelike beak [Fig. 303(E)] .... THYSANOPTERA
    Mouthparts in elongate beak [Fig. 336(B)] 35

35. Mouthparts usually arising at the anterior of head [Fig. 336(B)];
    filiform antennae; swimming legs may be present................. HEMIPTERA
    Mouthparts appearing to arise near or between base of
    prothoracic legs (Fig. 339); antennae filiform or setaceous;
    swimming legs absent.. HOMOPTERA

36. Labium longer than head when extended [Fig. 305(C)] .... ODONATA
    Labium not extendible, normal size.. 37

37. Prothorax reduced.. PSOCOPTERA
    Prothorax similar in size to other segments 38

38. Three (rarely 2) long "tails" at tip of abdomen;
    abdominal gills present (Fig. 311)....................... EPHEMEROPTERA
    Two "tails" or none at tip of abdomen [Fig. 302(E)]............................ 39

39. "Tails" (cerci) heavily sclerotized and forcepslike [Fig. 303(D)]...... DERMAPTERA
    "Tails" (cerci) not as above or absent.. 40

40. Fore basitarsi greatly enlarged (Fig. 326)......................... EMBIOPTERA
    Fore basitarsi not enlarged 41

41. Cerci long, one-half length or more of abdomen [Fig. 302(E)];
    gills may be present; saltatorial legs never present .... PLECOPTERA
    Cerci usually short (if long, then hind legs saltatorial) or absent; gills lacking....... 42

42. Front legs raptorial [Fig. 323(B)] .... MANTODEA
    Front legs not raptorial.. 43

43. Hind legs saltatorial [Fig. 302(D)] or front pair fossorial
    [Fig. 302(C)] .... ORTHOPTERA
    Legs not as above.. 43

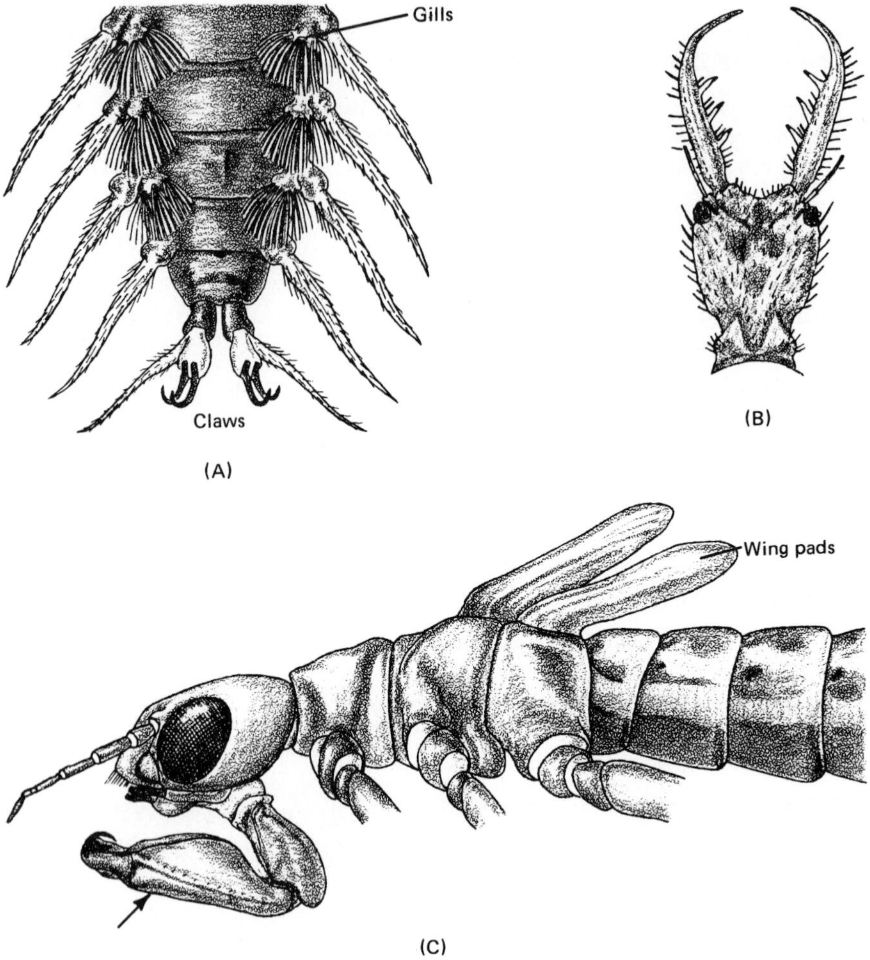

Figure 305. Diagnostic characters used in identifying immature insects. (A) lateral abdominal appendages, ventral view (Megaloptera); (B) mouthparts sickle-shaped and hollow (Neuroptera); (C) labium greatly enlarged but not extended; wing pads present (Odonata).

44. Body sticklike [Fig. 323(F)] .... PHASMIDA
    Body flattened dorsoventrally .... BLATTARIA

45. Short cerci present (Fig. 324) 46
    Cerci absent (Fig. 328) 49

46. Fore basitarsi enlarged (Fig. 326) .... EMBIOPTERA
    Fore basitarsi not enlarged 48

47. Antennae moniliform [Fig. 47(D)] .... ISOPTERA
    Antennae filiform [Fig. 47(A)] 48

48. Hind legs saltatorial [Fig. 302(D)] .... ORTHOPTERA
    Hind legs not saltatorial .... PHASMIDA

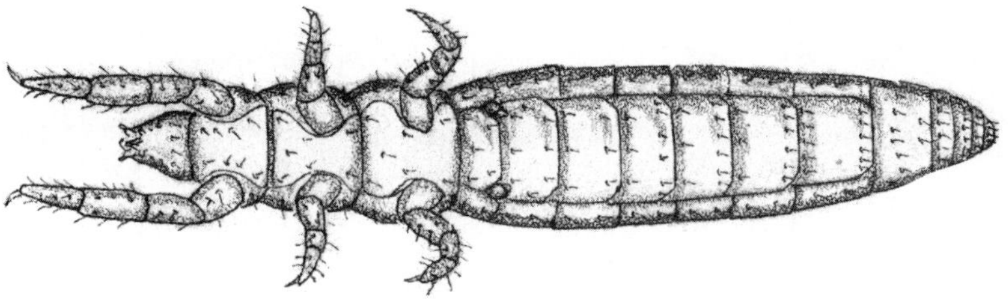

Figure 306. A proturan.

49. Antennae long and filiform (Fig. 328) . . . . PSOCOPTERA
Antennae short (Fig. 329). . . . . MALLOPHAGA

PROTURA

Proturans (Fig. 306) represent one of the most atypical of the insects and have some characteristics seen in centipedes and millipedes. These whitish and delicately sclerotized insects are normally less than 2 mm in length, lack antennae and compound eyes, are wingless, have short legs with a single claw, possess a 12-segmented abdomen in the adult stage with segments 1 to 3 having a pair of small appendages or *styli,* and lack cerci.

The gonopore is located on the eleventh abdominal segment, and small external genitalia are present. Spiracles are usually absent. Mouthparts are *entognathic;* that is, mandibles and maxillae are located in specialized pockets in the head. Malpighian tubules are represented by six minute papillae.

These apterygote insects are usually found under stones, in peat, or in organically enriched soil where they feed on fungi. The first pair of legs are held upward in antenna-like fashion when walking.

Eggs are deposited in damp soil and under logs, stones, or debris. Growth is by anamorphosis, the only such instance in insects. The first instar has 9 abdominal segments with a new segment added at each molt until 12 are attained. Ecdysis requires a transverse split at the posterior border of the head, a feature seen only in Collembola, centipedes, and millipedes.

The lack of antennae, the 12-segmented abdomen, the particular placement of the gonopore, and the anamorphic development are found in no other insect. The styles and entognathic mouthparts are found only in the most primitive insects.

About 280 species are known in the world with about 20 found in North America. They are categorized into three families. None are of economic importance to humans.

DIPLURA

Adult campodeids and japygids (Fig. 307), the former having filiform cerci and the latter forcepslike cerci, vary in length from 3 mm to more than 50 mm. These small apterygote insects lack compound eyes, ocelli, wings, external genitalia, and usually Malpighian tubules. Mouthparts are entognathic and chewing. Their antennae are elongate and

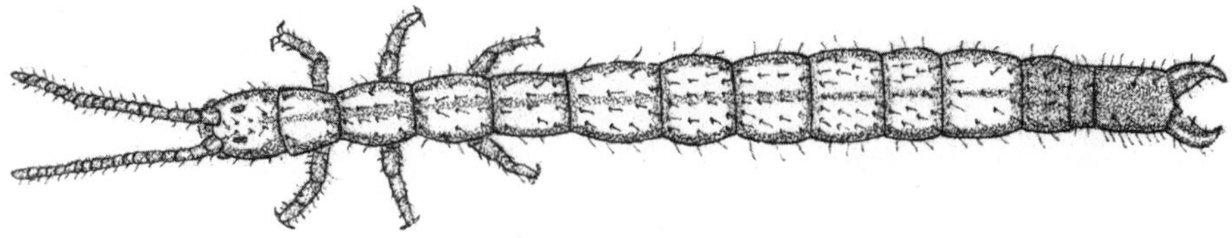

Figure 307. A dipluran, family Japygidae.

moniliform, and each segment has musculature, a noninsectan trait seen also in Collembola. Legs have a single tarsal segment but possess two claws. The abdomen has 11 segments, and styli with eversible sacs at their base are found on abdominal segments 2 to 7.

Eggs are deposited in moist soil under leaves, rotting logs, and litter, and they are often guarded by the female. As in springtails, no embryonic membranes are produced during embryology. Growth is epimorphic, and immatures resemble the adults except for size. Up to 30 molts have been reported for large species of *Campodea.*

More than 670 species have been described (64 in the United States) and are placed into three families. Both herbivores and carnivores have been observed with the former more common. No economic pests are known.

COLLEMBOLA

Another somewhat atypical insect is the springtail (Fig. 308). These arthropods are minute, seldom exceeding 6 mm in length. They are easily recognized by their small size, absence of wings, six-segmented abdomen, ventral tube, or *collophore,* on first abdominal segment, and the usual presence of an abdominal forked *furcula,* or spring; the latter, when present, is used to propel the organism forward. Springtail antennae are normally four-segmented, each segment having individual musculature; the possession of such true segmentation in the antennae is a trait found also in Diplura, millipedes, and centipedes. Their eyes are represented by lateral groups of from one to eight stemmata. Tarsi are single segmented. Spiracles are absent or located in the neck, a characteristic not seen in other insects. Other unique characteristics include a lack of corpora allata, gonads possessing a lateral germarium, Malpighian tubules and external genitalia being absent, and the presence of an Organ of Tömösvary, a structure just behind the antennae and found only in this order and in Diplura among insects. Springtail mouthparts are endognathic, a trait of many apterygotes.

Eggs are deposited in soil and undergo holoblastic cleavage; the latter trait is not found in other insects except for certain hymenopteran parasites. Ametabolous development occurs; except for size, all instars are similar in shape, and because external genitalia are absent, diagnosis of adults is based on whether or not the individuals reproduce. Molting often continues after adulthood is reached, a trait few insects demonstrate. A transverse ecdysial split, similar to Protura, is located at the posterior border of the head and permits the shedding of the old head capsule.

Between five and ten families have been proposed to include the approximately 3,500 species (approximately 700 in North America). Springtails are normally abun-

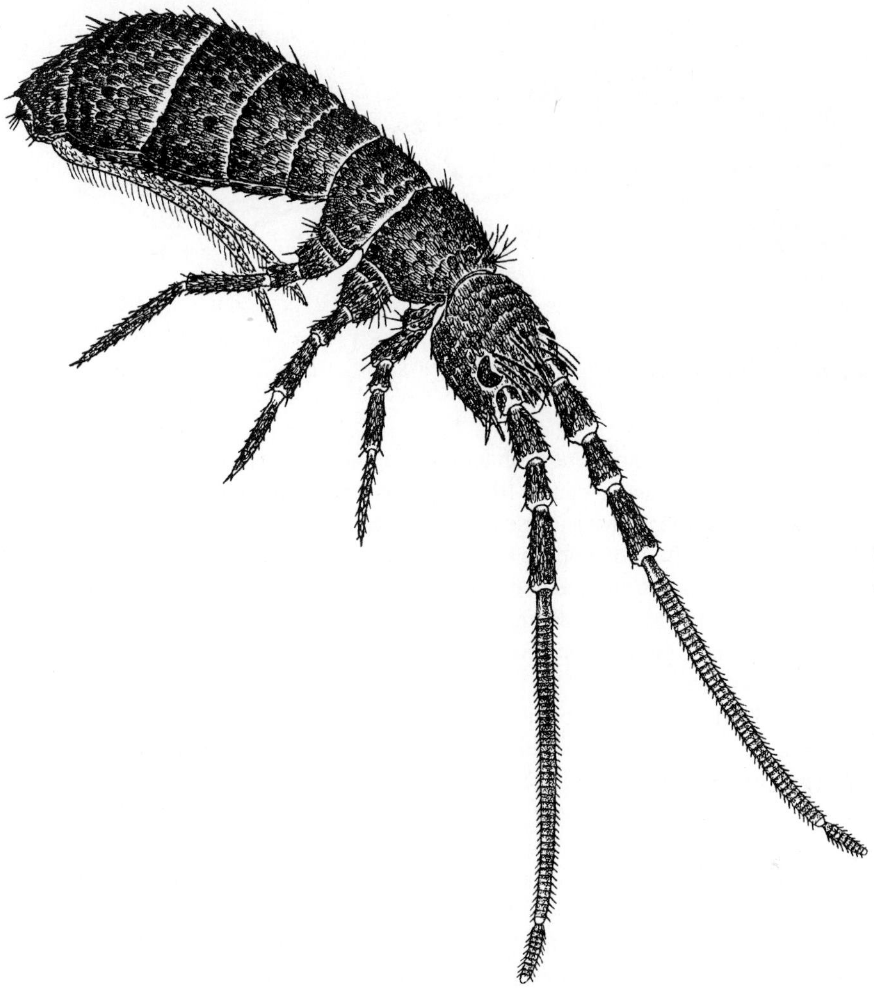

Figure 308. A springtail.

dant in soil or leaf litter and feed mainly as scavengers or on algae. They become pests only rarely. The lucern flea on alfalfa, a Puerto Rican species of Collembola transmitting disease to sugar cane, and species damaging mushrooms and house and greenhouse plant seedlings are examples of specific damage caused by springtails.

ARCHEOGNATHA

Bristletails (Fig. 309) have cylindrical bodies and are wingless. Their large compound eyes usually meet dorsally. Antennae are elongate and filiform, mouthparts are chewing but weak, and ocelli are present. The thorax is arched and styli may be present on

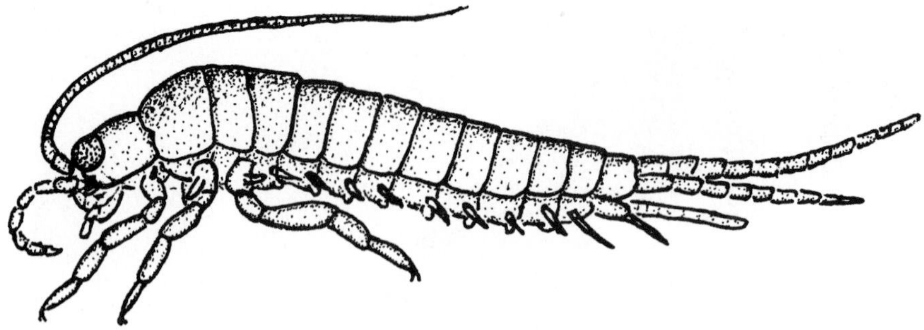

Figure 309. A bristletail.

the mesothorax, metathorax, and abdominal segments 2 to 9. Three terminal "tails" are present with the medial caudal filament considerably longer than the lateral cerci. An appendicular ovipositor is present in adult females.

Eggs are deposited in crevices or holes dug by the ovipositor. Both nymphs and adults are omnivorous and feed on lichens, algae, and organic debris. These insects are nocturnal, can move rapidly, and can jump to escape predators. Development is ametabolous, and molting can occur after reaching the adult stage. No economic species are known.

About 390 species are known (35 in North America), and they are grouped into two families with most species being Machilidae. This order is sometimes considered to be part of Thysanura.

THYSANURA

Silverfish (Fig. 310) are primitively wingless and flattened insects. Compound eyes are small and well separated or absent, ocelli are absent, and antennae are elongate and filiform. Their mouthparts are of a weak chewing type, and the abdomen possesses three elongate terminal tails (lateral two are cerci) of about equal length and usually diverging from the median caudal filament. Scales often cover the body, and *styli* or vestigial (single-segmented) abdominal legs are often restricted to segments 7 to 9. A well-developed appendicular ovipositor is present in females.

Metamorphosis is lacking and young resemble adults. Development may require several years to complete, and molting continues after adulthood is reached. A few species are associated with humans and are considered to be pest species. These insects can move rapidly.

About 370 species are known (13 in North America). These species are classified into four families, most of which are placed into the family Lepismatidae. The firebrat (*Thermobia domestica*) is found near furnaces and steam pipes. The silverfish (*Lepisma saccharina*) (Fig. 310) seeks cooler areas that have higher moisture. Bookbindings, wallpaper paste, flour, and organic debris form the diet of these two pests.

Figure 310. A silverfish (*Lepisma saccharina*).

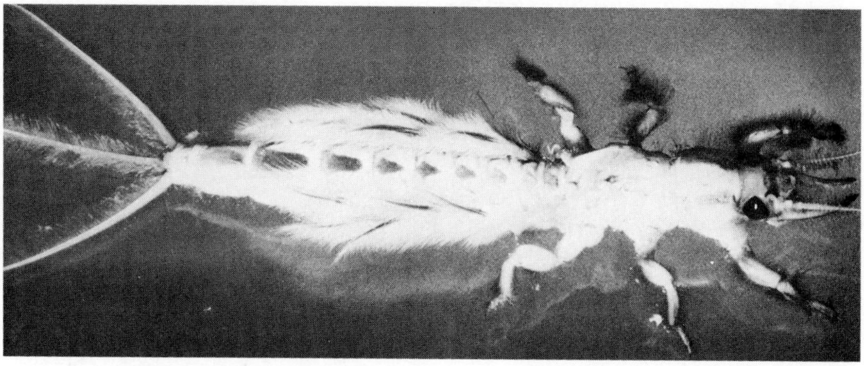

Figure 311. Mayfly nymphs are aquatic. This nymph (*Hexagenia limbata*) uses its tusklike mandibles for digging in the mud.

EPHEMEROPTERA

Mayflies (Figs. 311, 312) are from 4 mm to 50 mm in length and may be identified by their setaceous antennae and long multisegmented cerci. A median caudal filament may be observable. Two pairs of wings are usually present (hind pair usually reduced and sometimes lacking), have numerous veins, and are normally held vertically over the thorax at rest. Wing venation is considered to be the most primitive of all pterygote insects (Edmunds and Traver, 1954). Compound eyes are large and may be divided in

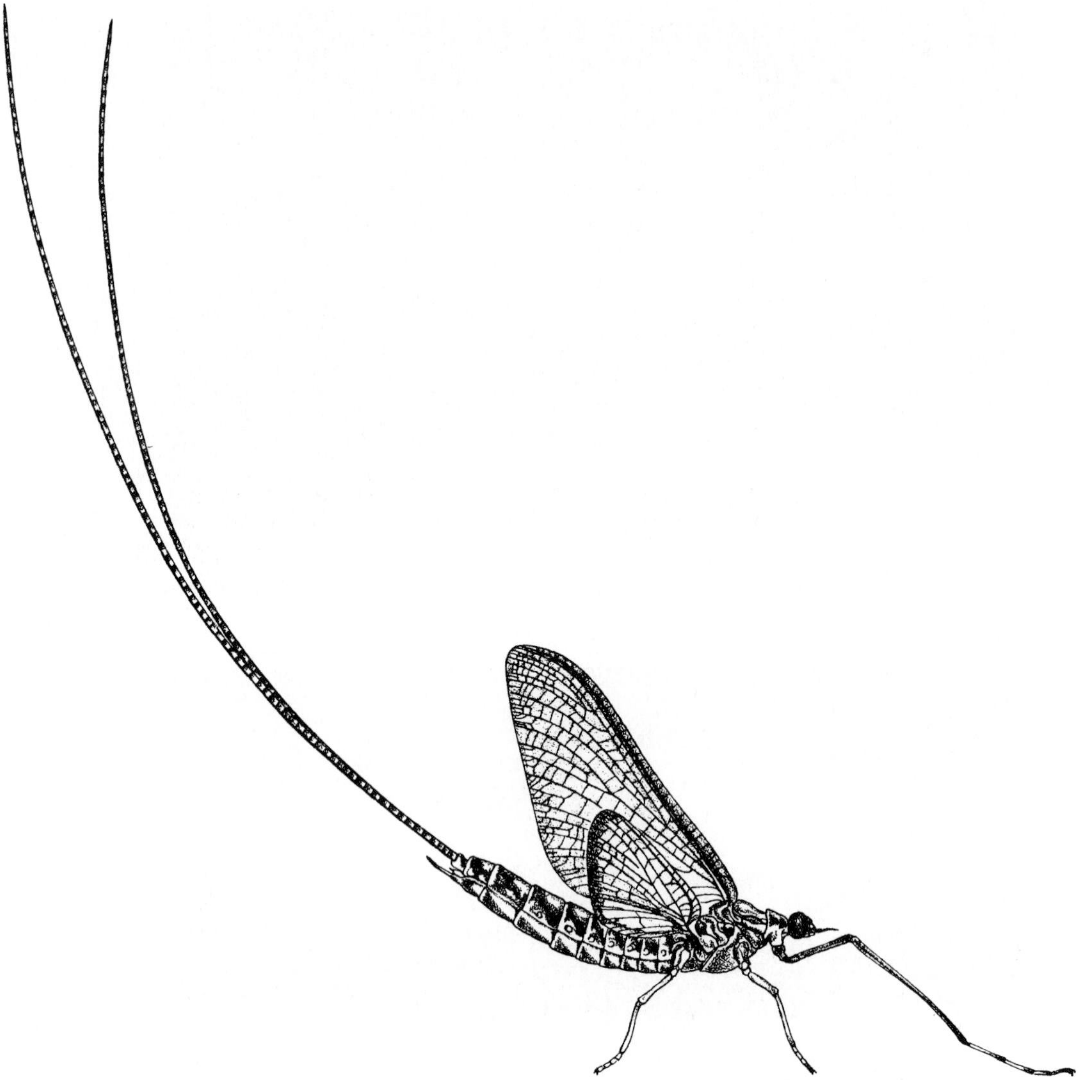

Figure 312. An adult mayfly (*Hexagenia limbata*). Note the wings in a vertical resting position.

males. Mouthparts are vestigial. Gonopores are paired in both sexes and open on the seventh abdominal segment in the female; the insectan genital chamber and median oviduct of females and the ejaculatory duct in males are lacking. Some species are parthenogenetic.

Metamorphosis is incomplete. Eggs [Fig. 91(A)] are deposited in water. Nymphs (Fig. 311) have abdominal gills and ordinarily have three "tails," two pairs of cerci, and a caudal filament. Some nymphs cling to rocks and are flattened and streamlined, whereas others are cylindrical and burrow (Fig. 311) or move about on the bottom. Specific habitat preferences are common. Food consists of aquatic organisms, usu-

Figure 313. Examples of Odonata. (A) darner dragonfly (Aeshnidae); (B) broad-winged damselfly (Calopterygidae); (C) common skimmer (Libellulidae); (D) broad-winged damselfly (Calopoterygidae).

ally plant and organic debris. After completing their immature development (1 to 3 years), nymphs leave the water and molt into a *subimago,* a winged immature stage unique to mayflies. Within hours to several days, another molt occurs, and the adult emerges (Fig. 312). Mating, often in swarms, takes place in flight. Eggs often are released when a rupture occurs in the female's body. Survival by adults is short (from several hours to rarely beyond 4 days) as indicated by the derivation of the order name (*ephemerous,* for living but a day).

Approximately 2,100 species are known (about 600 in North America), and these are categorized into as many as 17 families. This group forms and important source of food for some fish but is of not direct economic or medical importance to humans.

ODONATA

The Odonata (Fig. 313) are medium to large (wingspread from 25 mm to 190 mm) and have the following characteristics: large compound eyes, strong chewing mouthparts, setaceous antennae, elongate slender abdomen with small single-segmented cerci, and two pairs of long narrow wings with many veins, a notch, and an arculus [Fig. 302(F)].

Figure 314. An adult damselfly (*Agrion maculatum*).

Tarsi are three-segmented, and females often have an appendicular ovipositor. Two suborders exist in the United States. One, the Anisoptera, or dragonflies, hold their wings horizontally when at rest, have the thickest bodies, and are very active fliers, whereas the Zygoptera, or damselflies, hold their wings nearly vertical (Fig. 314) when at rest, have very slender bodies, and are less agile in flight.

Males have secondary copulatory organs on the second and third abdominal segments. During copulation the male grasps the female's "neck" with his claspers while she curves her abdomen up to his second segment (Fig. 246). Females then deposit eggs in or near water. Metamorphosis is incomplete. Nymphs are voracious predators and utilize an enlarged yet retractible labium for capturing prey (Figs. 138, 315). Damselfly nymphs (Fig. 316) have three terminal platelike gills, whereas oxygen exchange in dragonfly nymphs takes place internally in a musculated rectum. If the rectal muscles, used normally in exchanging water, contract strongly, water is forced out rapidly, and a form of "jet propulsion" occurs. After from 4 weeks to 4 years, nymphs crawl up emergent vegetation and metamorphose into adults.

The nearly 5,000 species (more than 400 in North America) are separated into nine families. Their value to humans is difficult to assess because they are general feeders, and, although many mosquitoes and other insects detrimental to humans are consumed, many beneficial species are also eaten.

The following two keys, one to the adults and the other to the nymphs, will aid the beginning student in recognizing the suborders and major families inhabiting the United States.

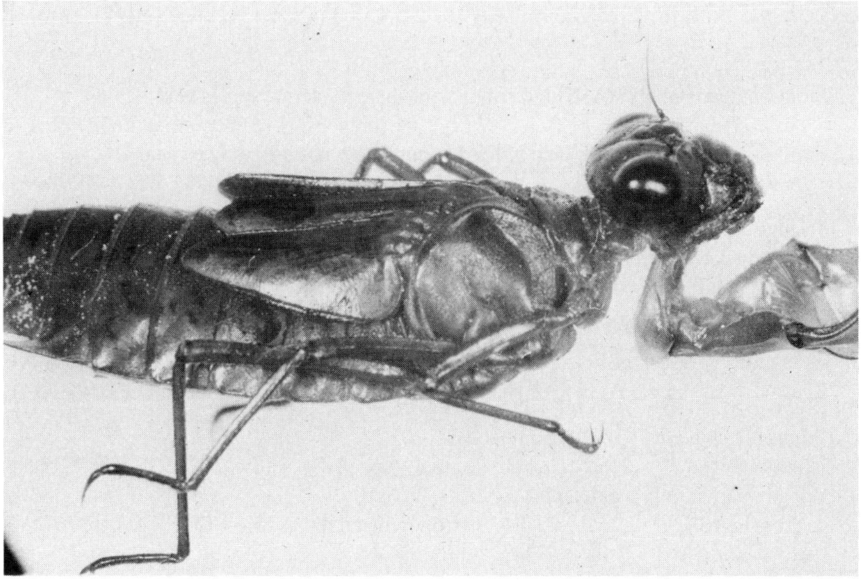

Figure 315. A dragonfly nymph. Note the enlarged labium and the external wing pads.

Figure 316. A damselfly nymph. Note the three terminal tracheal gills, two of which are abnormal and represent partial regeneration stages.

KEY TO SUBORDERS AND COMMON FAMILIES OF ADULT ODONATA

1. Stout-bodied [Fig. 313(A)]; hind wings with sharp posterior angles near base [Fig. 313(C)] . . . . (dragonflies) Suborder ANISOPTERA, 2
   Frail insects [Fig. 313(B)]; hind wings lacking sharp posterior angles near base [Fig. 313(D)] . . . . (damselflies) Suborder ZYGOPTERA, 4

2. Triangle of hind wing nearer arculus [Fig. 302(F)] than in front wing and shaped differently . . . . (skimmers) LIBELLULIDAE
   Triangle about equidistant from arculus in both front and hind wings and of similar shape 3

3. Eyes meeting on top of head . . . . (darners) AESCHNIDAE
   Eyes widely separated on top of head . . . . (clubtails) GOMPHIDAE

4. Wings not stalked; 10 or more antenodal cross veins . . . . (broad-winged damselflies) CALOPTERYGIDAE
   Wings stalked; only 2 antenodal cross veins . . . . (narrow-winged damselflies) COENAGRIONIDAE

KEY TO THE COMMON FAMILIES OF NYMPHAL ODONATA

1. Three large platelike gills present at tip of abdomen (Fig. 316) . . . . Suborder ZYGOPTERA, 2
   External gills absent (Fig. 315) . . . . Suborder ANISOPTERA, 3

2. Basal antennal segment long (Fig. 316) . . . . CALOPTERYGIDAE
   Basal antennal segment similar to distal segments in length . . . . COENARGIONIDAE

3. Antennae four-segmented . . . . GOMPHIDAE
   Antennae six- or seven-segmented 4

4. Distal portion of labium spoon-shaped (Fig. 315) . . . . LIBELLULIDAE
   Distal portion of labium flat . . . . AESHNIDAE

BLATTARIA

Cockroaches (Figs. 95, 297, 317) are small to large insects (up to about 65 mm) with dorsoventrally flattened bodies. The head is often covered by an enlarged pronotum. The compound eyes are usually large, the many segmented filiform antennae are long, and the chewing mouthparts are hypognathous. A thickened tegmina covers the membranous hind wings, which possess a large anal lobe. Micropterous, brachypterous, and macropterous morphs may be present in the same species (often as a sexual dimorphism, see Fig. 297), whereas some species are micropterous or apterous. Legs are long and cursorial, with five-segmented tarsi, and movements can be rapid. Other characteristics include the following: an appendicular ovipositor is hidden beneath a subgenital plate, the male genitalia are completely asymmetrical and concealed as in the female, and the multisegmented cerci are short to moderate in length.

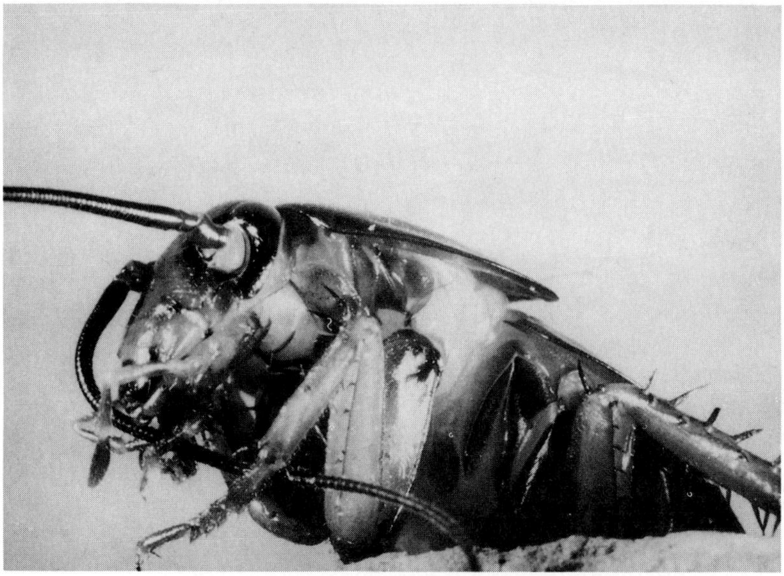

Figure 317. An American cockroach (*Periplaneta americana*) preening an antenna.

Eggs are often oviposited within a sclerotized oötheca (Fig. 95), although they may be retained internally in a special brood chamber in ovoviparous and viviparous species. Embryology and nymphal development are often extended, some requiring more than a year to reach adulthood. Nymphs and adults are often general feeders (omnivorous). In some species adults can survive up to 4 years.

Cockroaches are considered to be of economic importance. Although household food stuffs are eaten and roaches have been found with disease producing organisms, placing them into the pest category represents the psychological impact insects have on us (see Chapter 12).

About 3,500 species are known worldwide, mainly from the tropics, and only 67 are found in North America. Cockroaches are classified into from five to eight families, and sometimes they are grouped with the mantids into the order Dictyoptera or into the order Orthoptera by others.

MANTODEA

The mantids (sometimes referred to as praying mantids) are medium to large insects, and sometimes are flattened but mostly are slender and sticklike (Fig. 318). The head is highly articulated and can be rotated over 300°. The eyes are comparatively large and with many ommatidia. Mouthparts are chewing, hypognathic, and well modified for carnivory. The filiform antennae are long and slender. The prothorax is greatly elongated (but not covering the head) and holds the projecting raptorial forelegs [Fig. 323(B)] as if in prayer. Wings are two pairs with the front pair as narrowed tegmina in males, whereas the wing development in females is often reduced to absent.

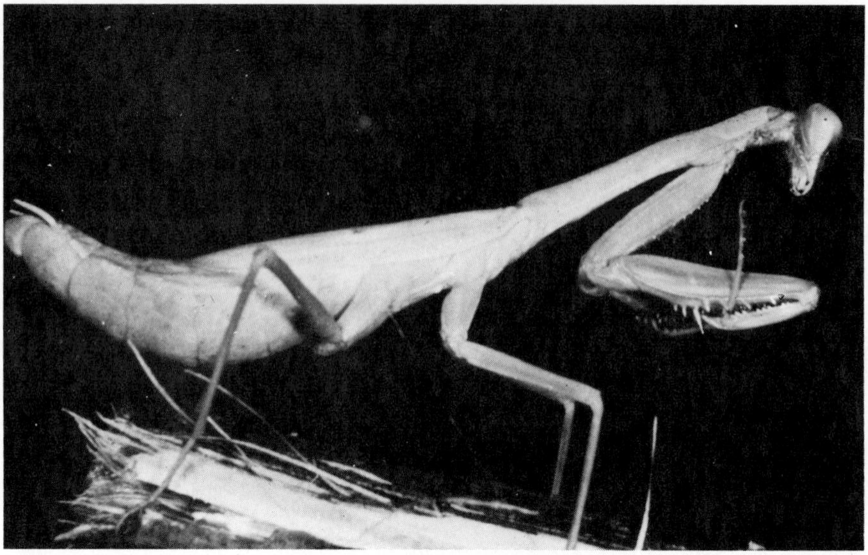

Figure 318. The Carolina mantis (*Stagomantis carolina*).

The mesothoracic and metathoracic legs are slender and have five-segmented tarsi. The abdomen is capable of great expansion during feeding or with the presence of eggs, and it carries both a hidden appendicular ovipositor or asymmetrical male genitalia and variably segmented cerci.

Eggs (varying from 10 to about 400) are deposited within a papery-textured oötheca and are the overwintering stage for temperate species. Both nymphs and adults are cryptically colored and suddenly ambush prey by grasping it with the forelegs (Fig. 288). This order is considered beneficial because pests are often eaten; however, they are generalists in food preference and can eat beneficial insects just as easily.

Nearly 2,000 species are known in the world (about 20 in North America), and these are grouped into eight families, only one of which, Mantidae, is found in North America. Some specialists combine the mantids with the cockroaches into the order Dictyoptera, whereas others group them into the Orthoptera.

PHASMIDA

These are moderate to large insects in length (to 30 cm) and are commonly called walkingsticks (Fig. 319) because of their often slender shape [Fig. 323(F)]. A few species, however, resemble leaves and are green. These insects have prognathic chewing mouthparts, compound eyes, and antennae varying from short to long and filimentous. The prothorax is small, and their legs are usually long and slender, normally with three- to five-segmented tarsi. Wings are usually absent in most nontropical species, but they may be present with the first pair shortened into tegmina and the hindwings normal and functioning for flight in some tropical males. Cerci are

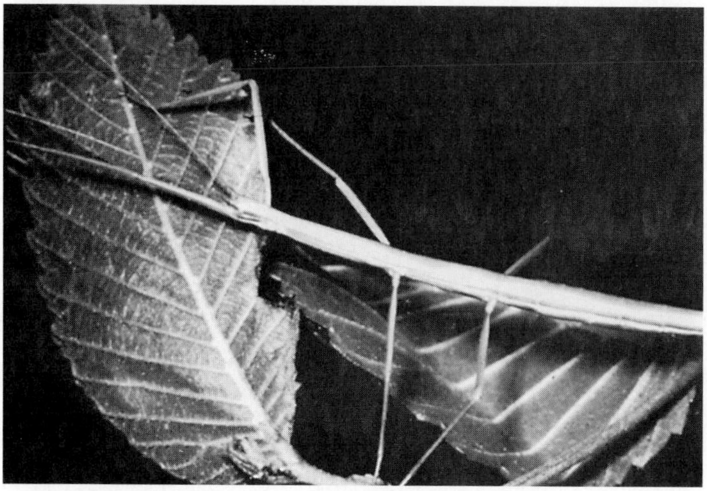

Figure 319. A walkingstick (*Diapherimera persimilis*).

short and single segmented. An appendicular ovipositor is present but concealed. Auditory or stridulatory apparatus are absent.

Large numbers (100 to 300) of eggs that resemble seeds are oviposited singly on vegetation or dropped to the ground. Hatching may be delayed up to 3 years. Development is hemimetabolous, and both nymphs and adults are herbivorous. Individuals often demonstrate reflex immobility and autotomy when disturbed. Facultative parthenogenesis is not uncommon.

Most species are not of economic importance except when rare outbreaks occur. If such an outbreak were to take place, severe defoliation of the host plant could occur.

More than 2,000 species are known (31 found in North America), and they are grouped into four families, only one of which, Phasmatidae, occurs in North America. This order may be grouped into the Orthoptera in some classifications.

ORTHOPTERA

Orthoptera (Figs. 320, 321, 322) include insects that have greatly enlarged hind legs, chewing mouthparts, one- to four-segmented tarsi, and large compound eyes. Wings are usually present and have many veins and are modified with the forewings often narrowed and thickened into a *tegmen,* whereas the hind wings are broad, membranous, and folded fanwise under the mesothoracic pair (Fig. 55). Flight is mainly through action of the hind wings. Stridulation or sound production by scraping (Fig. 167) is a means of attracting mates. An appendicular ovipositor is common and often measures as long as the abdomen. Cerci are often short. Antennae commonly are elongate and multisegmented. Size ranges from 12 mm to more than 120 mm in length.

Egg laying is variable. Most eggs are deposited in the soil, but others are deposited in or on vegetation (Fig. 176). Metamorphosis is incomplete. Most Orthoptera are herbivorous, but a few are carnivorous or omnivorous.

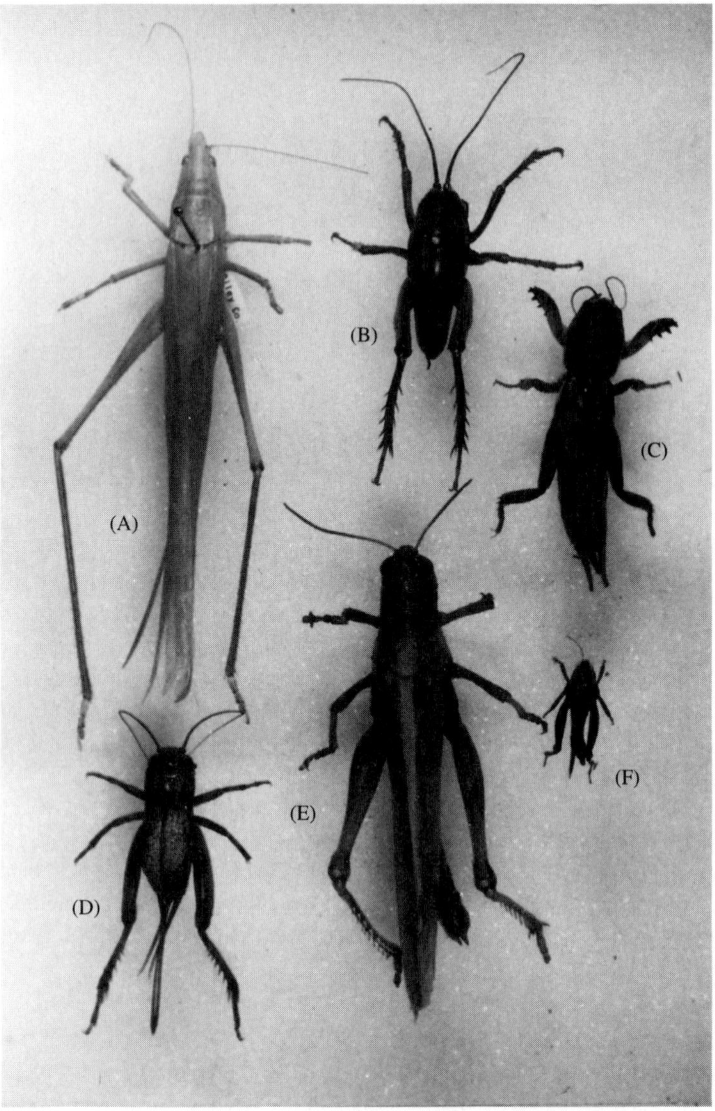

Figure 320. Examples of the common adult Orthoptera. (A) long-horned grasshopper (Tettigoniidae); (B) camel cricket (Gryllacrididae); (C) mole cricket (Gryllotalpidae); (D) cricket (Gryllidae); (E) short-horned grasshopper (Acrididae); (F) pigmy locust (Tetrigidae).

Some species are of economic importance. A few short-horned grasshoppers have been pests of crops throughout recorded history, especially in the temperate and arid regions of the world, and many areas in the United States conduct annual surveys to anticipate population increases that might reach economic thresholds. Migrating forms are of especial importance when populations are high. In some years the Mormon cricket, a wingless long-horned grasshopper of the intermountain West, has been very destructive because of its large size, gregariousness, and migratory behav-

Figure 321. A long-horned grasshopper (*Scudderia texensis*).

Figure 322. A field cricket (*Acheta assimilis*).

ior. Another example, the migratory locust, has been discussed previously in Chapter 6 on migrations and in Chapter 10 on detrimental aspects to plants. Field crickets may damage seedlings in truck crops.

More than 19,000 species are known (1,800 in North America), and they are placed into 14 to 27 families depending upon the classification system. The following key will help the beginner to identify the more common families in the United States.

KEY TO COMMON FAMILIES OF ADULT AND NYMPHAL ORTHOPTERA

1. Pronotum extending backward to near tip of abdomen [Fig. 323(A)] or beyond in adult, covering entire thorax in nymph . . . . (pigmy locusts) TETRIGIDAE
   Pronotum not extending behind thorax [Fig. 323(D), (E)], at least metanotum exposed in nymph 2

2. Hind tarsi one-segmented or absent. . . . . (pigmy locusts) TRIDACTYLIDAE
   Hind tarsi three- or four-segmented 3

3. Front tibiae and femora greatly enlarged for digging, with large black
   toothlike processes [Fig. 302(C)] . . . . (mole crickets) GRYLLOTALPIDAE
   Front tibiae and femora not greatly enlarged for digging 4

4. Tarsi three-segmented [Fig. 323(C)] 5
   Tarsi four-segmented 6

5. Antennae about as long as pronotum . . . . (short-horned grasshoppers) ACRIDIDAE
   Antennae much longer than pronotum . . . . (crickets) GRYLLIDAE

6. Wings present in adults; if absent, tympanum present
   on fore tibia (Fig. 169) in all but early instars;
   wing pads in most nymphs. . . . . (long-horned grasshoppers) TETTIGONIIDAE
   Wings absent in adults; tympanum absent; no wing pads in nymphs 7

7. Mesonota and metanota similar to pronotum
   [Fig. 323(D)]. . . . . (cave and camel crickets) GRYLLACRIDIDAE
   Pronotum longer than both mesonota and metanota
   and shieldlike. . . . . (sand crickets) STENOPELMATIDAE

ISOPTERA

Termites vary from 2 mm to 12 mm in length except for physogastric queens. They are characterized by a prognathic head, moniliform antennae with from 9 to 30 segments, chewing mouthparts, short and stout legs with four-segmented tarsi normal, one- to eight-segmented short cerci, and an absence of wings except for the reproductive caste (Fig. 324). Wings, when present, are longer than the body and are membranous. Fore and hind wings are similar in shape and size. All termites are social, and their castes and metamorphosis are discussed in detail in Chapter 7.

Approximately 2,120 species are known (45 in North America), and they are separated into from four to nine families. The most primitive family, Mastotermitidae, contains a single tropical species, *Mastotermes darwiniensis,* which has an oötheca deposited by the queen and has many other cockroach characteristics. Although 80 percent of the termite species of the world are in the Termitidae, subterranean termites (Rhinotermitidae) and drywood termites (Kalotermitidae) are the most common in the United States. Often termed "white ants," this group is of great biological and economic importance. In the tropics and in forests their feeding recycles nutrients and aids in soil development. In other instances, however, their eating is in direct conflict with humans. Because Isoptera feed upon paper, wood, and other similar cellulose goods, they cause considerable damage. In many regions of the United States nearly every house is or has been infested in the past 10 years unless the soil was treated during building. Some of the more important pest species in the United States are the subterranean *Reticulitermes flavipes* (central and eastern states), *R. hesperus* (Pacific states), *R. tibialis* (west of Mississippi River), and *Coptotermes formosanus* (Gulf states); the drywood *Incisitermes minor* (western states), *I. snyderi* (southeastern and Gulf states), and *Cryptotermes brevis* (Gulf states).

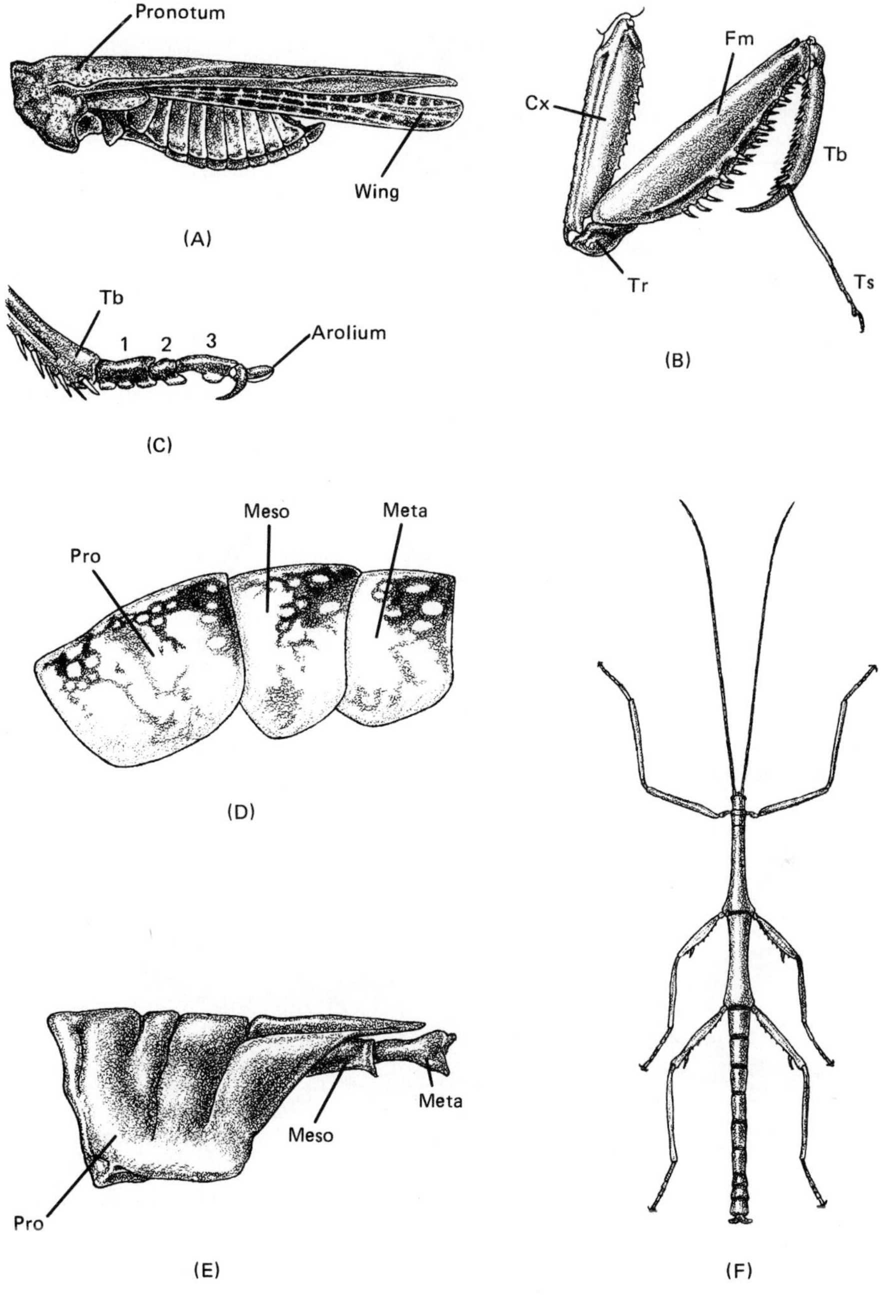

Figure 323. Some diagnostic characteristics of orthopteroid insects. (A) pronotum extending beyond tip of abdomen (Orthoptera: Tetrigidae); (B) grasping leg (Mantodea); (C) tarsus three-segmented (Orthoptera: Acrididae); (D) thoracic nota similar (Orthoptera: Gryllacrididae); (E) thoracic nota dissimilar (Orthoptera: Acrididae); (F) body sticklike (Phasmida).

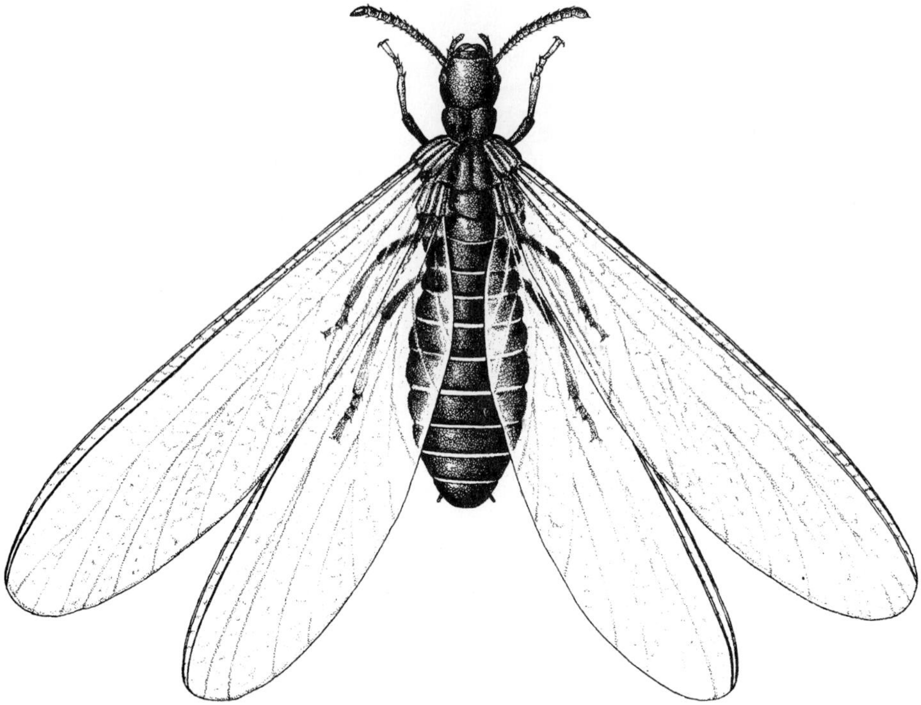

Figure 324. Reproductive termites possess two pairs of equal-sized wings. These wings are lost after the dispersal flight.

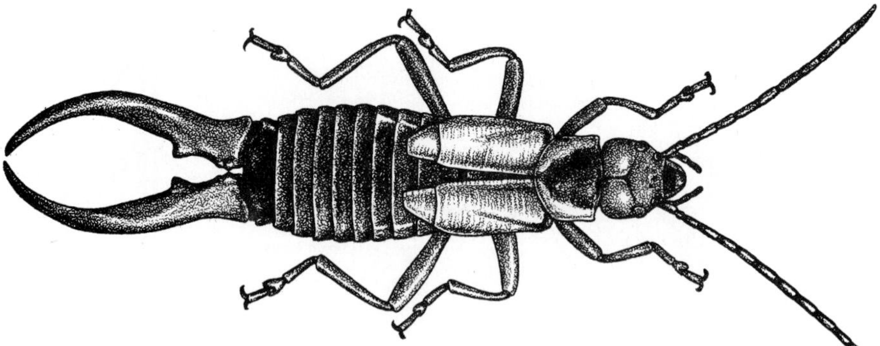

Figure 325. An earwig (*Forficula aurcularia*).

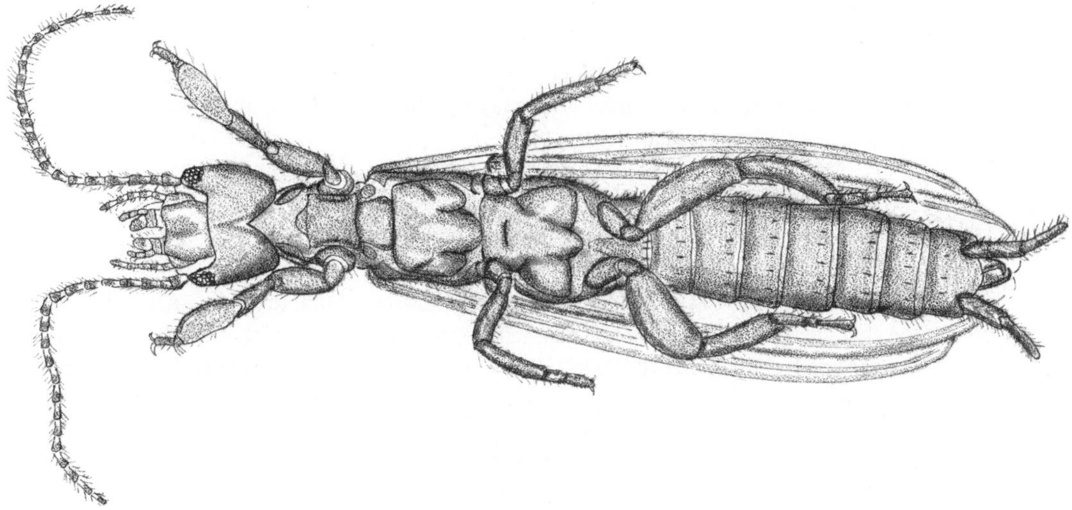

Figure 326. A male webspinner.

DERMAPTERA

Earwigs (Fig. 325) are medium-sized insects (from 5 mm to 35 mm) with characteristically enlarged unsegmented forcepslike cerci. They have prognathic chewing mouthparts, filiform antennae, usually with well-developed compound eyes, and long legs with three-segmented tarsi. The forewings are thickened into tegmina and are short. The semicircular hind wings fold fanlike longitudinally and twice transversely to fit under the reduced tegmina when not used in flight. About 20 to 80 eggs are often deposited in burrows, and, unlike most insects, there is maternal care for them through cleaning and protection. Metamorphosis is incomplete. Nymphs feed, as do the adults, on a wide variety of materials from dead to living plants and animals. Activity is normally at night.

Approximately 1,200 species are known, but only 23 have been found in North America. Four families exist. Although a few species occasionally feed on cultivated plants, generally they are of little economic importance.

EMBIOPTERA

Webspinners (Fig. 326) have chewing mouthparts, filiform antennae, and two-segmented cerci. Two pairs of nearly equal-sized wings are usually present in adult males, but females are apterous. Legs are short and tarsi three-segmented. These gregarious insects are easily recognized by the enlarged basitarsi of their forelegs, which contain the many silk glands used in constructing the silken galleries in which they dwell. Each gallery contains a female and her brood. Both eggs and nymphs develop within the protective webs and are cared for by the female. Food is of plant or microbial origin, although cannibalism exists in some species. Males die shortly after mating, but the female initiates a web of her own.

About 200 species are known (13 in the United States), and most are tropical in distribution. Three families have been classified using primarily male wing and genital characteristics as diagnostic features.

PLECOPTERA

Stoneflies (Fig. 327) are medium to large insects (from 12 mm to 65 mm in length) and are characterized by long filiform antennae, prognathic chewing mouthparts, and moderate to small compound eyes. Normally two pairs of wings are present with many veins and with similar textures; the forewings are narrower than the hind wings, and both fold back flat over the abdomen when not in flight. Also present are two- to three-segmented tarsi and long and multisegmented cerci. An appendicular ovipositor is absent.

Metamorphosis is incomplete. Eggs are mostly oviposited in well-aerated moving water, although a few exist in deep water lakes. Nymphs [Fig. 327(B)] have long antennae, chewing mouthparts, and two long cerci. Gills, when present, are on the thorax, neck, or first two to three abdominal segments. As the name *stonefly* implies, nymphs are commonly found on or under stones. Some species are herbivorous, but others are scavengers or carnivorous. Most North American species are univoltine, but maturation may take up to 4 years, at which time the nymphs leave the water and molt on nearby vegetation. Emergence from the water usually occurs in the summer, but some species normally accomplish this in the fall and winter.

The nearly 1,600 species (580 in North America) are classified into 6 to 10 families. They form an important food for many fish but are of no direct medical or economic importance to humans other than serving as good indicators of water quality.

PSOCOPTERA

Psocids are small (from 1 mm to 7 mm) and have the following diagnostic characteristics: compound eyes large or reduced, chewing mouthparts, long filiform antennae, slender legs with two- or three-segmented tarsi, and no cerci. Wings may be lacking or present. When developed, the hind wings are smaller than the front pair; both pairs flex rooflike over the abdomen when not in use and have few veins. An appendicular ovipositor is present but concealed by the eighth sternum.

Eggs are laid singly or in groups, either in buildings or outdoors. The outdoor species, including nymphs of all winged species, are found on tree trunks, in litter, on algae or lichens, and on vegetation in general. Many species are gregarious. Food consists of plant fragments including fungi, algae, and pollen. Indoor species are wingless and are called *booklice* (Fig. 328). Food includes the paste of bookbindings, glue, starch, grain, and cereal products; these insects often become pests. Six nymphal instars are common. Adults utilize the same food as the nymphs.

The 2,600 world species (257 in North America) are divided into 15 to 35 families.

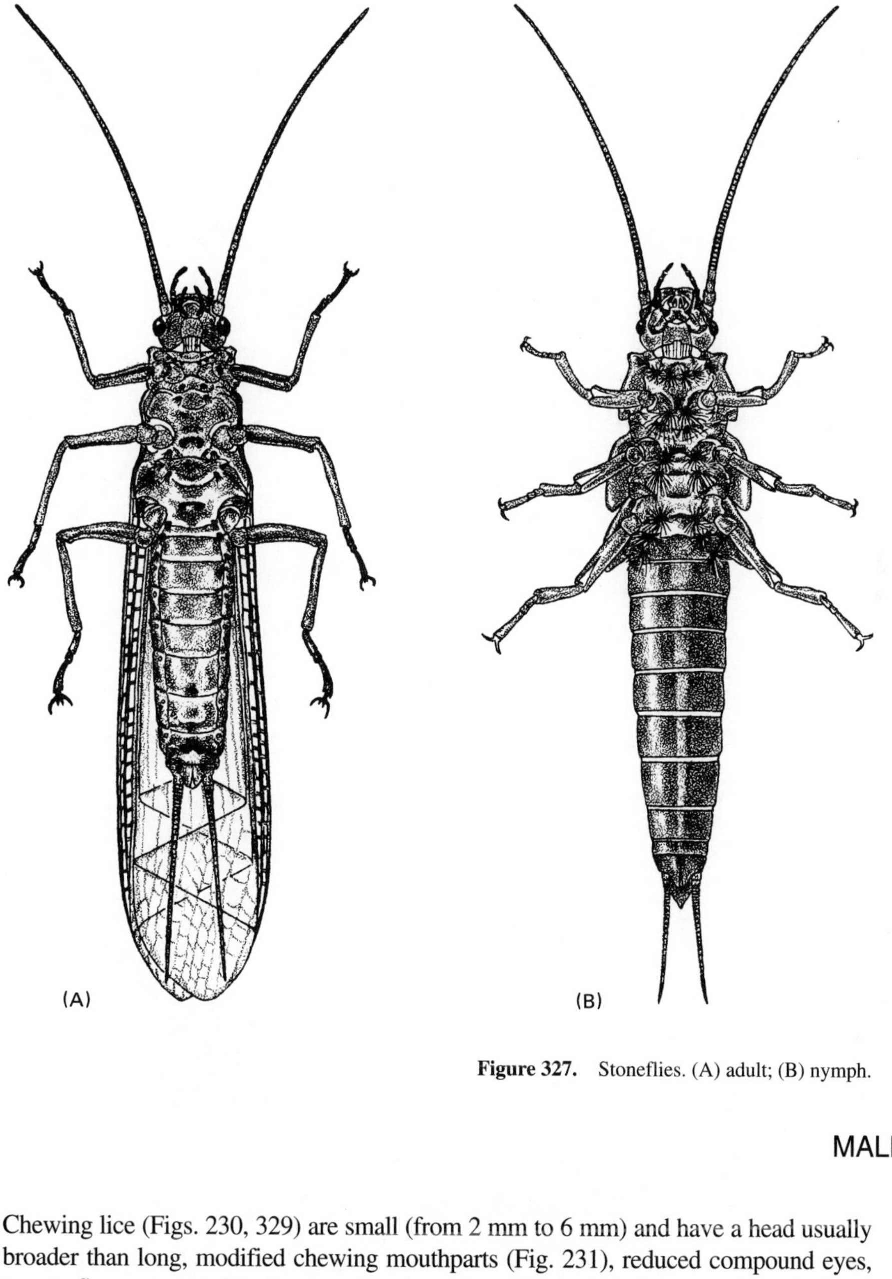

Figure 327. Stoneflies. (A) adult; (B) nymph.

MALLOPHAGA

Chewing lice (Figs. 230, 329) are small (from 2 mm to 6 mm) and have a head usually broader than long, modified chewing mouthparts (Fig. 231), reduced compound eyes, two- to five-segmented tarsi, no cerci, and no wings. The body is flattened dorsoventrally.

Eggs are fastened to feathers or hair of the host. Metamorphosis is incomplete. Both nymphs and adults ingest dead skin, feathers, hair, or scabs. Under high population pressures the dermal skin layer also may be attacked, particularly around wounds.

There are more than 2,800 species (700 in North America), and these are divided into six families. Most chewing lice infest birds, although a few utilize mammals as a host.

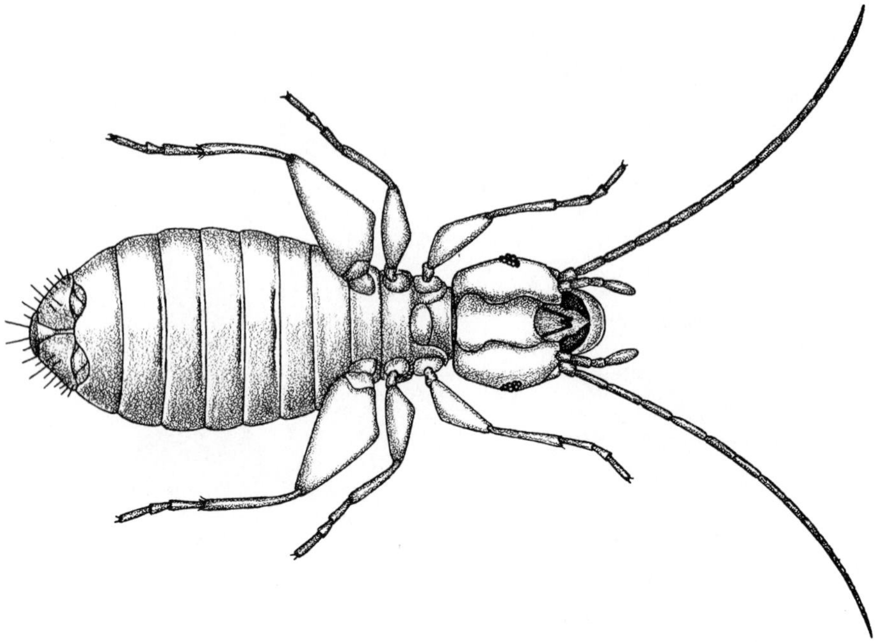

Figure 328. An adult booklouse (*Liposcelis tericolus*).

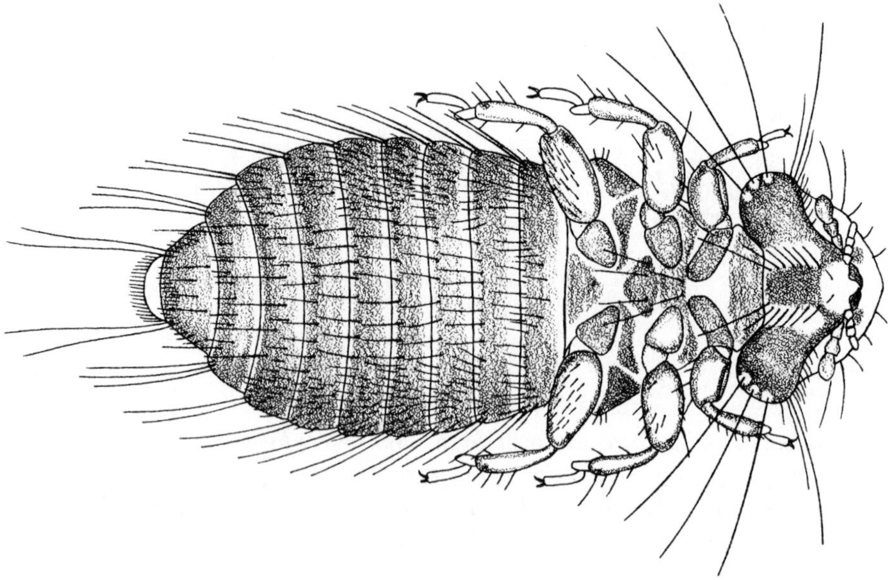

Figure 329. An adult chewing louse.

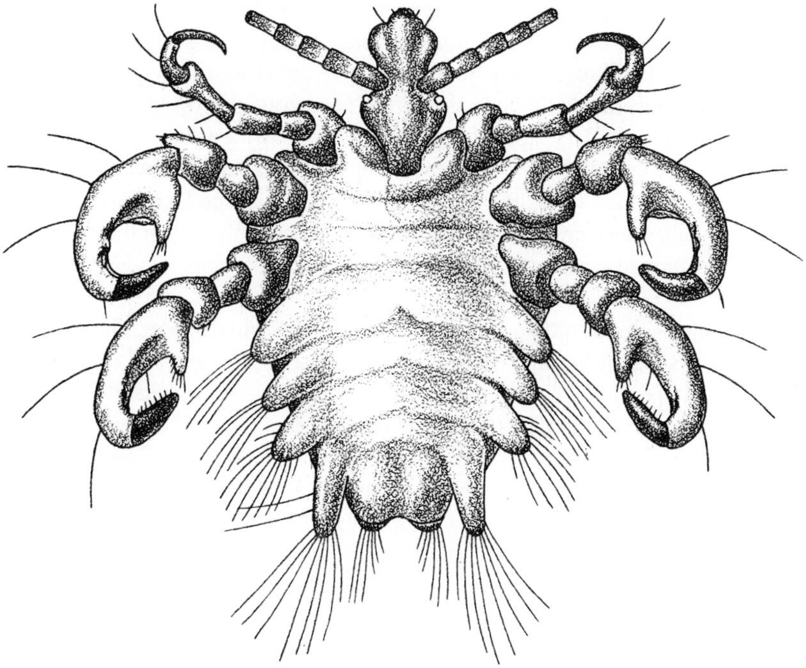

Figure 330. An adult sucking louse. This species, *Pthirus pubis,* is normally found in the pubic region of humans.

Host specificity is marked, and transfer from one host to another normally occurs only between two birds of the same species as the birds mate or nest. If a host dies, the louse fauna usually perishes. This order is of economic importance when domestic animals become infested; more than 40 species are known to parasitize poultry, and seven species are very common. A few species infest livestock (horses, cattle, and goats). Loss of weight and lowered egg production, in the case of birds, are two common results of infestations.

ANOPLURA

Sucking lice (Fig. 330) are minute to small (from 0.4 mm to 6.5 mm) and may be characterized by their narrower than long head, two- to five-segmented antennae, piercing-sucking mouthparts that are retracted into the head, greatly reduced eyes, absence of wings and cerci, and dorsoventrally flattened body. The legs are short, and the single tarsus and claw are modified into a grasping organ by being apposed to a tibial lobe (Fig. 233). The abdomen has only nine segments and lacks an ovipositor.

Sucking lice feed on blood, and their entire life cycle is spent on mammalian hosts. About 65 percent of all mammal species are parasitized by one or more louse species. Metamorphosis is incomplete. Eggs are glued singly to the hair of the host (Fig. 234). Development to the adult requires three nymphal instars.

A high degree of host specificity and preference for specific regions on the host are recognizable. The human louse, *Pediculus humanus,* infests humans, and whether it feeds on the head or body region has direct influence on its morphology and behavior (these two varieties, head and body lice, were once considered separate species). About 26 percent of secondary students in industrialized Britain were found to have head lice in 1970, and epidemics are becoming common today in many Western countries. Adults appear about 9 days after hatching from the egg. The crab louse, *Pthirus pubis* (Figs. 232, 330), another species found in man, is found mainly in the pubic and perianal region of humans and is currently epidemic, paralleling the incidence of venereal diseases.

Some systematists group Anoplura and Mallophaga into a single order, Phthiraptera, but these should be separate because of the work of Kim and Ludwig (1978). About 500 species are known (76 in North America), and these are usually separated into from three to nine families. Two families (Pediculidae, Pthiridae) include lice found on humans. A third family (Echinopthiriidae) infests only seals and walruses. The remaining families parasitize mainly rodents, rabbits, and hoofed animals. Sucking lice are important to humans in three ways: (1) several species infest livestock, (2) two species parasitize man and are irritating, and (3) human disease may be transmitted, including one type of relapsing fever, trench fever, and epidemic typhus. The latter two diseases are normally associated with wars.

THYSANOPTERA

Thrips (Fig. 331) are minute to small (from 0.5 mm to 14 mm) and have long and narrow bodies, large compound eyes, six- to nine-segmented antennae, mouthparts modified for sucking in a cone-shaped beak [Fig. 303(E)]. Other characteristics include one- to two-segmented tarsi, and an abdomen that is often pointed and lacking cerci. Wings are either absent or are long and narrow with long fringes or hair. Some species may have a protrusible bladder at the apex of each tarsus and an appendicular ovipositor.

From 25 to 200 eggs are laid either singly or in clusters on or beneath the surface of leaves. Metamorphosis appears intermediate between incomplete and complete. The first two instars lack external wing pads, and although the instars resemble adults, they are often called *larvae.* The remaining instars have external wing pads, are inactive, and are defined as *prepupa* and *pupae.* Most active stages feed on plants, but a few species are carnivorous. Parthenogenesis occurs in several species.

The 6,000 species (about 700 in North America) are separated into two principle suborders (Terebrantia, Tubulifera) and five families. Although a few species are predaceous, most feed on a wide range of flowers and cultivated crops and cause considerable damage. Examples include tobacco, onion, gladiolus, and citrus. A few plant viruses are transmitted by thrips.

HEMIPTERA

The true bugs (Figs. 332, 333) vary in length up to 100 mm, compound eyes are usually large, and antennae are from three- to five-segmented and often longer than the

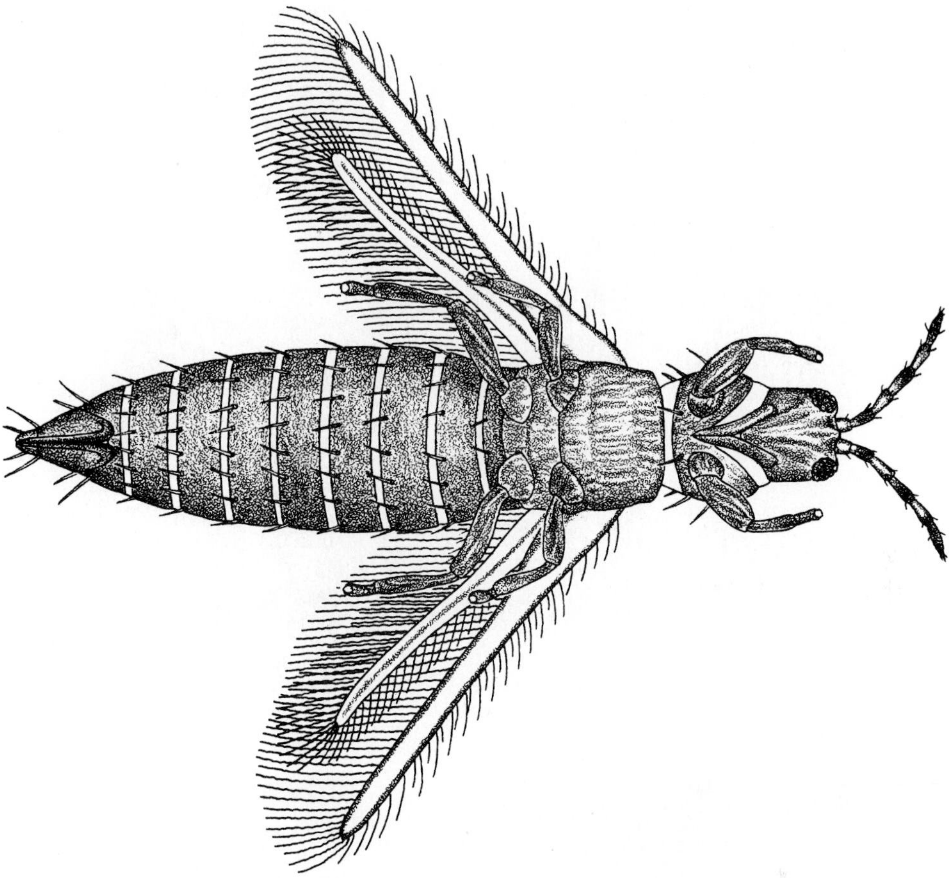

Figure 331. A female thrips (*Frankliniella tritici*).

head in terrestrial forms but are short in aquatic species. Mouthparts are piercing-sucking with the segmented beak arising from the anterior portion of the head and extending posteriorly along the sternum (Fig. 334). Tarsi are one- to three-segmented, and cerci and appendicular ovipositor are absent. Two pairs of wings are usually present and held flat over the body when at rest, but they may be short or absent in some species. The front pair of wings is usually thickened at the base and membranous apically to form a *hemelytron.* The hind wings are membranous and slightly shorter than the hemelytra. Located between the wings is an enlarged *scutellum.*

Metamorphosis is incomplete. Eggs are deposited in the habitat in which development occurs; many nymphs and adults are terrestrial, but a significant number are aquatic. Food is liquid, either sap or blood, and varies from the common herbivorous to carnivorous to parasitic. A summary of preferences for habitat and diet by family is shown in Table 12.

A number of true bugs are of economic importance. Pests of cultivated plants include: (1) Coreidae, the squash bug, *Anasa tristis* (squash and pumpkins); (2) Miridae,

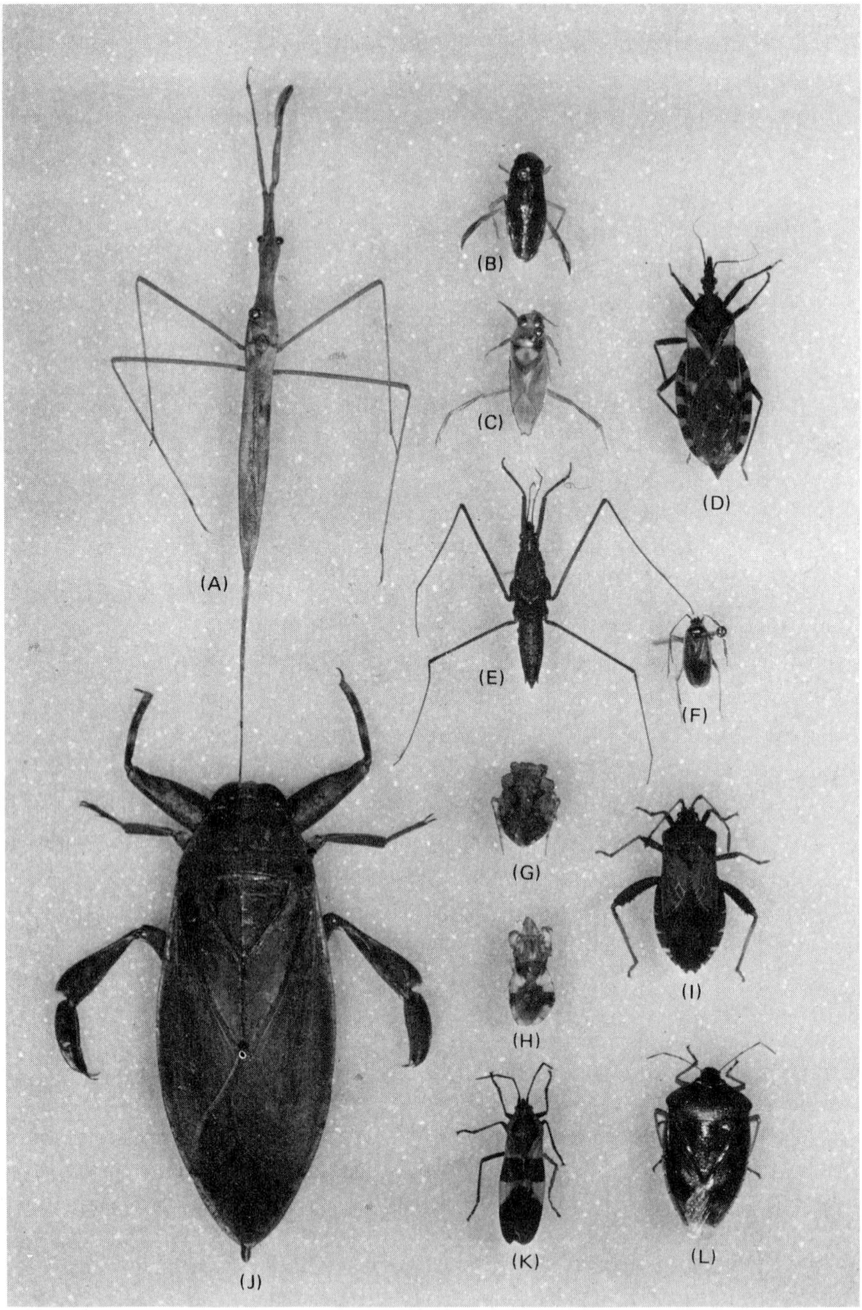

Figure 332. Examples of common Hemiptera. (A) water scorpion (Nepidae); (B) water boatman (Corixidae); (C) backswimmer (Notonectidae); (D) assassin bug (Reduviidae); (E) water strider (Gerridae); (F) plant bug (Miridae); (G) toad bug (Gelastocoridae); (H); ambush bug (Phymatidae); (I) coreid bug (Coreidae); (J) giant water bug (Belostomatidae); (K) lygaeid bug (Lygaeidae); (L) stink bug (Pentatomidae).

Figure 333. An assassin bug (*Arilus cristatus*). This species is referred to as a "wheel bug" because of the notched tergal region. It is a predator on many insects and has a painful bite.

TABLE 12. Major Families of Hemiptera Listing Their Diet and Habitat

Family	Diet	Habitat
Belostomatidae	Predator	Aquatic
Cimicidae	Parasite	Terrestrial
Coreidae	Herbivore, some scavengers	Terrestrial
Corixidae	Herbivore, scavenger	Aquatic
Cydnidae	Herbivore	Terrestrial
Gelastocoridae	Predator	Near water
Gerridae	Predator	Surface film of water
Lygaeidae	Herbivore, some predators	Terrestrial
Miridae	Herbivore	Terrestrial
Nabidae	Predator	Terrestrial
Naucoridae	Predator	Aquatic
Nepidae	Predator	Aquatic
Notonectidae	Predator	Aquatic
Pentatomidae	Herbivore, some predators	Terrestrial
Phymatidae	Predator	Terrestrial
Reduviidae	Predator, some parasites	Terrestrial
Scutelleridae	Herbivore	Terrestrial
Tingidae	Herbivore	Terrestrial
Veliidae	Predator	Surface film of water

the tarnished plant bug, *Lygus lineolaris* (fruit and flowers) and the cotton fleahopper, *Psallus seriatus* (cotton); (3) Lygaeidae, the chinch bug, *Blissus leucopterus* (wheat, oats, and other grains); and (4) Pentatomidae, the harlequin bug, *Murgantia histrionica* (cabbage and turnips) and the southern green stink bug, *Nezara viridula* (wide variety of crops). Some species of assassin bugs are naturally infected with Chagas' disease;

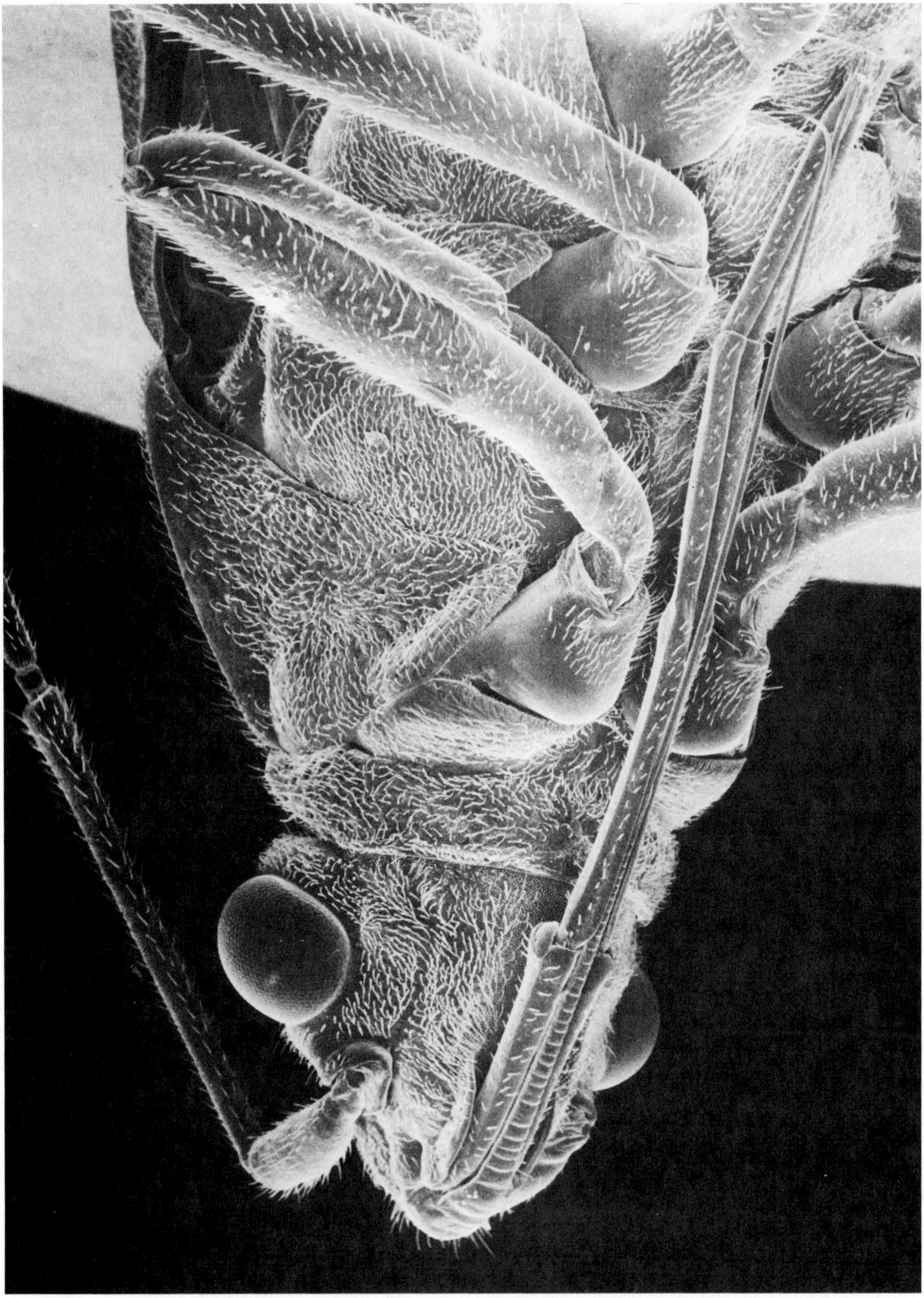

Figure 334. Scanning electron microscope photograph of the piercing-sucking mouthparts of the boxelder bug, *Leptocoris trivittatus*.

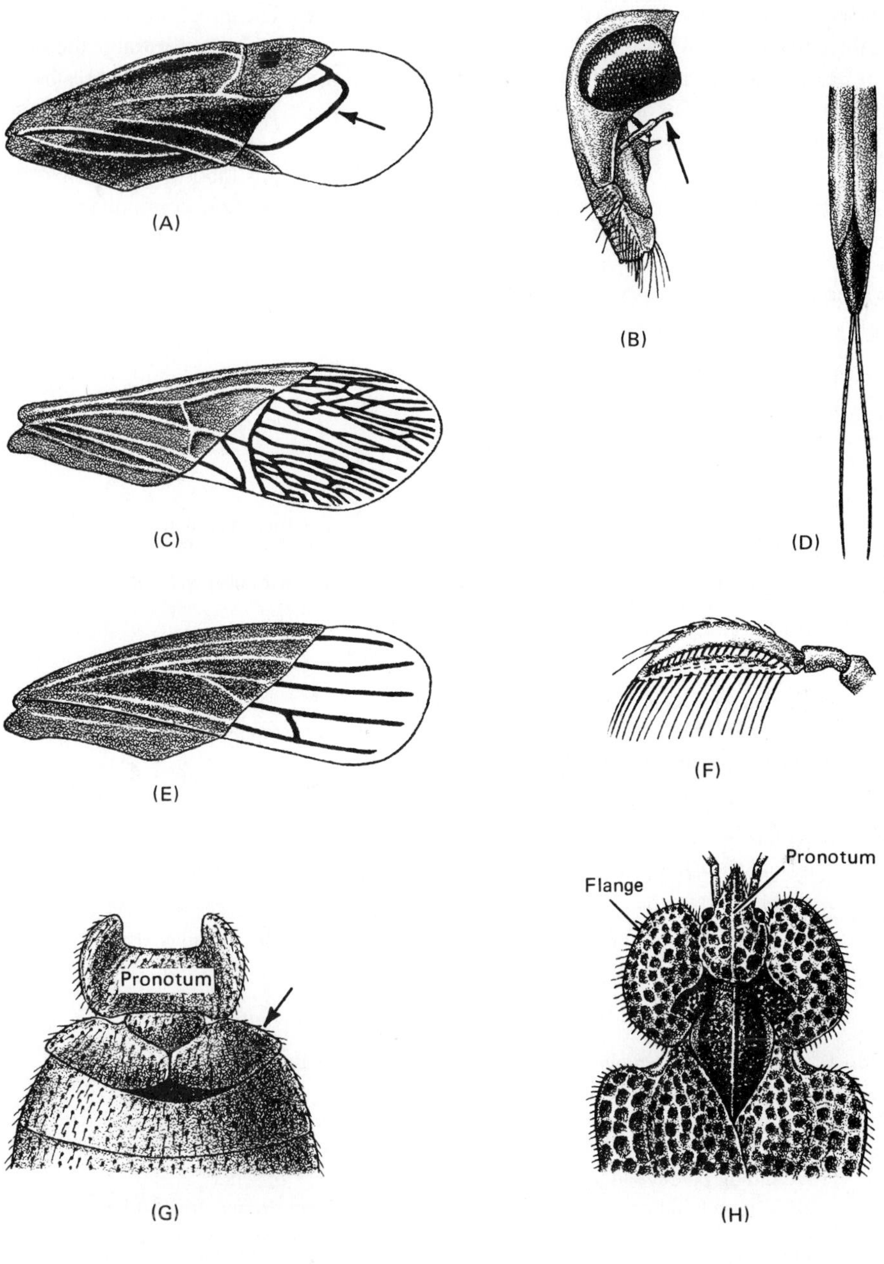

Figure 335. Diagnostic characters of Hemiptera. (A) loop veins in hemelytron (Miridae); (B) Antenna reduced (Corixidae); (C) many veins in membrane of hemelytron (Coreidae); (D) long caudal appendages (Nepidae); (E) few veins in membrane of hemelytron (Lygaeidae); (F) foreleg spoonlike (Corixidae); (G) wings scalelike (Cimicidae); (H) thorax with lacelike pattern (Tingidae).

most of these bugs belong to the genus *Triatoma* (Fig. 236). The infection may be transmitted to humans by rubbing the protozoan organism in *Triatoma* feces through the skin by scratching. Because of modern insecticides, bedbugs (Fig. 235) have nearly been eliminated as a pest of humans. A few species of Hemiptera, such as the giant water bug and backswimmer, bite severely and should be avoided.

This large order contains 44 families with 25,000 species (about 4,000 in North America). The following key will expedite identification of the more common families.

KEY TO NYMPHS AND ADULTS OF THE COMMON FAMILIES OF HEMIPTERA

1. Hind leg without tarsal claws 2
   Hind leg with tarsal claws 3

2. Front pair of legs spoon-shaped [Fig. 335(F)], lacking tarsal claws. . . . . (water boatman) CORIXIDAE*
   Front pair of legs not spoon-shaped, with tarsal claws . . . . (backswimmer) NOTONECTIDAE*

3. Antennae easily seen, not reduced 7
   Antennae reduced and difficult to locate [Fig. 335(B)] 4

4. Abdomen with long slender caudal appendage [Fig. 335(D)] . . . . (water scorpion) NEPIDAE*
   Abdomen lacking long slender caudal appendage 5

5. Hind legs fringed with long hair 6
   Hind legs not fringed with long hair; toadlike shape . . . . (toad bugs) GELASTOCORIDAE

6. Eyes protruding; fore tarsi with claws . . . . . . (giant water bugs) BELOSTOMATIDAE*
   Eyes not protruding; fore tarsi lacking. . . . . (creeping water bugs) NAUCORIDAE*

7. Claws of at least prothoracic legs subapical in position. 8
   Claws in normal apical position 9

8. Prothoracic coxae widely separated from mesothoracic coxae; hind femur extending far beyond tip of abdomen [Fig. 332(E)]. . . . . (water striders) GERRIDAE*
   All coxae usually equally separated; if not, then hind femur not extending far beyond tip of abdomen and fanlike fringe of setae present on middle tarsi . . . . (broad-shouldered water striders) VELIIDAE*

9. Adults wingless or front wings reduced to short scales [Fig. 335(G)]; nymphs and adults broad and flattened . . . . (bedbugs) CIMICIDAE
   Adults with wings fully developed; body not usually broadly flattened (if broad, then antennae five-segmented or body of nymph spinose) 10

10. Pronotum with broad flanges [Fig. 335(H)]; lacelike pattern in wings and pronotum; nymphs with large spines . . . . (lace bugs) TINGIDAE
    Not as above 11

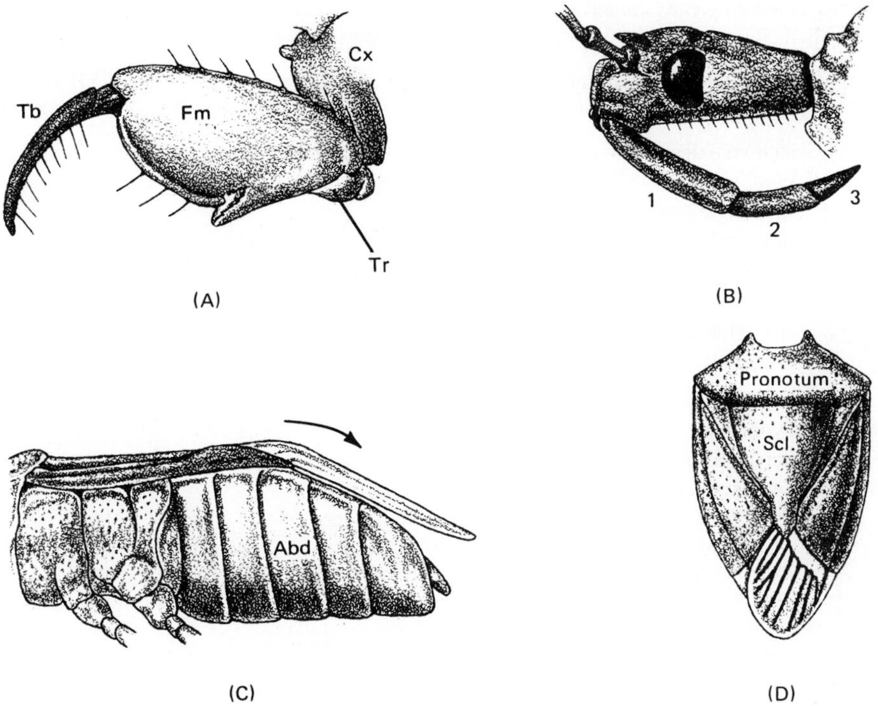

Figure 336. Diagnostic characteristics of Hemiptera. (A) thick femur (Phymatidae); (B) beak three-segmented (Reduviidae); (C) wings tilted downward (Miridae); (D) scutellum greatly enlarged (Pentatomidae).

11. Front femora cigar-shaped and long, with dense fine hair and 4 or 5 evenly spaced long black hairs on ventral surface; front tibiae with 2 rows of black spines on posterior surface . . . . (damsel bugs) NABIDAE
    Not as above 12

12. Adults with loop veins in membrane of hemelytra [Fig. 335(A)] and/or raptorial front legs; nymphs with either raptorial front legs *or* 0 to 1 pair of minute gland openings on abdominal terga 13
    Adults lacking both loop veins in membrane of hemelytra [Fig. 335(C)] and raptorial front legs; nymphs lacking raptorial front legs but with 2 or 3 pairs of tergal gland openings on abdomen 14

13. Femur of front legs very thick, triangular-shaped [Fig. 336(A)] . . . . (ambush bugs) PHYMATIDAE
    Femur of front legs not thick or triangular 15

14. Head elongate, beak usually short and curved, three-segmented [Fig. 336(B)], front wings in adult not tilted downward . . . . (assassin bugs) REDUVIIDAE
    Head not elongate; beak essentially straight, four-segmented; front wings in adult tilted at distinct angle posteriorly to abdomen [Fig. 336(C)] . . . . (plant bugs) MIRIDAE

15. Scutellum greatly enlarged in adults [Fig. 336(D)]; antennae five-segmented 16
    Scutellum not greatly enlarged in adults; antennae four-segmented 17

16. Tibia with many strong spines . . . . (cydnid bugs) CYDNIDAE
    Tibia lacking strong spines. 18

17. Scutellum extending to posterior tip of abdomen in adults;
    pronotum without lateral toothlike projections;
    first tergal gland opening on abdomen
    of nymph far apart . . . . (shield-backed bugs) SCUTTELLERIDAE
    Scutellum not extending to posterior tip of abdomen [Fig. 336(D)];
    pronotum with lateral toothlike projections;
    first tergal gland opening on abdomen
    of nymph close together. . . . . (stink bugs) PENTATOMIDAE

18. Front wing membrane in adults with 4 to 6 veins [Fig. 335(E)];
    nymphs with small tergal gland openings,
    not on raised areas. . . . . (lygaeid bugs) LYGAEIDAE
    Front wing membrane in adults with more than 10 veins
    [Fig. 335(C)]; tergal gland openings in nymphs large,
    usually on raised areas. . . . . (coreid bugs) COREIDAE

*Aquatic.

HOMOPTERA

The Homoptera order (Fig. 337) includes cicadas (Fig. 338), leafhoppers, treehoppers, spittlebugs, aphids, psyllids, and scales. Sizes range from minute to large (from 0.3 mm to 80 mm) with United States species, except the cicadas, measuring less than 12 mm. Diagnostic characteristics include large compound eyes, piercing-sucking mouthparts with the beak arising near the prothoracic legs and extending posteriorly along the sternum, setaceous or filiform antennae, one- to three-segmented tarsi, wings absent or two pairs (each of uniform texture) folded either rooflike or flat over the body. Cerci are absent. Females often possess an appendicular ovipositor [Fig. 63(B)]. Sound producing structures may be present.

These insects undergo incomplete metamorphosis, but life cycles may be very complex, involving winged and wingless stages. Many aphids reproduce parthenogenetically and are viviparous during the summer but oviparous when fall arrives. Scales and aleyrodids undergo a metamorphosis intermediate between incomplete and complete since a pupal stage occurs in most males. Completion of life cycles varies from less than a week in some aphids to 17 years in a few cicadas (Figs. 339, 340). With the exception of cicadas, development takes place above ground.

All Homoptera are herbivorous and include many pests. Species of leafhoppers, whiteflies, scales, and aphids are especially detrimental to a wide variety of crops. Leafhoppers and aphids (Fig. 341) transmit over two thirds of the known viral plant diseases of economic importance. Some species have become symbiotic with ants.

These insects may be consolidated as a suborder in Hemiptera in some classifications or may be elevated to two or three orders in others. More than 40,000 (about

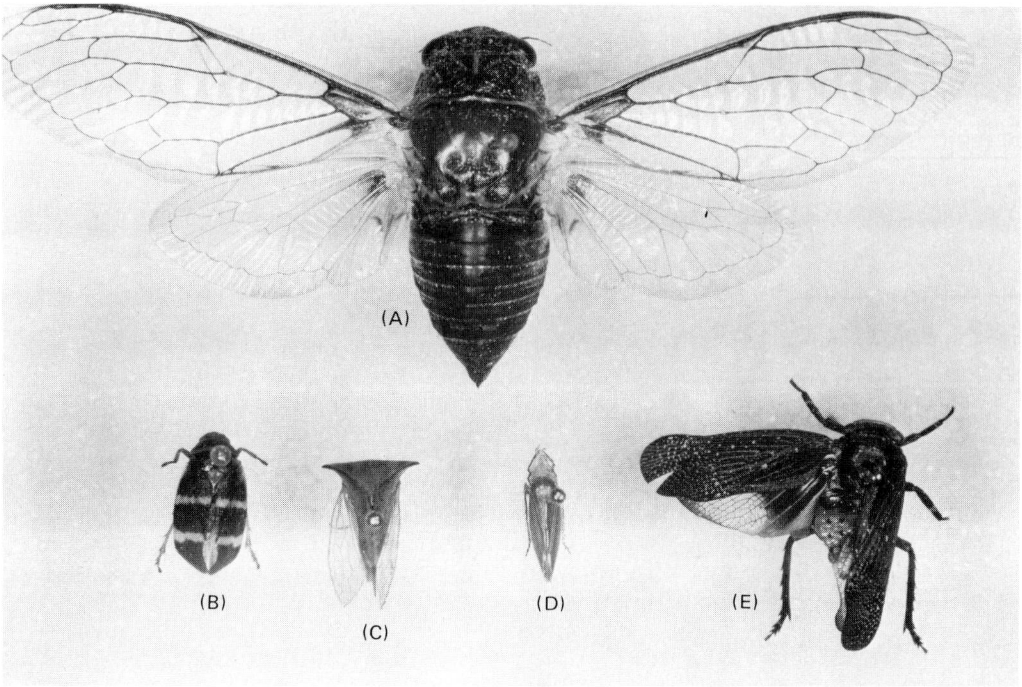

Figure 337. Examples of Homoptera. (A) cicada (Cicadidae); (B) spittlebug (Cercopidae); (C) treehopper (Membracidae); (D) leafhopper (Cicadellidae); (E) planthopper (Fulgoridae).

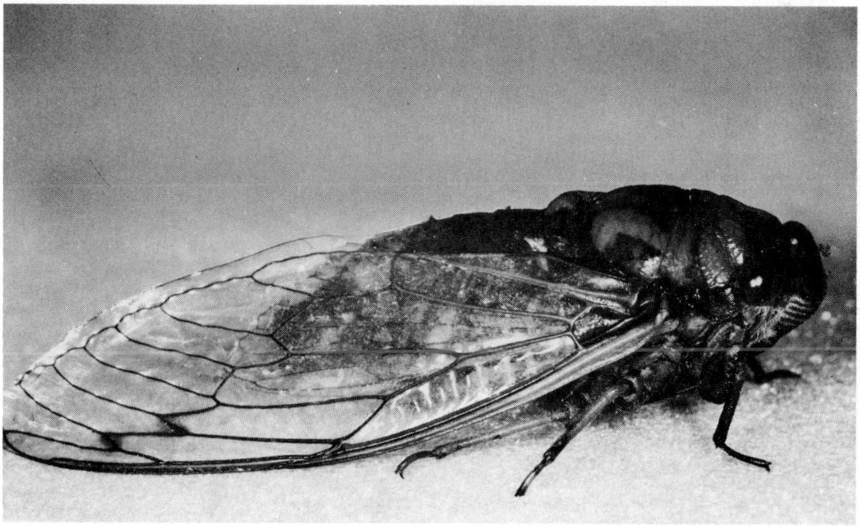

Figure 338. An adult cicada (*Tibicen pruinosa*).

Figure 339. A cicada nymph (*Tibicen pruinosa*) with its large digging front pair of legs.

Figure 340. The emergence of an adult cicada (*Tibicen pruinosa*) from the exoskeleton of the last instar nymph.

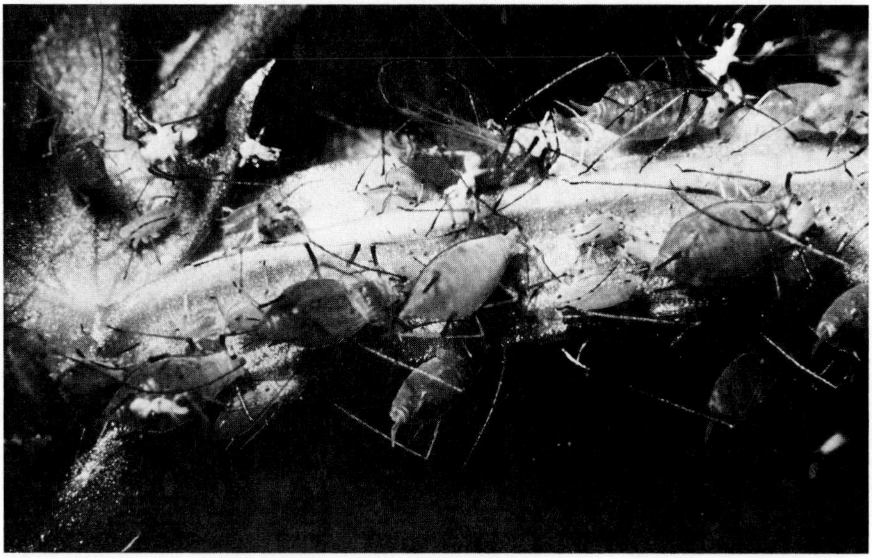

Figure 341. An aphid (*Macrosiphum pisi*). Both winged and wingless individuals are present.

7,000 in North America) species are known, and these are grouped into about 32 families (varies with classification). The most common families may be determined by using the following key.

KEY TO THE COMMON NYMPHS AND ADULTS OF THE HOMOPTERA

1. Setaceous antennae [Fig. 342(A)]; tarsi three-segmented 2
   Filiform antennae; tarsi one- or two-segmented 12

2. Adults larger than 1/2 in. (1.27 cm); 3 ocelli; nymphs
   with digging front legs . . . . (cicadas) CICADIDAE
   Adults shorter than 1/2 in. (1.27 cm); 2 ocelli; legs never for digging 3

3. Pronotum enlarged dorsally [Fig. 342(C)]; dorsal thoracic
   and abdominal spines often in nymphs . . . . (treehoppers) MEMBRACIDAE
   Pronotum not enlarged dorsally 4

4. Antennae arise beneath or behind eyes [Fig. 342(D)] 5
   Antennae arise between eyes 11

5. Large flattened movable spur at hind tibial apex
   [Fig. 343(A)]. . . . . (delphacid planthoppers) DELPHACIDAE
   No such spur present 6

6. Second tarsal segment of hind leg small, its apex rounded
   with single tooth or none; adult body often wedge-shaped 7
   Second tarsal segment of hind leg long, its apex truncate or emarginate
   and with row of small spines; adult body not wedge-shaped 9

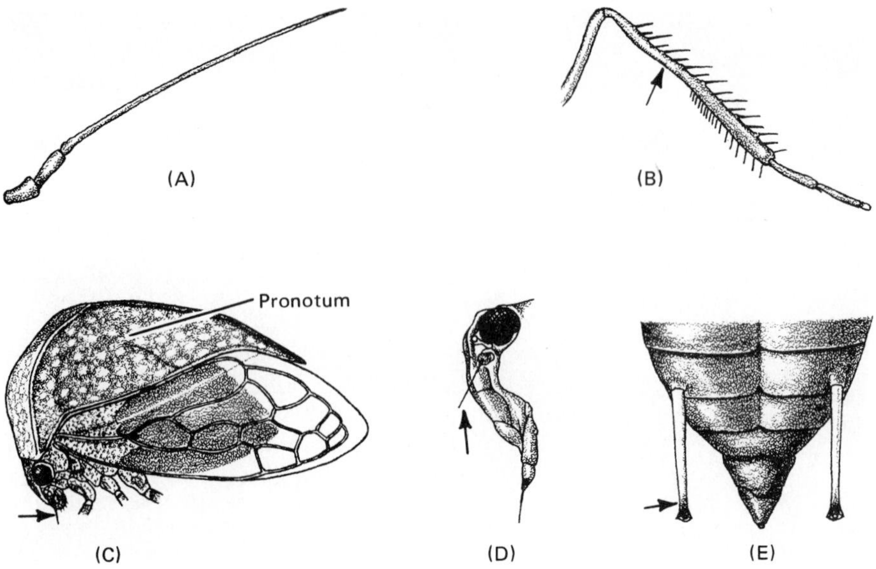

Figure 342. Diagnostic characters of Homoptera. (A) setaceous antenna (Cicadellidae); (B) hind tibia with row of spines (Cicadellidae); (C) pronotum enlarged dorsally (Membracidae); (D) antenna below eye (Flatidae); (E) cornicles (Aphididae).

7. Hind tibiae with stout lateral spines [Fig. 343(A)]. 8
   Hind tibiae lacking stout lateral
   spines. . . . . (acanaloniid planthoppers) ACANALONIIDAE

8. Wings held nearly vertical; costal area of wings with numerous
   cross veins; nymphal wax glands present on abdominal
   terga when viewed caudally . . . . (flatid planthoppers) FLATIDAE
   Wings not held vertically; costal area of wings lacking numerous
   cross veins; nymphal wax glands absent . . . . (issid planthoppers) ISSIDAE

9. Head either prolonged into elongate point or with 2 or 3 median carinae
   [Fig. 343(B)]. . . . . (dictyopharid planthopper) DICTYOPHARIDAE
   Head not as above 10

10. Front wings overlap terminally when folded; body flattened;
    wax-bearing glands in nymphal abdominal terga
    moderate-sized and widely spaced. . . . . (achilid planthoppers) ACHILIDAE
    Front wings not overlapping terminally when folded; wax-bearing glands
    large and nearly touching [Fig. 343(C)] . . . . (cixiid planthoppers) CIXIIDAE

11. Hind tibia with 1 or more rows of numerous
    small spines [Fig. 342(B)] . . . . (leafhoppers) CICADELLIDAE
    Hind tibia only with 0 to 2 stout lateral spines. . . . . (spittlebugs) CERCOPIDAE

12. Abdomen usually with elongate cornicles [Fig. 342(E)];
    body rounded to pear-shaped . . . . (aphids) APHIDIDAE
    Abdomen lacking or with reduced cornicles; body shape various,
    usually covered with waxy material 13

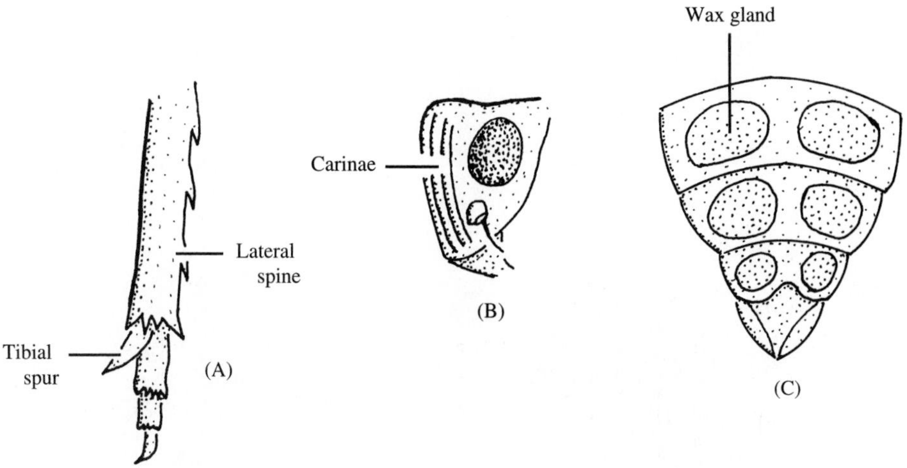

Figure 343. Diagnostic characteristics of Homoptera. (A) hind tibia with lateral spines and tibial spur (Delphacidae); (B) lateral carinae (Dictyopharidae); (C) dorsal view of abdomen with nymphal wax glands (Cixiidae).

13. Antennae three- to seven-segmented 14
    Antennae ten-segmented . . . . (psyllids) PSYLLIDAE

14. Tarsi two-segmented 15
    Tarsi one-segmented or legs absent. . . . . scales, superfamily COCCOIDEA

15. Antennae three-segmented . . . . (phylloxerans) PHYLLOXERIDAE
    Antennae six- or seven-segmented. 16

16. Antennae six-segmented; wings membranous in winged forms with front wings longer than hind wings; nymphs covered with white wooly or waxy material . . . . (wooly aphids) ERIOSOMATIDAE
    Antennae seven-segmented; wings opaque and whitish, all of equal length; nymphs often covered by black material . . . . (whiteflies) ALEYRODIDAE

MEGALOPTERA

Dobsonflies [Fig. 344(A)] and alderflies are characterized by having filiform antennae, large widely separated compound eyes, prognathic chewing mouthparts that may be greatly elongate in some males, long legs with five-segmented tarsi, and no cerci. Their two pairs of different-sized wings have numerous wing veins and cross veins (particularly at the leading edge of the front wings), and an anal fold is present in the hind wing.

Metamorphosis is complete. Up to 3,000 eggs are often deposited in compact masses on streamside vegetation or rocks. Hatching larvae drop into the water and often locate themselves in rapids. The larvae possess lateral abdominal appendages and gills (Fig. 345), have strong chewing mouthparts, and are voracious predators. The pupal stage (Fig. 139) is short in duration and takes place outside of the water in a

Figure 344. Examples of some minor insect orders. (A) Megaloptera (dobsonfly); (B) Neuroptera (green lacewing); (C) Neuroptera (ant lion); (D) Trichoptera (caddisfly); (E) Mecoptera (scorpionfly).

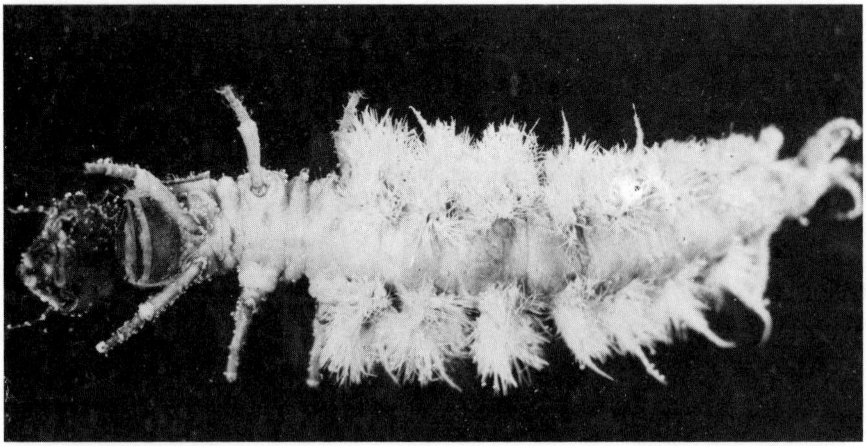

Figure 345. A larval dobsonfly (*Corydalus cornutus*), often termed *hellgrammite.* Its aquatic existence may be deduced by the abdominal gills. Food consists of mayfly and stonefly nymphs and caddisfly larvae.

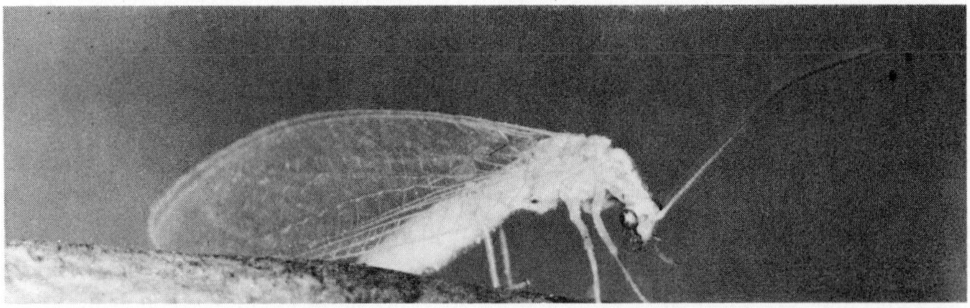

Figure 346. An adult green lacewing (*Chrysopa carnea*).

TABLE 13. Major Families of Neuroptera Listing Adult and Larval Diet and Larval Habitat

Family	Adult Diet	Larval Diet	Larval Habitat
Ascalaphidae	Predator	Predator	Vegetation or soil
Chrysopidae	Predator	Predator	Vegetation
Hemerobiidae	Predator	Predator	Vegetation
Mantispidae	Predator	Parasite	Wasps and spiders
Myrmeleontidae	Predator if feeds	Predator	Sandy soil

concealed cell under a stone, log, or other hiding place. Development from egg to adult requires 1 to 5 years. Adults rarely feed.

Approximately 260 species are known (43 in North America), and these species are normally placed in two families. The adults of Corydalidae (dobsonflies) are large (over 400 mm), and the larvae have terminal hooks at the end of the abdomen [Fig. 305(A)], whereas the adult Sialidae (alderflies) are less than 370 mm in length, and the larvae have an elongate terminal segment without hooks. This order is sometimes grouped into the Neuroptera.

No species are of direct economic or medical importance to humans, but their presence normally indicates good water quality.

NEUROPTERA

Neuroptera [Figs. 162, 344(B), (C)] are characterized by having large compound eyes widely separated, antennae usually filiform, chewing mouthparts, long legs with five-segmented tarsi, no cerci, and two pairs of usually many-veined membranous wings that are similar in size and are usually held rooflike over the body when at rest (Fig. 346).

Metamorphosis is complete. Larvae are usually carnivorous and terrestrial. The mandible and maxilla on each side fuse to form hollow tubes [Fig. 305(B)] for sucking fluids from prey. Some feed in the soil and may produce cone-shaped depressions in loose sand to trap insects (Figs. 250, 347), and others are found on vegetation (Fig. 251). The rarely seen mantispid larvae are often parasitic on ground spider eggs. A summary of adult diet and larval habitat and diet by family may be seen in Table 13.

Approximately 5,000 species are known (about 300 in North America), and these are placed into 11 to 15 families. Some authors include the Megaloptera and Raphidiodea (snakeflies) within this order. A few species are beneficial to humans, including green lacewings that feed on such pests as aphids and other Homoptera. None appears to be detrimental.

The following keys, one for the adults and the other for larvae, will permit the identification of the more common families of Neuroptera.

KEY TO THE COMMON FAMILIES OF ADULT NEUROPTERA

1. Antennae clubbed [Fig. 344(C)] 2
   Antennae not clubbed (Fig. 346) 3

2. Antennae more than half as long as first pair of wings . . . . (owlflies) ASCALAPHIDAE
   Antennae less than half the length of first pair of wings [Fig. 344(C)] . . . . (ant lions) MYRMELEONTIDAE

3. Raptorial forelegs (Fig. 162) . . . . (mantispids) MANTISPIDAE
   Forelegs similar to other legs 4

4. Wings and body usually greenish (overwintering forms and pinned specimens often turn yellow); costal wing veins not divided . . . . (green lacewings) CHRYSOPIDAE
   Wings and body usually brown to black; some costal wing veins branched or forked . . . . (brown lacewings) HEMEROBIIDAE

KEY TO THE COMMON FAMILIES OF LARVAL NEUROPTERA

1. Abdomen with long lateral projections or appendages; posterior of head bilobed . . . . ASCALAPHIDAE
   Abdomen usually lacking such projections or appendages; if present, then posterior of head not bilobed 2

2. Teeth on mandible [Fig. 347(B)] . . . . MYRMELEONTIDAE
   Teeth absent on mandible 3

3. Trumpet-shaped empodium present on each leg . . . . CHRYSOPIDAE
   Trumpet-shaped empodium absent on each leg . . . . HEMEROBIIDAE

Figure 347. Ant lion larvae are found in sandy soil. (A) the cone-shaped depressions are constructed to facilitate the prey slipping to the bottom where the larva waits. (B) the larva with its sickle-shaped mouthparts used in capturing and sucking fluids from the prey.

COLEOPTERA

Beetles (Figs. 348, 349) vary in length from 0.25 mm to 150 mm and include the largest insects by biomass known (to 40 g, 20 times that of the smallest shrew). Beetles are characterized by having chewing mouthparts and large compound eyes and by lacking cerci and an appendicular ovipositor. Their thickened forewings or *elytra,* which meet at the midline when folded, are a representative characteristic of the order. The membranous hindwings are the primary flight organs (Fig. 56) and fold up under the elytra when not in use, thereby protecting them from damage and permitting beetles to inhabit many rugged environments without endangering their flight capability. Closed elytra also produce a cavity above the abdominal terga in which air can be stored and utilized because spiracles open into this chamber. Antennae are highly variable. Tarsi are primitively five-segmented, but the number of segments may be less; these differences are used in identification. Bodies are normally stout and sclerotization is extensive (Figs. 350, 351). Sexual dimorphism is sometimes obvious (Fig. 352), but more often sex can be determined only with difficulty. Some of the largest and smallest insects are beetles.

Metamorphosis is complete. Larvae are highly diverse in structure (Figs. 353, 354), even though the head is always well sclerotized and the mouthparts are chewing. Some species are heavily sclerotized (click beetles, ground beetles), whereas others have poor sclerotization (scarabs and weevils). Larvae of many beetles, such as predaceous diving beetles, whirligig beetles, and water scavenger beetles, inhabit water, but most are terrestrial. A few are legless (borers). Pupae are exarate. Several years may be required for development.

A summary of diet and habitats of the more common beetles is given in Table 14.

The larval stage is often the most destructive, but many adults are also of economic importance. Stored products are attacked by the rice weevil (*Sitophilus oryzae*), yellow mealworm (*Tenebrio molitor*), confused flour beetle (*Tribolium confusum*), red flour beetle (*Tribolium castaneum*), carpet beetle (*Anthrenus scrophulariae*), and the larder beetle (*Dermestes lardarius*). The boll weevil (*Anthonomus grandis*) is a great pest on cotton, and the many rootworm species (*Diabrotica* sp.) can cause much damage to corn. Vegetables may be greatly affected by the striped cucumber beetle (*Acalymma vittatum*), Colorado potato beetle (*Leptinotarsa decemlineata*), Mexican bean beetle (*Epilachna varivestis*), and the bean weevil (*Acanthoscelides obtectus*). The Japanese beetle (*Popillia japonica*) is one of the more serious pests of plants in the eastern United States. The larvae feed on roots of grasses and the adults on leaves of a variety of plants. It should be noted, however, that a few species, including the lady beetles, are highly beneficial as predators of other detrimental insects such as aphids.

The order Coleoptera includes approximately 40 percent of all species of insects and 30 percent of all animals. An estimated 350,000 species are known in the world (about 30,000 in North America), and they are classified into about 160 families (about 120 in the United States). Many families contain only a few species, whereas others have many. More than two thirds of the world's beetles are classified into the following eight families: Curculionidae (50,000), Chrysomelidae (35,000), Staphylinidae (30,000), Cerambycidae (26,000), Carabidae (25,000), Scarabaeidae (20,500), Tenebrionidae (20,000), and Buprestidae (13,000). The following two keys will be useful in identifying

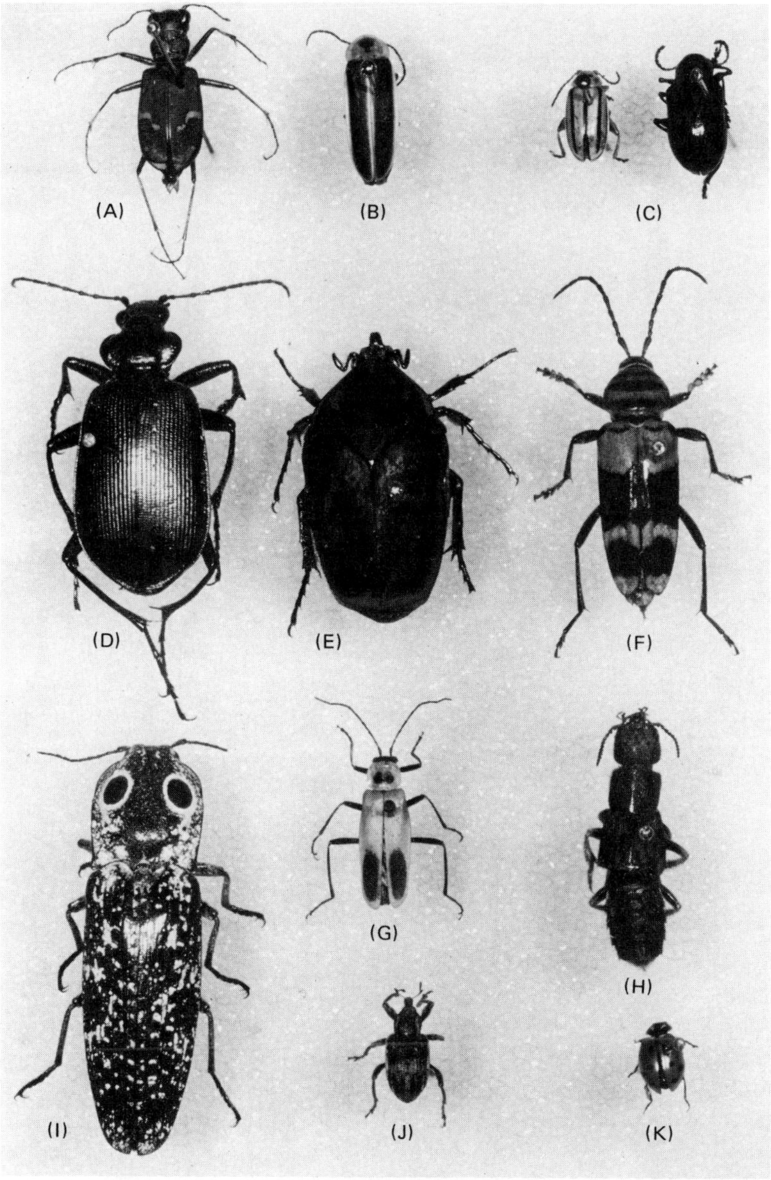

Figure 348. Examples of Coleoptera. (A) tiger beetle (Cicindelidae); (B) firefly (Lampyridae); (C) leaf beetles (Chyrsomelidae); (D) ground beetle (Carabidae); (E) scarab beetle (Scarabeidae); (F) long-horned beetle (Cerambycidae); (G) soldier beetle (Cantharidae); (H) rove beetle (Staphylinidae); (I) click beetle (Elateridae); (J) weevil (Curculionidae); (K) lady beetle (Coccinellidae).

Figure 349. Examples of Coleoptera. (A) carrion beetle (Silphidae); (B) metallic wood borer (Buprestidae); (C) passalid beetle (Passalidae); (D) blister beetle (Meloidae); (E) stag beetle (Lucanidae); (F) darkling beetle (Tenebrionidae); (G) predaceous diving beetle (Dytiscidae); (H) water scavenger beetle (Hydrophilidae); (I) whirligig beetle (Gyrinidae); (J) dermestid beetle (Dermestidae).

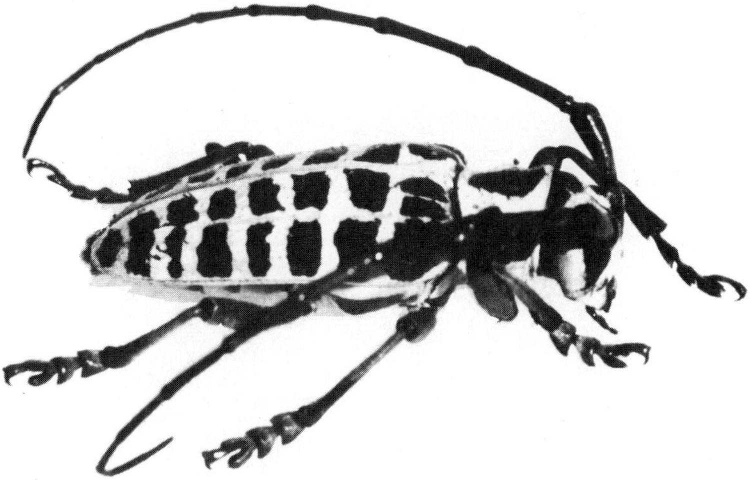

Figure 350. Cottonwood borer adult (*Plectrodera scalator*).

Figure 351. Ground beetle adult (*Harpalus caliginosus*).

Figure 352. Dimorphic appearance of dung beetle adults. Female is at the left and male at the right.

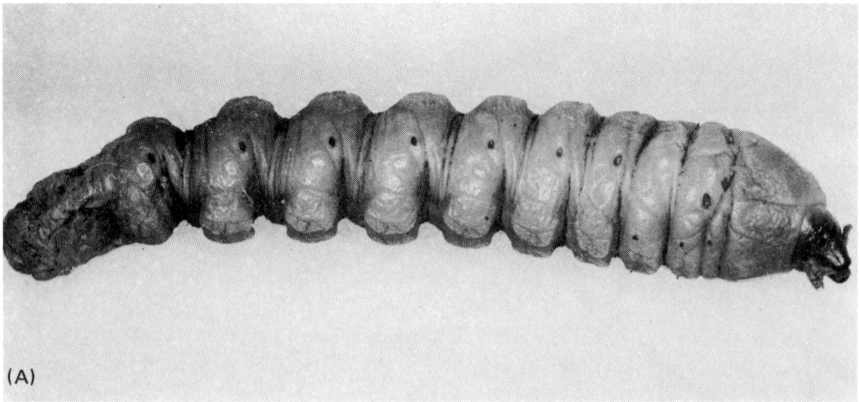

(A)

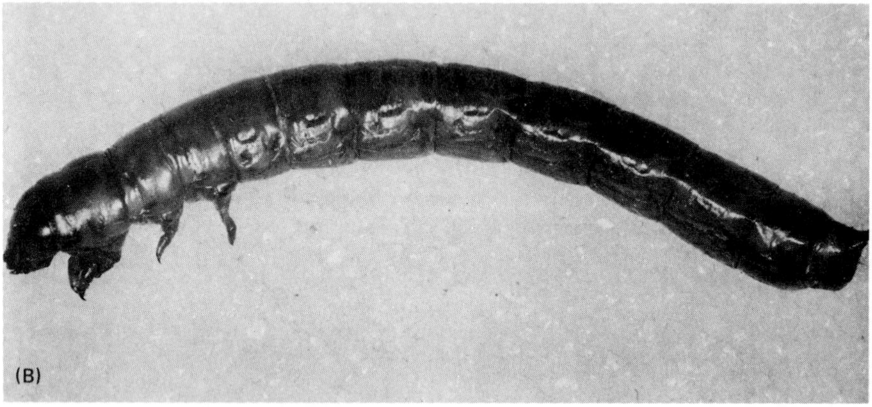

(B)

(C)

Figure 353. Examples of beetle larvae. (A) Cerambycidae; (B) Tenebrionidae; (C) Scarabeidae.

(A)

(B)

(C)

Figure 354. Examples of beetle larvae. (A) Carabidae; (B) Chrysomelidae; (C) Hydrophilidae.

adults and larvae of these and other common families of this order (more than 80 percent of the species of United States beetles are classified into these families). Less common beetles will key out incorrectly because they will not be included within the key; however, their absence reduces the difficulty in identifying the common forms.

KEY TO THE COMMON FAMILIES OF ADULT COLEOPTERA

1. Each eye divided into dorsal and ventral halves [Fig. 355(D)];
   second and third pair of legs short . . . . (whirligig beetles) GYRINIDAE*
   Eyes not divided; legs not short 2

2. Large platelike hind coxae that cover most of
   abdomen . . . . (crawling water beetles) HALIPLIDAE*
   Coxae not greatly enlarged. 3

TABLE 14. Families of Coleoptera Listing Adult and Larval Diet and Habitat

Family	Adult Diet	Adult Habitat	Larval Diet	Larval Habitat
Bostrichidae	Unknown	Vegetation, fungi	Wood	Wood borers
Bruchidae	Unknown	Vegetation	Seeds	Seed containers
Buprestidae	Herbivore, pollen	Vegetation	Wood	Wood borers
Cantharidae	Nectar, pollen	Vegetation, flowers	Predator, some herbivores	Litter and soil
Carabidae	Predator, some herbivores	Soil, under rocks	Predator	Soil, under rocks
Cerambycidae	Nectar, pollen	Vegetation, flowers	Wood	Wood borers
Chrysomelidae	Leaf matter	Vegetation	Leaves	Leaves, some roots
Cicindellidae	Predator	Soil surface	Predator	Soil in burrows
Cleridae	Predator, nectar	Foliage, bark	Predator, some omnivores	Under bark
Coccinellidae	Predator, few herbivores	Leaves	Predator, few herbivores	Vegetation
Cucujidae	Predator or seed eater	Vegetation, bark	Predator, seeds, wood, fungus	Under bark, seeds, storage bins
Curculionidae	Herbivore	Vegetation	Herbivorous	On or within vegetation, soil
Dermestidae	Scavenger	Vegetation, hidden areas	Scavenger	On or within dead matter, stored products
Dytiscidae	Predator	Water	Predator	Water
Elateridae	Unknown or nonfeeding	Variety of sites	Herbivore, some predators	Soil or in wood
Elmidae	Herbivore	Water	Herbivore, some predators and scavengers	Water
Gyrinidae	Scavenger	On water surface	Predator	Water
Haliplidae	Herbivore	Water	Herbivore	Water

Continued

TABLE 14. Families of Coleoptera Listing Adult and Larval Diet and Habitat—*Cont'd*

Family	Adult Diet	Adult Habitat	Larval Diet	Larval Habitat
Helodidae	Unknown	Water	Microorganisms	Water
Histeridae	Predator	Soil, dead animals and plants, animal burrows	Predator	Under bark and dead matter, rodent nests
Hydrophilidae	Scavenger, herbivore	Water	Predator	Water
Lampyridae	Nonfeeding, some predation	Vegetation	Predator	Under debris, rocks, soil
Lucanidae	Vegetation exudates	Vegetation, under debris or logs	Wood	In logs and stumps
Lycidae	Predator	Vegetation	Predator	Under bark, debris
Melandryidae	Unknown	Vegetation, under bark or fungi	Predator or fungi eaters	Under bark, fungi, or in dry wood
Meloidae	Herbivore or nectar feeder	Vegetation, flowers	Parasite	Grasshopper eggs, bee nests
Mordellidae	Flowers and nectar	Flowers	Predator	In wood burrows
Passalidae	Wood scavenger	In logs	Wood	In decaying wood
Psephenidae	Herbivore	Stream banks	Algae	Streams
Ptinidae	Unknown	Stored products	Scavenger	Stored products, dead plants and animals
Pyrochroidae	Unknown	Vegetation, bark	Predator	Under bark
Scarabaeidae	Herbivore, scavenger	Vegetation, soil, dung, flowers	Herbivore, dung feeders, rotting wood	On leaves, in soil or dung
Scolytidae	Herbivore, tree exudates	Under bark	Wood	Under bark
Silphidae	Scavenger	Under carrion	Scavenger, few predators and herbivores	Under carrion usually
Staphylinidae	Scavenger, some predators	Soil, under carrion or rocks	Scavenger, some predators and herbivores	Soil, under rocks and logs
Tenebrionidae	Scavenger, some seed eaters	Soil, under rocks, stored products, wood	Scavenger	Soil, under rocks, stored products, wood

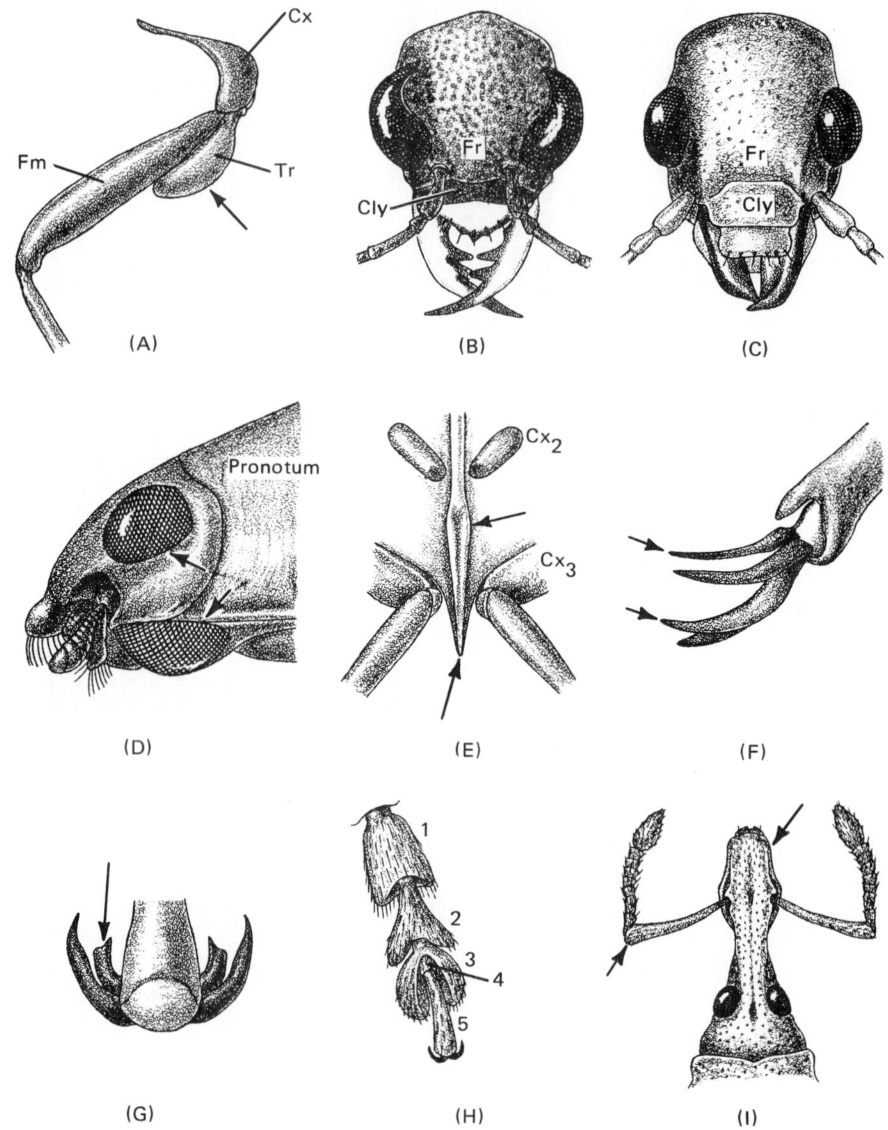

Figure 355. Diagnostic characteristics of adult Coleoptera. (A) hind trochanter with greatly enlarged lobes (Cicindelidae); (B) clypeus wide (Cicindelidae); (C) clypeus narrow (Carabidae); (D) eye divided into two halves (Gyrinidae); (E) sternal keel (Hydrophilidae); (F) split claws (Meloidae); (G) toothed claw (Cerambycidae); (H) tarsal formula appears to be four but is actually five (Chrysomelidae); (I) head prolonged, antenna elbowed (Curculionidae).

3. Legs modified for swimming [Fig. 61(D)]. 4
Legs not obviously modified for swimming 5

4. Sternal keel present [Fig. 355(E)] . . . . . (water scavenger beetles) HYDROPHILIDAE*
Sternal keel absent. (predaceous diving beetles) DYTISCIDAE*

5. Head usually prolonged into a snout [Fig. 355(I)];
antennae elbowed . . . . (weevils) CURCULIONIDAE
Head not prolonged into a snout; antennae usually not elbowed. 6

6. Less than 10 mm long; enlarged tibiae;
black and shiny . . . . (hister beetles) HISTERIDAE
Not with above combination of characters. 7

7. Tarsal formula 5-5-5. 18
Tarsal formula either less *or* appearing to be less than 5-5-5;
i.e., one segment may be greatly reduced [Fig. 355(H)] 8

8. Tarsal formula 5-5-4. 9
Tarsal formula not 5-5-4. 14

9. Each tarsal claw split longitudinally and appearing
to be double [Fig. 355(F)]. . . . . (blister beetles) MELOIDAE
Tarsal claws not split longitudinally [Fig. 355(G)] 10

10. Prothoracic coxal cavities closed
posteriorly. . . . . (darkling beetles) TENEBRIONIDAE
Prothoracic coxal cavities open 11

11. Antennae filiform. 12
Antennae pectinate to plumose . . . . (fire-colored beetles) PYROCHROIDAE

12. First tarsal segment from tibia on hind leg
elongated . . . . (false darkling beetles) MELANDRYIDAE
First tarsal segment from tibia on hind leg not elongated 13

13. Body flattened; reddish or yellowish in color. . . . . (flat bark beetles) CUCUJIDAE
Body wedge-shaped; black or mottled gray
in color . . . . (false darkling beetles) MELANDRYIDAE

14. Tarsal formula appears to be 3-3-3
(actually 4-4-4) . . . . (lady beetles) COCCINELLIDAE
Tarsal formula appears to be 4-4-4 [Fig. 355(H)] (actually 5-5-5) 15

15. Antennae long, usually longer than body; eyes notched
[Fig. 356(A)]. . . . . (long-horned beetles) CERAMBYCIDAE
Antennae much shorter than length of body 16

16. Antennae with distinct elbow [Fig. 356(E)]
and capitate. . . . . (bark beetles) SCOLYTIDAE
Antennae lacking elbow. 17

17. Third tarsal segment bilobed [Fig. 355(H)] . . . . (leaf beetles) CHRYSOMELIDAE
Third tarsal segment normal in shape. . . . . (checkered beetles) CLERIDAE

18. Hind trochanters with greatly enlarged lobes [Fig. 355(A)] 19
Hind trochanters without such enlarged lobes 20

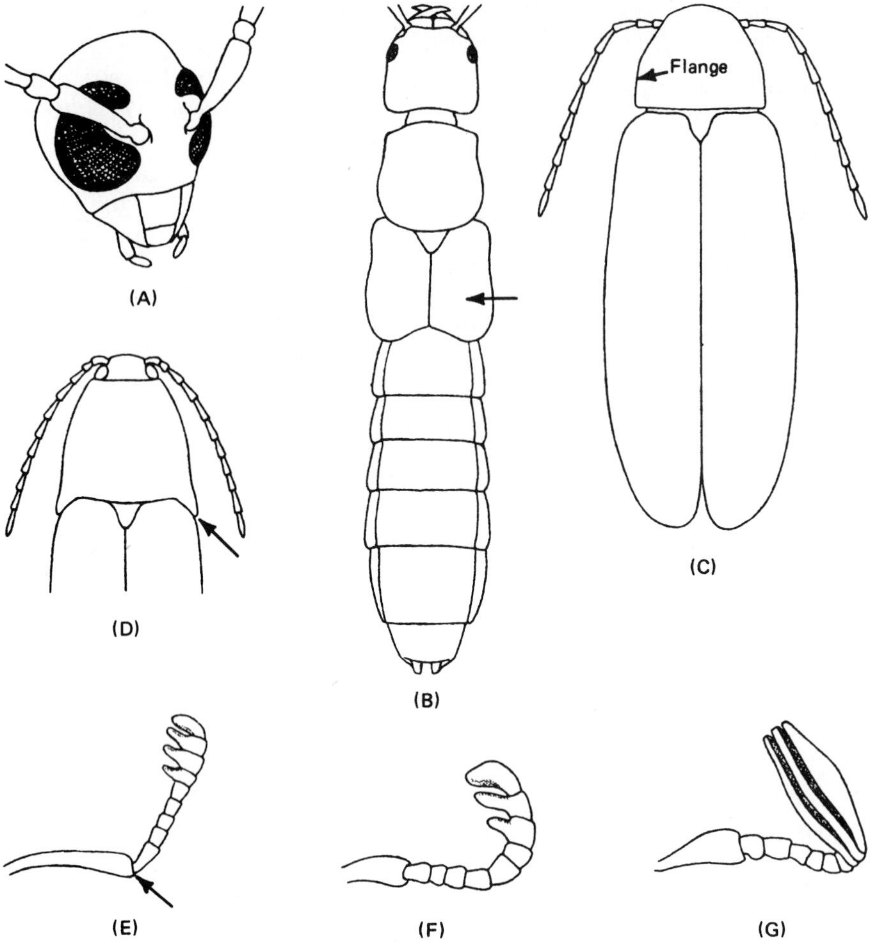

Figure 356. Diagnostic characteristics of adult Coleoptera. (A) notched eye (Cerambycidae); (B) elytra short and with five or more terga exposed (Staphylinidae); (C) pronotum enlarged and covering head (Lampyridae); (D) pronotum with sharp posterior angles to flange (Elateridae); (E) antenna elbowed but poorly lamellate (Lucanidae); (F) antenna not elbowed, poorly lamellate (Passalidae); (G) antenna not elbowed, highly lamellate (Scarabaeidae).

19. Clypeus wider than distance between antennal sockets [Fig. 355(B)]; head usually broader than prothorax .... (tiger beetles) CICINDELIDAE
    Clypeus narrower than distance between antennal sockets [Fig. 355(C)]; head usually narrower than prothorax............. (ground beetles) CARABIDAE

20. Elytra short, 3 or more terga exposed [Fig. 356(B)] 21
    Elytra covering the abdomen or only 1 or 2 segments exposed.................. 22

21. Three terga usually exposed........................ (carrion beetles) SILPHIDAE
    Five or more terga exposed [Fig. 356(B)] .... (rove beetles) STAPHYLINIDAE

22. Pronotum enlarged and with wide lateral flanges;
    head difficult to see from dorsal view [Fig. 356(C)] 33
    Pronotum without such widened flanges; head may be difficult
    to see from dorsal view in Bostrichidae 23

23. Pronotum prolonged posteriorly into long sharp
    angles at lateral corners [Fig. 356(D)] .... (click beetles) ELATERIDAE
    Pronotum without long sharp lateral angles [Fig. 356(C)] 24

24. Antennae elbowed [Fig. 356(E)]; mandibles greatly
    enlarged in males .... (stag beetles) LUCANIDAE
    Antennae not elbowed [Fig. 356(F), (G)] 25

25. Antennae highly lamellate [Fig. 356(G)] .... (scarab beetles) SCARABAEIDAE
    Antennae not lamellate or only poorly so [Fig. 356(F)] 26

26. Large, black, shiny [Fig. 349(C)];
    often "horn" on head .... (passalid beetles) PASSALIDAE
    Not as above 27

27. Body lightly sclerotized, especially elytra 28
    Body well sclerotized and hard 29

28. Elytra with reticulate sculpturing and ridges .... (net-winged beetles) LYCIDAE
    Elytra lacking such sculpturing .... (soldier beetles) CANTHARIDAE

29. Head deflected ventrally, usually not visible
    from dorsal view .... (false powderpost beetles) BOSTRICHIDAE
    Head not so positioned, although head may be difficult to see 30

30. Rounded body with long legs; legs with yellow hair .... (spider beetles) PTINIDAE
    Not as above 31

31. Most of ventral surface with dense white hair .... (dermestid beetles) DERMESTIDAE
    Little if any ventral white hair 32

32. Antennae not clubbed .... (metallic wood borers) BUPRESTIDAE
    Antennae clavate [Fig. 47(K)] .... (carrion beetles) SILPHIDAE

33. Antennae clubbed .... (carrion beetles) SILPHIDAE
    Antennae filiform .... (fireflies) LAMPYRIDAE

*Aquatic.

KEY TO THE MOST COMMON FAMILIES OF LARVAL COLEOPTERA

1. Thoracic legs absent [Fig. 353(A)] or reduced to either clawless
   segmented or nonsegmented rudiments 2
   Thoracic legs present, with claws [Fig. 353(C)] 6

2. Prothorax distinctly flattened dorsoventrally and much wider
   than abdominal segments .... (flatheaded wood borers) BUPRESTIDAE†
   Prothorax rounded and not greatly wider than other segments 3

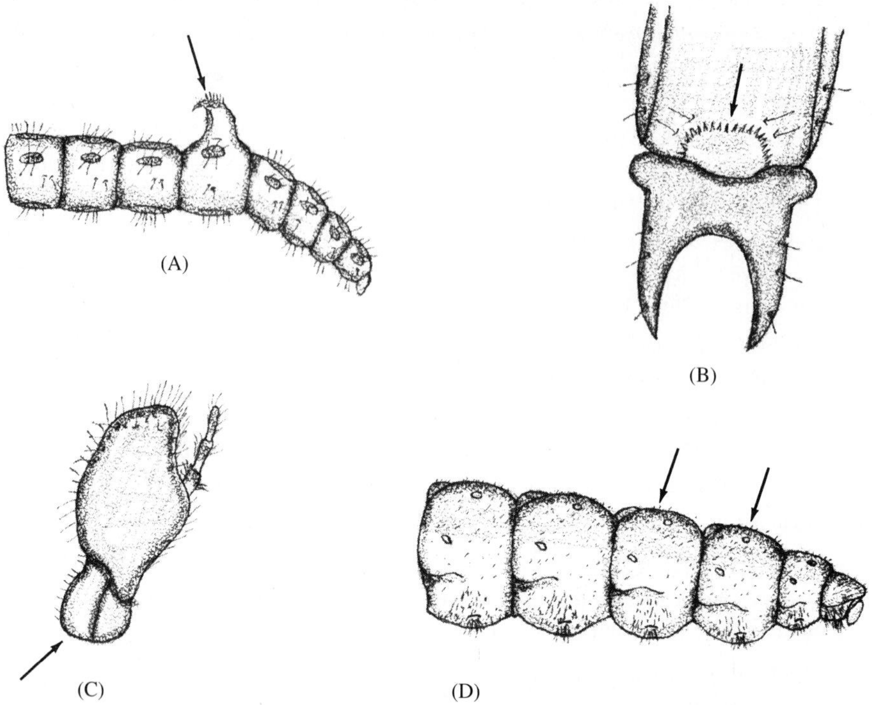

Figure 357. Diagnostic characteristics of Coleoptera larvae. (A) dorsal hooks (Cicindelidae); (B) aspergites (Pyrochroidae); (C) cardo divided into two parts (Melandryidae); (D) dorsal glandular openings (Cantharidae).

3. Mouthparts prognathous [Fig. 354(A)] 5
   Mouthparts hypognathous [Fig. 353(C)] 4

4. Ninth abdominal segment with unpaired spine [Fig. 67(D)] or paired solid urogomphi . . . . MORDELLIDAE†
   Ninth abdominal segment lacking spine or urogomphi . . . . SCOLYTIDAE and CURCULIONIDAE

5. Pronotum distinctly sclerotized; pronounced narrowing between abdominal segments [Fig. 353(A)] . . . . CERAMBYCIDAE†
   Not as above . . . . BRUCHIDAE

6. Legs five-segmented 7
   Legs four-segmented 11

7. Dorsal hooks on fifth abdominal segment [Fig. 357(A)]. . . . . CICINDELIDAE
   Dorsal hooks on fifth abdominal segment lacking. 8

8. Two pairs of gills on ninth abdominal segment. . . . . GYRINIDAE*
   Gills on ninth abdominal segment lacking. 9

9. Tenth abdominal segment long and tapering, *or* 2 pairs of long segmented filaments present per abdominal segment . . . . HALIPLIDAE*
   Tenth abdominal segment shortened 10

10. Mouthparts of suctorial type (mandibles hollowed, lacking teeth) . . . . . DYTISCIDAE*
    Mouthparts not suctorial (mandibles solid, usually with a tooth). . . . . CARABIDAE

11. Body C-shaped [Fig. 353(C)]; exoskeleton thin without abdominal sclerites. 12
    Body straight or nearly so (some curvature from alcohol may occur);
    exoskeleton variable 17

12. Metathoracic legs reduced to 2 segments or absent. . . . . PASSALIDAE†
    Metathoracic legs normally developed [Fig. 353(C)] 13

13. Mouthparts prognathic 14
    Mouthparts hypognathic. 15

14. Antennae three-segmented; labial palpi two-segmented . . . . BOSTRICHIDAE†
    Antennae two-segmented; labial palpi single-segmented . . (seed beetles) BRUCHIDAE

15. Antennae short, about length of labrum [Fig. 354(B)]. . . . . CHRYSOMELIDAE
    Antennae long, often length of head 16

16. Anus transverse or V-shaped. . . . . (grubs) SCARABAEIDAE
    Anus longitudinal or Y-shaped . . . . LUCANIDAE

17. Antennae as long as abdomen and filiform, multisegmented . . . . HELODIDAE*
    Antennae short, 1 to 5 segments. 18

18. Body flattened into disc shape (Fig. 128);
    ventral gills present . . . . (water pennies) PSEPHENIDAE*
    Body not as above 19

19. Labrum distinct sclerite 20
    Labrum absent or indistinct 31

20. Body distinctly flattened 21
    Body round to slightly depressed due to flattened terga 23

21. Eighth abdominal segment twice the length of seventh, with crescent-shaped
    ventral row of aspergites [Fig. 357(B)] . . . . PYROCHROIDAE†
    Eighth abdominal segment shortened or lacking aspergites 22

22. Terga with elongate lateral flanges . . . . SILPHIDAE
    Terga lacking such elongate flanges . . . . (flat bark beetles) CUCUJIDAE†

23. Body with long plumose setae, especially posteriorly. . . . . DERMESTIDAE
    Body lacking such long setae. 24

24. Body with warts or branched spines . . . . COCCINELLIDAE
    Body lacking warts or branched spines; if spines or warts present,
    mandible flattened and tips palmate 25

25. Solid urogomphi present [Fig. 67(D)] 26
    Urogomphi absent or segmented, not solid 28

26. Mandible with a molar (Fig. 34) 27
    Mandible lacking a molar. . . . . CLERIDAE

27. Cardo divided into 2 parts
    [Fig. 357(C)] . . . . (melandryid bark beetles) MELANDRYIDAE†
    Cardo not in 2 parts . . . . TENEBRIONIDAE

28. Urogomphi two-segmented . . . . SCAPHIDIIDAE
    Urogomphi absent 29

29. Body heavily sclerotized [Fig. 353(B)] . . . . (falsewire worms) TENEBRIONIDAE
    Body with many separate sclerites or membranous 30

30. Antennae extremely reduced; no dorsal swellings
    for locomotion . . . . (rootworms, leaf beetles) CHRYSOMELIDAE
    Antennae readily seen; swellings on dorsum of abdominal
    segments 1–6 . . . . MELANDRYIDAE

31. Head covered dorsally by pronotum . . . . (glowworms) LAMPYRIDAE
    Head partially or not covered by pronotum 32

32. Thorax and abdomen heavily sclerotized dorsally and ventrally;
    unsegmented urogomphi may be present . . . . (wireworms) ELATERIDAE
    Thorax and abdomen with much membranous areas 33

33. Tenth segment narrow and elongate; one- to three-segmented
    urogomphi present . . . . STAPHYLINIDAE
    Tenth segment reduced or lacking 34

34. Glandular openings on thoracic and most abdominal segments
    [Fig. 357(D)] . . . . CANTHARIDAE
    Glandular openings absent 35

35. Five to six pairs of ocelli; antennae three- to four-segmented, slender
    [Fig. 354(C)] . . . . HYDROPHILIDAE*
    Zero to one pair of ocelli; antennae short stubs . . . . LYCIDAE

*Aquatic.
†Primarily under bark or in wood.

HYMENOPTERA

The order Hymenoptera includes bees, wasps, ants, and sawflies (Figs. 358, 359). Sizes range from 0.21 mm to 65 mm in length, excluding the appendicular ovipositor (Fig 360). Characteristics include filiform antennae, chewing or chewing-lapping mouthparts, large compound eyes except for ants, long legs with five-segmented tarsi, cerci minute or absent, and wings absent (Fig. 361) or two pair that are long and narrow with fused venation. The mesothorax is enlarged for flight efficiency with the forewings becoming the major flight organs and the hind wings reduced and coupled to the fore wings by *hamuli* (Fig. 59). The second abdominal segment is often narrowed (Apocrita) to form a slender waist or *petiole*. The ovipositor may be modified into a saw, drill, or sting (Fig. 64). Some species are social and have caste differences (see Chapter 7 on social insects). Because of their stings and distastefulness to vertebrates, many species have served as models for numerous mimics (see Chapter 6 on mimicry).

Metamorphosis is complete. Larvae are either caterpillar-like (have thoracic and abdominal legs) as in the sawflies (Fig. 362) or vermiform (legless) in the remaining species (Fig. 363). Caterpillars are herbivorous, but vermiform larvae vary in diet from

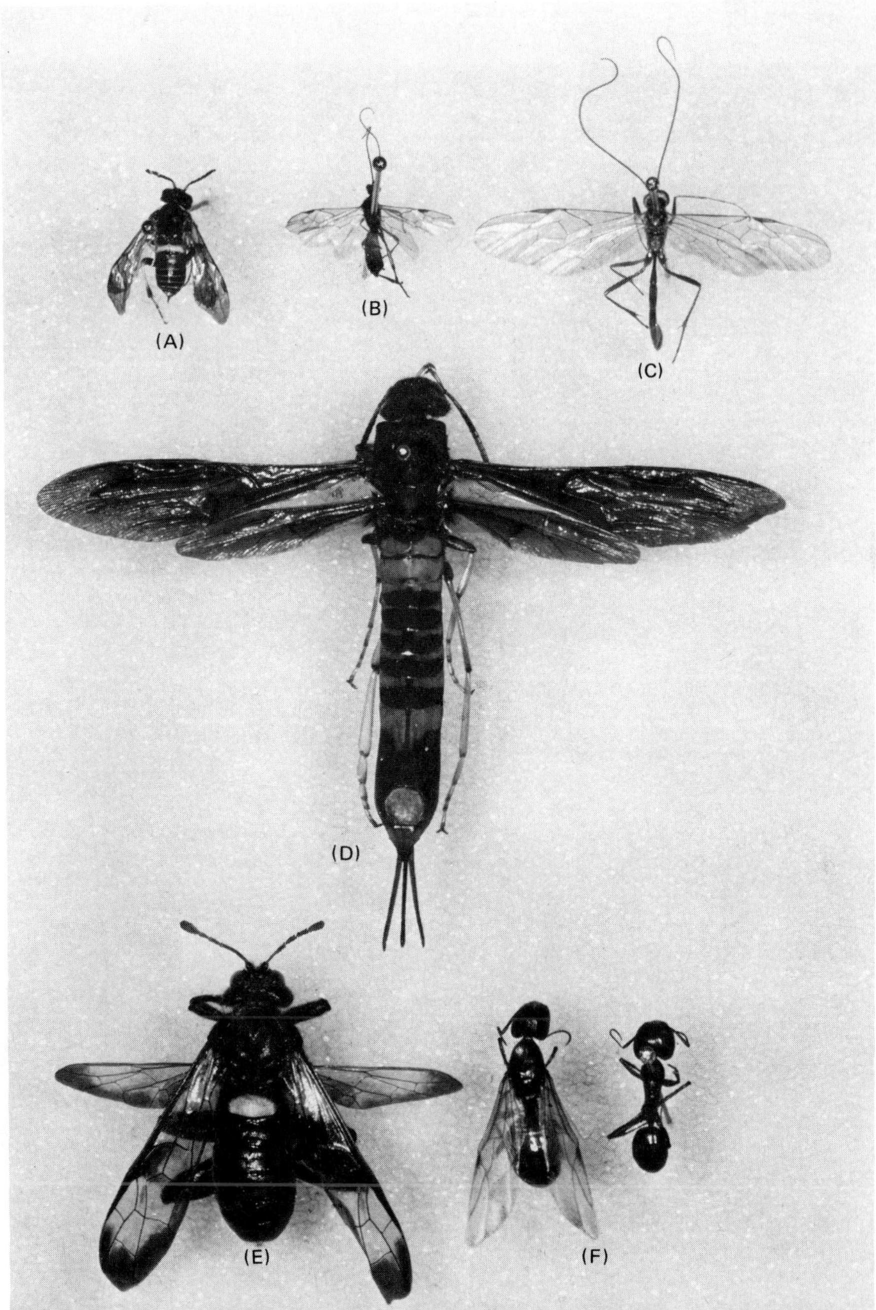

Figure 358. Examples of Hymenoptera. (A) sawfly (Tenthredinidae); (B) braconid wasp (Braconidae); (C) ichneumonid wasp (Ichneumonidae); (D) horntail (Siricidae); (E) cimbicid sawfly (Cimbicidae); (F) ants (Formicidae).

Figure 359. Examples of Hymenoptera. (A) vespid wasp (Vespidae); (B) velvet ant (Mutillidae); (C) honeybee (Apidae); (D) sweat bee (Halictidae); (E) leafcutting bee (Megachilidae); (F) spider wasp (Pompilidae); (G) bumblebee (Apidae); (H) sphecid wasp (Sphecidae); (I) carpenter bee (Apidae).

Figure 360. A horntail (*Tremex columba*).

Figure 361. A female velvet ant (*Dasymutilla occidentalis*). Larvae feed on other wasp larvae.

parasitic to scavengers. Many wasps deposit eggs on hosts, and the larvae develop as parasitoids. In the remaining species, adults either feed the apodous larvae daily or mass-provision nests with sufficient food for the larvae to complete development. No connection exists between the midgut and hindgut in legless larvae, so wastes build up into a *meconium* until after pupation when the connection is established. Pupae are exarate and sometimes enclosed within a cocoon (Fig. 203). Adult males result from parthenogenesis (in all species studied). The diet of many adults is unknown (Table 15).

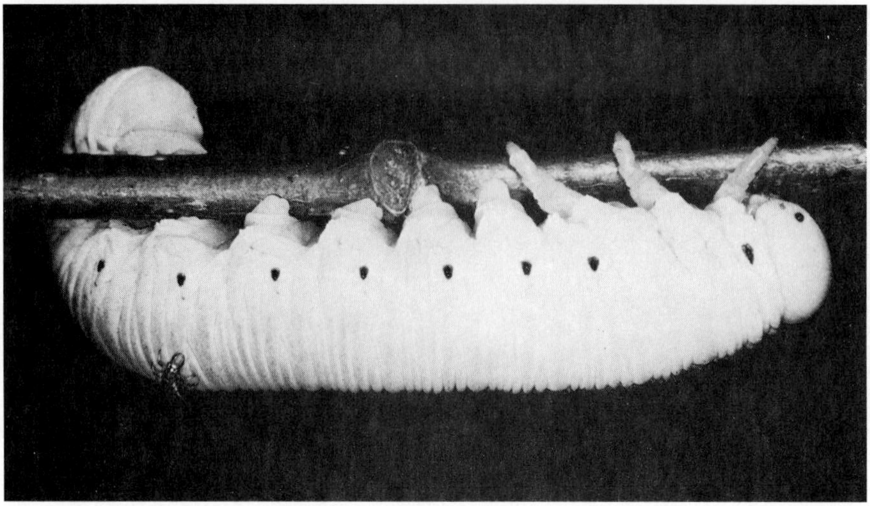

Figure 362. The elm sawfly (*Cimbex americana*) larva.

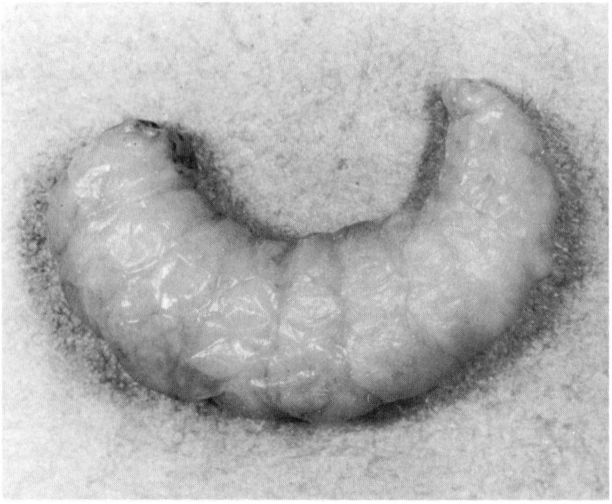

Figure 363. A bumblebee larva (*Bombus americanorum*).

Relatively few species are considered pests. Wasps (Fig. 215), bees (Fig. 225), and some ants such as the southern fire ant (*Solenopsis xyloni*) are well known for their stings, and serious reactions do occur (see Chapter 10). In addition to stings and allergic reactions, the Hymenoptera also may affect humans by being major pests of crops. Of conspicuous economic importance are the pearslug (*Caliroa cerasi*), larch sawfly (*Pristiphora erichsonii*), wheat stem sawfly (*Cephus cinctus*), wheat jointworm (*Harmolita tritici*), and the Texas leafcutter and ant (*Atta texana*). Beneficial species should be also noted. Some of the major pollinators, such as the honeybee (*Apis mellifera*), and many of the natural biological control agents of other insects are

TABLE 15. Major Families of Hymenoptera Listing Adult and Larval Diet

Suborder	Family	Adult Diet	Larval Diet
Symphyta	Argidae	Unknown	Leaves
	Cimbicidae	Unknown	Leaves
	Diprionidae	Unknown	Leaves
	Pamphiliidae	Unknown	Leaves
	Siricidae	Unknown	Wood
	Tenthredinidae	Unknown	Leaves
Apocrita	Andrenidae	Pollen, nectar	Pollen, nectar
	Apidae	Pollen, nectar	Pollen, nectar
	Braconidae	Unknown, nectar	Parasitic
	Chalcididae	Unknown, nectar	Parasitic
	Chrysididae	Unknown, pollen, nectar	Parasitic
	Cynipidae	Unknown, nectar	Plant gall tissues
	Eulophidae	Unknown, nectar	Parasitic
	Formicidae	Wide variety	Wide variety
	Halictidae	Pollen, nectar	Parasitic
	Ichneumonidae	Unknown, nectar	Parasitic
	Megachilidae	Pollen, nectar	Leaves
	Mutillidae	Unknown, predator	Arthropods*
	Pompilidae	Pollen, nectar	Arthropods*
	Pteromalidae	Unknown, arthropod tissues	Parasitic
	Sphecidae	Pollen, nectar	Arthropods*
	Vespidae	Arthropod tissue, nectar	Masticated arthropod tissue

*Larval diet of arthropods is provided by adults, and so could be classified as either predaceous or parasitic.

Hymenoptera. Ichneumonid, braconid, and chalcid wasps, for example, deposit their eggs on many aphids and lepidopteran caterpillars, which are pests of humans.

The order may be subdivided into two suborders, the primitive Symphyta and the highly specialized and diverse Apocrita. The 110,000 species (17,400 in North America) are usually divided into 71 families, of which about more than 50,000 species in 40 families are defined as being parasites of other insects. The following key to adults permits identification of the more common or conspicuous families in the United States.

KEY TO THE ADULTS OF THE COMMON FAMILIES OF ADULT HYMENOPTERA

1. Slender waist (petiole) between abdominal segments 1 and 2 absent [Fig. 364(A)]. . . . . Suborder SYMPHYTA, 2
   Slender waist (petiole) between abdominal segments 1 and 2 present [Fig. 364(B)] . . . . Suborder APOCRITA, 6

2. Antennae three-segmented; terminal segment elongated and often U-shaped in males . . . . (argid sawflies) ARGIDAE
   Antennae more than three-segmented 3

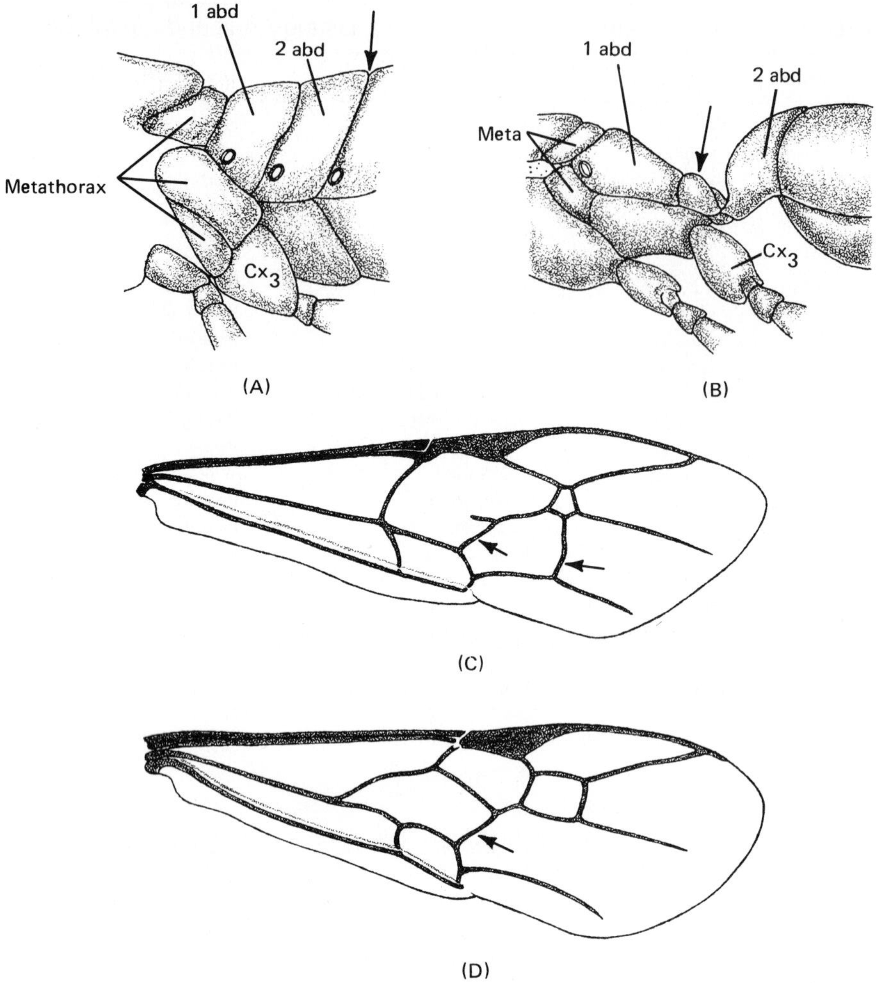

Figure 364. Diagnostic characteristics of adult Hymnoptera. (A) abdominal segments not narrowed (Tenthredinidae); (B) abdominal segments narrowed to form a waist or petiole (Sphecidae); (C) two recurrent veins present in forewing (Ichneumonidae); (D) one recurrent vein in forewing (Braconidae).

3. Antennae clubbed [Fig. 358(E)] . . . . (cimbicid sawflies) CIMBICIDAE
   Antennae filiform 4

4. Usually 1 in. (2.54 cm) or more in length. . . . . (horntails) SIRICIDAE
   Usually less than 1/2 in. (1.27 cm) in length 5

5. Antennae 13- or more segmented; serrate. . . . . (conifer sawflies) DIPRIONIDAE
   Antennae normally nine-segmented;
   filiform . . . . (tenthredinid sawflies) TENTHREDINIDAE

6. Minute in size (smaller than 8 mm) 7
   Small to large (larger than 8 mm); if minute, then node
   or spine on petiole segment [Fig. 365(C)] 10

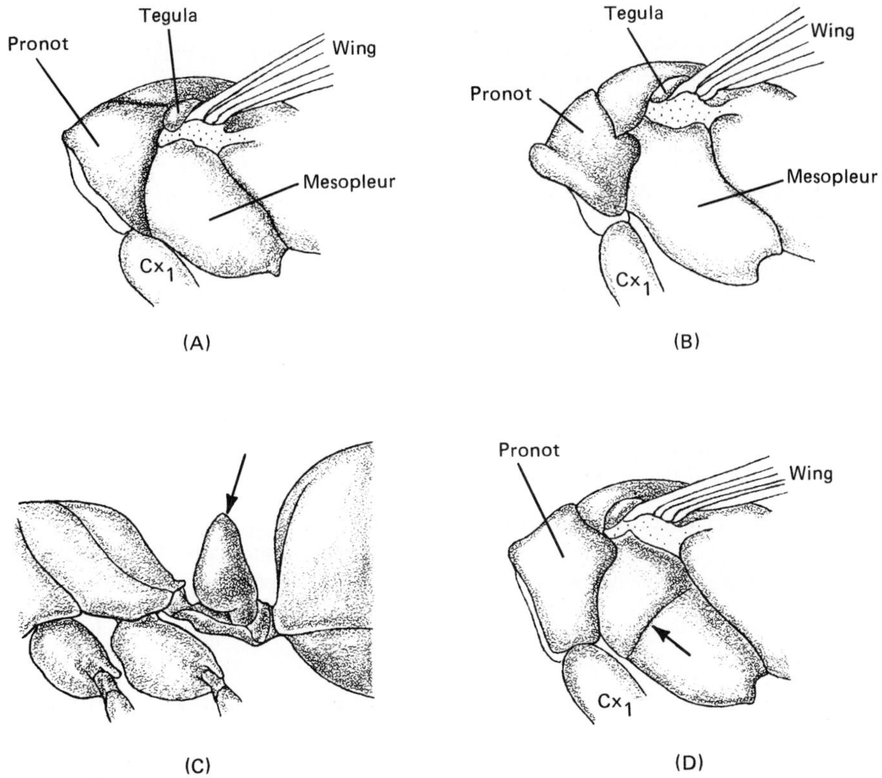

Figure 365. Diagnostic characteristics of adult Hymenoptera. (A) pronotum extending to tegula (Vespidae); (B) pronotum not extending to tegula (Sphecidae); (C) petiole with node or spine (Formicidae); (D) mesopleuron with transverse suture (Pompilidae).

7. Wing venation reduced to single anterior vein; antennae elbowed 8
   Wing venation reduced, but several veins present;
   antennae not elbowed. . . . . (gall wasps) CYNIPIDAE

8. Hind femora greatly enlarged for grasping . . . . (chalcid wasps) CHALCIDIDAE
   Hind femora not greatly enlarged. 9

9. Tarsi five-segmented; ovipositor located anteriorly near middle
   of abdomen. . . . . (pteromalid wasps) PTEROMALIDAE
   Tarsi four-segmented; ovipositor not as above. . . . . . . (eulophid wasps) EULOPHIDAE

10. Antennae long, with 16 or more segments. 11
    Antennae with fewer than 16 segments 12

11. One recurrent vein [Fig. 364(D)] . . . . (braconids) BRACONIDAE
    Two recurrent veins [Fig. 364(C)]. . . . . (ichneumons) ICHNEUMONIDAE

12. Pronotum extending posteriorly to tegula [Fig. 365(A)];
    may be wingless or hairy and slender (Fig. 361). 13
    Pronotum not extending posteriorly to tegula; posterior edge of pronotum
    lobelike [Fig. 365(B)]; winged; may be hairy and stout-bodied 18

13. Segment in petiole with a dorsal bump or spine [Fig. 365(C)]; antennae elbowed; usually wingless . . . . (ants) FORMICIDAE
    Petiole without bumps or spines [Fig. 364(B)] 14

14. Usually covered by conspicuous dense hair (Fig. 361) 15
    Dense hair absent 16

15. Hind coxae contiguous or nearly so; female wingless . . . . . (velvet ants) MUTILLIDAE
    Hind coxae broadly separated . . . . (scoliid wasps) SCOLIIDAE

16. Mesopleuron divided with straight transverse suture [Fig. 365(D)]; hind femora often extend to or beyond tip of abdomen . . . . (spider wasps) POMPILIDAE
    Not as above 17

17. Wings folded longitudinally when flexed . . . . (vespid wasps) VESPIDAE
    Wings not folded longitudinally when flexed . . . . (tiphiid wasps) TIPHIIDAE

18. With branched and plumose hair present on body . . . . (bees) 20
    Without branched and plumose hair 19

19. Coarsely sculptured; abdomen somewhat hollowed out ventrally . . . . (cuckoo wasps) CHRYSIDIDAE
    Sculpturing absent; abdomen not hollowed . . . . (sphecid wasps) SPHECIDAE

20. Thorax usually metallic green . . . . (mining or sweat bees) HALICTIDAE
    Thorax not metallic green 21

21. Abdomen with ventral pollen basket . . . . (leafcutting bees) MEGACHILIDAE
    Abdominal pollen basket absent 22

22. Two subantennal sutures . . . . (andrenid bees) ANDRENIDAE
    One subantennal suture . . . . (honeybees, bumblebees, carpenter bees, and stingless bees) APIDAE

KEY TO THE COMMON FAMILIES OF LARVAL HYMENOPTERA

1. Segmented thoracic legs present (Fig. 362) . . . . Suborder SYMPHYTA, 2
   Thoracic legs reduced to unsegmented lobes or absent (Fig. 363) 6

2. Prolegs present (Fig. 362) 3
   Prolegs absent . . . . (webspinning sawflies) PAMPHILIIDAE

3. Ventral pad under claw [Fig. 366(A)] . . . . (argid sawflies) ARGIDAE
   Ventral pad lacking 4

4. Antennae one- to two-segmented; usually more than 30 mm when grown . . . . (cimbicid sawflies) CIMBICIDAE
   Antennae three- to five-segmented (some may be reduced); less than 30 mm when grown 5

5. Antennae three-segmented; terminal segment peglike [Fig. 366(B)] . . . . (conifer sawflies) DIPRIONIDAE
   Antennae four- to five-segmented; terminal segment usually not peglike . . . . (sawflies) TENTHREDINIDAE

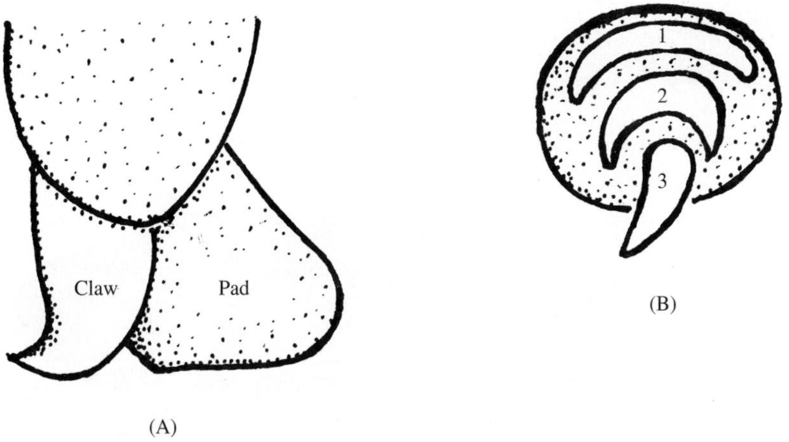

Figure 366. Diagnostic characteristics of larval Hymenoptera. (A) clawed pad at tip of abdominal proleg (Argidae); (B) antennae with two reduced segments and peglike terminal segment (Diprionidae).

6. Median sclerotized process from terminal segment . . . . Suborder SYMPHYTA, SIRICIDAE
   Median process absent . . . . Suborder APOCRITA*

*Families not presently keyable for beginning students.

MECOPTERA

Scorpionflies [Fig. 344(E)] are medium-sized (from 10 mm to 25 mm) and have the following important characteristics: chewing mouthparts are located at the end of a snoutlike head (Fig. 367), antennae are filiform and long, compound eyes are large and well separated, legs are long and tarsi are five-segmented, cerci are simple in males and are two-segmented in females, and two pairs of membranous wings (lacking in some species) are of similar size and shape and held in a rooflike position when at rest. Male genitalia are usually enlarged, bulblike, and held scorpionlike in repose above the abdomen.

Metamorphosis is complete. Eggs are normally deposited singly or in mass on or in the soil, and the caterpillar-like larvae feed as predators or as scavengers on dead and decaying materials. Pupation takes place in a cell in the soil or in damp debris. Adults are omnivorous and feed on such a variety of food as pollen, nectar, juices from decaying fruit, and dead or live insects. They are most often located in moist, cool habitats. One group, the snow scorpionflies, are active throughout the year.

About 500 species (80 in North America) are known, and these are relegated to four to nine families. Species are of no economic or medical importance to humans.

Figure 367. A female scorpionfly (*Panorpa* sp.).

TRICHOPTERA

Caddisflies [Figs. 344(D), 368] are small to moderate in size (from 1.5 mm to 40 mm) and slender and fragile appearing. The compound eyes are large, mouthparts are chewing but poorly developed except for the four- to five-segmented maxillary palpi, antennae are long and filiform, legs are long and tarsi are five-segmented, cerci are one- or two-segmented, and two pairs of membranous wings are held rooflike over the abdomen. Hind wings are wider than the forewings, and few cross veins are present in either set of wings. Hair is common on the wing surface as well as on the body.

Metamorphosis is complete. Eggs are deposited in gelatinous secretions in water or on objects at the water's edge. Larvae have small peglike antennae, chewing mouthparts, and are poorly sclerotized except for the head, pronotum and a few other body sclerites. They have numerous abdominal gills and a pair of terminal hooks for aiding in locomotion and holding onto rocks or to a portable case (Fig. 369). These cases are distinctive with each species using characteristic materials and shape. Those species that lack cases produce a web of silk that traps organic matter passing in the current. A few species are carnivorous and some scrape algae from the substrate. Pupation takes place in the case. The exarate pupa, equipped with sharp mandibles that will be lost after metamorphosis, cuts its way free of this case, swims to the shore, and molts into the adult. Adults are nonfeeders and are usually found on vegetation near fresh water. Most are nocturnal or crepuscular.

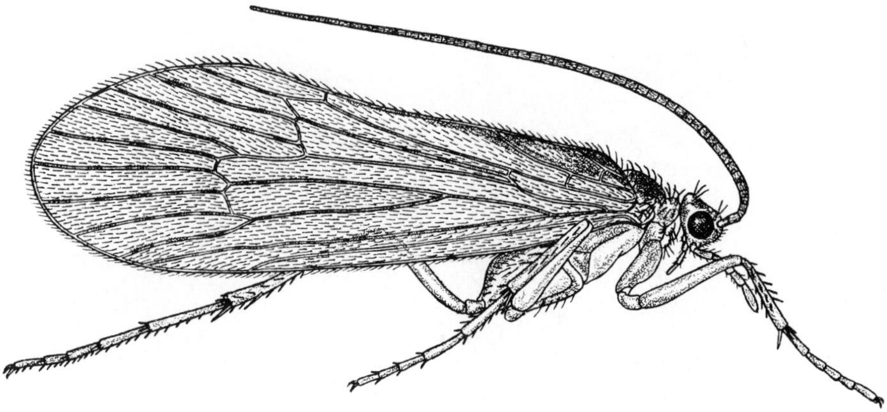

Figure 368. A caddisfly adult.

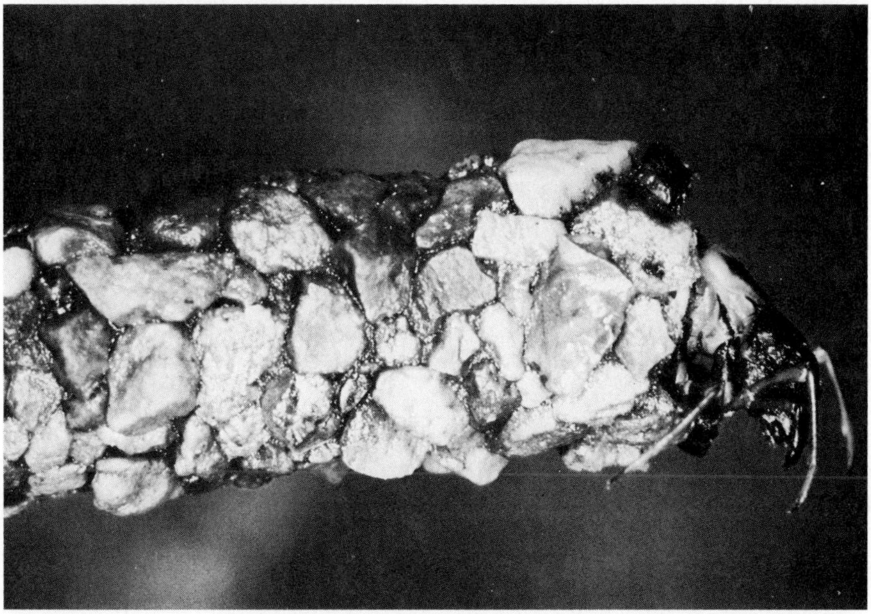

Figure 369. A larval caddisfly with its pebble case.

Figure 370. Examples of moths. (A) Sphingidae; (B) Pyralidae; (C) Yponomeutidae; (D) Notodontidae; (E) Arctiidae; (F) Noctuidae; (G) Geometridae.

About 6,000 species are known (1,300 in North America), and these are often divided into 18 families. Although they are part of many aquatic food chains, caddisflies cannot be considered to be of direct economic importance.

LEPIDOPTERA

Moths (Fig. 370) and butterflies (Fig. 371) vary in wingspread from 5 mm to 270 mm. This conspicuous order may be identified by siphoning mouthparts coiled under their head (Fig. 372), vestigial or absent maxillary palpi, large labial palpi, large compound eyes, long legs with five-segmented tarsi, absence of cerci, two pairs of membranous wings clothed with overlapping scales. Some species have apterous females and winged males. Lepidopteran bodies are covered with hair and scales. Color is the result not only of pigments in the hair and scales but also from structural ridges and layers that reflect light differentially to cause iridescence. Much mimicry exists among species. Antennae vary greatly and are useful in identification.

Figure 371. Examples of butterflies and skippers. (A) Nymphalidae; (B) Papilionidae; (C) Hesperiidae; (D) Lycaenidae; (E) Danaidae; (F) Pieridae; (G) Libytheidae.

Figure 372. A butterfly.

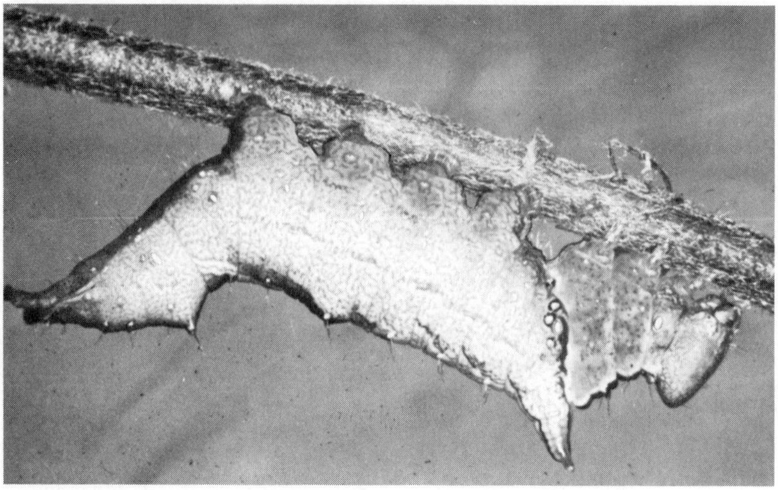

Figure 373. This lepidopteran caterpillar (*Schizura unicornis*) feeds on a wide variety of plants including apple, cherry, plum, willow, elm, oak, locust, and dogwood.

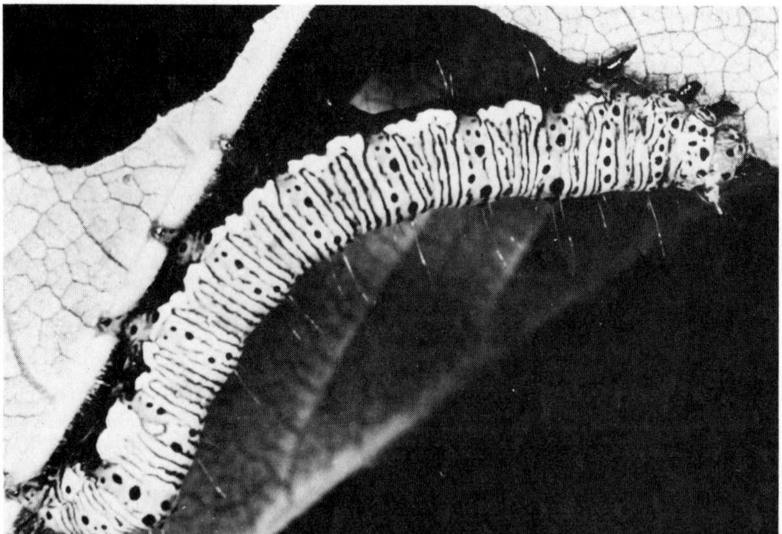

Figure 374. The eight-spotted forester moth larva (*Alypia octomaculata*) has a narrow food preference and feeds mainly on grape and Virginia creeper.

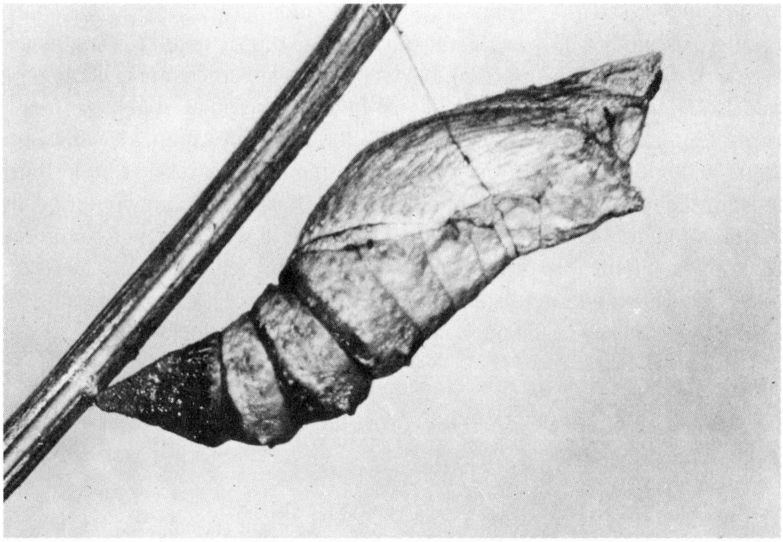

Figure 375. An obtect pupa of the black swallowtail, *Papilio polyxenes.*

Metamorphosis is complete. Eggs are often deposited on host plants or other food sources. Larvae (Figs. 373, 374), called *caterpillars,* have chewing mouthparts and up to five pairs of abdominal prolegs (located on abdominal segments 3 to 6 and 10) provided with hooks or *crochets* (Fig. 68). Lepidopteran caterpillars rely on cryptic coloration to hide or are aposematic and brightly colored. A few have stinging or urticating hair (Fig. 274). Cocoons may be spun by moth larvae, but most butterflies lack this structure. Pupae are normally obtect (Fig. 375). The dominant feeding stage is the larva, and, although some species are borers, most larvae utilize leaves as food. When adults feed, they usually ingest nectar, but some also take in fluids from rotting fruit and feces, sap exuding from plants, and honeydew. A few species feed on animal tears. Some Lepidoptera do not feed as adults but utilize only stored fat from the larval stage as an energy source.

Many species of Lepidoptera are of economic or medical importance. Forests are frequently damaged by the gypsy moth (*Lymantria dispar*), eastern tent caterpillar (*Malacosoma americanum*), and the forest tent caterpillar (*Malacosoma disstria*). Caterpillars that damage fruit trees include the peach tree borer (*Synanthedon existiosa*), peach twig borer (*Anarsia lineatella*), Oriental fruit moth (*Grapholitha molesta*), and the codling moth (*Laspeyresia pomonella*). The following are serious pests of grain: European corn borer (*Ostrinia nubilalis*), southwestern corn borer (*Diatraea grandiosella*), corn earworm (*Helicoverpa zea*), fall armyworm (*Spodoptera frugiperda*), armyworm (*Pseudaletia unipuncta*), several species of cutworms, and the sugar cane borer (*Diatraea saccharalis*). Cotton is damaged by the pink bollworm (*Pectinophora gossypiella*), and tobacco is frequently eaten by the tobacco hornworm (*Manduca sexta*) and the tobacco budworm (*Helicoverpa virescens*). The imported cabbage worm (*Artogeia rapae*) is a pest on cabbage and related plants. Bagworms (*Thyridopteryx ephemeraeformis*) are common on evergreens. Stored products are often infested with Indian-meal moth (*Plodia interpunctella*), Angoumois grain moth (*Sitotroga cerealella*), Mediterranean flour moth (*Anagasta kuhniella*), and two species of clothes moths (*Tineola bisselliella* and *Tinea pellionella*). Several species are urticating, including the puss caterpillar (*Megalopyge opercularis*) and the Io moth caterpillar (*Automeris io*).

This large order includes 120,000 species (11,300 in North America), which are grouped into 77 families. About 12 to 15 of these families are termed butterflies, whereas most of the remainder are called moths. Identification of adults relies heavily on wing venation, and because wings are covered with scales, viewing requires special bleaching techniques. To avoid this difficulty for the beginner, a key has been prepared that includes only those distinctive families that are easily recognizable by color and body shape. A key to the more common larvae is also included.

KEY TO THE MOST EASILY RECOGNIZABLE FAMILIES OF ADULT LEPIDOPTERA

1. Wingless . . . . (females of certain moths) 21
   Wings present 2

2 Antennae clubbed (Fig. 372) 3
   Antennae not clubbed . . . . (moths) 9

3. Terminal hook usually projecting from clubbed antenna [Fig. 376(A)] . . . . (skippers) HESPERIIDAE
   Terminal hook lacking . . . . (butterflies) 4

4. Prothoracic legs much reduced [Fig. 376(C)] 5
   Prothoracic legs normal size 6

5. Antennae void of scales and club reduced in size . . (milkweed butterflies) DANAIDAE
   Antennae with scales. . . . . (brushfooted butterflies) NYMPHALIDAE*

6. Large, wingspread usually more than 3 in. (7.5 cm); hind wings with "tails" [Fig. 371(B)]; prothoracic legs bearing tibial epiphyses [Fig. 376(D)]. . . . . (swallowtails) PAPILIONIDAE
   Not as above. 7

7. Labial palpi long and projecting far forward [Fig. 376(G)] . . . . (snout butterflies) LIBYTHEIDAE
   Labial palpi normal size 8

8. Margin of eye notched where antenna located [Fig. 376(E)]; color usually metallic in sheen. . . . . (blues, coppers, etc.) LYCAENIDAE
   Not with above characteristics . . . . (whites and sulphurs) PIERIDAE

9. Wings fringed with numerous long hair; minute to small size . . . . MICROLEPIDOPTERA
   Wings not fringed with numerous long hair. 10

10. Eyespots usually on wings; wingspread usually more than 3 in. (7.6 cm) . . . . (giant silkworm moths) SATURNIIDAE
    Eyespots usually absent; if present, then antennae not pectinate to tip [Fig. 376(B)] 11

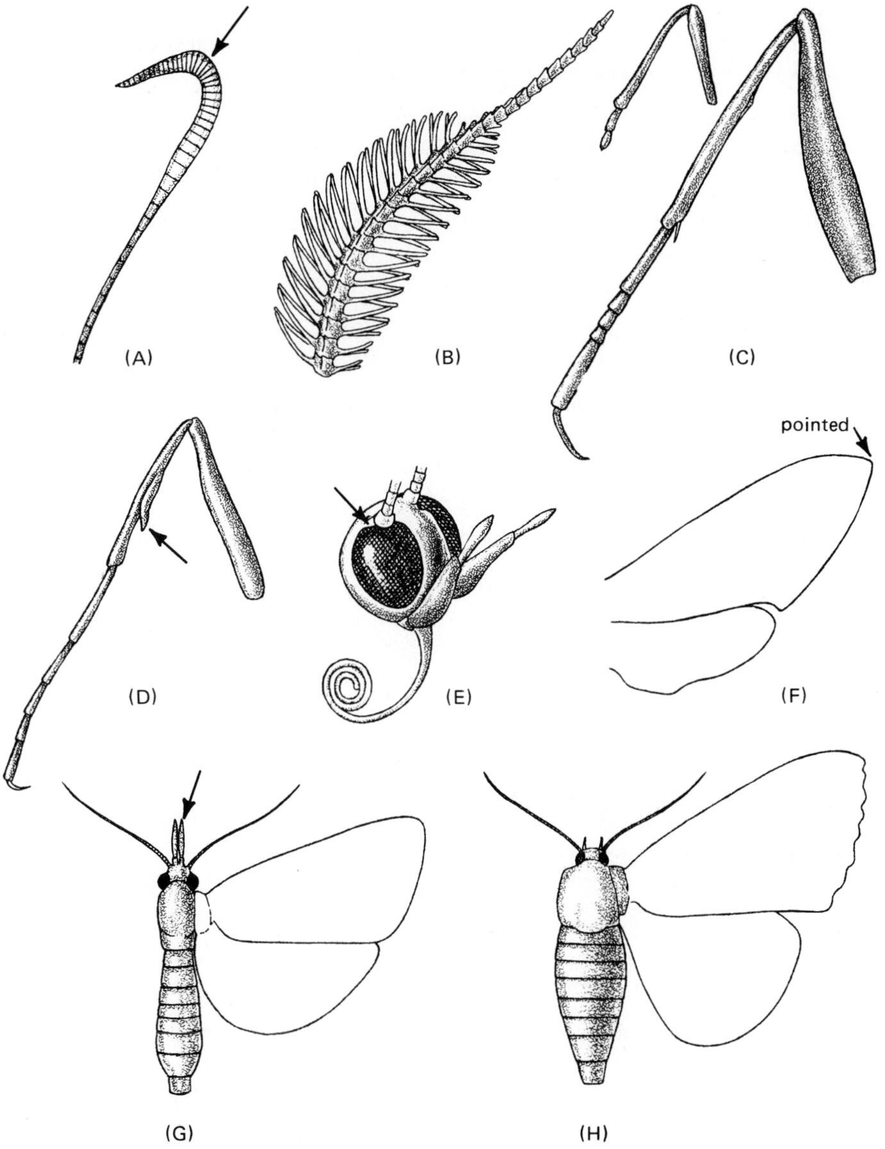

Figure 376. Diagnostic characteristics in Lepidoptera. (A) clubbed antenna with a hook (Hesperiidae); (B) antenna not pectinate to tip (Saturniidae); (C) prothoracic leg much smaller than mesothoracic (Nymphalidae); (D) tibial epiphysis (Papilionidae); (E) notched eye (Lycaenidae); (F) hind wing reduced in width, fitting into notch in pointed forewing (Sphingidae); (G) hind wing normal in width, beaklike palpi (Pyralidae); (H) hind wing normal in width, body large (Noctuidae).

11. Antennae pectinate but not obviously to tip [Fig. 376(B)] . . . . (royal moths) SATURNIIDAE
    Antennae either pectinate to tip (Fig. 166) *or* not conspicuously pectinate 12

12. Hind wings reduced in width, often fitting into indented posterior margin of front wings [Fig. 376(F)] 13
    Hind wings normal; no indented posterior margin in front wings [Fig. 376(H)] 15

13. With either large black spots or mottled without distinct bands; abdomen often shiny . . . . (carpenterworm moths) COSSIDAE
    Not as above 14

14. Antennae pectinate [Fig. 47(E)] . . . . (male bagworm moths) PSYCHIDAE
    Antennae not obviously pectinate . . . . (sphinx or hawk moths) SPHINGIDAE

15. Abdomen colored other than brown, often with bands or spots; front wings often white *or* brightly striped . . . . (tiger moths) ARCTIIDAE
    Abdomen light yellow-brown to dark brown 16

16. Body large when compared to wing size [Fig. 376(H)] 18
    Body slender and small when compared to wing size 17

17. Large beaklike palpi extending forward of head [Fig. 376(G)] . . . . (pyralid moths) PYRALIDAE
    Palpi not large, although may be seen from dorsal view . . . . (geometrid moths) GEOMETRIDAE

18. Antennae distinctly pectinate 19
    Antennae usually not distinctly pectinate 20

19. Anterior base of hindwing not expanded; frenulum present; proboscis reduced . . . . (tussock moths) LYMANTRIIDAE
    Anterior base of hind wing greatly expanded; frenulum absent; proboscis well developed . . . . (tent caterpillar moths) LASIOCAMPIDAE

20. Cubitus of front wing appears three-branched; often have wavy lines on forewings . . . . (notodontid moths) NOTODONTIDAE
    Cubitus of front wing appears four-branched . . . . (noctuid moths) NOCTUIDAE

21. Found within case made by larva (Fig. 154) . . . . (bagworm moths) PSYCHIDAE
    Not located within case 22

22. Body slender . . . . (cankerworm moths) GEOMETRIDAE
    Body stout . . . . (tussock moths) LYMANTRIIDAE

*Includes Satyridae and Heliconiidae.

KEY TO THE MOST COMMON FAMILIES OF LARVAL LEPIDOPTERA

1. Prolegs well developed and with crochets (Fig. 377) 2
   Prolegs and crochets absent . . . . (slug caterpillars) LIMACODIDAE

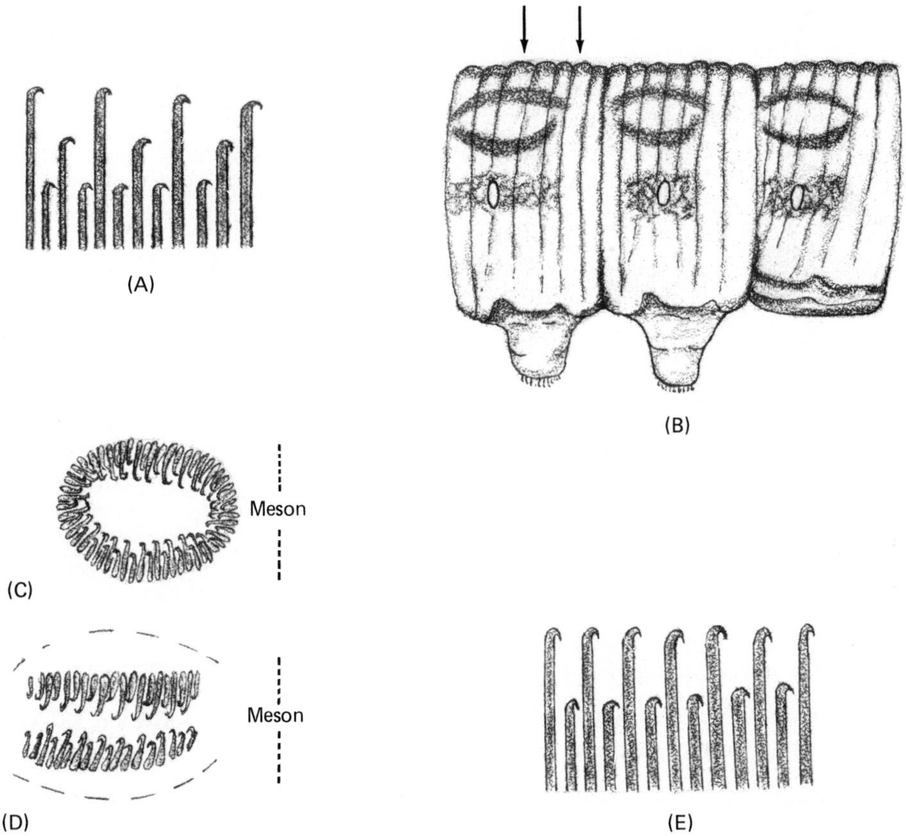

Figure 377. Diagnostic characteristics of larval Lepidoptera. (A) triordinal crochets (Nymphalidae); (B) annulets (Sphingidae); (C) crochets in ellipse (Pyralidae); (D) crochets in transverse bands; (E) biordinal crochets (Saturniidae).

2. Median horn or scolus on dorsum of eighth abdominal segment
   (may be lost in a few late instar Sphingids but a scar remains) 3
   Median horn or scolus absent, no scar present 7

3. Body with 2 or more horns or enlarged tubercles (scoli) (Fig. 112) 4
   Body with single horn (or scar from loss) 6

4. Crochets biordinal [Fig. 377(E)] 5
   Crochets triordinal [Fig. 377(A)] . . . . NYMPHALIDAE

5. Ninth abdominal segment lacking single dorsal horn;
   anal plate smooth . . . . SATURNIIDAE
   Ninth abdominal segment with middorsal horn or scolus;
   anal plate with small spines . . . . SATURNIIDAE

6. Abdominal segments with 6 to 8 annulets [Fig. 377(B)] . . . . SPHINGIDAE
   Abdominal segments with 3 or less annulets . . . . BOMBYCIDAE

7. Eversible osmeterium present on dorsum of prothorax (Fig. 195) . . . . PAPILIONIDAE
   Osmeterium lacking 8

8. Body with many long setae 9
   Body with either obscure short setae or few long setae 13

9. Crochets biordinal [Fig. 377(E)] . . . . LASIOCAMPIDAE
   Crochets uniordinal 10

10. Eversible middorsal glands on sixth and seventh abdominal segments; dense brush of setae on several abdominal segments . . . . LYMANTRIIDAE
    Not as above 11

11. Anal prolegs reduced (Fig. 192) . . . . NOTODONTIDAE
    Anal prolegs not reduced 12

12. Body setae usually not plumose . . . . NOCTUIDAE
    Body setae usually plumose . . . . ARCTIIDAE

13. Fleshy filaments on mesothorax and usually on eighth abdominal segment (Fig. 193) . . . . DANAIDAE
    Fleshy filaments absent on mesothorax and on eighth abdominal segment 14

14. Less than 5 pairs of prolegs (including anals) 15
    Five pairs of prolegs present 16

15. Prolegs absent or vestigial on fifth abdominal segment . . . . GEOMETRIDAE
    Prolegs present on fifth abdominal segment . . . . NOCTUIDAE

16. Crochets arranged in a single band 17
    Crochets in circle, ellipse [Fig. 377(C)], or as transverse bands [Fig. 377(D)] 22

17. Anal prolegs reduced or absent; horns or odd-shaped terga often present (Fig. 373) . . . . NOTODONTIDAE
    Not as above 18

18. Distinct lateral lobe interrupting crochets . . . . LYCAENIDAE
    Lateral lobe lacking 19

19. Crochets uniordinal . . . . NOCTUIDAE
    Crochets biordinal [Fig. 377(E)] or triordinal [Fig. 377(A)] 20

20. Distinct scoli (enlarged tubercles) on head or body (Fig. 112) 21
    Scoli lacking . . . . PIERIDAE

21. Crochets biordinal [Fig. 377(E)] . . . . SATURNIIDAE
    Crochets triordinal [Fig. 377(A)] . . . . NYMPHALIDAE

22. Crochets usually in complete circle [Fig. 377(C)] 25
    Crochets in incomplete circle or transverse bands [Fig. 377(D)] 23

23. Prothoracic spiracle oval and on horizontal axis . . . . PSYCHIDAE
    Prothoracic spiracle rounded or oval, on perpendicular axis 24

24. Crochets in transverse bands [Fig. 377(D)] . . . . SESIIDAE
    Crochets in ellipse formation . . . . NOCTUIDAE

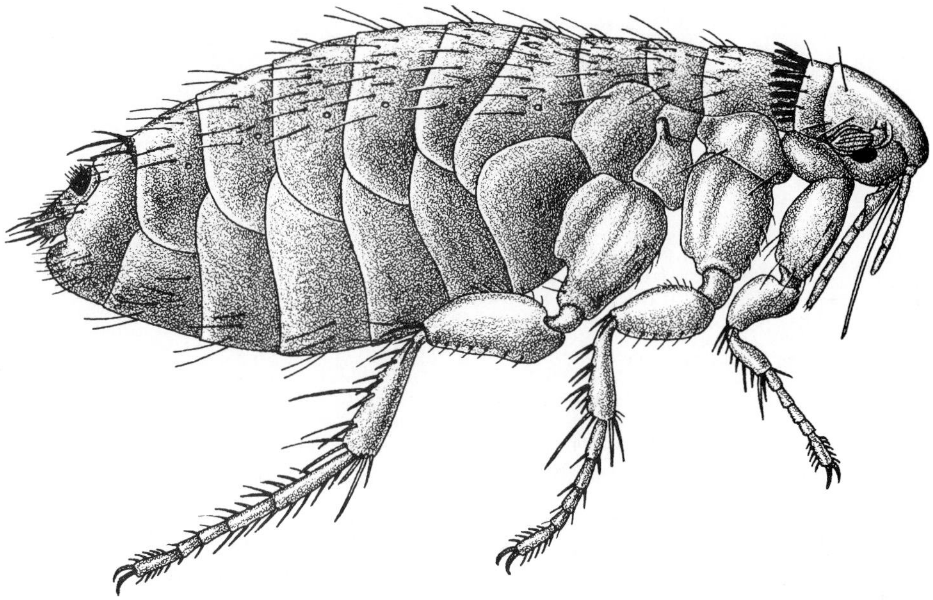

Figure 378. An adult male flea (*Thrassis pansus*).

25. Head large, much wider than prothorax . . . . HESPERIIDAE
    Head narrower than prothorax 26

26. Two setae in plate anterior to prothoracic spiracle . . . . PYRALIDAE
    Three setae in plate anterior to prothoracic spiracle . . . . TORTRICIDAE

SIPHONAPTERA

Fleas (Figs. 238, 378) are minute to small (from 0.8 mm to 5 mm) and have the following characteristics: compound eyes are absent or each is represented by a single ommatidium, antennae are short and can be folded into grooves in the head, mouthparts are piercing-sucking [Fig. 239(B)], coxae are long and tarsi are five-segmented, cerci are small and in a single segment, and wings are absent. The body is laterally flattened [Fig. 239(A)]; posteriorly projecting rows of strong spines may be present, especially on the gena and pronotum where they may be enlarged to form combs or *ctenidia* (Fig. 298).

Metamorphosis is complete. Eggs are oviposited on the host or more often in the host's nest; in the former case, eggs fall off prior to hatching. The legless larva (Fig. 379) feeds upon such organic matter as may be available, including fecal material from adult fleas that contains blood residues. There are usually three larval instars. Pupation is in silken cocoons. Adults feed on blood from either birds or mammals, the latter being more common since more than 70 percent of the flea species parasitize rodents. Host specificity exists but is not as well developed as in lice. Some species are predominantly on the host, but if the host has a nest, many species of fleas leave the host

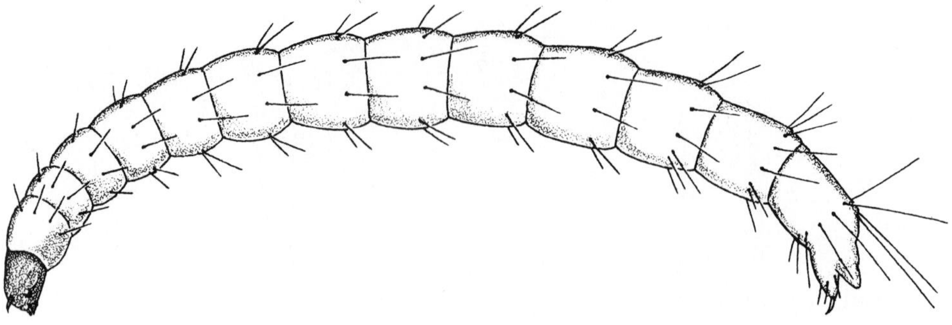

Figure 379. A larval flea.

during nonfeeding periods. These fleas, commonly called *nest fleas,* have their highest population in the nest, and scientific surveys only of hosts will fail to reveal the true flea population.

The 2,100 species (260 in North America) are classified into 9 to 15 families depending on the classification used. Beyond mere irritation, fleas are of medical importance to humans through disease transfer. Fleas are vectors of plague (bubonic form) and of endemic or murine typhus. Several species of tapeworm can, but not commonly, infect humans after utilizing the flea as an intermediate host. In the tropics the chigoe flea attaches itself to humans and can initiate severe lesions around the toe nails and feet. Fleas also can become pests to such domesticated animals as dogs, cats, and, in rare cases, dairy cattle.

DIPTERA

True flies (Figs. 380, 381, 382) are a diverse group and are minute to medium-sized (from 0.5 mm to 50 mm) with wingspreads ranging from 1 mm to 100 mm. All are characterized by having only one pair of flight wings; the second pair is modified into halteres (Figs. 57, 58). A great variety of mouthparts have evolved, including piercing-sucking [Figs. 36(D), 382], cutting-sponging [Fig. 36(A)], and the most common type, sponging [Figs. 36(B), 383]. Labial palpi are absent. Compound eyes are very large, antennae have many shapes, the mesothorax is disproportionately large, legs are long and tarsi are five-segmented, and cerci are vestigial to absent. A tubular ovipositor is often present (Fig. 65).

Metamorphosis is complete. Larvae are normally legless and vermiform (Figs. 384, 385), often without a well-developed head capsule [Fig. 385(C)].

Great differences in diet exist. Many habitats are exploited (Table 16), but they are nearly always with high moisture. Pupae may be active and nonfeeding (mosquitoes), but most pupae are sessile within puparia (Fig. 115). A significant number of adults utilize blood or nectar for food.

Numerous species are of economic and medical importance to humans. Injurious to plants as larvae are the Hessian fly (*Mayetiola destructor*) on wheat, sorghum midge (*Contarinia sorghicola*) on sorghum, and the Mediterranean fruit fly (*Ceratitis capitata*)

Figure 380. Examples of Diptera. (A) blow fly (Calliphoridae); (B) tachina fly (Tachinidae); (C) flesh fly (Sarcophagidae); (D) flower fly (Syrphidae); (E) mosquito (Culicidae); (F) midge (Chironomidae); (G) deer fly (Tabanidae); (H) house fly (Muscidae).

Figure 381. Examples of Diptera. (A) crane fly (Tipulidae); (B) horse fly (Tabanidae); (C) bee fly (Bombyliidae); (D) bee fly (Bombyliidae); (E) robber fly (Asilidae); (F) robber fly (Asilidae).

Figure 382. A robber fly (*Protacanthella cacopiloga*).

Figure 383. A flesh fly (*Sarcophaga bullata*).

Figure 384. Examples of Diptera larvae. (A) Tipulidae; (B) Tabanidae; (C) Syrphidae.

on a wide number of fruits. Many larvae are parasitic on livestock, including the screwworm (*Cochliomyia hominivorax*), cattle grubs (*Hypoderma lineatum* and *H. bovis*), sheep bot fly (*Oestrus ovis*), and common horse bot fly (*Gasterophilus intestinalis*). Adults that may ingest blood from domesticated animals include the horn fly (*Haematobia irritans*), stable fly (*Stomoxys calcitrans*), and many species of genera of horse flies, deer flies, black flies, and mosquitoes. Of prime importance to humans are the many diseases transmitted by mosquitoes, tsetse flies, and black flies, some of which are summarized in Table 6. A few species, primarily in the families Syrphidae and Tachinidae, are of benefit to humans because of their parasitism of other insects.

More than 100,000 species (17,130 in North America) are described and separated into 105 families. The most conspicuous families, which include about 80 percent of the United States' species, may be identified by using the following keys.

KEY TO THE COMMON FAMILIES OF ADULT DIPTERA

1. Flattened dorsoventrally; usually wingless; coxae distantly separated 32
   Not as above 2
2. Antennae filiform to plumose, with at least 7 segments [Fig. 386(A)] 3
   Antennae with fewer than 7 segments [Fig. 386(B), (C)] 11

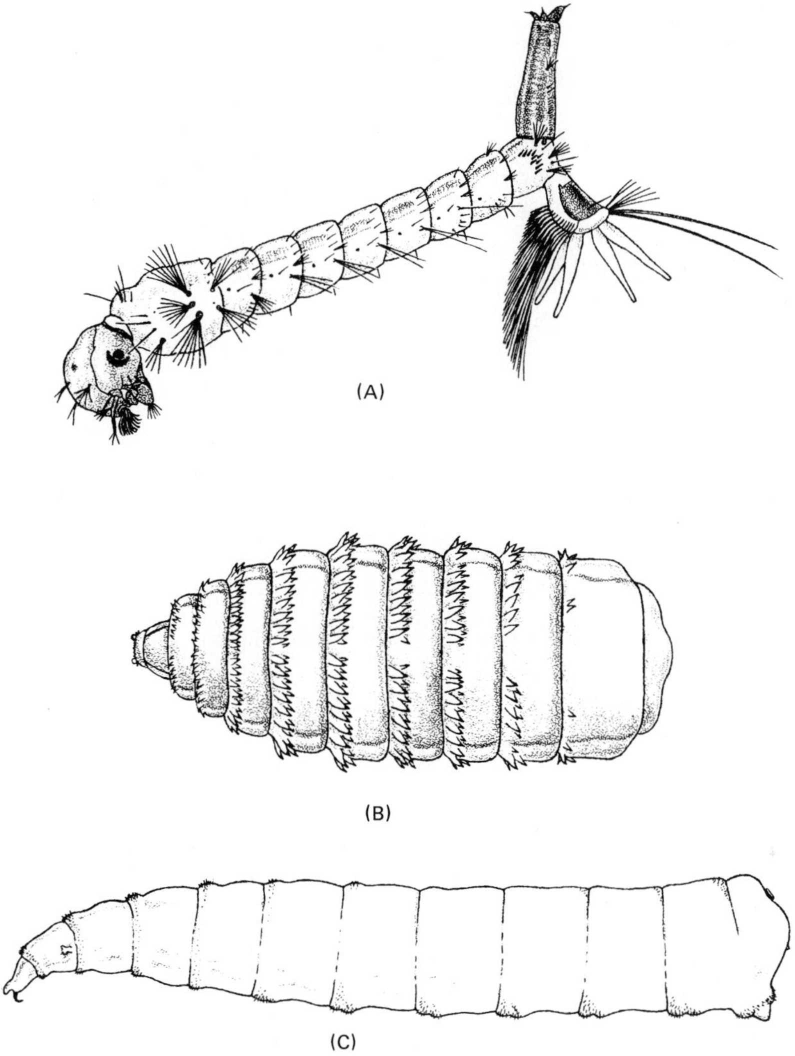

Figure 385. Examples of Diptera larvae. (A) mosquito (*Aedes* sp.) (B) horse bot fly (*Gasterophilus* sp.); (C) house fly (*Musca* sp.).

3. Minute size (6 mm or less) 4
   Small to large 6

4. Body and wings hairy . . . . (moth flies) PSYCHODIDAE
   Body and wings not obviously hairy 5

5. Legs very long; costa continuing around
   wing tip . . . . (gall midges) CECIDOMYIIDAE
   Legs not very long; costa ending near wing tip . . . .
   . . . . (black flies) SIMULIIDAE or (biting midges) CERATOPOGONIDAE

TABLE 16. Major Families of Diptera Listing Adult and Larval Diet and Larval Habitat

Suborder	Family	Adult Diet	Larval Diet	Larval Habitat
Nematocera	Bibionidae	Unknown, scavenger	Scavenger, some herbivores	Decomposing plants and soil
	Cecidomyiidae	Usually none, some nectar	Scavenger, herbivore or parasites	Galls, excreta, insects, soil
	Ceratopogonidae	Blood, some parasites on insects	Scavenger, predator	Mud, water, detritus
	Chironomidae	Unknown, none or honeydew	Plankton, scavenger	Water
	Culicidae	Nectar, blood	Plankton, some predators	Water
	Dixidae	Unknown	Plankton	Water
	Mycetophilidae	Unknown	Fungi, few predators	Fungi, decay, soil
	Psychodidae	Blood	Scavenger	Decaying plants, soil
	Simuliidae	Blood	Plankton	Streams
	Tipulidae	Unknown, none	Scavenger, herbivore, predator	Water, soil
Brachycera	Asilidae	Predator, some nectar	Predator, some scavengers	Soil, rotting wood
	Bombyliidae	Nectar	Parasitic on invertebrates	Soil on host
	Dolichopodidae	Nectar, predator	Predator, some scavengers	Mud, wood, vegetation
	Empididae	Predator, some nectar	Predator, some scavengers	Soil, decaying wood, water
	Rhagionidae	Predator, some blood	Predator, some scavengers	Soil and litter, some water
	Stratiomyidae	Nectar	Scavenger, some herbivores	Water, rotting wood
	Tabanidae	Blood, nectar	Predator, some scavengers	Water, rotting logs, moist soil

Continued

TABLE 16. Major Families of Diptera Listing Adult and Larval Diet and Larval Habitat—*Cont'd*

Suborder	Family	Adult Diet	Larval Diet	Larval Habitat
Cyclorrhapha	Anthomyiidae	Predator, nectar, dung, or none	Herbivore, some scavengers	Plants, soil, water
	Calliphoridae	Nectar, scavenger, or none	Scavenger, some parasites of invertebrates and vertebrates	Decaying animals, excreta, animal hosts
	Chloropidae	Honeydew, some parasites	Herbivore or dung, some predators and parasites	Grass stems, detritus
	Drosophilidae	Scavenger, some nectar	Scavenger or fungivore, and parasites on invertebrates	Soil, decaying matter, hosts
	Ephydridae	Scavenger	Scavenger, herbivores, some predators	Water usually
	Gasterophilidae	None	Parasite	Digestive system of horses
	Hippoboscidae	Ectoparasitic	Viviparous	Within mother
	Muscidae	Scavenger, blood, nectar, some predators	Scavenger, some parasites and predators	Decaying matter, some soil, birds' nests
	Nycteribiidae	Ectoparasitic on bats	Viviparous	Within mother
	Oestridae	None	Parasite	Mammals
	Otitidae	Unknown, none	Scavenger, some herbivores	Decaying materials
	Pipunculidae	Unknown	Parasites	Planthoppers, leafhoppers
	Pyrgotidae	Unknown	Parasite	June beetles
	Sarcophagidae	Scavenger, nectar, none	Scavenger, some parasites	Decaying matter, animal hosts
	Sciomyzidae	Unknown, scavenger	Predator	Water
	Streblidae	Ectoparasitic on bats	Viviparous	Within mother
	Syrphidae	Nectar, pollen	Scavenger, predator, some herbivores	Water, decaying matter, vegetation
	Tachinidae	None, nectar	Parasitic on invertebrates	Host
	Tephritidae	Rotting fruit, bacteria, and yeast	Herbivore	Fruit, some seeds, vegetation

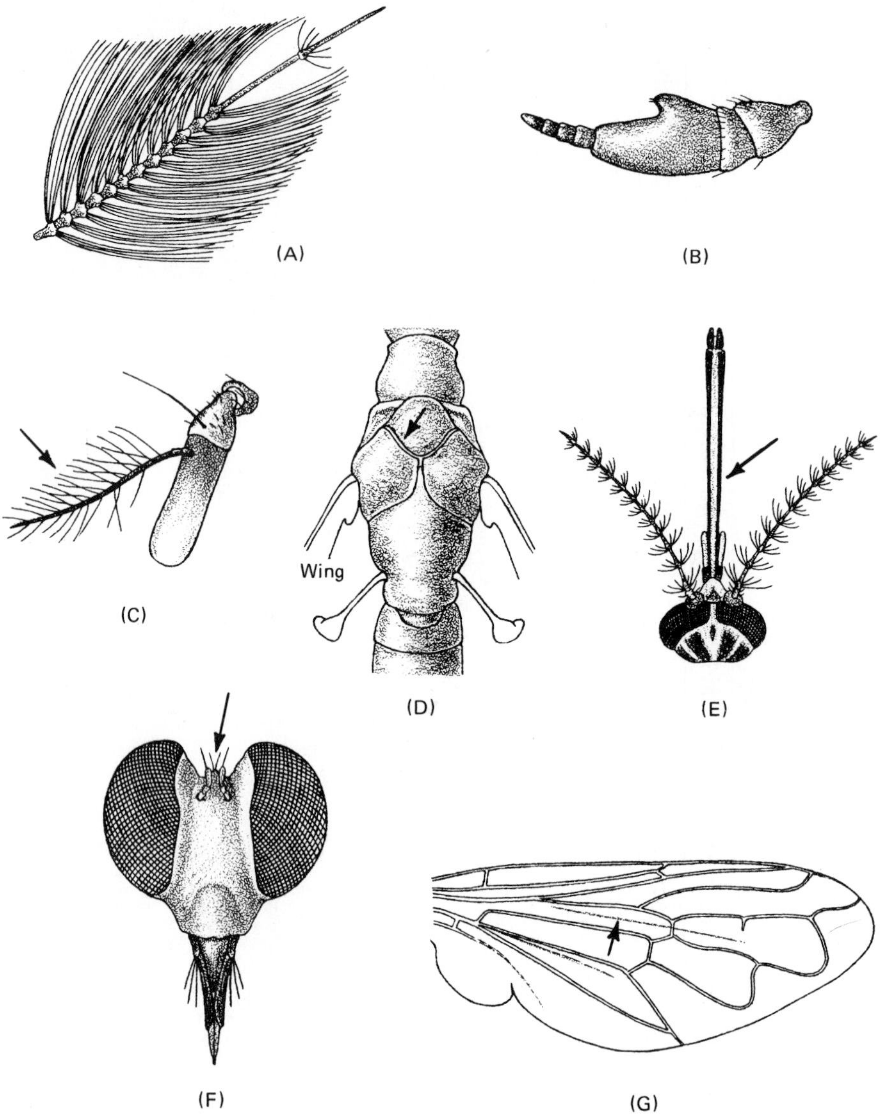

Figure 386. Diagnostic characteristics of adult Diptera. (A) antenna with more than seven segments and plumose (Culicidae); (B) antenna with less than seven segments, arista absent (Tabanidae); (C) antenna with less than seven segments, arista present and plumose to tip (Calliphoridae); (D) mesonotum with V-shaped suture (Tipulidae); (E) long proboscis (Culicidae); (F) vertex indented (Asilidae); (G) spurious vein present in wing (Syrphidae).

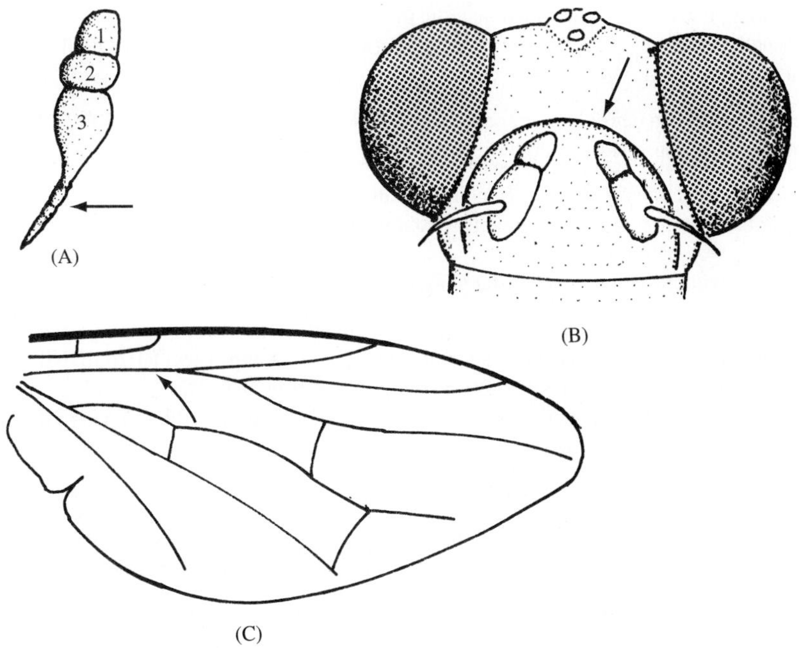

Figure 387. Diagnostic characteristics of adult Diptera. (A) third antennal segment rounded and terminal style present (Rhagionidae); (B) frontal suture (Drosophilidae); (C) subcostal vein bent abruptly toward costa (Tephritidae).

6. Slender bodies and long legs (very fragile appearing) 7
   Bodies not slender 10

7. Mesonotum with V-shaped suture above front edge of wings [Fig. 386(D)] . . . . . (crane flies) TIPULIDAE
   Mesonotum without V-shaped suture 8

8. With long beak [Fig. 386(E)]; scales on wing veins . . . . . (mosquitoes) CULICIDAE
   Beak short or absent; scales on wings absent 9

9. Coxae long . . . . . (fungus gnats) MYCETOPHILIDAE
   Coxae not long . . . . . (midges) CHIRONOMIDAE

10. Antennae arising above level of the middle of the eye; humpbacked appearance . . . . . (black flies) SIMULIIDAE
    Antennae arising at or below lower edge of eye . . . . . (March flies) BIBIONIDAE

11. Arista absent [Fig. 387(A)], although terminal stylus may be present; frontal suture absent 12
    Arista (large spine midway on antennae) present [Fig. 386(C)]; frontal suture usually present [Fig. 387(B)] 17

12. Vertex deeply indented between eyes [Fig. 386(F)]; no metallic coloring . . . . . (robber flies) ASILIDAE
    Vertex not indented or metallic colored 13

13. Legs long; body slender and often metallic in color;
    usually small . . . . (long-legged flies) DOLICHOPODIDAE
    Not as above 14

14. Body usually with dense hair; empodium bristlelike . . . . . . (bee flies) BOMBYLIIDAE
    Body without dense hair; empodium pulviliform 15

15. Antennae often diverge sharply from one another after initial
    parallel direction of basal segments; costal wing vein
    ending well before wing tip . . . . (soldier flies) STRATIOMYIDAE
    Antennae without sharp angle; costal wing vein extending around wing tip. 16

16. Third antennal segment usually rounded, with terminal style
    [Fig. 387(A)] head rounded from dorsal view. . . . . (snipe flies) RHAGIONIDAE
    Third antennal segment elongate or with enlarged lobe [Fig. 386(B)];
    head triangular from dorsal view . . . . (horse and deer flies) TABANIDAE

17. Frontal suture absent 18
    Frontal suture present [Fig. 387(B)]. 19

18. Spurious vein present in wing [Fig. 386(G)] . . . . (flower flies) SYRPHIDAE
    Spurious vein absent in wing . . . . (big headed flies) PIPUNCULIDAE

19. Small (3–4 mm) with reddish eyes . . . . (vinegar flies) DROSOPHILIDAE
    Not as above 20

20. Wings with distinct patterns. 21
    Wings lacking distinct patterns 24

21. Subcosta wing vein bent abruptly toward costa
    [Fig. 387(C)] . . . . (fruit flies) TEPHRITIDAE
    Subcosta wing vein only slightly curved toward costa 22

22. Ocelli absent . . . . (pyrgotid flies) PYRGOTIDAE
    Ocelli present 23

23. Yellow to brown in color. . . . . (marsh flies) SCIOMYZIDAE
    Black or metallic green in color . . . . (otitid flies) OTITIDAE

24. Either specialized wing lobes at posterior base of wing or mouthparts vestigial 26
    Basal wing lobes absent; mouthparts well developed 25

25. Costa broken once (just prior to R_1 vein;
    arista bare) . . . . (chloropid flies) CHLOROPIDAE
    Costa broken twice (at R_1 and at humoral cross vein) . . . . . (shore flies) EPHYDRIDAE

26. Mouthparts vestigial or absent; medium to large flies 27
    Mouthparts functional 28

27. Light yellow in color resembling honeybee . . . . (horse bot flies) GASTEROPHILIDAE
    Dark-colored . . . . (bot and warble flies) OESTRIDAE

28. Ridge or distinct bulge beneath and behind scutellum
    [Fig. 388(A)]; arista bare . . . . (tachina flies) TACHINIDAE
    No distinct ridge or bulge as above [Fig. 388(B)];
    arista at least partly plumose [Fig. 386 (C)] 29

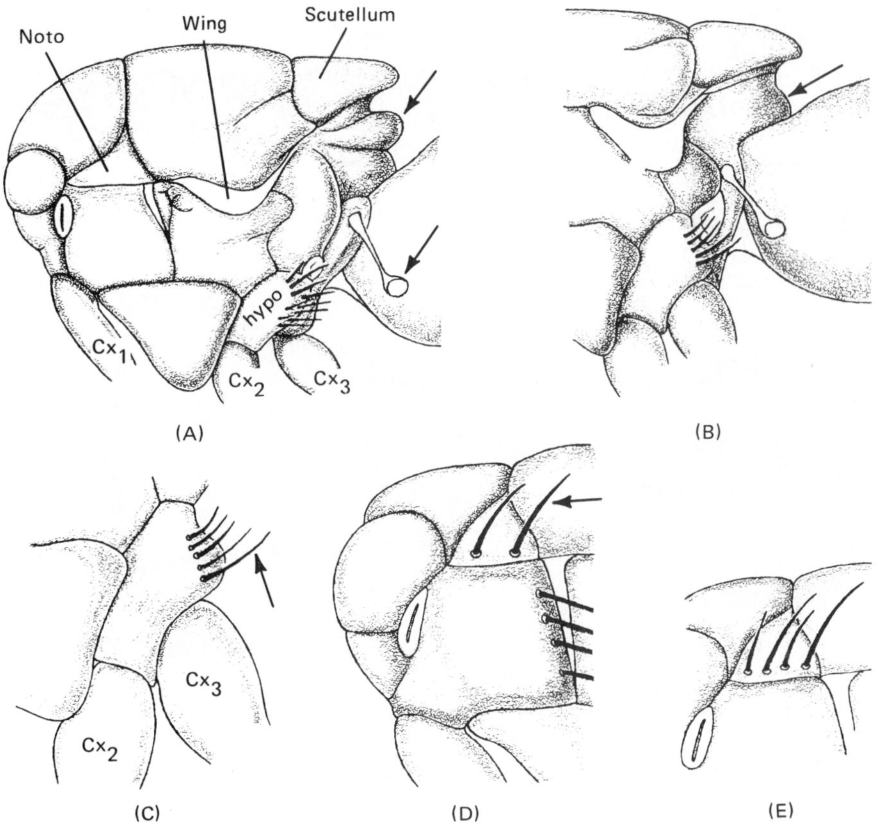

Figure 388. Diagnostic characteristics of adult Diptera. (A) upper arrow indicates distinct ridge behind scutellum, lower arrow points to haltere (Tachinidae); (B) distinct ridge behind scutellum absent (Calliphoridae); (C) hypopleuron with distinct row of bristles (Calliphoridae); (D) notopleuron with two large bristles (Calliphoridae); (E) notopleuron with four large bristles (Sarcophagidae).

29. Hypopleuron without row of bristles 30
    Hypopleuron with distinct row of bristles [Fig. 388(C)] 31

30. Underside of scutellum lacking fine setae . . . . (muscid flies) MUSCIDAE
    Underside of scutellum with fine setae . . . . (anthomyiid flies) ANTHOMYIIDAE

31. Notopleuron with 2 large bristles [Fig. 388(D)]; arista plumose tip [Fig. 386(C)]; often metallic colored . . . . (blow flies) CALLIPHORIDAE
    Notopleuron with 4 large bristles [Fig. 388(E)]; arista plumose only in basal half; color usually gray with dark patterns . . . . (flesh flies) SARCOPHAGIDAE

32. Head small and folds posteriorly into groove . . . . (bat flies) NYCTERIBIIDAE
    Head not as above 33

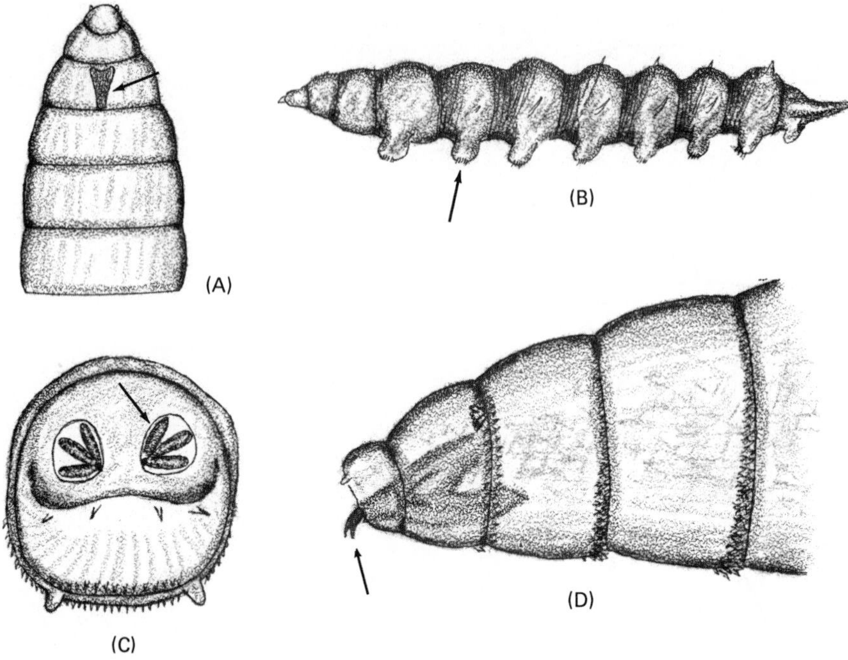

Figure 389. Diagnostic characteristics of Diptera larvae. (A) sclerotized ventral sclerite (Cecidomyiidae); (B) prolegs (Anthericidae); (C) posterior spiracles (Calliphoridae); (D) paired mouthhooks (Calliphoridae).

33. Eyes absent or minute; observable neck present. . . . . (bat flies) STREBLIDAE
    Eyes moderate, round; head partially retracted
    into thorax . . . . (louse flies) HIPPOBOSCIDAE

KEY TO COMMON FAMILIES OF LARVAL DIPTERA

1. Body deeply cleft into 7 areas; 6 median ventral
   suckers present . . . . BLEPHAROCERIDAE*
   Not as above 2
2. Narrow sclerotized ventral plate on thorax [Fig. 389(A)];
   body 5 mm or less in length. . . . . CECIDOMYIIDAE
   Not as above 3
3. Head capsule well developed, may be retracted into thorax [Fig. 384(A)];
   mandibles opposable 4
   Head capsule incomplete or absent [Fig. 385(C)]; mouthparts parallel in action 11
4. Diameter thorax much greater than abdomen;
   terminal breathing tube usually present [Fig. 385(A)]. . . . . CULICIDAE*
   Thorax not enlarged or only slightly; breathing tube absent 5

5. Large mouth fans present [Fig. 304(E)]; posterior abdominal segment with single sucker surrounded by hooks . . . . SIMULIIDAE*
   No conspicuous mouth fans or sucker as above. 6

6. First 2 abdominal segments with prolegs . . . . DIXIDAE*
   First 2 abdominal segments lacking prolegs 7

7. Prolegs on prothorax and terminal segment . . . . CHIRONOMIDAE*
   Prolegs absent on either prothorax or terminal segment; if both present, many abdominal filaments present. 8

8. Head distinct and not retracted. 9
   Head retracted into prothorax [Fig. 384(A)] . . . . TIPULIDAE*

9. Body very slender and head longer than broad *or* with long fleshy filaments on many segments . . . . CERATOPOGONIDAE*
   Body not as above 10

10. Body terminating with short breathing tube . . . . PSYCHODIDAE
    Body lacking breathing tube. . . . . BIBIONIDAE

11. Prolegs present, with crochets [Fig. 389(B)] 12
    Prolegs absent or with small creeping ridges [Fig. 385(C)]. 18

12. Long terminal breathing tube or tubes present [Fig. 384(C)] 13
    Terminal breathing tubes absent or short, fleshy filaments may be present 14

13. Abdominal breathing tube unbranched [Fig. 384(C)] . . . . SYRPHIDAE*
    Abdominal breathing tube branched at tip . . . . EPHYDRIDAE*

14. Largest prolegs in middle of body. . . . . TABANIDAE*
    Prolegs equal in length or terminal pair longest. 15

15. Terminal pair of prolegs long, others reduced . . . . EMPIDAE*
    Prolegs of equal length; if terminal longest, then others also well developed. 16

16. Two or more caudal fleshy filaments. 17
    Single caudal breathing tube. . . . . EPHYDRIDAE*

17. Terminal prolegs enlarged . . . . EMPIDIDAE*
    All prolegs similar in size [Fig. 389(B)] . . . . ANTHERICIDAE, RHAGIONIDAE*

18. Parasites of livestock 19
    Not parasites of livestock. 21

19. Most segments with large posteriorly directed spines [Fig. 385(B)]. . . . . (bot flies) GASTEROPHILIDAE
    Large spines lacking. 20

20. Body ovoid; caudal spiracular slits with various patterns . . . . (cattle grubs) OESTRIDAE
    Body cylindrical; each spiracle with 3 distinct slits . . . . (screwworms) CALLIPHORIDAE

21. Large maxillae with short palpi 22
    Maxillae small or mouthhooks present [Fig. 389(D)] 24

22. Caudal spiracles in pocket or at end of elongate terminal segment;
    body surface often rough or striated 23
    Caudal spiracles on eighth abdominal segment and visible . . . . . . ASILIDAE

23. Head fixed in position; body usually
    with roughened surface . . . . STRATIOMYIDAE*
    Head retractable; body smooth or longitudinally striated;
    segmental swellings often present [Fig. 384(B)] . . . . . . TABANIDAE*

24. Caudal spiracles small, on separate lobes; fleshy lobes may be present
    on body segments. 25
    Caudal spiracles not on separate lobes but flush
    with body surface or in a depression 26

25. Most segments with fleshy protuberances. . . . . MUSCIDAE
    Most segments lacking fleshy protuberances . . . . . . DROSOPHILIDAE, others

26. Caudal spiracles small, inconspicuous 27
    Caudal spiracles large, conspicuous [Fig. 389(C)] 28

27. Three pairs of prominent tubercles ventral to spiracles . . . . . . ANTHOMYIIDAE
    Not as above. . . . . SYRPHIDAE

28. Single mouthhook present. . . . . MUSCIDAE
    Two mouthhooks present [Fig. 389(D)]. 29

29. Caudal spiracles in deep depression;
    spiracular slits nearly vertical . . . . SARCOPHAGIDAE
    Caudal spiracles in shallow depression; spiracular slits horizontal
    to 45° angle or in wavy pattern 30

30. Parasites in insects; spiracular slits straight or wavy . . . . . . TACHINIDAE
    Saprozoans; spiracular slits only straight [Fig. 389(C)] . . . . . . CALLIPHORIDAE

*Aquatic.

14 Making an Insect Collection

WHERE AND HOW TO COLLECT

Before starting to make a collection, one should be aware that rare and endangered insects, like other rare and endangered plants and animals, are protected by law. It is the responsibility of the *collector* to be aware of these restrictions. Also written permission must be obtained prior to collecting in wildlife reserves, land tracts managed by the Nature Conservancy, or state and local parks. Collecting in national parks is forbidden by law.

To secure a representative collection of insects during any season of the year, one must be constantly alert for different species. A collection built from specimens picked up at random or over a very short period is never representative of the species that occur in a locality. Therefore, make definite collecting trips to various habitats and localities to secure the best material. Large numbers of insects may be caught by turning over stones, logs, rubbish, and leaves. Sweepings in alfalfa, grass, and weed patches will reveal entirely different species. Butterflies may be caught by dropping a net (Fig. 390) over them when they alight. Water insects may be seined using a water net. Various insects are attracted to sap exuding from stumps or tree trunks, which furnish good collecting in the very early spring. Soil insects may be located by placing humus and leaf matter in a berlese funnel or by using different sieves to sort material by size.

The list of habitats given below is by no means an exhaustive one, but it does include sites where representatives of orders may be found.

Apterygota (including springtails and bristletails). Under old paper, leaves, on old books, and in flour mills. Especially common in damp places such as under boards or decaying vegetation.

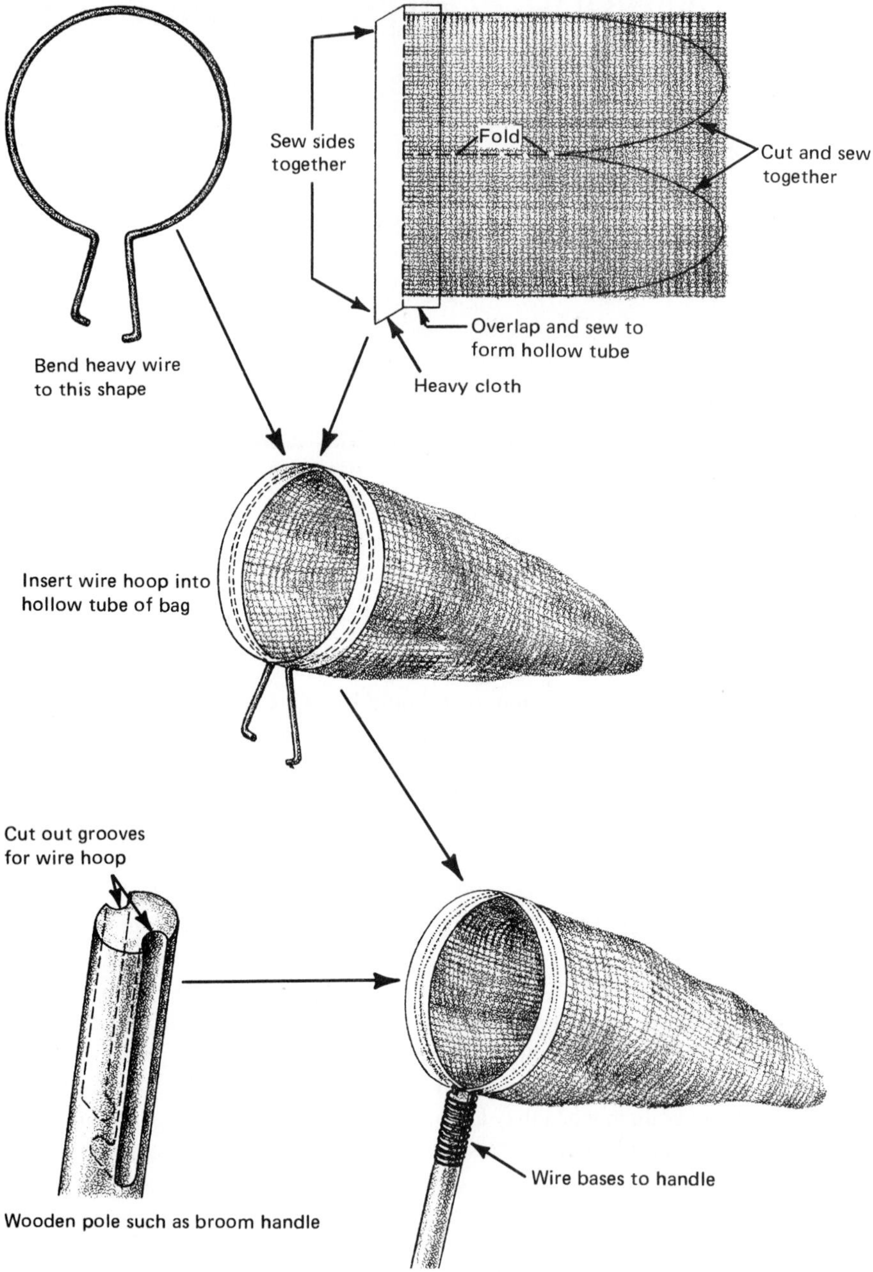

Figure 390. Constructing a simple insect net.

Ephemeroptera (mayflies), ***Odonata*** (damselflies, dragonflies), ***Plecoptera*** (stoneflies), and ***Trichoptera*** (caddisflies). Adults near streams and ponds resting on rocks, reeds, and bushes, or near lights. Immatures in streams, ponds, and other water habitats.

Orthoptera (grasshoppers, crickets, etc.). In long grass, under bark, under logs and rocks, at edges of fields, at lights, in trees, and in damp basements.

Blattaria (cockroaches). Under bark, rocks, and rotting wood and in houses and damp basements, garages, and garbage deposits.

Mantodea (mantids). On vegetation and near lights.

Phasmida (walkingsticks). On trees and shrubs.

Isoptera (termites). At the base of fence posts, under and in rotting wood, under dung in pastures, in weed stems and trees, and in discarded boxes and newspaper.

Psocoptera (psocids and booklice). In old books or papers and on tree trunks.

Anoplura (sucking lice). On livestock and rodents.

Mallophaga (chewing lice). Generally on birds.

Hemiptera (true bugs). In grass and weed patches, under rocks, near or in water, under bark and leaves, at lights, and in gardens.

Homoptera (aphids, leafhoppers, etc.). Around shrubbery, grass, weed patches, and many crops.

Thysanoptera (thrips). Common in most flowers.

Coleoptera (beetles). In trees and shrubbery, under rocks, beneath bark, in rotting logs, under decaying matter, in water, at lights, and around flowers.

Megaloptera (alderflies, dobsonflies). Vegetation around streams, in streams as larvae, and around lights.

Neuroptera (lacewings, etc.). On vegetation, at lights, and some in soil during larval stage.

Mecoptera (scorpionflies). Near streams and in dense woods and vegetation.

Lepidoptera (butterflies, skippers, and moths). At flowers, on trees, at lights, and on many types of vegetation as larvae.

Diptera (flies, etc.). At lights, on flowers, near water, about livestock and fecal material, and near decaying matter.

Hymenoptera (bees, wasps, ants, etc.). On flowers, near mud banks, at lights, under stones and wood, on shrubbery and trees, around barns and houses, under bridges, and in soil.

KILL BOTTLE

A suggested design of a kill bottle is provided in Figure 391. Any size bottle can be used, but the most useful sizes are half-pint and pint wide-mouth jars. Thick-walled bottles are best because they are less likely to break. Prepare a relatively thick solution of plaster of Paris and pour in over the layer of sawdust that was previously

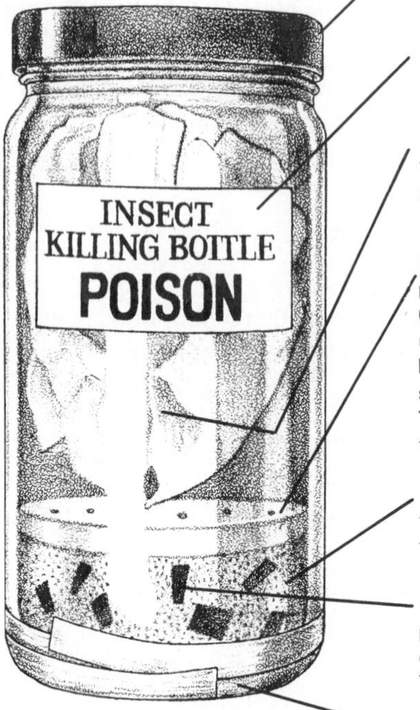

Figure 391. One way of constructing a kill bottle.

compacted into the jar. As the solution begins to set, thrust a pencil through the layer at several points to form ducts for the fumigant to penetrate. Allow the mixture to dry for about a week with the lid completely off. Fumigant, such as ethyl acetate, may be added after the plaster is dry. A layer of fumigant 1 cm deep will be completely absorbed in a standard pint kill bottle. DO NOT INHALE THE FUMIGANT. LABEL THE BOTTLE "POISON." A few strips of blotting paper placed in the completed bottle will absorb excess moisture and keep the interior dry.

As soon as hard-bodied insects are collected, they should be placed in the kill bottle. Allow them to remain for approximately 1 hour to ensure death. During a collecting trip in which many kinds and numbers of insects are being collected, it is wise to have several bottles or to remove the more delicate insects to prevent damage by larger insects. Fumigants are lost when the bottle is opened and closed; therefore, add them periodically, particularly prior to collecting trips and at other times when insects are not killed within 2 to 3 minutes. Large butterflies, moths, beetles, and dragonflies may be placed in a separate bottle with a piece of cotton saturated with fumigant, or they may be injected with alcohol or formalin by using a syringe.

Soft-bodied insects, especially immature stages, should *not* be placed in the kill bottle but should be killed by inserting them into or injecting them with alcohol. Although ethyl or isopropyl alcohols are adequate for preserving these fragile insects for future observations, other preservatives are available and are preferred if the specimen is to be placed in a research collection (see subsequent section on liquid presentation).

RELAXING

If, for any reason, collected insects cannot be mounted until they are dry, relaxing them will prevent breakage. **This step is vital**. A relaxing jar is easily constructed by placing 5 cm of sand in the bottom of a glass jar. The sand is then saturated with water, to which a few drops of formalin or carbolic acid crystals have been added to prevent mold growth, and a piece of blotting paper is placed on the sand surface to receive the specimens. The lid is then closed for 1 to 3 days, depending on the size of the insects (the larger the size, the longer the period). Be careful, because too long a period results in fungal growth and loss of the specimen.

PINNING

Much of the appearance of the insect and ease in viewing diagnostic characteristics depends on the neatness and care with which pinning is performed. Figure 392 shows where the special insect pin is to be inserted in some of the major orders. Pin all Coleoptera through the right elytron, one-quarter the distance from the base of the wing. Hemiptera and Homoptera should be pinned through the scutellum. Other orders should be pinned through the mesothorax. Specimens that are too small and delicate to be pinned normally are mounted either on cardboard points or on micropins. If points are used, the insect is touched by the point to which a small dab of shellac, nail polish, or glue has been added and is positioned to the left of the insect pin with the head forward (Fig. 392). Micropins are short, very delicate headless pins that should pierce the insect from the ventral side. Care should be taken so that the pin does not penetrate the dorsum. The other side of the micropin is in turn connected to the insect pin through bits of cork or pith.

It is best to spread the wings of moths and butterflies. First, pin the insect through the mesothorax and insert the pin in the crease of the pinning board (Fig. 393). Move the forewing anteriorly until the back margin forms a right angle with a line parallel with the long axis of the body. Hold the wing in position temporarily with a pin behind the large anterior veins. Move the hind wings forward to their proper position. Next, cut small strips of paper and place them across the wings and pin around the wing outline (exercise care not to penetrate the wings). The temporary pins may now be removed. Three to five days may be required, especially for large specimens, to permit thorough drying in areas with high humidity.

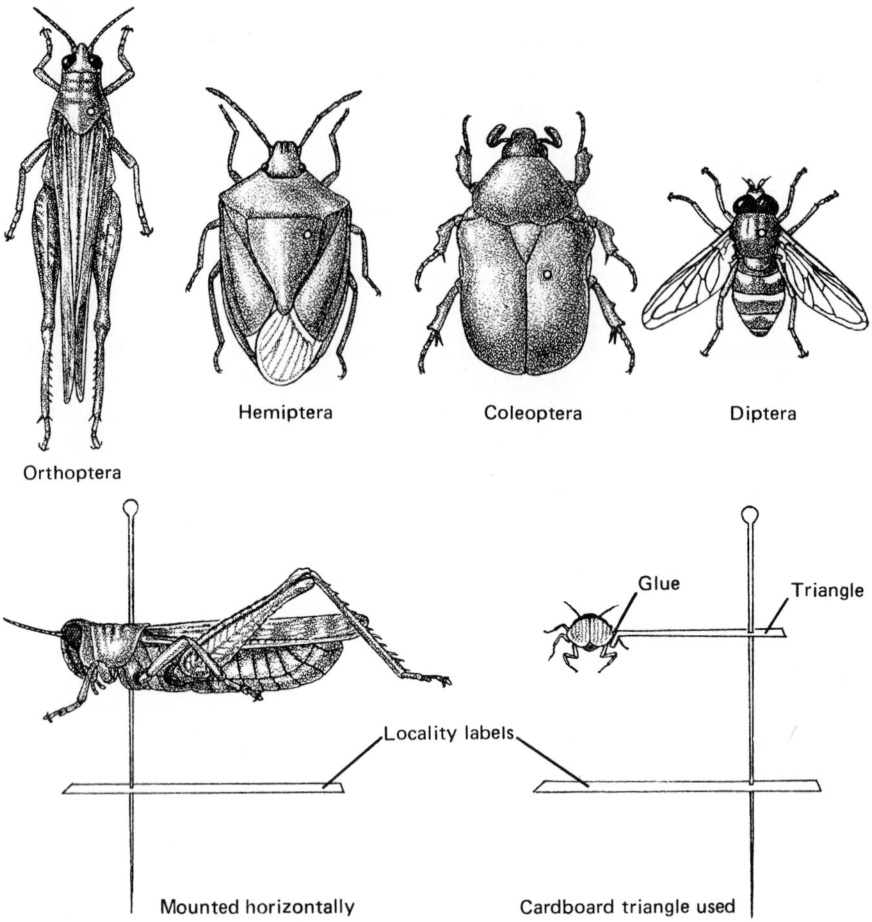

Figure 392. Pin positions for insects.

LABELING

All specimens should bear collection site and date of capture on the *locality label* (Fig. 392). Be accurate, because incorrect information is worse than no data. If you do not know the exact date, include only what is definite. Dates should be written in one of the following styles: 10 Aug. 1977, 10.VIII.1977, or VIII.10.1977. Never use 8/10/77 or 10/8/77. A suggested form is:

KANSAS: Riley Co.
Manhattan
VIII.10.1977

Labels placed on insect pins should not exceed 18 mm by 5 mm in size. Names of collectors are placed on a second label beneath the locality label. Photographic col-

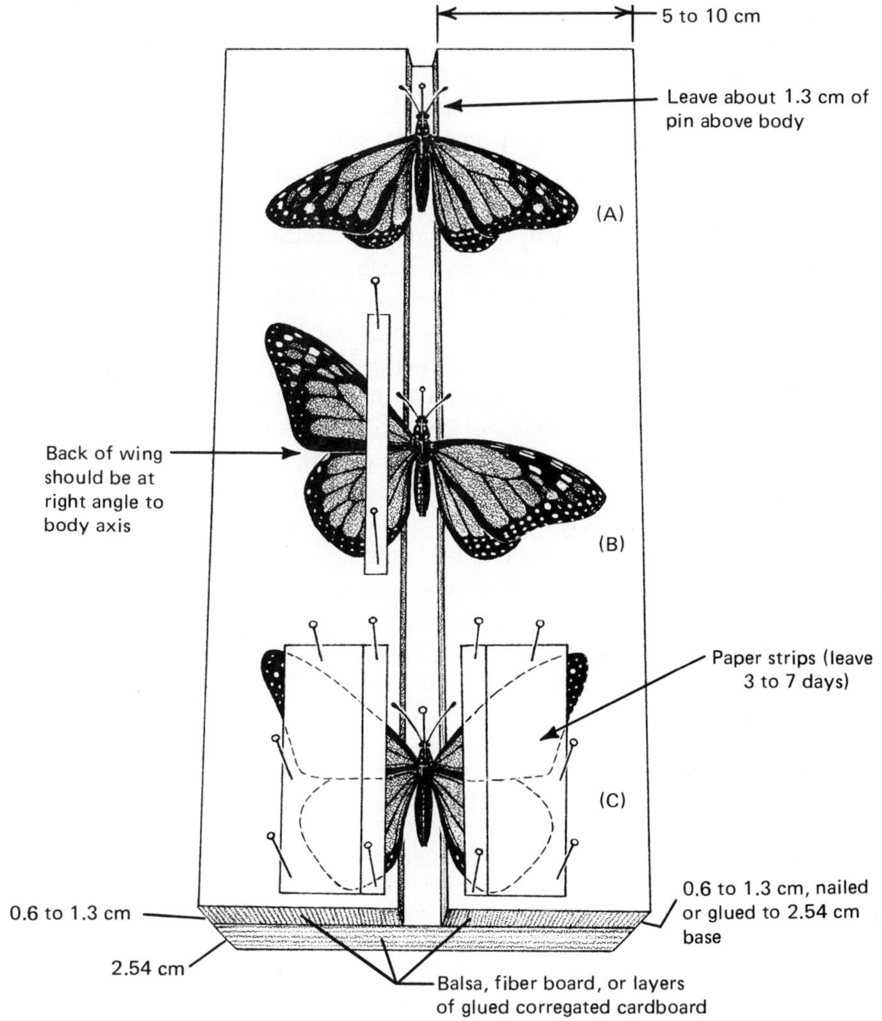

Figure 393. Spreading board for drying insect wings in a horizontal position.

lecting labels may be secured by filling a typewritten sheet with columns of the various data desired. This sheet is then reduced by photography and printed. Computers can be programmed to produce the necessary labels.

DISPLAY AND STORAGE

Collections become meaningful when specimens can be studied or displayed. Museum and many private collections are normally housed in wooden or steel cabinets containing glass-covered museum drawers. Each drawer, in turn, contains unit trays (cardboard trays lined with polyethylene foam pinning bottoms) that permit the pinned specimens to be easily inserted or removed as needed. Each tray contains a single species, and the trays are usually arranged alphabetically by species within a genus, by genus within a family, and by family within an order. Cabinets are usually arranged in phylogenetic order for easy reference. Even with tight-fitting drawer tops and cabinet doors, insect pests, such as dermestid beetles, often gain entry; therefore, a repellent or fumigant (mothballs, insecticide) must be used, and drawers must be inspected routinely for signs of insect damage (powdery residues under pinned specimens that indicate pest feeding). Damage is least likely when the cabinets are maintained in rooms with low temperatures and moisture and where specimens are fumigated prior to storage.

For the beginner, expensive cabinets and museum drawers often are not realistic; therefore, how can specimens be stored and permit viewing? The least expensive method is by using cigar boxes or Schmitt boxes, but these lack glass coverings and increase the likelihood of accidental damage. However, these boxes can be stored in appropriate containers or areas, but frequent inspections for pests are recommended. A second possibility is the use of glass-topped display boxes other than museum drawers. Such boxes are less likely to be available from biological supply houses; however, they can be made by local cabinet makers or by the collector to the size desired. Again, a polyethylene foam pinning bottom is inserted to retain the pinned specimens, and a repellent or insecticide can be added to prevent insect pest damage. Computer-generated or hand-printed order or family labels can be pinned into the boxes in appropriate positions for displaying the diversity of the specimens.

LIQUID PRESERVATION

Many insects, particularly immatures, are soft-bodied and cannot be adequately preserved by pinning because shrinkage and body distortion will make identification difficult or impossible. Such organisms are best preserved by killing and maintaining them in liquids.

The most widely used preservative is 70 to 80 percent isopropyl or ethyl alcohol to which glycerine is sometimes added (about 1 percent or less of total volume). Alcohol should be replaced 2 to 3 days after preserving an arthropod because it will be diluted by body fluids. Such a collection may be maintained indefinitely if alcohol levels are checked periodically. Glycerine keeps the specimens pliable even if the alcohol should evaporate from the container, but it does clear specimens with time. Unfortunately some shrinkage always occurs, and colors do not remain true, especially the greens.

Special liquid preservatives have been developed to maintain body rigidity and prevent darkening of the specimens. The one that seems most satisfactory for larvae to the author is as follows:

Kerosene	30 parts
Glacial acetic acid	20 parts
95% ethyl alcohol	90 parts
Isobutyl alcohol	30 parts
Dioxane (not dioxin)	10 parts

The fluids are mixed in a descending order of the listing. Kerosene may be reduced 50 percent if very "thin-skinned" specimens such as dipteran larvae are to be collected (to prevent bursting as they straighten). For best results, specimens should be placed directly into the solution while alive. Immatures with gills should not be preserved in this fluid as gills tend to become distorted. Insects with thick exoskeletons also are best preserved in other fluids; if adults, they should be pinned. After several days to a week, the fluid should be removed and should be replaced with 70 to 80 percent isopropyl or ethyl alcohol for permanent storage, so that fats (these give many larvae their natural white color) will not leach from the specimen into the solution.

Collection and storage are usually done in vials. Patent-lipped vials with neoprene stoppers are best because alcohol evaporation is minimal, but they are expensive, and so shell caps or screw caps are often substituted. However, long-term storage in the latter types is risky because tight seals are uncommon. The level of alcohol should be monitored regularly and replaced as needed; rapid loss may require the replacement of the vial and cap. The most popular vial sizes are 2 or 4 drams. Vial racks are handy for storage and permit ready access to the specimens; vials should not be placed into collection boxes with pinned specimens because they often roll about and can destroy much of the collection.

Each vial should contain a label identifying the collection site, date, and collector as minimal data. A *separate* label with identification and the identifier's name may be added later. Labels should be made in India ink or using soft pencil; *never* use ballpoint pen because its ink is soluble in alcohol. Photographic or computer-produced labels are handy but practical only if large collections are made.

If specimens become dry through evaporation of alcohol from the vial, they may be reconditioned. An often-used, but somewhat inferior, method is soaking specimens in a weak solution of potassium hydroxide (less than 5 percent) until normal shape is restored. Washing in weak acetic acid and then transferring to 25, 50, and finally 75 percent alcohol completes the process. Acetic acid neutralizes any remaining KOH, and the slow exposure to increased alcohol concentrations prevents collapse. Another and better technique involves boiling the specimen in 85 percent alcohol and then transferring it to cool alcohol of the same concentration. This latter method is usually easier and safer and produces more reliable results than the KOH technique.

Glossary

Abdomen: The major region of the body posterior to the thorax or cephalothorax.

Adaptability: Capacity for evolutionary change.

Adaptive radiation: Evolution and spread of a population of organisms into different ecological niches.

Aerobic respiration: The oxidative process of releasing energy within the cell.

Aeropyle: Specialized pores on the egg surface for oxygen uptake.

Aggregation: Grouping by a species resulting from some attractive stimulus.

Alimentary canal: The tube between the mouth and anus in which food movement, digestion, and absorption occur.

Allele: The alternate form of a gene, having the same locus in homologous chromosomes.

Allelochemical: Substance that mediates an interaction between individuals of different species.

Allomone: Allelochemical giving the emitter an adaptive advantage.

Ametabolism: Development in insects where no metamorphosis occurs.

Amino acid: Any organic acid containing an amino radical (NH_2); the building stones of proteins.

Anaerobic respiration: Releasing energy within the cell without oxygen.

Anamorphic development: Development by adding segments to the body.

Androconia: Patch of secretory scales in Lepidoptera.

Annulets: Rings within a segment produced by constrictions.

Antennae: A pair of segmented sensory appendages on the head.

Antenodal cross veins: Cross veins from the costa, in the region from the wing base to the node in Odonata.

Anus: The posterior opening of the alimentary canal.

Apolysis: Freeing of the epidermis from the old cuticle.

Apophysis: Elongate process of the exoskeleton.

Aposematic: Brightly colored insects producing bad taste or experience to vertebrate predators.

Apterous: Lacking wings.

Arculus: First cross vein between radius and cubitus in Odonata wings.

Arista: Large bristle on the third segment of the antennae of specialized flies.

Arthropoda: A phylum of animals, all of whom have segmented bodies and legs and exoskeletons.

Asexual reproduction: Reproduction not involving the fusion of nuclei from different individuals.

Aspergites: Small spinelike protuberances, usually in rows.

ATP: Adenosine triphosphate; a molecule having high-energy bonding through a phosphate group.

Basitarsus: The tarsal segment attached to the tibia.

Batesian mimicry: Palatable species protected from predators by resembling aposematic models.

Bilateral symmetry: Symmetry in which the body can be divided into halves that are mirror images of one another.

Biomass: The amount of living matter.

Biome: A group of communities of plants having a similar life form.

Biordinal: A row of crochets with two alternating lengths.

Blastoderm: The cells produced through cleavage of the zygote.

Blastokinesis: Movement or migration of the embryo within the egg.

Brachypterous: With reduced wing length, yet capable of flight.

Carbamates: A group of synthetic organic insecticides that are cholinesterase inhibitors.

Cardo: The first segment of the maxilla.

Carnivore: An eater of animals.

Cephalothorax: A body region having the characteristics of both a head and a thorax.

Cercus (pl. cerci): Appendage of the eleventh abdominal segment.

Chelicerae: Appendages of the third body segment that become the major feeding structures in Arachnida.

Chitin: A nitrogenous polysaccharide with an empirical formula of $C_8H_{13}0_5N_n$.

Chlorinated hydrocarbons: Synthetic insecticides that act both as contact and stomach poisons; most have long life and degrade slowly.

Circadian rhythm: Physiological rhythm with an approximate 24-hour cycle.

Classification: The process of dividing and subdividing into a system of categories.

Clypeus: Sclerite beneath the frons.

Coarctate pupa: Pupa remains within the last larval exoskeleton, the puparium.

Community: All plant and animals living in a specific region.

Complete metamorphosis: Holometabolic development; immatures (larvae) do not resemble the adults, and a pupal stage is present.

Compound eye: Optic organ formed from ommatidia.

Cornicles: A pair of dorsal tubes on the abdomen of some Homoptera.

Corpus allatum: Endocrine gland located behind the brain which produces JH, or the juvenile hormone.

Coxa: Basal segment of a leg.

Crepuscular: Active during low periods of light such as twilight.

Crochets: Sclerotized hooks found on the prolegs of some holometabolous immatures.

Cryptic coloration: Coloration of organism that blends with background.

Cuticle: The secreted part of the integument.

Density-dependent: Factor whose influence varies with the density of the population.

Density-independent: Factor whose influence does not vary with the number of individuals in a population.

Diapause: A state of hibernation or suspended development in insects.

Digestion: Enzymatic hydrolysis of food into absorbable-size molecules.

Dimorphic: Having two different forms or sizes.

Dioecious: Condition in which male and female reproductive organs are in separate individuals.

Diploid: Possessing a double set of chromosomes ($2n$).

Diurnal: Active during daytime hours.

Diversity: Measure of variety of species that takes into account the relative abundance of each species.

Dorsum: Upper part of body or back.

Ecdysial suture: Weakened area of integument that splits during molting to permit the old exoskeleton to be shed.

Ecdysis: The process of shedding an exoskeleton.

Ecdysone: A hormone produced in the prothoracic glands that activates epidermal cells to secrete a new exoskeleton during the process of molting; also causes differentiation of adult tissues when not inhibited by the juvenile hormones.

Eclosion: Process of an individual's emerging or hatching from an egg.

Ecological efficiency: Percent of energy in biomass produced by a trophic level that is incorporated into the biomass by the next higher trophic level.

Ecological niche: The way of life of an organism, including food, habitat requirements, and so forth.

Economic injury level (EIL): Pest population level at which economic incremental damage equals the cost of preventing the damage.

Economic threshold (ET): Pest population density that controls measures need to be applied to prevent EIL from being reached.

Ecosystem: The living and nonliving factors and their interactions in an area.

Ectoderm: The outer germ layer of an early embryo.

Ectoparasite: A parasite that feeds on the surface of the host.

Ectotherm: An organism that does not maintain body temperatures above ambient; often referred to as *cold-blooded* or *poikilothermic*.

Elytra (sing. elytron): Thickened and hardened front wings of beetles.

Empodium: A median lobe or process originating between the claws.

Endocrine: A ductless gland producing hormones that are released into the blood for distribution.

Endoderm: The innermost layer of cells produced in an early embryo that becomes part of the lining of the alimentary canal.

Endoparasite: A parasite located within the body of its host.

Endotherm: An organism that maintains body temperature above ambient temperature.

Entomophobia: The fear of insects and closely related animals.

Environment: The total surroundings of an organism.

Enzyme: An organic catalyst.

Epimorphic development: Development whereby immatures possess the same number of body segments as adults.

Estivation: Reduction of biological activity by an organism in response to periods of hot or dry weather.

Exarate pupa: Pupa in which the appendages are not appressed to the body.

Exocrine: Tissue or gland that produces and releases pheromones or allomones.

Exoskeleton: A skeleton on the outside of the body.

Exuviae: The cast exoskeleton after molting has occurred.

Femur (pl. femora): Third segment of the leg from the body.

Fertilization: The fusion of an egg and sperm.

Food chain: The direction of energy flow between species that feed upon one another.

Frons: Area between eyes bordered above by the vertex and below by the clypeus.

Furca: Forked endosternal process in thorax of higher insects.

Galea: Outer distal lobe of the maxilla.

Gamete: A mature reproductive cell, either sperm or ovum.

Ganglion (pl. ganglia): A concentration of motor nerve cell bodies and association neurons performing nervous coordination.

Gena (pl. genae): Cheek area of insect head below compound eye.

Gene pool: The entire genetic complement of a population.

Genetic drift: Change in allele frequency as a result of random occurrences, variations in fecundity, and mortality in a population.

Genotype: The sum total of all genes found in an organism.

Germarium: Terminal portion of gonad tubule containing either oögonia or spermatogonia.

Haltere: Modified metathoracic wing used as balancing organ during flight.

Haploid: Possessing a single set of chromosomes (n).

Hemelytra: Front pair of wings of true bugs (Hemiptera); the basal part is thickened, and the distal part is membranous.

Hemimetabolous: Designating development that produces gradual changes in form during growth from an immature to adulthood.

Hemocoel: Blood cavity not entirely lined by mesoderm.

Herbivore: An organism that consumes living plant material or sap.

Heterotroph: Describing organisms that are unable to synthesize their own food.

Heterozygous: Containing two forms (alleles) of a gene, one from each parent.

Holometabolism: Development that produces drastic changes in form during growth from an immature to adulthood.

Homozygous: Containing identical alleles of a gene.

Hyperosmotic: Having a salt concentration greater than that of the surrounding medium.

Hypognathic: Condition where chewing mouthparts are located beneath the head capsule.

Hypopharynx: Tonguelike structure anterior to labium in chewing mouthparts.

Hypopleuron: Pleural plate located directly above and between the last two pairs of legs in adult flies.

Hyposmotic: Having a salt concentration less than that of the surrounding medium.

Imaginal disc: Cluster of undifferentiated cells in larvae that will give rise to adult structure.

Incomplete metamorphosis: Hemimetabolic development; the immatures (nymphs) resemble the adults, and a pupal stage is lacking.

Inquiline: A symbiotic organism associated with social insects. Such organisms are usually commensals, but some are carnivores or parasites.

Insecticide: Substance that is more lethal to insects than to other organisms.

Instar: The insect from one molt to the next.

Integument: The body wall or exoskeleton.

Intima: The extension of the exoskeleton into the internal cavities of the body for lining the foregut, hindgut, and tracheae.

Irritability: The ability to respond to stimuli.

Juvenile hormone: Any of a group of hormones produced in corpora allata that maintains functioning of immature genes and inhibits metamorphosis.

Kin selection: Selection operating between closely related individuals to produce cooperation.

***K*-selection:** Favoring population densities near carrying capacity (*K*).

Labellum (pl. labella): Modified tip of the labium of some flies.

Labium (pl. labia): Fused third pair of appendages in insect mouthparts.

Labrum: Sclerite beneath the clypeus; serves as upper lip in insects that have chewing mouthparts.

Lacinia: Inner distal lobe of maxilla.

Lamellate: Platelike lateral processes from antennal club segments.

Larva: Immatures stage in life cycle; often restricted to specific forms such as feeding immatures in insects that have complete metamorphosis.

Larvapod: Abdominal appendage of immature insects.

Life cycle: The sequence of events from egg to adult.

Macropterous: With normal length wings.

Mandibles: Appendages of the fourth body segment that become the first pair of mouthparts in mandibulate species.

Maxillae: Second pair of feeding appendages in mandibulate species.

Meconium: Excretory material stored by the pupa and released upon metamorphosis to an adult.

Meiosis: Process of reducing the number of chromosomes from diploid to haploid in gamete formation.

Mesenteron: The middle portion of the alimentary canal that initially forms as a blind sac.

Mesoderm: The embryonic group of cells forming between the ectoderm and endoderm.

Meson: Imaginary plane dividing the body into right and left halves.

Mesothorax: The second or middle segment in the insect thorax.

Metabolism: The many chemical reactions and concomitant energy changes that occur within living cells.

Metamere: A somite or segment.

Metamorphosis: Change in shape from one stage to another in the life cycle.

Metathorax: Third segment in thorax of insect.

Micropterous: With greatly reduced wings.

Mimicry: Superficial resemblances between two organisms that result in protection for at least one organism, the mimic.

Molting: The process of shedding an exoskeleton in arthropods.

Mutualism: Symbiotic relationship requiring both organisms for survival.

Mutation: Process that results in gene changes from one heritable allele to another heritable allele.

Natural selection: Differential survival and reproduction of individuals.

Neoptera: Insects capable of flexing or folding their wings.

Neuron: A nerve cell, one highly specialized for conducting cellular impulses.

Nocturnal: Pertaining to the night hours.

Notum (pl. nota): The dorsal plate or tergum of each thoracic segment.

Nymph: The immature stage in incomplete metamorphosis.

Obtect pupa: Pupa that has appendages closely appressed to the body.

Occiput: Portion of head behind the vertex, eyes, and gena.

Ocellus (pl. ocelli): Photoreceptor located between compound eyes in adults and nymphs.

Ommatidium (pl. ommatidia): Visual unit of a compound eye.

Omnivore: An organism that has a broad diet including both plant and animal material.

Oötheca: An egg pod secreted by the accessory glands.

Organic phosphates: A group of cholinesterase-inhibiting insecticides that are usually short-lived; some are highly toxic to mammals.

Osmeterium: An eversible process from the prothorax of papilionid larvae that is used to repel predators through its odors.

Osmoregulation: Regulation of the salt and water concentration in cells and organisms.

Oviparous: Egg laying with hatching outside the body of the female.

Oviposition: Placing of eggs in suitable position and habitat.

Ovipositor: Egg-laying organ.

Ovoviviparous: Eggs hatch within the female's body.

Paedogenesis: Reproduction by immature insects.

Paleoptera: Insects that are not capable of flexing or folding their wings.

Palpus (pl. palpi): Segmented appendage on maxillae and labium.

Paranotum (pl. paranota): Notal flange in fossil insects.

Parasite: An organism living at the expense of another.

Parasitoid: A parasite that eventually kills its host.

Parthenogenesis: Development from an unfertilized egg.

Pedipalpi: Second pair of appendages in Arachnida; homologous to mandibles.

Peritrophic membrane: Structure generated within the mesenteron that protects the lining cells from the abrasive action of ingested materials.

pH: Symbol denoting relative concentration of hydrogen ions in solution; a scale of relative acidity and alkalinity.

Pharate: An individual insect after apolysis and prior to molting.

Pheromones: Chemical substances used for communication between individuals of the same species.

Phoresy: The process of being transported from one area to another without using the host for food.

Photoperiodism: Seasonal responses to changes in length of daylight.

Phylum (pl. phyla): A category of classification comprised of closely related classes.

Physogastric: Body swollen with eggs.

Plastron: Structure to obtain dissolved oxygen.

Pleural sulcus: Groove running dorsally from coxa; often referred to as *pleural suture*.

Pleuron: Lateral plate in an insect segment.

Plumose: Featherlike, usually referring to the antennae.

Pollination: Transfer of male gametophyte (pollen) to female flower or female part of flower.

Polyethism: Differences in behavior between individuals and castes in social insects.

Polymorphism: Variability into distinctive phenotypes within a series population, often resulting in distinctive groupings or morphs.

Population: All organisms sharing a gene pool.

Predator: Organism that requires feeding on and killing more than one other animal to complete its development and reproduce.

Proctodeum: Posterior division of alimentary canal formed from invagination.

Prognathous: The position of mouthparts directed forward to plane of body.

Proleg: A fleshy abdominal leg.

Pronotum: The dorsal plate in the prothorax.

Prothorax: First segment of insect thorax.

Pseudergate: The last instar apterous worker immature in lower termites that is capable of molting into any caste.

Ptilinum: Eversible sac in some dipterous adults that is used to rupture the puparium.

Pupa: Intermediate stage between larva and adult in complete metamorphosis.

Puparium: Covering of pupa in some dipterous species; represents last larval exoskeleton.

Raptorial: Pertaining to predation.

Resistance: The ability of an organism to withstand control measures.

Royal jelly: Substance secreted by pharyngeal glands of worker honeybees that is high in protein and sugars.

***r*-selection:** Favoring rapid population growth at low density.

Sclerite: Plate in body wall formed from either sclerotization or calcification.

Sclerotization: Process of hardening integument by tanning.

Scolus: A large sclerotized process from the body wall.

Scutellum: A posteriorly located triangular-shaped subdivision of a notum.

Search image: Behavior that enables predators to increase efficiency in locating prey by selecting certain shapes and colors most likely to represent food.

Semiochemical: Chemical released to the environment by one organism that affects interactions with other individuals.

Seta (pl. setae): Hairlike process from integument.

Sibling: One of two or more individuals having common parents.

Solitary: Living alone.

Speciation: Separation of one population into two or more reproductively isolated independent evolutionary units.

Spermatophore: Gelatinous-covered or sclerotized packets containing sperm.

Spermiogenesis: Process of transforming spermatids into functional sperm cells.

Spinneret: A structure from which silk exudes and is spun.

Spiracle: Opening into the tracheal system.

Spurious vein: A linear thickening in the wing that resembles a true vein.

Stage: A definite period in the development of an arthropod.

Stemmata: Photoreceptors in larval insects and some ametabolous adults; often referred to as *lateral ocelli.*

Stipes: Second segment in the maxilla.

Stomodeum: Anterior portion of alimentary canal formed from invagination.

Stylet: Mouthpart modified into a pointed or daggerlike structure.

Styli: Short tapering paired abdominal appendages in Apterygota.

Sulci: Linear braces, sometimes referred to as sutures.

Surface/volume ratio: A ratio permitting comparisons to be made on effects of size.

Symbiosis: Close association between two organisms.

Tagmosis: Specialization of body into regions.

Tarsus (pl. tarsi): Last or distal part of an insect leg.

Taxonomy: The system of classifying organisms.

Tegmen (pl. tegmina): Parchmentlike textured forewing of Orthoptera.

Tentorium: An endoskeletal brace that helps prevent collapse of the head capsule.

Tergum (pl. terga): Dorsal part of a body segment; often referred to as the *notum* in thoracic segments.

Tibia (pl. tibiae): Fourth segment of an insect leg.

Trachea (pl. tracheae): Tube used to transport air throughout the body.

Tracheole: The untapering terminal trachea in which gaseous exchange occurs.

Transverse band of crochets: Crochets arranged at right angle to body axis on proleg.

Trehalose: A disaccharide occurring in significant amounts in the blood.

Triordinal: A row of crochets with three alternating lengths.

Trochanter: Second leg segment from the body.

Tympanum (pl. tympana): Membrane that serves as an eardrum or covers the auditory organs.

Uniordinal: A row of crochets of the same length.

Urogomphus: A process or movable projection from the ninth abdominal tergum in beetle larvae.

Urticating setae: Setae capable of producing irritation in another organism.

Vertex: Top of the insect head.

Voltinism: The number of generations per year.

Wing pad: A double-walled wing primordium that develops externally in hemimetabolous insects and internally within pouches in holometabolous species.

Selected References

AHEARN, G. 1970. Control of water loss in beetles. *J. Exp. Biol.* 53:573–95.

AKRE, R. D. 1995. Our stinging friends? The ambivalent yellowjackets. *Amer. Entomol.* 41(1):21–29.

ALEXANDER, R. D. 1974. The evolution of social behavior. *Ann. Rev. Syst. Ecol.* 4:325–83.

______ and W. L. BROWN, JR. 1963. Mating behavior and the origin of insect wings. *Occas. Papers Mus. Zool. Univ. Mich.* 628:1–19.

ALVARADO, C. A., and L. J. BRUCE-CHWATT. 1962. Malaria. *Scientific Amer.* 206(5):86–98.

ANDREWARTHA, H. G. 1971. *Introduction to the Study of Animal Population,* 2nd ed. Chicago: Univ. Chicago Press, 275 pp.

______. 1984. *The Ecological Web: More on the Distribution and Abundance of Animals.* Chicago: Univ. Chicago Press, 506 pp.

ANGIER, N. 1981. Menace of the Medfly. *Discover* 2(9):16–21.

AOKI, S. 1982. In *The Biology of Social Insects,* M. D. Breed, C. D. Michener and H. E. Evans, eds., pp. 154–158. Westview Press.

ASKEW, R. R. 1971. *Parasitic Insects.* London: Heinemann Ed. Books, 316 pp.

ATSATT, P. R., and D. J. O'DOWD. 1976. Plant defense guilds. *Science* 193:24–29.

BAILEY, W. 1990. *Reproductive Behavior of Insects: Individuals and Populations.* Routledge, Chapman & Hall, 400 pp.

BAKER, H. G. 1968. Intrafloral ecology. *Ann. Rev. Entomol.* 13:385–414.

BARROWS, E. M., W. J. BELL, and C. D. MICHENER. 1975. Individual odor differences and their social functions in insects. *Proc. Nat. Acad. Sci.* 72:2824–28.

BARTHOLOMEW, T. M. 1977. Endothermy during terrestrial activity in large beetles. *Science* 195:882–83.

BATRA, S. W. T. 1982. Biological control in agroecosystems. *Science* 215:134–39.

______ and L. R. BATRA. 1967. The fungus garden of insects. *Scientific Amer.* 217(5):112–20.

BECK, S. D. 1980. *Insect Photoperiodism*, 2nd ed. New York: Academic Press, 387 pp.

BEIRNE, B. P. 1963. *Collecting and Preserving Insects.* Ottawa: Canada Dept. Agric. Publ. 932.

BERLESE, A. 1909. *Gli Insetti: Embriologia E. Morfologia.* Milano: Societa Editrice Libraria, 1004 pp.

BICKET-SMITH, S. J. R. 1984. Prolegs, legs, and wings of insects. *Entomonograph* 5:1–129.

BISHOP, J. A., and L. M. COOK. 1975. Moths, melanism, and clean air. *Scientific Amer.* 232(1):90–99.

BLEST, A. D. 1957. The evolution of protective displays in the Saturnoidea and Sphingidae (Lepidoptera). *Behavior* 11:257–309.

______. 1963. Longevity palatability, and natural selection in five species of new world saturnid moths. *Nature* 197:1183–86.

BLUM, M. S., and N. A. BLUM. 1979. *Sexual Selection and Reproductive Competition in Insects.* New York: Academic Press, 463 pp.

BODENHEIMER, F. 1951. *Insects as Human Food.* The Hague: W. Junk, 353 pp.

BOER, P. J., and G. R. GRADWELL. 1970. *Dynamics of Populations: Proceedings of the Advanced Study Institute on "Dynamics of Numbers in Populations."* Oosterbeck, Netherlands, Wageningen: Centre for Agri. Publ. & Documentation.

BORROR, D., and N. F. JOHNSON. 1989. *An Introduction to the Study of Insects,* 6th ed. Philadelphia: Saunders College Publ., 875 pp.

BORROR, D. J., and R. E. WHITE. 1970. *A Field Guide to the Insects of America North of Mexico.* Boston: Houghton Mifflin, 404 pp.

BOUDREAUX, H. B. 1979. *Arthropod Phylogeny with Special Reference to Insects.* New York: John Wiley & Sons, 320 pp.

BOUGHEY, A. S. 1972. *Fundamental Ecology.* San Francisco: Intext Educational Publ., 222 pp.

BOWERS, W. S., et al. 1976. Discovery of insect ant-juvenile hormones in plants. *Science* 193:542–47.

BROWER, L. P. 1969. Ecological chemistry. *Scientific Amer.* 220(2):22–29.

______. 1977. Monarch migration. *Nat. History* 8(6):40–53.

______, J. V. Z. BROWER, and C. T. COLLINS. 1963. Relative palatability and Mullerian mimicry among neotropical butterflies of the subfamily Heliconiinae. *Zoologica* 48:65–84.

BROWER, L. P., and S. C. GLAZIER. 1975. Localization of heart poisons in the monarch butterfly. *Science* 188:12–25.

BROWN, J. L. 1975. *The Evolution of Behavior.* New York: W. W. Norton & Co., 761 pp.

BRUES, C. T. 1972. *Insects, Food, and Ecology.* New York: Dover, 465 pp.

BUSH, G. L., et al. 1976. Screwworm eradication: Inadvertent selection for noncompetitive ecotypes during mass rearing. *Science* 193:491–93.

BUSHLAND, R. C. 1975. Screwworm research and eradication. *Bul. Entolmol. Soc. Amer.* 21(1):23–26.

BUTT, F. H. 1949. Embryology of the milkweed bug, *Oncopeltus fasciatus* (Hemiptera). *Cornell Univ. Ag. Expmt. Sta. Memoir* 283:1–43.

CALVERT, W. H., L. E. HEDRICK, and L. P. BROWER. 1979. Mortality of the monarch butterfly (*Danaus plexippus L.*): Avian predation at five overwintering sites in Mexico. *Science* 204:847–51.

CAMAZINE, S., and R. A. MORSE. 1988. The Africanized honeybee. *Amer. Scientist* 76(5): 464–71.

CAMHI, J. M. 1980. The escape mechanism of the cockroach. *Scientific Amer.* 243(6):158–72.

CARSON, R. 1962. *Silent Spring.* Greenwich, Conn.: Fawcett, 304 pp.

CASEY, T. M., and J. R. HEGEL. 1981. Caterpillar setae: Insulation for an ectotherm. *Science* 214:1131–32.

CHAI, P. 1987. Wing coloration of free-flying neotropical butterflies as a signal learned by a specialized avian predator. *Biotropics* 20(1):20–30.

CHAPMAN, R. F. 1982. *The Insects Structure and Function,* 3rd ed. Cambridge, Mass: Harvard Univ. Press.

CHENG, L., ed. 1976. *Marine Insects.* New York: American Elsevier, 581 pp.

CHERRY, R. H. 1985. Insects as sacred symbols in ancient Egypt. *Bul. Entomol. Soc. Amer.* 31(2):15–19.

CRANE, P. R., E. M. FRIIS, and K. R. PEDERSEN. 1995. The origin and early diversification of angiosperms. *Nature* 374(2):27–33.

CRESPI, B. J. 1992. Eusociality in Australian gall thrips. *Nature* 359:724–26.

CROZIER, R. H. 1982. On insects and insects: Twists and turns in our understanding of the evolution of eusociality, pp. 4–9. In *The Biology of Social Insects,* M. D. Breed, C. D. Michener, and H. E. Evans, eds. Boulder, Colo.: Westview Press.

CUMMINGS, K. W. 1973. Trophic relations of aquatic insects. *Ann. Rev. Entomol.* 18:183–206.

CSIOR. 1991. *Insects of Australia,* 2nd ed., 2 vols. New York: Cornell Univ. Press, 1500 pp.

DALY, H. V., J. T. DOYEN, and P. R. EHRLICH. 1978. *Introduction to Insect Biology and Diversity.* New York: McGraw-Hill, 564 pp.

DECK, E. 1975. Federal and state pesticide regulations and legislation. *Ann. Rev. Entomol.* 20:119–31.

DECKER, G. C. 1960. Insecticides in the 20th-century environment. *A.I.B.S. Bull.* 10(2):27–31.

DEFOLIART, G. R. 1975. Insects as a source of protein. *Bul. Entomol. Soc. Amer.* 21(3):161–63.

______. 1989. The human use of insects as food and as animal feed. *Bul. Entomol. Soc. Amer.* 35(1):22–35.

DEWILDE, J., and J. BEETSMA. 1982. The physiology of caste development in social insects. *Advances Insect Physiol.* 16:167–246.

DOUGLAS, M. M. 1981. Thermoregulatory significance of thoracic lobes in the evolution of insect wings. *Science* 211:84–85.

DOW, J. A. T. 1986. Insect midgut function. *Advances Insect. Physiol.* 19:188–306.

DOWNES, J. A. 1965. Adaptations of insects in the Arctic. *Ann. Rev. Entomol.* 10:257–74.

DOWNES, W. L. JR. 1987. The impact of vertebrate predators on early arthropod evolution. *Proc. Entomol. Soc. Wash.* 89(3):389–406.

DYER, F. C. and J. L. GOULD. 1981. Honey bee orientation: A backup system for cloudy days. *Science* 214:1041–42.

EBERHARD, M. J. W. 1975. The evolution of social behavior by kin selection. *Quarterly Rev. Biol.* 50(1):1–33.

______. 1978. Polygyny and the evolution of social behavior in wasps. *Jour. Kans. Entomol. Soc.* 51(4):832–56.

EDMUNDS, G. E. and J. R. TRAVER. 1954. The flight mechanics and evolution of the wings of Ephemeroptera, with notes on the archeotype insect wing. *Jour. Wash. Acad. Science* 44:390–99.

EDNEY, E. B. 1957. *The Water Relations of Terrestrial Arthropods.* Cambridge, England: Cambridge Univ. Press, 105 pp.

EDWARDS, C. 1969. Soil pollutants and soil animals. *Scientific Amer.* 220(4):88–99.

EHRLICH, P. R., and P. H. RAVEN. 1963. Butterflies and plants: A study in coevolution. *Evolution* 18(4):586–608.

ELZINGA, R. J. and A. B. BROCE. 1986. Labellar modifications of Muscamorpha flies. *Ann. Entomol. Soc. Amer.* 79(1):150–200.

______ and T. L. HOPKINS. 1994. Foregut microspines in four families of cockroaches (Blattaria). *Int. J. Insect Morphol. & Embryol.* 23(3):253–260.

______ and T. L. Hopkins. 1995. Microspine variation in hindgut regions of four families of cockroaches (Blattaria). *Int. J. Insect Morphol. & Embryol.* 24(2):203–11.

ENGLEMANN, F. 1970. *The Physiology of Insect Reproduction.* New York: Pergamon Press, 307 pp.

Entomological Society of America. 1989. *Common Names of Insects and Related Organisms,* Special Publ., 199 pp.

ERWIN, T. L. 1982. Tropical forests, their richness in Coleoptera and other arthropod species. *Coleopt. Bull.* 35:74–75.

EVANS, H. F. 1977. Extrinsic versus intrinsic factors in the evolution of insect sociality. *Bio. Science* 27(9):613–17.

______ and M. J. W. EBERHARD. 1970. *The Wasps.* Ann Arbor: Univ. Mich. Press, 265 pp.

EWING, A. W. 1990. *Arthropod Bioacustics.* New York: Cornell Univ. Press, 272 pp.

FERNALD, H. T., and H. H. SHEPHERD. 1942. *Applied Entomology,* 4th ed. New York: McGraw-Hill, 400 pp.

FERRON, O. 1978. Biological control of insect pests by entomogenous fungi. *Ann. Rev. Entomol.* 23:409–42.

FITZGERALD, T. D. 1976. Trail marking by larvae of the eastern tent caterpillar. *Science* 193:961–63.

______ and S. C. PETERSON. 1988. Cooperative foraging and communication in caterpillars. *Bioscience* 38(1):20–25.

FLETCHER, D. J. C., and K. G. ROSS. 1985. Regulation of reproduction in eusocial Hymenoptera. *Ann. Rev. Entomol.* 30:319–43.

FRAENKEL, G. S., and D. L. GUNN. 1940. *The Orientation of Animals.* Oxford: Clarendon Press, 352 pp.

FRAZER, J. F. D., and M. ROTHSCHILD. 1960. Defense mechanisms in warningly coloured moths and other insects. *Symposium,* 4:249–56. XI International Kongress fur Entomologie, Wien.

FRAZIER, C. A. 1969. *Insect Allergy: Allergic and Toxic Reactions to Insects and Other Arthropods.* St. Louis: Warren H. Green, 493 pp.

FUTUYMA, D. J., and M. SLATKIN. 1983. *Coevolution.* Sunderland, Mass.: Sunderland Assoc. 535 pp.

GARCIA-BELLIDO, A., and P. RIPOLL. 1978. The number of genes in *Drosophilia melanogaster. Nature* 273(566):399–400.

GHIRADELLA, H. 1984. Structure of iridescent lepidopteran scales: Variations on several themes. *Ann. Entomol. Soc. Amer.* 77(6):637–45.

GILBERT, L. E. 1976. Post mating female odor in *Heliconius* butterflies: A male contributed antiaphrodisiac? *Science* 193:419–20.

GILBERT, L. I. 1976. *The Juvenile Hormone.* New York: Plenum Press, 572 pp.

______ and E. FRIEDEN. 1981. *Metamorphosis, a Problem in Developmental Biology.* New York: Plenum Press, 578 pp.

GILLET, J. D. 1973. The mosquito: still man's worst enemy. *Amer. Scientist* 61(4):430–36.

GOLDEN, D. B. K., D. G. MARSH, A. KAGEY-SOBOTKA, L. FREIDHOFF, M. SZKLO, M. D. VALENTINE, and L. M. LICHTENSTEIN. 1989. Epidemiology of insect venom sensitivity. *JAMA* 262(2):240–45.

GORDON, R. M., and W. H. LUMSDEN. 1939. A study of the behavior of the mouthparts of mosquitoes when taking blood from living tissue; together with some observations on the ingestion of microfilariae. *Ann. Trop. Med. Parasitol.* 33:259–278.

GOSS, R. J. 1969. *Principles of Regeneration.* New York: Academic Press, 287 pp.

GOULD, S. J. 1989. *Wonderful Life.* New York: W. W. Norton, 347 pp.

HADLEY, N. F. 1972. Desert species and adaptations. *Amer. Scientist* 60(3):338–47.

HAGAN, K. S., R. H. DADD, and J. REESE. 1984. The food of insects, pp. 79–111. In *Ecological Entomology,* C. B. Huffaker and R. L. Rabb, eds. New York: John Wiley & Sons, Inc.

HAGEN, H. R. 1951. *Embryology of the Viviparous Insects.* New York: Ronald, 472 pp.

HAMILTON, K. G. A. 1971. The insect wing, Part I. Origin and development from notal lobes. *Kans. Entomol. Soc.* 44(4):421–33.

HAMILTON, W. D. 1964a. The genetic theory of social behavior, 1. *J. Theoret. Biol.* 7:1–16.

______. 1964b. The genetic theory of social behavior, 2. *J. Theoret. Biol.* 7:17–52.

HAVERTY, M. I. 1975. Natural wood preferences of desert termites. *Ann Entomol. Soc. Amer.* 68(3):533–36.

HAWKINS, F. 1970. The clock of the malaria parasite. *Scientific Amer.* 222(6):123–31.

HEINRICH, B. 1981. Insect Thermoregulation. New York: John Wiley & Sons, 328.

HENNIG, W. 1981. *Insect Phylogeny.* New York: John Wiley & Sons, 514 pp.

HICKINN, N. E., and R. EDWARDS. 1975. *Insect Factor in Wood Decay,* 3rd ed. London: Associated Business Programmes Ltd., 383 pp.

HINTON, H. E. 1963. The origin and function of the pupal state. *Proc. Royal Entom. Soc. London* 38:77–85.

______. 1969. Respiratory systems of insect egg shells. *Ann. Rev. Entomol.* 14:343–68.

______. 1976a. Plastron respiration in bugs and beetles. *Jour. Insect Physiol.* 22:1529–50.

______. 1976b. Enabling mechanisms. *Proc XV Internat. Cong. Entomol.,* pp. 71–83.

HOFFMANN, C. H. 1974. Critical issues face entomologists. *Bul. Entomol. Soc. Amer.* 20(4):316–18.

HÖLLDOBLER, B., and E. O. WILSON. 1990. *The Ants.* Cambridge, Mass.: Belknap Press, 732 pp.

HOLLING, C. S. 1965. The functional response of predators to prey density and its role in mimicry and population. *Mem. Entomol. Soc. Canada* 45:1–60.

HORRIDGE, G. A., ed. 1975. *The Compound Eye and Vision of Insects.* Oxford: Clarendon Press, 595 pp.

HOY, M. 1994. *Insect Molecular Genetics.* New York: Academic Press, 402 pp.

HYNES, H. B. N. 1970. The ecology of stream insects. *Ann. Rev. Entomol.* 15:25–42.

JACOBSON, M. 1972. *Insect Sex Pheromones.* New York: Academic Press, 382 pp.

______, and M. BEROZA. 1964. Insect attractants. *Scientific Amer.* 211(2):20–27.

JANZEN, D. H. 1970. Herbivores and the number of tree species in tropical forests. *Amer. Nat.* 104:501–28.

______. 1984. Weather-related color polymorphism of *Rothschildia lebeau* (Saturniidae). *Bull. Entomol. Soc. Amer.* 30(2):16–20.

JOHANNSEN, O. A., and F. H. BUTT. 1941. *Embryology of Insects and Myriapods.* New York: McGraw-Hill, 462 pp.

JOHNSON, C. G. 1963. The aerial migration of insects. *Scientific Amer.* 209(6):132–38.

______. 1966. A functional system of adaptive flight. *Ann. Rev. Entomol.* 11:233–53.

KETTLE, D. S. 1984. *Medical and Veterinary Entomology.* New York: John Wiley & Sons, 658 pp.

KETTLEWELL, H. B. D. 1959. Darwin's missing evidence. *Scientific Amer.* 200(3):48–53.

KIM, K. C., and H. W. LUDWIG. 1978. Phylogenetic relationships of parasitic Psocoptera and taxonomic position of the Anoplura. *Ann. Entomol. Soc. Amer.* 76(6):910–22.

KINGSOLVER, J. G., and M. A. R. KOEHL. 1994. Selective factors in the evolution of insect wings. *Ann. Rev. Entomol.* 39:425–51.

KIRCHNER, W. H., and W. F. TOWNE. 1994. The sensory basis of the honeybee's language. *Scientific Amer.* 270(6):74–80.

KNIPLING, E. F. 1959. Sterile-male method of population control. *Science* 130:902–4.

KOLATA, G. 1984. Scrutinizing sleeping sickness. *Science* 226:956–59.

KRISHNA, K., and F. M. WEESNER. 1969. *Biology of Termites,* vols. 1 and 2. New York: Academic Press, 598 and 643 pp.

KRISTENSEN, N. P. 1981. Phylogeny of insect orders. *Ann. Rev. Entomol.* 26:135–57.

KUKALOVA-PECK, J. 1978. Origin and evolution of insect wings and their relation to metamorphosis, as documented by the fossil record. *J. Morph.* 156:53–126.

______. 1983. Origin of the insect wing and wing articulation from the arthropodan leg. *Canad. J. Zool.* 61:1618–69.

______. 1986. New carboniferous Diplura, Monura, and Thysanura, the hexapod ground plan, and the role of thoracic side lobes in the origin of wings (Insecta). *Canad. J. Zool.* 65:2327–45.

LANE, R. P., and R. W. CROSSKEY. 1993. *Medical Insects and Arachnids.* London: Chapman & Hall, 723 pp.

LEE, K. E. and T. G. WOOD. 1971. *Termites and Soils.* New York: Academic Press, 251 pp.

LUCK, R. F., R. VAN DER BOSCH, and R. GARCIA. 1977. Chemical insect control—A troubled pest management strategy. *Bio. Science* 27(9):606–11.

LÜSCHER, M. 1961a. Air-conditioned termite nests. *Scientific Amer.* 205(1):138–45.

______. 1961b. Social control of polymorphism in termites. In *Insect Polymorphism,* pp. 57–67, Symposium no. 1, R. Ent. Soc. Lond.

______. 1976. *Phase and Caste Determination in Insects.* Oxford: Pergamon Press, 130 pp.

MARDEN, J. H., and M. G. KRAMER. 1994. Surface-skimming stoneflies: a possible intermediate stage in insect flight evolution. *Science* 266:427–30.

MARSHALL, E. 1981. Man vs. medfly: Some tactical blunders. *Science* 213:417–18.

MATHEWS, R. W., and J. R. MATHEWS. 1978. *Insect Behavior.* New York: John Wiley & Sons, 507 pp.

MATSUDA, R. 1965. *Morphology and Evolution of the Insect Head.* Ann Arbor, Mich.: Memoirs of the Amer. Entom. Instit. No. 4, 334 pp.

______. 1976. *Morphology and Evolution of the Insect Abdomen.* Oxford: Pergamon Press, 538 pp.

______. 1981. The origin of insect wings (Arthropoda: Insecta). *Int. J. Insect Morph. Embryol.* 10(5/6):387–98.

MATTSON, W. J., and N. D. ADDY. 1975. Phytophagus insects as regulators of forest primary production. *Science* 190:515–22.

MAUGH, T. H., II. 1982. Exploring plant resistance to insects. *Science* 216:722–23.

MAY, M. L. 1976. Thermoregulation and adaptation to temperature in dragonflies (Odonata: Anisoptera). *Ecolog. Monographs* 46(1):1–32.

MAY, R. M. 1988. How many species are there on earth? *Science* 241:1441–49.

MICHENER, C. D. 1958. The evolution of social behavior in bees. *Proc. Tenth Int. Congr. Entomol., Montreal* 2:441–47.

______. 1974. *The Social Behavior of the Bees,* Cambridge, Mass.: Harvard Univ. press, 404 pp.

______. 1975. The Brazilian bee problem. *Ann. Rev. Entomol.* 20:399–416.

______, and M. H. MICHENER. 1951. *American Social Insects.* New York: Van Nostrand, 267 pp.

MILNE, L. J., and M. MILNE. 1978. Insects of the water surface. *Scientific Amer.* 238(4):134–42.

NAULT, L. R., et al. 1976. Ant-aphid association: Role of aphid alarm pheromone. *Science* 192:1349–50.

NEVILLE, A. C. 1975. *Biology of the Arthropod Cuticle.* New York: Springer-Verlag, 448 pp.

NEWSOM, L. D. 1967. Consequences of insecticide use on nontarget organisms. *Ann. Rev. Entomol.* 12:323–46.

NIJHOUT, H. F., and D. E. WHEELER. 1982. Juvenile hormone and the physiological basis of insect polymorphisms. *Quart. Rev. Biol.* 57(2):109–28.

NOIROT, C. 1989. Social structure in termite societies. *Ethology Ecol. & Evol.* 1:1–17.

______, and J. M. PASTEELS. 1987. Ontogenetic development and evolution of the worker caste in termites. *Experientia* 43(8):851–60.

NUESCH, H. 1968. The role of the nervous system in insect morphogenesis and regeneration. *Ann. Rev. Entomol.* 13:27–44.

O'BRIEN, R. W., and M. SLAYTOR. 1982. Role of microorganisms in the metabolism of termites. *Aust. J. Biol. Sci.* 35:239–62.

OKOT-KOTBER, B. M. 1981. Instars and Polymorphism of castes in *Macrotermes michaelseni* (Isoptera, Macrotermitidae), *Social Insects* 28(3):233–46.

______. 1982a. Ergatoid reproductives in *Nasutitermes corniger* (Isoptera: Termitidae). *Int. J. Insect Morph Embryol.* 11(3/4):213–26.

______. 1982b. Correlation between larval weights, endocrine gland activities and competence period during differentiation of workers and soldiers in *Macrotermes michaelseni* (Isoptera: Termitidae). *J Insect Physiol.* 28(11):905–10.

______. 1983. Influence of group size and composition on soldier differentiation in female final instars of higher termite, *Macrotermes michaelseni. Phys. Entomol.* 8:41–47.

OSSIANNILSON, F. 1966. Insects in the epidemiology of plant viruses. *Ann. Rev. Entomol.* 11:213–32.

PAINTER, R. H. 1968. *Insect Resistance in Crop Plants.* Lawrence: Univ. Press of Kansas, 544 pp.

PAMPANA, E. J., and P. F. RUSSELL. 1955. *Malaria, a World Problem.* Geneva: World Health Organization, 72 pp.

PAPAGEORGIS, C. 1975. Mimicry in neotropical butterflies. *Amer. Scientist* 63:522–32.

PARRISH, H. M. 1963. Analysis of 460 fatalities from venomous animals in the U.S. *Jour. Med. Science,* 245(2):129–41.

PASTEELS, J. M., and J. C. GREGOIRE. 1983. The chemical ecology of defense in arthropods. *Ann. Rev. Entomol.* 28:263–89.

PASTEUR, G. 1982. A classificatory review of mimicry systems. *Ann. Rev. Ecol. Syst.* 13:169–99.

PEIGLER, R. S. 1993. Wild silks of the world. *Amer. Entomol.* 39(3): 151–61.

PIANKA, E. R. 1974. *Evolutionary Ecology.* New York: Harper & Row, 356 pp.

PIVNICK, K. A., and J. M. MCNEIL. 1987. Puddling in butterflies: sodium affects reproductive success in *Thymelicusa lineola. Physiol. Entomol.* 12:461–72.

PLAPP, F. W. 1976. Biochemical genetics of insecticide resistance. *Ann. Rev. Entomol.* 21:179–97.

POMERANTZ, C. 1959. Arthropods and psychic disturbances. *Bul. Entomol. Soc. Amer.* 5:65–67.

POWELL, J. H. 1949. *Bring Out Your Dead: The Great Plague of Yellow Fever in Philadelphia in 1793.* Philadelphia: Univ. Penn. Press, 304 pp.

PRICE, P. W., ed. 1975. *Evolutionary Strategies of Parasitic Insects and Mites.* New York: Plenum Press, 224 pp.

______. 1984. *Insect Ecology,* 2nd ed. New York: John Wiley & Sons, 600 pp.

PUTMAN, R. J. 1978. The role of carrion-frequenting arthropods in the decay process. *Ecological Entomol.* 3(2):133–39.

RASNITSYN, A. P. 1981. A modified paranotal theory of insect wing origin. *Jour. Morph.* 168:331–38.

REED, C. 1943. *Land Birds East of the Rockies.* Garden City, N.Y.: Doubleday.

RENTZ, D. C. 1972. The lock and key as an isolating mechanism in katydids. *Amer. Scientist* 60(6):750–55.

RETTENMEYER, C. W. 1963. Behavioral studies of army ants. *Univ. Kans. Science Bul.* 44(9):281–465.

______. 1970. Insect mimicry. *Ann. Rev. Entomol.* 15:43–74.

RICHARDS, O. W., and R. G. DAVIES. 1977. *Imm's General Textbook of Entomology,* 10th ed. London: Chapman & Hall, 123 pp.

RICKLEFS, R. E. 1973. *Ecology.* Newton, Mass: Chiron Press, 861 pp.

ROBERTSON, R. M., and H. REICHERT. 1982. Flight interneurons in the locust and the origin of insect wings. *Science* 217:177–79.

ROITBERG, B. D., and M. B. ISMAN. 1992. *Insect Chemical Ecology: An Evolutionary Approach.* New York: Routledge, Chapman & Hall, 359 pp.

ROGERS, H. T. 1981. Victory over Medfly. *Ag. Consult. Fieldman.* 37(6):17–18.

ROGERS, S. H., and H. WELLS. 1984. The structure and function of the bursa copulotrix of the monarch butterfly (*Danaus plexippus*). *Jour. Morph.* 180:213–21.

ROUSH, R., and B. TABASHNIK. 1990. *Pesticide Resistance in Arthropods.* New York: Routledge, Chapman & Hall, 352 pp.

RUDOLPH, D. 1982. Site, process, and mechanisms of active uptake of water vapour from the atmosphere in the Psocoptera. *J. Insect Physiol.* 28(3):205–12.

______. 1983. The water uptake system of the Pthiraptera. *J Insect Physiol.* 29(1):15–25.

RYAN, C. A. 1983. Insect-induced chemical signals regulating natural plant protection responses, pp. 43–59. In *Variable Plants and Herbivores in Natural and Managed Systems,* R. F. DENNO and M. S. MCCLURE, eds. New York: Academic Press.

SAILER, R. I. 1977. Our immigrant insect fauna. *Ann. Entomol. Soc. Amer.* 24(1):3–12.

SALT, R. W. 1961. Principles of insect cold-hardiness. *Ann. Rev. Entomol.* 6:55–74.

SCHALK, J. M., and R. H. RATCLIFFE. 1976. Evaluation of ARS program for alternative methods of insect control: Host plant resistance to insects. *Bul. Entomol. Soc. Amer.* 22(1):7–10.

SCHALLER, F. 1968. *Soil Animals.* Ann Arbor: Univ. Mich. Press, 144 pp.

______. 1971. Indirect sperm transfer by soil arthropods. *Ann. Rev. Entomol.* 16:406–46.

SCHOONHOVEN, L. M. 1968. Chemosensory bases of host plant selection. *Ann. Rev. Entomol.* 13:115–36.

SCHWALM, F. E. 1988. *Insect Morphogenesis.* Basel: Karger, 356 pp.

SEELEY, T. D. 1983. The ecology of temperate and tropical honeybee societies. *Amer. Scientist* 71:264–72.

SEELY, M. K. 1976. Fog basking by the Namib Desert beetle, *Onymacris unguicularis. Nature* 262:284–85.

______, and W. J. HAMILTON. 1976. Fog catchment sand trenches constructed by tenebrionid beetles, *Lepidochora,* from the Namib Desert. *Science* 193:484–86.

SEXTON, O. J. 1960. Experimental studies of artificial Batesian mimics. *Behaviour,* 15:244–52.

SILBERGLIED, R. E. 1979. Communication in the ultraviolet. *Ann. Rev. Ecol. Syst.* 10:373–98.

SILVERSTEIN, R. M. 1981. Pheromones: Background and potential for use in insect pest control. *Science* 213:1326–32.

SKAIFE, S. H. 1954. The black-mound termite of the cape, *Amitermes atlanticus* Fuller. *Transactions of the Royal Soc.of South Africa* 34(1):251–71.

SMITH, R. L. 1979. Repeated copulation and sperm precedence: Paternity assurance for a male brooding water bug. *Science* 206: 1029–31.

SNODGRASS, R. E. 1935. *Principles of Insect Morphology.* New York: McGraw-Hill, 667 pp.

______. 1958. Evolution of arthropod mechanisms. *Smithsonian Misc. Coll.,* 138(2):1–77.

______. 1965. *A Textbook of Arthropod Anatomy.* New York: Stechert-Hafner, 363 pp.

SOUTHWOOD, T. R. E. 1977. Entomology and mankind. *Proc. XV Internat. Cong. Entomol.,* pp. 35–51.

SPIVAK, M., D. J. C. FLETCHER, and M. D. BREED. 1991. *The "African Honeybee."* Boulder, Colo.: Westview Press, 435 pp.

STAHL, G. B. 1975. Insect growth regulators with juvenile hormone activity. *Ann. Rev. Entomol.* 20:417–60.

STEELMAN, C. D. 1976. Effects of external and internal arthropods on domestic livestock production. *Ann. Rev. Entomol.* 21:155–78.

STEHR, F. W. 1987. *Immature Insects,* vol. 1. Dubuque, Iowa: Kendall/Hunt Publ. Co., 754 pp.

______. 1991. *Immature Insects,* vol. 2. Dubuque, Iowa: Kendall/Hunt Publ. Co., 996 pp.

STORK, N. E. 1988. Insect diversity: Facts, fiction, and speculation. *Biolog. Jour. Linnean Soc.* 35:321–37.

TAUBER, M. J., C. A. TAUBER, and S. MASAKI. 1986. *Seasonal Adaptations.* New York: Oxford Univ. Press, 411 pp.

TAYLOR, C. E., and C. CONDRA. 1980. *r*- and *K*-selection in *Drosophila pseudoobscura. Evolution* 34(6):1183–93.

TEAL, J. M. 1971. Community metabolism in a temperate cold spring. *Ecological Monog.* 27:283–302.

TOONG, U. C., D. A. SHOOLEY, and F. C. BAKER. 1988. Isolation of insect juvenile hormone III from a plant. *Nature* 333:170–71.

TU, A. T. 1984. Handbook of Natural Toxins, vol. 2, In *Insect Poisons, Allergens, and Other Invertebrate Venoms.* New York: Marcel Dekker, Inc., 732 pp.

TUSKES, T. M., and L. P. BROWER. 1978. Overwintering ecology of monarch butterfly, *Danaus plexippus L.,* in California. *Ecological Entomol.* 3(2):141–53.

URQUHART, F. A. 1976. Found at last: the monarch's winter home. *Nat. Geographic* 150(2):160–73.

VINSON, S. B. 1976. Host selection by insect parasitoids. *Ann. Rev. Entomol.* 21:109–34.

VON FRISCH, K. 1962. Dialects in the language of the bees. *Scientific Amer.* 207(2): 78–87.

______. 1971. *Bees—Their Vision, Chemical Senses, and Language.* Ithaca, N.Y.: Cornell Univ. Press, 157 pp.

WALDRON, W. G. 1962. The role of the entomologist in delusionary parasitosis (entomophobia). *Bul. Entomol. Soc. Amer.* 8(2):81–83.

WALOFF, Z. 1966. *The Upsurges and Recessions of the Desert Locust Plague: An Historical Survey.* London: Anti-Locust Research Center, 111 pp.

WHEELER, D. E. 1986. Developmental and physiological determinants of caste in social Hymenoptera: evolutionary implications. *Amer. Nat.* 128(1):13–34.

WHITTEN, J. M. 1976. Definition of insect instars in terms of 'apolysis' or 'ecdysis'. *Ann. Entomol. Soc. Amer.* 69(3):556–59.

WHITTEN, J. M., and G. G. FOSTER. 1975. Genetic methods of pest control. *Ann. Rev. Entomol.* 20:461–76.

WIGGLESWORTH, V. B. 1984. *The Principles of Insect Physiology,* 8th ed. New York: E. P. Dutton, 544 pp.

WIKLUND, C., and T. JÄRVI. 1982. Survival of distasteful insects after being attacked by naive birds: A reappraisal of the theory of aposematic coloration evolving through individual selection. *Evolution* 36(5):998–1002.

WILLIAMS, C. M. 1947. Physiology of insect diapause. II. Interaction between the pupal brain and prothoracic glands in metamorphosis of the giant silkworm, *Platysamia cecropia. Biol. Bull.* 93(2):89–98.

______. 1952. Physiology of insect diapause. IV. The brain and prothoracic glands as an endocrine system in the Cecropia silkworm. *Biol. Bull.* 103(1):120–38.

______. 1958. The juvenile hormone. *Scientific Amer.* 198(2):67–74.

WILLIAMS, R. E., et al. 1985. *Livestock Entomology.* New York: John Wiley & Sons, 335 pp.

WILLMER, P. G. 1982. Microclimate and the environmental physiology of insects. *Advances Insect Physiol.* 16:1–58.

WILSON, E. O. 1971. *The Insect Societies.* Cambridge, Mass.: Belknap Press of Harvard Univ. Press, 548 pp.

______. 1985. The sociogenesis of insect colonies. *Science* 228 (4707): 1489–95.

______. 1987. The arborial ant fauna of Peruvian Amazon forests: A first assessment. *Biotropica* 19(3):245–51.

WILSON, D. S. 1989. Reviving the superorganism. *J. Theor. Biol.* 136:337–56.

WINSTON, M. L. 1987. *The Biology of the Honey Bee.* Cambridge, Mass.: Harvard Univ. Press, 281 pp.

WOLFGANG, W. J., and L. M. RIDDIFORD. 1981. Cuticular morphogenesis during continuous growth of the final instar of a moth. *Tissue Cell* 13(4):757–72.

ZUMPT, F. 1965. *Myiasis in Man and Animals in the Old World.* London: Butterworth, pp. 1–4.

Index

Numbers in italic print represent illustration-bearing pages. Many items in this index are defined in the glossary.